This book studies the geometric theory of polynomials and rational functions in the plane. Any theory in the plane should make full use of the complex numbers and thus the early chapters build the foundations of complex variable theory, melding together ideas from algebra, topology and analysis.

In fact, throughout the book, the author introduces a variety of ideas and constructs theories around them, incorporating much of the classical theory of polynomials as he proceeds. These ideas are used to study a number of unsolved problems, bearing in mind that such problems indicate the current limitations of our knowledge and present challenges for the future. However, theories also lead to solutions of some problems and several such solutions are given including a comprehensive account of the geometric convolution theory.

This is an ideal reference for graduate students and researchers working in this area.

CAMBRIDGE STUDIES IN
ADVANCED MATHEMATICS 73

COMPLEX POLYNOMIALS

Already published

1 W.M.L. Holcombe *Algebraic automata theory*
2 K. Petersen *Ergodic theory*
3 P.T. Johnstone *Stone spaces*
4 W.H. Schikhof *Ultrametric calculus*
5 J.-P. Kahane *Some random series of functions, 2nd edition*
6 H. Cohn *Introduction to the construction of class fields*
7 J. Lambek & P.J. Scott *Introduction to higher-order categorical logic*
8 H. Matsumura *Commutative ring theory*
9 C.B. Thomas *Characteristic classes and the cohomology of finite groups*
10 M. Aschbacher *Finite group theory*
11 J.L. Alperin *Local representation theory*
12 P. Koosis *The logarithmic integral I*
13 A. Pietsch *Eigenvalues and s-numbers*
14 S.J. Patterson *An introduction to the theory of the Riemann zeta-function*
15 H.J. Baues *Algebraic homotopy*
16 V.S. Varadarajan *Introduction to harmonic analysis on semisimple Lie groups*
17 W. Dicks & M. Dunwoody *Groups acting on graphs*
18 L.J. Corwin & F.P. Greenleaf *Representations of nilpotent Lie groups and their applications*
19 R. Fritsch & R. Piccinini *Cellular structures in topology*
20 H. Klingen *Introductory lectures on Siegel modular forms*
21 P. Koosis *The logarithmic integral II*
22 M.J. Collins *Representations and characters of finite groups*
24 H. Kunita *Stochastic flows and stochastic differential equations*
25 P. Wojtaszczyk *Banach spaces for analysts*
26 J.E. Gilbert & M.A.M. Murray *Clifford algebras and Dirac operators in harmonic analysis*
27 A. Frohlich & M.J. Taylor *Algebraic number theory*
28 K. Goebel & W.A. Kirk *Topics in metric fixed point theory*
29 J.E. Humphreys *Reflection groups and Coxeter groups*
30 D.J. Benson *Representations and cohomology I*
31 D.J. Benson *Representations and cohomology II*
32 C. Allday & V. Puppe *Cohomological methods in transformation groups*
33 C. Soulé et al *Lectures on Arakelov geometry*
34 A. Ambrosetti & G. Prodi *A primer of nonlinear analysis*
35 J. Palis & F. Takens *Hyperbolicity, stability and chaos at homoclinic bifurcations*
37 Y. Meyer *Wavelets and operators I*
38 C. Weibel *An introduction to homological algebra*
39 W. Bruns & J. Herzog *Cohen-Macaulay rings*
40 V. Snaith *Explicit Brauer induction*
41 G. Laumon *Cohomology of Drinfield modular varieties I*
42 E.B. Davies *Spectral theory and differential operators*
43 J. Diestel, H. Jarchow & A. Tonge *Absolutely summing operators*
44 P. Mattila *Geometry of sets and measures in euclidean spaces*
45 R. Pinsky *Positive harmonic functions and diffusion*
46 G. Tenenbaum *Introduction to analytic and probabilistic number theory*
47 C. Peskine *An algebraic introduction to complex projective geometry I*
48 Y. Meyer & R. Coifman *Wavelets and operators II*
49 R. Stanley *Enumerative combinatorics I*
50 I. Porteous *Clifford algebras and the classical groups*
51 M. Audin *Spinning tops*
52 V. Jurdjevic *Geometric control theory*
53 H. Voelklein *Groups as Galois groups*
54 J. Le Potier *Lectures on vector bundles*
55 D. Bump *Automorphic forms*
56 G. Laumon *Cohomology of Drinfeld modular varieties II*
57 D.M. Clarke & B.A. Davey *Natural dualities for the working algebraist*
59 P. Taylor *Practical foundations of mathematics*
60 M. Brodmann & R. Sharp *Local cohomology*
61 J.D. Dixon, M.P.F. Du Sautoy, A. Mann & D. Segal *Analytic pro-p groups, 2nd edition*
62 R. Stanley *Enumerative combinatorics II*
64 J. Jost & X. Li-Jost *Calculus of variations*
68 Ken-iti Sato *Lévy processes and infinitely divisible distributions*
71 R. Blei *Analysis in integer and fractional dimensions*
72 F. Borceux & G. Janelidze *Galois theories*
73 B. Bollobás *Random graphs*
74 R.M. Dudley *Real analysis and probability*

COMPLEX POLYNOMIALS

T. SHEIL-SMALL
University of York

CAMBRIDGE UNIVERSITY PRESS
Cambridge, New York, Melbourne, Madrid, Cape Town, Singapore, São Paulo

Cambridge University Press
The Edinburgh Building, Cambridge CB2 8RU, UK

Published in the United States of America by Cambridge University Press, New York

www.cambridge.org
Information on this title: www.cambridge.org/9780521400688

First published 2002

A catalogue record for this publication is available from the British Library

Library of Congress Cataloguing in Publication data

Sheil-Small, T. (Terence), 1937–
Complex polynomials / T. Sheil-Small.
p. cm. – (Cambridge studies in advanced mathematics ; 75)
Includes bibliographical references and index.
ISBN 0 521 40068 6
1. Polynomials. 2. Functions of complex variables. I. Title. II. Series.
QA161.P59 S45 2002
512.9′42 – dc21 2001052690

ISBN 978-0-521-40068-8 hardback

Transferred to digital printing 2007

Contents

	Preface	*page* xi
	List of notation	xix
1	The algebra of polynomials	1
	1.1 Complex polynomials	1
	1.2 The number of zeros of a real analytic polynomial	4
	1.3 Real analytic polynomials at infinity	13
2	The degree principle and the fundamental theorem of algebra	22
	2.1 The fundamental theorem of algebra	22
	2.2 Continuous functions in the plane	26
	2.3 The degree principle	31
	2.4 The degree principle and homotopy	40
	2.5 The topological argument principle	43
	2.6 The coincidence theorem	47
	2.7 Locally 1–1 functions	56
	2.8 The Borsuk–Ulam theorem	79
3	The Jacobian problem	81
	3.1 The Jacobian conjecture	81
	3.2 Pinchuk's example	90
	3.3 Polynomials with a constant Jacobian	105
	3.4 A topological approach	118
	3.5 The resultant and the Jacobian	124
4	Analytic and harmonic functions in the unit disc	125
	4.1 Series representations	125
	4.2 Positive and bounded operators	138
	4.3 Positive trigonometric polynomials	144
	4.4 Some inequalities for analytic and trigonometric polynomials	151

4.5	Cesàro means	156
4.6	De la Vallée Poussin means	160
4.7	Integral representations	167
4.8	Generalised convolution operators	168
5	Circular regions and Grace's theorem	172
5.1	Convolutions and duality	172
5.2	Circular regions	178
5.3	The polar derivative	184
5.4	Locating critical points	186
5.5	Critical points of rational functions	190
5.6	The Borwein–Erdélyi inequality	193
5.7	Univalence properties of polynomials	196
5.8	Linear operators	203
6	The Ilieff–Sendov conjecture	206
6.1	Introduction	206
6.2	Proof of the conjecture for those zeros on the unit circle	207
6.3	The direct application of Grace's theorem	208
6.4	A global upper bound	213
6.5	Inequalities relating the nearest critical point to the nearest second zero	216
6.6	The extremal distance	221
6.7	Further remarks on the conjecture	223
7	Self-inversive polynomials	228
7.1	Introduction	228
7.2	Polynomials with interspersed zeros on the unit circle	232
7.3	Relations with the maximum modulus	238
7.4	Univalent polynomials	241
7.5	A second necessary and sufficient condition for angular separation of zeros	249
7.6	Suffridge's extremal polynomials	251
8	Duality and an extension of Grace's theorem to rational functions	263
8.1	Linear operators and rational functions	263
8.2	Interpretations of the convolution conditions	270
8.3	The duality theorem for $T(1, \beta)$	275
8.4	The duality theorem for $T(m, \beta)$	282
8.5	The duality principle	286
8.6	Duality and the class $T(\alpha, \beta)$	289
8.7	Properties of the Kaplan classes	293
8.8	The class $S(\alpha, \beta)$	296

8.9 The classes $T_0(\alpha, \beta)$ 300
8.10 The class $T(2, 2)$ 302
9 Real polynomials 304
9.1 Real polynomials 304
9.2 Descartes' rule of signs 317
9.3 Strongly real rational functions 319
9.4 Critical points of real rational functions 323
9.5 Rational functions with real critical points 325
9.6 Real entire and meromorphic functions 326
10 Level curves 350
10.1 Level regions for polynomials 350
10.2 Level regions of rational functions 353
10.3 Partial fraction decomposition 355
10.4 Smale's conjecture 358
11 Miscellaneous topics 370
11.1 The *abc* theorem 370
11.2 Cohn's reduction method 372
11.3 Blaschke products 373
11.4 Blaschke products and harmonic mappings 377
11.5 Blaschke products and convex curves 382
11.6 Blaschke products and convex polygons 392
11.7 The mapping problem for Jordan polygons 402
11.8 Sudbery's theorem on zeros of successive derivatives 407
11.9 Extensions of Sudbery's theorem 413
References 416
Index 421

Preface

Polynomials play an important role in almost all areas of mathematics. From finding solutions to equations, finding the number of solutions of equations, understanding the role of critical points in determining the geometric behaviour of the distribution of values, determining the properties of geometric curves and much else, polynomials have wielded an enormous influence on the development of mathematics since ancient times. This is a book on polynomials in the plane with a special emphasis on the geometric theory. We can do justice to only a very small part of the subject and therefore we confine most of our attention to the study of a number of specific problems (some solved and some unsolved). However, as the book is directed towards graduate students and to a broad audience of scientists and mathematicians possessing a basic knowledge of complex variable theory, we have concentrated in the earlier chapters on building some theoretical foundations, melding together algebraic, topological and analytic ideas.

Description of the chapters.
Chapter 1. **The algebra of polynomials.** The chapter deals with two important issues concerning real analytic polynomials – a topic normally regarded as being part of classical algebraic geometry. The first is a straightforward account of Bézout's theorem in the plane. We prove a version of the theorem which is independent of any algebraic restrictions, and therefore immediately applicable and useful to the analyst. The second topic concerns the calculation and properties of asymptotic values at infinity. We give details of an algorithm, with which one can calculate all asymptotic values and the tracts within which they lie. The method is not quite computable, since it involves finding the exact zeros of certain polynomials. However, we establish two important facts: (1) all finite limiting values at infinity are asymptotic values; (2) the repetition

property of asymptotic values. This work will be applied in our study of the Jacobian problem in chapter 3.

Chapter 2. **The degree principle and the fundamental theorem of algebra.** The first half of this chapter is effectively a mini-course in plane topology, and in fact is based on undergraduate lectures given at York c. 1980. The degree principle, as stated and proved, is a fundamental rotation theorem in plane topology and deserves to be as familiar to all mathematicians as is Cauchy's theorem, to which it is closely analogous. Its corollary, the argument principle, is a powerful general result for locating certain types of 'singularity' of continuous functions and immediately places the classical argument principle for analytic or meromorphic functions as a topological rather than an analytic result. The statement of the principle is virtually unchanged from the classical statement, providing only that one has a clear concept of topological multiplicity. Some theorems, such as Rouché's theorem, have topological analogues, which include the Brouwer fixed point theorem. We are also able to prove, not only the fundamental theorem of algebra, but existence of a minimum number of zeros for more general polynomial-type functions. Included here are the complex-valued harmonic polynomials (with a mild restriction). We show that a harmonic polynomial of degree n has at least n zeros and, using the algebra of chapter 1, at most n^2 zeros; we include Wilmshurst's elegant example of a harmonic polynomial with exactly n^2 zeros. We also begin in this chapter our study of locally 1–1 mappings of the plane, including the important theorem that the global injectivity depends entirely on the behaviour at infinity. We introduce a second topological method 'continuation of the inverse function' using the monodromy theorem. The final part of the chapter contains a detailed study of harmonic polynomials on the critical set.

Chapter 3. **The Jacobian problem.** The problem of when it is possible to solve uniquely a set of simultaneous algebraic equations is among the most beautiful problems of classical mathematics and remains only partially resolved. This chapter is a 'bottom up' and elementary account of the two-dimensional case, in which we consider the sufficiency of the non-vanishing of the Jacobian. In line with chapters 1 and 2 we adopt a quite specific approach by examining the possibility and nature of a polynomial mapping having asymptotic values at infinity. For real variables the central question has been resolved by S. Pinchuk, who has constructed a polynomial mapping of degree 25 with a positive Jacobian, but which is not globally 1–1. We give the details of his elegant construction and then by explicitly calculating the asymptotic values are able to examine the geometric and topological behaviour of the mapping.

Topologically Pinchuk's example turns out to be as simple as such a mapping can be and lends credence to the possibility that below degree 25 the main conjecture is true (this also seems likely from the simplicity of the original algebraic construction). We consider this question and obtain some estimates for the degree of counter-examples. As the author's algebraic capacity and resources are limited, we suggest that resolving this question is a suitable topic for a PhD research thesis.

For complex variables the Jacobian is constant, a fact which has strong algebraic implications. There is a massive literature on this problem (see e.g. [1], [5], [66]), much of it algebraically very advanced. The conjecture is true up to degree 100 and also under various additional conditions on the degrees of the components of the mapping. The general conjecture is unresolved. We continue with our elementary approach using asymptotic values, obtaining a few of the known results and, we hope, leading the student to a greater understanding of the complexity of the problem and the probable need for more advanced methods. We also include a topological approach to the problem, which certainly turns up some interesting and strange features.

Chapter 4. **Analytic and harmonic functions in the unit disc.** In this chapter we begin our study of convolutions and convolution operators, which is one of the central themes of the book. A knowledge of general real analysis and the theory of Fourier series will be useful here. The main themes are the annihilation properties of convolution operators, the representation of general linear operators, the structure preserving properties of bounded and positive operators and the construction of approximate identities. We also introduce a class of multiplication operators and use these to give very simple proofs of some classical polynomial inequalities, which we are able to generalise. We include an account of variation diminishing transformations and give Pólya and Schoenberg's proof that the de la Vallée Poussin means form a variation diminishing approximate identity. The chapter concludes with a brief account of generalised convolution operators, including a coefficient inequality due, essentially, to Rogosinski. As a whole non-trivial class of such operators is established in chapter 8, we have here a family of (unstated) theorems and inequalities.

Chapter 5. **Circular regions and Grace's theorem.** This is our main chapter on the classical theory 'Geometry of the zeros', of which the centrepiece is the beautiful theorem of Grace. Our approach to the theorem and its proof is unorthodox and is, in part, motivated by the need to set the scene for the powerful generalisations of chapter 8. We begin by introducing a general concept of duality defined in terms of convolutions. This is related to the functional analysis concept but is centred around a quite different class of functions: analytic functions with no zeros in the unit disc. Grace's theorem is stated as a second dual theorem and is closely related to G. Szegö's version expressed in terms of linear functionals. The proof given is due to the author, but is based on an ingenious idea due to St. Ruscheweyh, which we have expressed as the 'finite product lemma'. This lemma will form a central plank in the later convolution theory. After transforming Grace's theorem into a theorem for general circular regions we define apolarity and prove Grace's apolarity theorem and Walsh's theorem on symmetric linear forms. We then relate the theory to the more conventional geometric theory, proving a number of inequalities and giving an account of a number of Walsh's geometric theorems, including his celebrated 'two circles theorem'. We also include an account of the recent generalisation of Bernstein's theorem to rational functions due to Borwein and Erdélyi. At the expense of a slightly weaker result, we give a quite different proof using the methods of this chapter and also obtaining some other inequalities. We then establish the circle of theorems surrounding the Grace–Heawood theorem including the interesting 'bisector' theorem of Szegö. The chapter concludes with an extension of Grace's theorem for general linear operators.

Chapter 6. **The Ilieff–Sendov conjecture.** This problem, still unsolved, is one of those problems which create a mathematical itch. The question whether, given a polynomial with all its zeros in the unit disc, there is necessarily a critical point within unit distance of a given zero, is an innocent sounding proposition, which one would expect to be able to resolve quite quickly. Indeed Q.I. Rahman, who has contributed some of the most attractive ideas to the problem, once had the experience of mentioning this problem in a lecture attended by engineers, who laughed at the silliness of the question. He therefore offered a prize to any who could solve the problem by the end of the conference. For the remainder of the conference groups of engineers could be seen trying to answer this trivial question. Of course, Rahman kept his money. The problem is now approaching its fortieth birthday. However, let not age deter the bright young newcomer. Simple problems do sometimes have simple solutions, but nevertheless last a long time. The author has twice had the experience of producing simple solutions to old problems – one lasted over forty years, the other seventy-seven! (Both are in this book). However, a word of warning: simple solutions are rarely simply found and can cost a great deal of time and effort. Most (all?) mathematicians spend 99% of their time failing to solve the problems in which they are interested. The author would be delighted with a 1% success rate: the solution to the seventy-seven year old problem reduced to a short paragraph of verbal reasoning, but took eight years to discover.

For the Ilieff–Sendov conjecture, we concentrate on presenting some of the general ideas and methods, which have been obtained. A recent result is in an interesting and informative paper of J.E. Brown and G. Xiang [12], who establish the conjecture for polynomials up to degree 8 and further for polynomials with at most eight distinct zeros.

Chapter 7. **Self-inversive polynomials.** These are polynomials with their zeros either on the unit circle or distributed in pairs conjugate to the circle. After a few preliminaries our main attention is focused on polynomials whose zeros lie on the circle. We establish some inequalities for integral means and coefficients relating these to the maximum modulus on the circle. We also establish an interspersion theorem for the ratio of two polynomials whose zeros are interspersed on the circle. However, the bulk of the chapter is an exposition of T.J. Suffridge's startling theory of polynomials whose zeros lie on the circle and are separated by a minimum specified gap. He shows that such polynomials can be used to approximate some geometrically defined classes of analytic functions in the unit disc. Furthermore the polynomials can be classified according to a number of different criteria. However, his deepest and most significant result is his convolution theorem for such polynomials. This is undoubtedly one of the most beautiful results in the entire theory of polynomials and deserves to be much better known and understood. Unfortunately, because of the length and complexity of the proof, we are able to give only a partial account of his main argument. It would be a useful research project for someone to take apart and, if possible, simplify his methods. We make one suggestion for a possible simplification of one part of his theorem.

Chapter 8. **Duality and an extension of Grace's theorem to rational functions.** This chapter is the heart of our development of the convolution theory and contains the deepest results in the book. The central result is that the Kaplan classes (defined in chapter 7) lie in second dual spaces for very simple classes of functions. Part of the importance of duality theory lies in its connection with convexity and extreme point theory, and we explain this. Furthermore, the Kaplan classes form very general classes of non-vanishing analytic functions in the disc, and therefore the theorems and inequalities obtained give results of considerable generality and power. The theory is a development, taking place over a number of years, of the work of St. Ruscheweyh and the author arising out of their proof of the Pólya–Schoenberg conjecture: that the convolution of two convex univalent mappings of the disc remains convex univalent, and the chapter includes this result. However, we have emphasised in this book the importance to this development of the ideas and results of earlier authors: Szegö, Pólya and Schoenberg, Suffridge. Although the main theorems are not theorems on polynomials, nevertheless they arise naturally out of the polynomial theory, as we show, and in any case include a whole range of results for various types of finite and infinite products. Certain structure preserving classes arise in this theory and identifying these explicitly is a central problem. The author posed to his research student S. Robinson the problem of finding polynomial members. Robinson undertook a substantial amount of computer work and by means of some brilliant pattern spotting proposed certain specific polynomials. Although he had never heard of them, these turned out to be the generating polynomials for the Cesàro means! We have included his conjectures as beautiful unsolved problems.

Chapter 9. **Real polynomials.** This chapter centres on some questions of a quite specific type, namely estimating the number of real, or non-real, zeros or critical points of various types of real polynomials and rational functions. This constitutes a return to topological and geometric thinking typical of chapter 2. After introducing the Jensen circles we give a detailed discussion of the unresolved conjecture of Craven, Csordas and Smith (recently nicknamed the Hawaii conjecture, as the three authors are all from the University of Hawaii), concerned with relating the number of non-real zeros of a real polynomial to the number of real critical points of the logarithmic derivative. As this conjecture relates closely to the topological structure formed by the level curves on which the logarithmic derivative is real, it is a problem of fundamental interest in understanding the structure of real polynomials.

After a discussion of strongly real rational functions (rational functions which are real only on the real axis), we establish a general theorem on critical points. We then generalise this theory to real entire and meromorphic functions, concluding with a proof of Wiman's conjecture relating the genus of a real entire function of finite order possessing only real zeros to the number of non-real zeros of the second derivative. This is an account of the author's proof of the conjecture, in which we have simplified some of the original arguments in order to avoid the use of advanced methods and to make the argument self-contained.

Chapter 10. **Level curves.** Level curves have appeared many times earlier in the book. Here we establish two of the main facts: the relationship between polynomials in level regions and Blaschke products in the unit disc, and for rational functions the important Riemann–Hurwitz formula relating for a level region the number of zeros, critical points and components of the boundary.

The final part of the chapter centres around the fascinating Smales' conjecture, still unsolved after twenty years and originally formulated by S. Smale as a result he would like to be true for its usefulness in the theory of computer complexity. We have included our account in a chapter on level curves, because of the geometric orientation which study of the problem constantly arouses. In particular the conjecture raises questions concerning the relationship between the shape of level curves and the number of critical points (inside or near by). Although Hilbert long ago showed that the level curve of a polynomial could have any shape (specifically, given two Jordan curves, one inside the other thus creating an annular region beween the curves, one can construct a polynomial possessing a level curve lying inside the annular region and containing the inner of the two curves in its interior), nevertheless this says nothing about the critical points. The recent book by Borwein and Erdélyi [9] gives substantial coverage of the metric theory of level curves or 'lemniscates'.

Chapter 11. **Miscellaneous topics.** This chapter contains one extended piece of theory plus some isolated and striking results. Mason's *abc* theorem and its corollary Fermat's last theorem for polynomials were taken from a popular work on mathematics by Ian Stewart [60] simply for their attractiveness. We have expressed the theorem as a particular case of a result giving a lower bound on the number of points at which a rational function can take on specified values. Cohn's lemma is a well-known and useful result on finding the number of zeros of a polynomial in the unit disc. Blaschke products have been prominent in many places in the book. Here we give an extended account of a variety of features. Firstly we include Walsh's beautiful analogue of the Gauss–Lucas theorem and the connection with non-euclidean geometry. We then move on to a detailed account of the connections between Blaschke products and certain harmonic mappings of the disc, particularly the harmonic extension into the disc of a step function on the circle, of mappings onto convex curves and of mappings onto polygons. Included here are a proof of a conjecture of H.S. Shapiro on the Fourier coefficients of an n-fold mapping of the circle, a proof of the Rado–Kneser–Choquet theorem on the univalence of the harmonic extension of a mapping of the circle onto a convex curve and a proof of the finite valence of the analytic part of the harmonic extension for n-fold mappings. Also included is a statement of the as yet unsolved rotation conjecture of the author. The final topic of the book is Sudbery's proof of Popoviciu's conjecture on the number of distinct zeros of the successive derivatives of a polynomial. We extend Sudbery's result by considering the number of zeros of just some of the derivatives.

Concluding remarks. Prerequisites for this book are graduate level real analysis, basic topology and algebra and some complex variable theory, preferably up to and including the Riemann mapping theorem. We have made every effort to avoid advanced material and very technical knowledge in the hope that the work will then be available to a wide audience. We have included a scattering of problems for the reader; however, the book has not been written as a textbook; it is more a presentation of ideas, methods and especially proofs. The interested reader should certainly consult the classic works on polynomials, notably Marden [30], Walsh [69], Pólya and Szegö [41] and Levin [25], as well as more recent works: the book by Borwein and Erdélyi [9] and the forthcoming work of Rahman and Schmeisser cover numerous topics not mentioned in this book, including substantial amounts of work connected with approximation theory. Iteration theory is nowadays an important subject in its own right and is very fully covered in a number of works. Finally, the author has uncovered many errors in his various readings of the text. Inevitably many will remain undetected. The author apologises for these and emphasises that such errors are the author's errors and should not be attributed to the writers being discussed.

Notations

$\mathbb{C}$	set of complex numbers
$\mathbb{R}$	set of real numbers
$\mathbb{U}$	the open unit disc: $\{z \in \mathbb{C}: \|z\| < 1\}$
$\Re z$	real part of complex number z ($= x$ if $z = x + iy$, with x, y real)
$\Im z$	imaginary part of z ($= y$)
$\overline{z}$	complex conjugate of complex number z ($= x - iy$)
$\overline{A}$	topological closure of set A
$K(\alpha, \beta)$	Kaplan class: chapter 7, 7.4.5
Π_α	chapter 7, (7.107)
$S(\alpha, \beta)$	class of functions f of form $f = g/h$, where $g \in \Pi_\alpha$ and $h \in \Pi_\beta$
H^+	upper half-plane: $\{z: \Im z > 0\}$
H^-	lower half-plane: $\{z: \Im z < 0\}$
iff	abbreviation for 'if, and only if'
$\mathcal{P}(z)$	the Poisson function: $\mathcal{P}(z) = (1 - \|z\|^2)/\|1 - z\|^2$ for $\|z\| < 1$ (chapter 4, (4.13))
$\mathcal{H}$	class of complex-valued functions harmonic in $\mathbb{U}$
$\mathcal{A}$	class of functions analytic in $\mathbb{U}$
$\mathcal{A}_n$	class of complex polynomials of degree not exceeding n

1
The algebra of polynomials

1.1 Complex polynomials

1.1.1 Definitions

A **complex polynomial** is a function of the form

$$P(z) = \sum_{k=0}^{n} a_k z^k, \tag{1.1}$$

where the a_k are complex numbers not all zero and where z is a complex variable. We also use the terms **analytic** polynomial (reflecting the fact that the polynomial is an analytic function) and **algebraic** polynomial (since the polynomial contains only algebraic operations on the variable z). If $a_n \neq 0$ the polynomial is said to have **degree** n. In particular, a polynomial of degree 0 is, by definition, a non-zero constant. The function which is identically zero is often regarded as being a polynomial of degree $-\infty$. When the a_k are all real numbers, the polynomial $P(z)$ is called a **real polynomial**. Observe that $P(z)$ is a real polynomial iff

$$\overline{P(\overline{z})} = P(z) \tag{1.2}$$

for all $z \in \mathbb{C}$. From this it follows that, if $P(a) = 0$ then $P(\overline{a}) = 0$; therefore either a is real or P has the **conjugate pair** of zeros a and $\overline{a}$.

1.1.2 Number of zeros

Lemma *A complex polynomial of degree n has **at most** n zeros.*

The proof of this is an entirely elementary fact from algebra; this is to be contrasted with the stronger **fundamental theorem of algebra** (see chapter 2) which states that a polynomial of degree $n \geq 0$ has exactly n zeros; the usual proofs of this use methods of analysis or topology (it is not a result which follows purely from the algebraic field property of the complex numbers). The proof of the weaker statement is by induction. The result is trivial when $n = 0$. Assume the statement proved for polynomials of degree $n-1$, where $n \geq 1$, and let P be a polynomial of degree n given say by (1.1). Either P has no zeros and there is nothing further to prove, or $\exists a \in \mathbb{C}$ such that $P(a) = 0$. We then have

$$\frac{P(z)}{z-a} = \frac{P(z) - P(a)}{z-a} = \sum_{k=0}^{n} a_k \frac{z^k - a^k}{z-a} = \sum_{k=1}^{n} a_k \sum_{j=0}^{k-1} z^{k-1-j} a^j, \tag{1.3}$$

and the last expression is clearly a polynomial of degree $n-1$. Therefore by the induction hypothesis $P(z)/(z-a)$ has at most $n-1$ zeros and so $P(z)$ has at most n zeros. The result follows by induction.

This result has a number of important consequences.

1.1.3

Uniqueness theorem *If $P(z)$ and $Q(z)$ are polynomials of degree not exceeding n and if the equation*

$$P(z) = Q(z) \tag{1.4}$$

is satisfied at $n+1$ distinct points, then $P = Q$.

For otherwise, $P - Q$ is a polynomial of degree not exceeding n with $n+1$ zeros. We deduce the next result.

1.1.4

Theorem: Lagrange's interpolation formula *Let $z_1, z_2, \ldots, z_{n+1}$ be $n+1$ distinct points and let $w_1, w_2, \ldots, w_{n+1}$ be arbitrary complex numbers (not necessarily distinct but not all zero). Among all polynomials of degree not exceeding n there is a unique polynomial $P(z)$ such that*

$$P(z_k) = w_k \quad (1 \leq k \leq n+1). \tag{1.5}$$

This has the representation

$$P(z) = \sum_{k=1}^{n+1} w_k \frac{Q(z)}{Q'(z_k)(z - z_k)}, \tag{1.6}$$

where $Q(z) = \prod_{k=1}^{n+1}(z - z_k)$.

By the previous result there is at most one such polynomial; on the other hand the polynomial so represented is immediately seen to have the desired property.

1.1.5

An alternative formulation of Lagrange's formula is the following.

Corollary *Let $z_1, z_2, \ldots, z_{n+1}$ be $n+1$ distinct points and let $Q(z) = \prod_{k=1}^{n+1}(z - z_k)$. If $P(z)$ is a polynomial of degree not exceeding n, then*

$$P(z) = \sum_{k=1}^{n+1} P(z_k) \frac{Q(z)}{Q'(z_k)(z - z_k)}. \tag{1.7}$$

In other words we have an explicit method for determining the values of a polynomial of degree n in terms of its values at $n + 1$ known points.

1.1.6

Example Let $Q(z) = z^{n+1} - 1 = \prod_{k=0}^{n}(z - \omega_k)$, where $\omega_k = e^{2\pi ik/(n+1)}$ are the $(n + 1)$th roots of unity. Then, if $P(z)$ is a polynomial of degree not exceeding n, we have

$$P(z) = \frac{1}{n+1} \sum_{k=0}^{n} P(\omega_k) e_n(\overline{\omega}_k z), \tag{1.8}$$

where $e_n(z) = 1 + z + \cdots + z^n$.

1.1.7 Representation for harmonic polynomials

A **harmonic polynomial** $T(z)$ is a function of the form $T(z) = \overline{Q(z)} + P(z)$, where Q and P are analytic polynomials, and so is a complex-valued harmonic function in $\mathbb{C}$ (the complex plane). T can be represented in the form

$$T(z) = \sum_{k=-n}^{n} a_k r^{|k|} e^{ik\theta} \quad (z = re^{i\theta}). \tag{1.9}$$

T has **degree** n if either a_n or a_{-n} is non-zero. Note that a harmonic polynomial is the sum of a polynomial in the variable $\overline{z}$ and a polynomial in the variable z. The polynomial is uniquely determined by the $2n + 1$ coefficients a_k. These coefficients also determine uniquely the **trigonometric polynomial**

$$T(e^{i\theta}) = \sum_{k=-n}^{n} a_k e^{ik\theta}. \tag{1.10}$$

Indeed, given this latter expression the coefficients can be recovered from the formula

$$a_k = \frac{1}{2\pi}\int_0^{2\pi} e^{-ik\theta}T(e^{i\theta})d\theta \quad (-n \le k \le n). \tag{1.11}$$

We can apply Lagrange's interpolation formula to obtain a representation for a harmonic polynomial as follows. Let $\zeta_k = e^{2\pi ik/(2n+1)}$ $(0 \le k \le 2n)$ and set

$$D_n(z) = \overline{z}^n + \cdots + \overline{z} + 1 + z + \cdots + z^n. \tag{1.12}$$

Then, *if $T(z)$ is a harmonic polynomial of degree not exceeding n*, we have

$$T(z) = \frac{1}{2n+1}\sum_{k=0}^{2n} T(\zeta_k)D_n(\overline{\zeta}_k z). \tag{1.13}$$

This follows easily by applying example 1.1.6 to $e^{in\theta}T(e^{i\theta})$, which is a polynomial of degree at most $2n$ in the variable $e^{i\theta}$. We obtain the above formula on the unit circle. Since both sides of the equation are harmonic polynomials, it follows that the equality holds for all $z \in \mathbb{C}$.

1.2 The number of zeros of a real analytic polynomial

1.2.1

Definition A **real analytic** polynomial is an expression of the form

$$P(x, y) = \sum_{j=0}^{m}\sum_{k=0}^{n} a_{j,k}x^j y^k \tag{1.14}$$

where the coefficients $a_{j,k}$ are real or complex numbers and where x and y are real variables. The **degree** of the term $a_{j,k}x^j y^k$ is $j + k$ provided that $a_{j,k} \ne 0$, and the **degree** of P is the largest of the degrees of the individual terms. Every algebraic polynomial and every harmonic polynomial is real analytic and their degrees earlier defined agree with the above definition.

1.2.2

Bézout's theorem *Let $P(x, y) = u(x, y) + iv(x, y)$ (with u and v real) be a complex-valued real analytic polynomial, where u has degree m and v has degree n, and suppose that u and v are relatively prime (i.e. contain no non-trivial common factors). Then P has at most mn zeros in $\mathbb{C}$.*

Proof This is an algebraic result which is proved using linear techniques. Firstly, we may note that $P(0, y)$ is not identically zero, for otherwise x is a common factor of u and v. We then write u and v in the form

$$u(x, y) = \sum_{k=0}^{m} u_k(x) y^k, \qquad v(x, y) = \sum_{k=0}^{n} v_k(x) y^k, \tag{1.15}$$

where u_k are real polynomials in x of degree at most $m-k$ and v_k are real polynomials in x of degree at most $n-k$. Also, after an affine transformation of the variables, we may assume that u_m and $v_n \neq 0$. (**Affine transformations** are first degree real analytic polynomial mappings, which are bijective mappings of $\mathbb{C}$; for an account of these see chapter 2, section 2.7.21.) Now the equation $P = 0$ holds iff $u = 0$ and $v = 0$ simultaneously. Let us assume this is the case at a particular pair (x, y). We use the **method of Sylvester** to eliminate the quantity y from the two equations $u(x, y) = 0$, $v(x, y) = 0$. The method is to observe that the following $m + n$ equations hold:

$$y^\mu u(x, y) = 0 \quad (0 \leq \mu \leq n-1); \qquad y^\nu v(x, y) = 0 \quad (0 \leq \nu \leq m-1). \tag{1.16}$$

Here $y^0 = 1$ by definition. Each of these equations can be interpreted as the vanishing of a linear combination of the $m+n$ quantities $1, y, y^2, \ldots, y^{m+n-1}$. This implies the vanishing of the determinant of coefficients:

$$D(x) = \begin{vmatrix} u_0(x) & u_1(x) & u_2(x) & 0 \\ 0 & u_0(x) & u_1(x) & u_2(x) \\ v_0(x) & v_1(x) & v_2(x) & 0 \\ 0 & v_0(x) & v_1(x) & v_2(x) \end{vmatrix} = 0. \tag{1.17}$$

This example is the case $m = n = 2$. The determinant $D(x)$ is a polynomial in x. We will show that (a) $D(x)$ *does not vanish identically* and (b) $D(x)$ *has degree at most* mn. When these facts are established, it will follow that there are at most mn values of x for which the determinant can vanish. For each such x there are only finitely many values of y such that $P(x, y) = 0$. Thus we obtain at most finitely many points (x, y) for which $P(x, y) = 0$. Because of this we may now perform an affine transformation on (x, y) so that, after the transformation to new variables x', y', the resulting polynomial has for each fixed x' at most one zero in y'. (This is easily seen geometrically; e.g. a small rotation will do.) It follows immediately from the above applied to the new coordinates that P has at most mn zeros altogether, proving the theorem.

To prove (b) we let $w_{i,j}$ denote the terms of the $(m+n) \times (m+n)$ matrix of coefficients and observe that each term is a polynomial in x whose degree satisfies

$$\text{degree of } w_{i,j} \leq \begin{cases} m - j + i & (1 \leq i \leq n), \\ -j + i & (n+1 \leq i \leq m+n). \end{cases} \tag{1.18}$$

Here we have taken the degree of the identically zero polynomial to be $-\infty$. Now the value of the determinant is given by the formula

$$|w_{i,j}| = \sum(-1)^{\sigma} \prod_{i=1}^{m+n} w_{i,\phi(i)} \tag{1.19}$$

where the sum ranges over all permutations $\phi(i)$ of the numbers $1, 2, \ldots, m+n$, and where $\sigma = 0$ for even permutations and $\sigma = 1$ for odd permutations. It follows that the degree of the determinant cannot exceed

$$\sum_{i=1}^{n}(m - \phi(i) + i) + \sum_{i=n+1}^{m+n}(-\phi(i) + i) = mn + \sum_{i=1}^{m+n}(-\phi(i) + i) = mn, \tag{1.20}$$

since $\sum \phi(i) = \sum i$, as $\phi(i)$ runs through the numbers i in some order. This proves (b).

To prove (a) we assume that the determinant vanishes identically and show that this implies that u and v have a common factor, contradicting the hypotheses. Firstly, we observe that the vanishing of the determinant is equivalent to the vanishing of the determinant of the transpose matrix. Secondly, we observe that the elements of the matrix are polynomials in x and therefore belong to the field of rational functions in x. Because the determinant vanishes for all x, we may interpret this as the statement that the determinant of this matrix consisting of elements of this field vanishes, as all the operations required to evaluate the determinant are algebraically defined over any field. Therefore from the algebraic theory of determinants we can assert the existence of elements α_k, β_k of the field not all zero and satisfying the set of equations

$$u_0\alpha_0 + v_0\beta_0 = 0, \qquad u_1\alpha_0 + u_0\alpha_1 + v_1\beta_0 + v_0\beta_1 = 0, \qquad \ldots \text{etc.} \tag{1.21}$$

or generally

$$\sum_{k=r-m}^{r} u_{r-k}\alpha_k + \sum_{k=r-n}^{r} v_{r-k}\beta_k = 0 \quad (0 \leq r \leq m+n-1), \tag{1.22}$$

where $\alpha_k = 0$ for $k \geq n$ and $\beta_k = 0$ for $k \geq m$. We now set

$$A(y) = \sum_{k=0}^{n-1} \alpha_k y^k, \qquad B(y) = -\sum_{k=0}^{m-1} \beta_k y^k \tag{1.23}$$

and note that by multiplying out and equating coefficients to zero the above set of equations is equivalent to the single equation

$$uA - vB = 0. \tag{1.24}$$

Since not both A and B are zero, it follows from this equation that neither A nor B is the zero element, as otherwise one of u and v is zero, contradicting the hypothesis. Now the coefficients of A and B are rational functions in x; therefore multiplying through by a suitable polynomial in x, we obtain an equation of the same form with A and B polynomials in both x and y. Furthermore, we may divide out the equation by any common factor of A and B. Thus we obtain an equation of the above form with A and B relatively prime polynomials, where the degree of A in y is smaller than the degree of v in y and the degree of B in y is smaller than the degree of u in y. We will show that $f = v/A = u/B$ is a polynomial in x and y, which is then the required common factor of u and v. Consider first the case when B is constant in the variable y, and so a pure polynomial in x. B can be factorised as a product of irreducible polynomials; hence it is enough to show that each irreducible factor R of B is a factor of u. Expanding u in powers of y, this is true iff every coefficient of u is divisible by R. Write $u = u_1 + u_2$, where u_1 consists of those terms in the expansion of u whose coefficients are divisible by R, and u_2 consists of the remaining terms. Similarly, write $A = A_1 + A_2$. Since R is not a factor of A, $A_2 \neq 0$. Thus we have $vB = Au = A_1u_1 + A_1u_2 + A_2u_1 + A_2u_2$, from which we see that A_2u_2 is divisible by R. This implies that $u_2 = 0$, as otherwise the lowest coefficient of this product in y is divisible by R, but also is the product of a coefficient of A_2 and a coefficient of u_2, which is not divisible by R. Thus B divides u in this case.

Returning to the general case, from the usual division algorithm for polynomials we can write

$$\frac{v}{A} = f + \frac{\rho}{A}, \qquad \frac{u}{B} = g + \frac{\sigma}{B} \tag{1.25}$$

where f, g, ρ and σ are polynomials in y with rational coefficients in x, and where ρ has y-degree smaller than A and σ has y-degree smaller than B. Hence as $y \to \infty$, ρ/A and $\sigma/B \to 0$ and so $f - g \to 0$. This clearly implies that $f = g$. We therefore obtain $\rho/A = \sigma/B$ and so either (i) $\rho = \sigma = 0$ or (ii) $A/\rho = B/\sigma$. In case (ii) we see that we are back to the same problem for A and B that we had for u and v, but A and B have smaller degrees. Therefore we may use an induction argument to deduce that A and B have a common factor, contradicting our legitimate assumption. It follows that case (i) holds and so f is a common factor of u and v. The coefficients of f are rational functions in x, but by multiplying through by the smallest polynomial p in x to cancel out the denominators of the coefficients, we find as above that p divides both A and B. Since A and B are relatively prime, f is a polynomial in both x and y. This proves (a) and so completes the proof of Bézout's theorem.

1.2.3

Bézout's theorem can be used to put a bound on the number of zeros of a real analytic polynomial even in the absence of algebraic information on the polynomial. For example it may often be the case that on analytic or topological grounds one can establish the finiteness of the number of zeros. In this case we have the following result.

Theorem *Let $P(x, y) = u(x, y) + iv(x, y)$ be a complex-valued real analytic polynomial, where u has degree $m \geq 0$ and v has degree $n \geq 0$. Then P has at most mn isolated zeros in $\mathbb{C}$. In particular, if P has at most finitely many zeros, then P has at most mn zeros in $\mathbb{C}$.*

Proof The following argument is an adaptation of a method given by Wilmshurst [72]. If $m = 0$, then u is a non-zero constant and so P has no zeros; similarly if $n = 0$. If $m = 1$, then either (i) u divides v or (ii) u and v are relatively prime. In case (ii) Bézout's theorem shows that P has at most n zeros; in case (i) P vanishes exactly when u vanishes, which is on a line. Hence P has no isolated zeros. Thus the theorem follows in this case. We proceed by induction and assume the theorem is true for polynomials $U + iV$, where U has degree smaller than m and V has degree smaller than n. If u and v are relatively prime, then Bézout's theorem gives the result. Therefore we may assume that u and v have highest common factor p say, where p is a real polynomial in (x, y) of degree $r \geq 1$. Then u/p and v/p are relatively prime polynomials of degrees $m - r$ and $n - r$ respectively, and so P/p has at most $(m - r)(n - r)$ zeros. The remaining zeros of P are the zeros of p. Now p is a real-valued continuously differentiable function in the plane. It follows from the implicit function theorem (see chapter 2) that at a zero c of p either (i) p vanishes on a curve passing through c or (ii) both partial derivatives p_x and p_y are zero at c. In case (i) c is a non-isolated zero of P. Thus if c is an isolated zero of P, which is also a zero of p, then c is an isolated zero of p, and so from case (ii) the polynomial $Q = p + ip_x$ has an isolated zero at c. Since p has degree r and p_x has degree $r - 1$, it follows by the induction hypothesis that Q has at most $r(r - 1)$ isolated zeros. Hence P has at most $(m - r)(n - r) + r(r - 1) < mn$ isolated zeros. The result follows by induction.

1.2.4 A complex approach

It is clear that Bézout's theorem will be equally valid for a pair of complex polynomials $u(z, \zeta)$, $v(z, \zeta)$ of complex variables z and ζ. From the two equations $u(z, \zeta) = 0$ and $v(z, \zeta) = 0$ we can use Sylvester's method to eliminate the variable ζ, obtaining a determinant $D(z)$, which is an algebraic polynomial in z whose zeros contain those z for which there exists a pair (z, ζ) satisfying the two equations. The maximum degree of D is the product of the degrees of u and v. We can apply this result to our original problem of finding the zeros of a complex-valued real analytic polynomial P. P is a polynomial in the real variables x and y; by making the substitutions $x = \frac{1}{2}(z + \overline{z})$, $y = \frac{1}{2i}(z - \overline{z})$, P becomes a polynomial $P(z, \overline{z})$ in the 'variables' z and $\overline{z}$; in other words a real analytic polynomial can be written in the form of a finite sum

$$P(z, \overline{z}) = \sum_{j,k} a_{j,k} \overline{z}^j z^k, \tag{1.26}$$

where the $a_{j,k}$ are complex numbers and the degree of P is the maximum of $j + k$ over terms where $a_{j,k} \neq 0$. This is clearly the restriction of $P(z, \zeta) = \sum_{j,k} a_{j,k} \zeta^j z^k$ to $\zeta = \overline{z}$. To apply the Sylvester method of elimination we require two equations. However, $P(z, \overline{z}) = 0$ iff $\overline{P(z, \overline{z})} = 0$; thus we eliminate ζ from the two equations

$$P(z, \zeta) = 0, \qquad \overline{P(\overline{\zeta}, \overline{z})} = 0 \tag{1.27}$$

which take the form

$$\sum_{j,k} a_{j,k} \zeta^j z^k = 0, \qquad \sum_{j,k} \overline{a_{j,k}} z^j \zeta^k = 0. \tag{1.28}$$

We obtain the determinant equation $D(z) = 0$, where D is an algebraic polynomial in z whose degree does not exceed n^2, where n is the degree of P. D is identically zero iff the above pair of polynomials in z and ζ have a common factor. We leave it as an exercise to show that this holds iff, writing $P = u + iv$ (u, v real), u and v have a common factor. Thus in this formulation we have lost the distinction between the possible different degrees of u and v, as n is clearly the larger of these two degrees. On the other hand we have constructed an algebraic polynomial $D(z)$ whose set of zeros contains all the zeros of the original real analytic polynomial P. In fact, if (z, ζ) is any pair satisfying the above two equations, then $D(z) = 0$. In general it need not be the case that the ζ satisfies $\zeta = \overline{z}$, and therefore D is likely to have more zeros than P. However, there are certainly cases where D and P have the same zeros. Indeed this is the case if P is itself an algebraic polynomial: then $D = cP^n$, where c is a constant $\neq 0$. Of course, if we count multiplicities, then D has n^2 zeros, whereas P has only n zeros. We will give an example in chapter 2 (subsection 2.6.11) where D has n^2 simple zeros all of which are zeros of P.

It is worth recording here the form of the $2n \times 2n$ matrix of which D is the determinant. We have

$$P(z, \bar{z}) = \sum_{j,k} a_{j,k} \bar{z}^j z^k = \sum_j \Big(\sum_k a_{j,k} z^k \Big) \bar{z}^j = \sum_k \Big(\sum_j a_{j,k} \bar{z}^j \Big) z^k \tag{1.29}$$

and so

$$P(z, \bar{z}) = \sum_{j=0}^{n} b_j(z) \bar{z}^j, \qquad \overline{P(z, \bar{z})} = \sum_{j=0}^{n} c_j(z) \bar{z}^j, \tag{1.30}$$

where

$$b_j(z) = \sum_{k=0}^{n-j} a_{j,k} z^k, \qquad c_j(z) = \sum_{k=0}^{n-j} \overline{a_{k,j}} z^k. \tag{1.31}$$

The terms of the matrix are given by

$$\begin{aligned} w_{i,j} &= b_{j-i} \quad \text{for } i \le j \le n+i \quad \text{and} \quad 1 \le i \le n, \\ w_{i,j} &= c_{j-i+n} \quad \text{for } i-n \le j \le i \quad \text{and} \quad n+1 \le i \le 2n, \\ w_{i,j} &= 0 \quad \text{otherwise.} \end{aligned} \tag{1.32}$$

As recorded earlier these relations imply that the maximum degree of $w_{i,j}$ is given by

$$\text{degree of } w_{i,j} \le \begin{cases} n - j + i & (i \le j \le n+i \text{ and } 1 \le i \le n), \\ -j + i & (i - n \le j \le i \text{ and } n+1 \le i \le 2n). \end{cases} \tag{1.33}$$

Otherwise degree of $w_{i,j} = -\infty$ (so trivially satisfies the above inequalities). We then have

$$D(z) = \sum (-1)^\sigma \prod_{i=1}^{2n} w_{i,\phi(i)} \tag{1.34}$$

where the sum ranges over all permutations $\phi(i)$ of the numbers $1, 2, \ldots, 2n$, and where $\sigma = 0$ for even permutations and $\sigma = 1$ for odd permutations. Each product $\prod_{i=1}^{2n} w_{i,\phi(i)}$ has degree not exceeding $\sum_{i=1}^{n}(n - \phi(i) + i) + \sum_{i=n+1}^{2n}(-\phi(i) + i) = n^2$, with equality only if every term $w_{i,\phi(i)}$ in the product attains its maximum degree $n - \phi(i) + i$ for $1 \le i \le n$ or $-\phi(i) + i$ for $n + 1 \le i \le 2n$.

As an application of these remarks, suppose that P has coefficients $a_{j,k}$ where $a_{j,n-j} = 0$ for $1 \le j \le n$ and $a_{0,n} \ne 0$; in other words the only term of highest degree n in the expansion of P is the term in z^n. Then $b_0(z)$ has full degree n, and $c_0(z)$ has degree 0, but the remaining polynomials $b_j(z)$ $(1 \le j \le n)$ and $c_j(z)$ $(0 \le j \le n-1)$ have degree strictly smaller than $n - j$. It follows that the only permutation giving a product of degree n^2 is the case $\phi(i) = i$ $(1 \le i \le 2n)$. Thus $D(z)$ has exact degree n^2, and therefore P has at most n^2 zeros.

Collecting up the results of this subsection we have the next theorem.

1.2.5

Theorem *Let $P(z, \overline{z}) = u + iv$ (u and v real) be a complex-valued real analytic polynomial of degree n. Then the Sylvester resultant $D(z)$ of $P(z, \zeta)$ and $\overline{P(\overline{\zeta}, \overline{z})}$ is a polynomial in z of degree at most n^2. Every zero of $P(z, \overline{z})$ is a zero of $D(z)$; indeed, if $P(z, \zeta) = \overline{P(\overline{\zeta}, \overline{z})} = 0$, then $D(z) = 0$. $D(z)$ vanishes identically if, and only if, u and v have a non-trivial common factor. Finally, if*

$$P(z, \overline{z}) = Q(z, \overline{z}) + Az^n, \tag{1.35}$$

where Q is a real analytic polynomial of degree $< n$ and A is a non-zero constant, then $D(z)$ has exact degree n^2 and therefore P has at most n^2 zeros. Indeed, in this case, P attains every value w at most n^2 times and is therefore an n^2-valent mapping of the plane.

The final remark comes from applying the theorem to $P - w$. We shall see in chapter 2 that P of this form does actually attain every value w, and so is a surjective mapping of the plane.

1.2.6 The Sylvester resultant of $P - w$

In studying the valence of a mapping $P(z, \overline{z})$ it is natural to consider the Sylvester resultant for $P - w$, where w is an arbitrary complex number. The resultant will take the form $D(z, w, \overline{w})$, where this is an algebraic polynomial in z and a real analytic polynomial in w. Furthermore, if P has degree n, then as a real analytic polynomial in w, D has at most degree n. This follows from the fact that the only terms involving w in the matrix are the terms $b_0 - w$ and $c_0 - \overline{w}$, so w only appears with degree 1 in $w_{i,j}$ if $j = i$ for $1 \le i \le n$ and if $j = i - n$ for $n + 1 \le i \le 2n$. All other terms have degree 0 or $-\infty$ in w. For any permutation ϕ of the numbers $1, 2, \ldots, 2n$, there are at most n values i such that (i) $\phi(i) = i$ where $1 \le i \le n$ and (ii) $\phi(i) = i - n$ where $n + 1 \le i \le 2n$.

1.2.7 Evaluating the Sylvester resultant

Let $b_0, b_1, \ldots, b_n$ and $c_0, c_1, \ldots, c_n$ be the generating elements of the $2n \times 2n$ Sylvester matrix. Then the resultant can be written in the form

$$\sum_{\phi} (-1)^{\sigma} \prod_{i=1}^{n} b_{\phi(i)-i}\, c_{\phi(i+n)-i} \tag{1.36}$$

summed over all permutations ϕ of the numbers $1, 2, \ldots, 2n$ such that for $1 \le i \le n$

$$0 \le \phi(i) - i \le n, \qquad 0 \le \phi(i+n) - i \le n. \tag{1.37}$$

Note that, for each permutation ϕ, the sum of the subscripts of the bs and the cs is exactly n^2.

1.2.8 Further properties of the resultant

Let u and v be real analytic polynomials of degrees m and n respectively written in the form

$$u(x, y) = \sum_{k=0}^{m} u_k(x) y^k, \qquad v(x, y) = \sum_{k=0}^{n} v_k(x) y^k, \tag{1.38}$$

where $u_m \neq 0$ and $v_n \neq 0$. If the resultant is not identically zero, then the set of equations

$$\sum_{k=r-m}^{r} u_{r-k} \alpha_k + \sum_{k=r-n}^{r} v_{r-k} \beta_k = \lambda_r \quad (0 \le r \le m+n-1) \tag{1.39}$$

can be uniquely solved for α_k and β_k, where $\alpha_k = 0$ for $k \ge n$ and $\beta_k = 0$ for $k \ge m$. Here the λ_r are given arbitrary elements in the field of rational functions in x, and the solutions α_k and β_k are then also elements in this field. Writing

$$\Lambda(x, y) = \sum_{r=0}^{m+n-1} \lambda_r(x) y^r, \quad A(x, y) = \sum_{k=0}^{n-1} \alpha_k(x) y^k, \quad B(x, y) = \sum_{k=0}^{m-1} \beta_k(x) y^k, \tag{1.40}$$

the above equations are equivalent to the single equation

$$uA + vB = \Lambda. \tag{1.41}$$

In other words, *if the resultant of u and v does not vanish identically, then given Λ of degree in y at most $m + n - 1$ we can find a unique A of degree at most $n - 1$ in y and a unique B of degree at most $m - 1$ in y to satisfy this equation.*

In particular, taking $\Lambda = 1$, *we can find a unique p of degree at most $n - 1$ and a unique q of degree at most $m - 1$ such that*

$$up + vq = 1. \tag{1.42}$$

Here p and q are polynomials in y with coefficients which are rational functions in x. If $R(x)$ is the lowest common denominator of these rational coefficients, then we obtain the relation

$$uP + vQ = R \tag{1.43}$$

where $P = pR$, $Q = qR$ are now polynomials in both x and y. The polynomial $R(x)$ is a factor of the resultant: if we use the Sylvester matrix to solve the equation $up + vq = 1$ for p and q, then, applying Cramer's rule, the solutions for each of the coefficients of p and q (expanded as polynomials in y) will be the ratio of two determinants; the numerator is $\pm$ the determinant of one of the minors of the matrix corresponding to the first column; the denominator is the determinant of the matrix, i.e. the Sylvester resultant. Since u and v have polynomial coefficients, it follows that the Sylvester resultant is a common denominator of all the rational coefficients of p and q. Hence the lowest common denominator R is a factor of the resultant; if R is a proper factor, then there is a factor of the resultant which is a common factor of the determinants of each of the minors corresponding to the first column.

1.3 Real analytic polynomials at infinity

1.3.1 Resolving the singularity: the blow-up method

Let $P(x, y)$ be a real analytic polynomial of degree N. The limiting behaviour at infinity of P can be determined by performing a sequence of transformations, which we now describe. Firstly, after a suitable affine transformation we may assume that both the real and imaginary parts of $P(0, y)$ have full degree N. We then make the transformation

$$x \mapsto 1/x, \; y \mapsto y/x, \tag{1.44}$$

obtaining

$$R(x, y) = P\left(\frac{1}{x}, \frac{y}{x}\right) = \frac{Q(x, y)}{x^N}, \tag{1.45}$$

where $Q(x, y)$ is a polynomial of degree N such that $Q(0, y)$ has degree N. Any finite limiting value of P at infinity is then a limiting value of R as $x \to 0$ and y remains bounded. We expand Q in powers of x, obtaining

$$R(x, y) = \frac{\sum_{k=0}^{N} Q_k(y) x^k}{x^N}. \tag{1.46}$$

It is clear that $R(x, y) \to \infty$ as $x \to 0$ and y remains bounded, unless y tends to a real zero of $Q_0(y)$; for otherwise the term $Q_0(y)/x^N$ dominates the expression. Thus to determine the possible finite limiting values of R we will consider $x \to 0$ through positive values and $y \to a$, where a is a real zero of Q_0. We write

$$Q_k(y) = (y - a)^{r_k} S_k(y) \quad (0 \le k \le N), \tag{1.47}$$

where r_k is the multiplicity of the zero a of Q_k (if $Q_k(a) \neq 0$, then $r_k = 0$); if $Q_k \equiv 0$, for the purposes of this discussion we may take $r_k = +\infty$. Next, let us consider the substitution

$$y = a + tx^p, \tag{1.48}$$

where p is a positive number to be determined. Note that the mapping $(x, y) \mapsto (x, t)$ is a 1–1 mapping of the half-plane $x > 0$ onto itself. We obtain

$$R(x, y) = x^{-N}\left(\sum_{k=0}^{N} S_k(y)t^{r_k}x^{pr_k+k}\right). \tag{1.49}$$

Now observe that, for small positive p, $pr_0 < pr_k + k\,(1 \leq k \leq N)$ and the dominating term is $S_0(a)t^{r_0}/x^{N-pr_0}$. If $pr_0 = N$ and $pr_0 < pr_k + k\,(1 \leq k \leq N)$, then with the choice $p = N/r_0$ we obtain the limiting values $S_0(a)t^{r_0}$ (t real). If r_0 is odd, this is a simply described straight line through the origin; if r_0 is even, it is a doubly described ray with endpoint at the origin. Otherwise, we choose $p > 0$ to satisfy $pr_0 = pr_j + j$ for some j and $pr_0 \leq pr_k + k\,(1 \leq k \leq N)$; in other words

$$p = \min \frac{j}{r_0 - r_j}, \tag{1.50}$$

where the minimum is taken over those j $(1 \leq j \leq N-1)$ for which $r_0 > r_j$. We obtain

$$R(x, a + tx^p) = \frac{F(x, t)}{x^{N-pr_0}}, \tag{1.51}$$

where $F(x, t)$ is polynomial in t of degree at most N, and contains fractional powers of x – in fact is polynomial in x and x^p. Note also that $F(0, t)$ is a polynomial in t of degree r_0, which contains only those powers t^{r_j} for which $pr_0 = pr_j + j$. Suppose now that $p = \mu/\nu$, where μ and ν are relatively prime natural numbers. Then with the substitution $x = s^\nu$ we obtain

$$R(s^\nu, a + ts^\mu) = \frac{A(s, t)}{s^{\nu N - \mu r_0}}, \tag{1.52}$$

where $A(s, t)$ is a polynomial in s and t and $A(0, t) = F(0, t)$ has degree r_0. Furthermore, since the powers t^{r_j} occurring in $A(0, t)$ satisfy $pr_0 = pr_j + j$, we obtain

$$r_0 - r_j = \frac{\nu j}{\mu}, \tag{1.53}$$

and, since μ and ν are relatively prime and $r_0 - r_j$ is an integer, μ divides j and so $r_0 - r_j$ is an integral multiple of ν. It follows that $A(0, t)$ has the form

$$A(0, t) = t^\alpha B(t^\nu), \tag{1.54}$$

where α is a non-negative integer and $B(t)$ is a polynomial in t of degree ≥ 1.

1.3.2

Now the form of R after this transformation is of the same general type in the variables s and t as it was in the variables x and y, namely a polynomial in s and t divided by a power of s. Therefore it is open to us to repeat this reducing process. Indeed, it follows from our comparison principle (see 1.3.6 below) that any finite limiting value of R will be attained along some curve $y = a + tx^p$, with $x \to 0$ and t remaining bounded, where p is a positive and indeed rational number. On the other hand $R \to \infty$ along such curves, if p is smaller than the above minimum choice. Therefore, finite limiting values can only be attained for either this or larger values of p. This implies that with our chosen minimum value of p and the above transformation, limiting values of R (associated with the zero a) will be attained with $s \to 0$ and t remaining bounded (in fact $t \to 0$ if the limiting value corresponds to a larger value of p). As before, each limiting value will correspond to a real zero, b say, of the polynomial $A(0, t)$. If the multiplicity of the zero b is s_0, then we make a transformation $t = b + us^q$, where q is chosen according to the same minimum process with which we chose p. Now clearly $s_0 \leq r_0 \leq N$; if $s_0 = r_0$, then $A(0, t) = c(t - b)^{r_0}$, where c is a non-zero constant, and therefore $A(0, t)$ contains all powers of t from 0 to r_0. It follows that $\nu = 1$ and so the power of s in the denominator is at most $N - r_0$. We see, therefore, that, if we continue repeating the process, at each step either the degree of the leading term strictly decreases or the power of the denominator strictly decreases by an integer amount. Thus eventually the process will terminate in the following way. Using the original notation $Q(x, y)/x^N$ as a standard form, the final transformation will take the form $y = a + tx^p$, where $p = N/r_0$. As earlier described either this will lead to a line or ray of asymptotic values or this choice of p will coincide with the minimum choice $p = \min(j/(r_0 - r_j))$. Following the transformation as before we obtain a curve of asymptotic values $A(0, t)$, a polynomial in the real variable t of degree $r_0 \geq 1$. Indeed, as the denominator is eliminated at the final step, the sequence of transformations will send our original polynomial P to a new polynomial $A(s, t)$. It is clear that for the existence of asymptotic values it is both necessary and sufficient that there exist such a finite sequence of transformations from P to another polynomial A. If we perform our construction for every real zero of our leading term at every stage, we will obtain all possible asymptotic values with $x > 0$ tending to ∞. We can obtain in this way a maximum of $N =$ degree of P asymptotic value curves, which are polynomial images of the real axis. Of course, if at one or other stage the leading term has no real zeros, then the polynomial tends to ∞ along the resulting tract. In a similar way, to obtain the asymptotic values of P with $x < 0$ tending to $-\infty$, we simply apply the same reasoning to the function $(-1)^N Q(-x, y)/x^N$, again taking $x > 0$ and tending to 0. This gives another N possible asymptotic value curves.

1.3.3 Repetition of asymptotic values

However, as we shall see, each asymptotic value of P at ∞ is repeated along quite separate tracts, and in general each asymptotic value curve appears twice (except in one circumstance), either repeated or reversed. This is a consequence of the fact that, according to the above reasoning, the asymptotic values are attained as limits along rational algebraic curves of a particularly simple type. To be specific, if we put together the sequence of transformations from the initial $R(x, y)$ to the final polynomial $A(s, t)$, we obtain a single transformation of the form

$$x = s^q, \qquad y = \phi(s) + ts^\beta, \tag{1.55}$$

where q and β are natural numbers and where $\phi(s)$ is a real polynomial of degree $< \beta$; furthermore, the highest common factor of the powers of s appearing in the expression $\phi(s) + ts^\beta$ is 1. In short, we have

$$R(s^q, \phi(s) + ts^\beta) = A(s, t). \tag{1.56}$$

If $\phi(s) = \sum_{k=0}^{m} c_k s^k$, then the coefficients c_k are zeros of successive leading terms in our sequence of transformations. Note that the mapping $(s, t) \mapsto (x, y)$ is a 1–1 mapping of the half-plane $\{s > 0\}$ onto the half-plane $\{x > 0\}$. Indeed, this is clearly the case at each stage of the resolution process, which consists of blowing up a chosen zero into an entire line, the remainder of the line being pushed off to ∞. Furthermore, the mapping is 1–1 from $\{s < 0\}$ onto $\{x < 0\}$, if q is odd, and onto $\{x > 0\}$, if q is even. The curve $A(0, t)$ is a limiting curve of asymptotic values as $s \to 0$ through either positive values of s or negative values of s. If q is odd, the approach through negative values of s will correspond in the (x, y)-plane to a tract in $\{x < 0\}$ diametrically opposite the constructed tract in $\{x > 0\}$, and therefore the asymptotic value curve is repeated in a completely different portion of the plane. If q is even, the curve will again appear twice as limits along separate tracts in $\{x > 0\}$, unless $\phi(-s) = \phi(s)$.

This last possibility will occur if $\phi(s)$ is a polynomial in s^2 and β is odd. We will show that repetition of asymptotic values still occurs, though in a slightly strange way. Let d be the highest power of 2 which is a factor of q and the powers of s appearing in the expansion of ϕ. Then $x = s^{\sigma d}$, $y = \psi(s^d) + ts^\beta$, where $\psi(u)$ is a real polynomial in u and $\sigma d = q$. Now suppose that

$$\omega^d = 1 \quad \text{and} \quad \gamma = \overline{\omega}^\beta; \tag{1.57}$$

then

$$x = (\omega s)^{\sigma d}, \qquad y = \psi((\omega s)^d) + (\gamma t)(\omega s)^\beta \tag{1.58}$$

and so

$$A(\omega s, \gamma t) = A(s, t). \tag{1.59}$$

In particular, since d is even and β is odd, this holds when $\omega = -1, \gamma = -1$; i.e.

$$A(-s, -t) = A(s, t), \tag{1.60}$$

and in particular $A(0, -t) = A(0, t)$. It follows that our polynomial curve is an even function and so has the form $A(0, t) = B(t^2)$. The curve traced out is thus $B(t)$ from $+\infty$ to 0, followed by the reverse curve $B(t)$ from 0 to $+\infty$; i.e. we obtain half of $B(t)$ traversed twice. We show that in a separate tract the other half of $B(t)$ with $t < 0$ is similarly traversed. Firstly, note that either (i) $\sigma = q/d$ is odd, or (ii) σ is even. In case (i) we consider the tract given by

$$x = -s^{\sigma d} = (\lambda s)^{\sigma d}, \qquad y = \psi(-s^d) + ts^\beta = \psi((\lambda s)^d) + \eta t(\lambda s)^\beta, \tag{1.61}$$

where $\lambda^d = -1$ and $\eta = \overline{\lambda}^\beta$. Thus

$$R(x, y) = A(\lambda s, \eta t) \to A(0, \eta t) \text{ as } s \to 0. \tag{1.62}$$

Now $A(0, t) = A(0, \gamma t)$ and so the non-vanishing terms in the expansion of $A(0, t)$ involve only those powers t^k for which $\gamma^k = 1$; i.e. $\omega^{\beta k} = 1$. If $\omega = e^{2\pi i/d}$, this implies that d divides βk and so, as d is a power of 2 and β is odd, d divides k. Thus $A(0, t)$ has the form

$$A(0, t) = F(t^d), \tag{1.63}$$

where $F(t)$ is a real polynomial. Thus $A(0, \eta t) = F(\eta^d t^d) = F(-t^d)$. Thus with $s \to 0$ through positive values, $x \to 0$ through negative values; i.e. our tract lies in the negative half-plane and has limiting values the half-curve described by $F(t)$ with $t < 0$, as asserted. In case (ii) we consider the tract

$$x = s^{\sigma d} = (\lambda s)^{\sigma d}, \qquad y = \psi(-s^d) + ts^\beta = \psi((\lambda s)^d) + \eta t(\lambda s)^\beta, \tag{1.64}$$

and note that exactly the same reasoning applies, but our tract will now lie in the positive half-plane. However, in this case the polynomial $\psi(s)$ contains at least one odd power of s and so we obtain a distinct tract.

1.3.4

As we have seen, apart from the above case, the asymptotic value curve $A(0, t)$ appears twice corresponding to the distinct tract obtained by changing the variable s to $-s$. The parameter β determines whether the curve is repeated or reversed as the tract is traversed in the positive direction: the positive direction corresponds to y increasing. If β is odd, $s \mapsto -s$ gives the tract

$$x = (-s)^q, \qquad y = \phi(-s) - ts^\beta \tag{1.65}$$

and for $s > 0$, y is clearly a decreasing function of t. Therefore, as the tract is traversed positively, we obtain the limiting curve $A(0, -t)$; i.e. the curve is reversed. On the other hand, when β is even, $s \mapsto -s$ gives $y = \phi(-s) + ts^\beta$ is an increasing function of t, and so the limiting curve is $A(0, t)$; i.e. the curve is repeated in the same direction.

1.3.5 Inverting the transformation

The mapping $x = s^q$, $y = \phi(s) + ts^\beta$ has the inverse

$$s = x^{1/q}, \qquad t = x^{-\beta/q}(y - \phi(x^{1/q})) \tag{1.66}$$

and so

$$R(x, y) = A(x^{1/q}, x^{-\beta/q}(y - \phi(x^{1/q}))). \tag{1.67}$$

This gives

$$P(x, y) = A(x^{-1/q}, x^{\beta/q}(y/x - \phi(x^{-1/q}))), \tag{1.68}$$

and so

$$P(x^q, y) = A(x^{-1}, -x^\beta \phi(x^{-1}) + yx^{\beta-q}) = A(x^{-1}, \psi(x) + yx^\alpha), \tag{1.69}$$

where $\alpha = \beta - q$ and, since ϕ has degree $\beta - h$, where $h \geq 1$, $\psi(x) = -x^\beta \phi(1/x)$ is a polynomial of the form $\psi(x) = cx^h + \cdots$, where $c \neq 0$. This may be regarded as the inverse of the formula

$$A(s, t) = P(s^{-q}, \phi(s)s^{-q} + ts^\alpha). \tag{1.70}$$

From these relations we see that on the curve $y = x^q \phi(x^{-1})$ we have

$$P(x^q, y) = A(x^{-1}, 0), \tag{1.71}$$

and, if $q \geq \beta - h$, y remains bounded on this curve as $x \to 0$. Thus $A(x^{-1}, 0)$ remains bounded as $x \to 0$; i.e. $A(x, 0)$ is a bounded polynomial as $x \to \infty$. This implies $A(x, 0)$ is constant, and so $P(x^q, y)$ is constant on the curve. Therefore, unless the polynomial $P(x, y)$ is constant on some curve,

$$q < \beta - h < \beta, \tag{1.72}$$

and in particular

$$\beta \geq q+2 \quad \text{and} \quad \alpha > h. \tag{1.73}$$

To reiterate: knowing the tract equation $y = \phi(s)s^{-q} + ts^{\alpha}$, we obtain the polynomial $A(s,t)$ from $P(x,y)$; conversely, using the reverse tract equation $t = -s^{\beta}\phi(1/s) + ys^{\alpha}$, we can recover $P(x^q, y)$ from $A(s,t)$. However, if we work directly with $A(s,t)$, applying our process, but not assuming any prior knowledge of the asymptotic tracts, we will not necessarily obtain the above tract for t. The reason for this is that α need not be greater than the largest power of s in the expression $s^{\beta}\phi(1/s)$. No term s^{γ} with $\gamma \geq \alpha$ is needed in determining the asymptotic tract appropriate for $A(s,t)$ in the above direction; we will obtain a tract by deleting all terms involving those powers s^{γ} with $\gamma \geq \alpha$. On the other hand we can obtain precisely invertible transformations if ϕ has the form $\phi(x) = c_r x^r + \cdots$, where $c_r \neq 0$ and $r > q$; for then $\alpha > \deg(s^{\beta}\phi(1/s))$. Then the asymptotic tract is such that $y \to 0$ as $x \to \infty$. In this case, if the process is actually carried through, at each stage the value of p is always an integer and furthermore the zero is always unique and is a multiple zero of exact degree N; this is because the initial largest power of t is t^N, but also the final leading term is $P(0,y)$, which has exact degree N (by our initial assumption). Therefore the leading term has exact degree N at each stage. This phenomenon indicates that the possible polynomials $A(s,t)$ which can be obtained by applying the process to a polynomial P are of a rather special type. $A(s,t)$ has the asymptotic value curve $P(0,y)$ along a curve with $s \to \infty, t \to 0$; but all other asymptotic values of A, if any, will occur along curves where $s \to 0, t \to \infty$.

1.3.6 The comparison principle

We have omitted from our discussion of asymptotic values an important general principle, which establishes the algebraic nature of the asymptotic tracts and also the fact that, *for real analytic polynomials, the only finite limiting values at ∞ are asymptotic values*. The argument is easily sketched as follows.

Suppose we have a finite sum of terms, for which along a sequence the sum tends to a finite value, though some individual terms tend to ∞. We compare two terms by considering for a suitable global subsequence the ratio of the absolute values of the two terms. If the limit of the ratio is finite and non-zero, we say the terms are comparable. If the limit is zero, then the upper term of the ratio is smaller than the lower term; and *vice versa*. In this way we can order the groups of comparable terms and pick out the largest grouping in this ordering. It is clear that the largest group must contain at least two terms, for if it contained only one, then this term would dominate the sum, and so the sum would tend to ∞. For polynomial terms the comparison of two such terms gives a tract of the type described above in which the sequence must lie.

1.3.7 Solving algebraic equations and algebraic functions

An algebraic function $f(x)$ is a function satisfying an equation of the form

$$P(x, f(x)) = 0, \tag{1.74}$$

where $P(x, y)$ is a real analytic polynomial. In other words $y = f(x)$ is a solution of the equation $P(x, y) = 0$. Of course, at a given x the equation may have no solutions or several solutions, though no more than the degree of P. We are interested in continuous solutions for an interval of values of x. We can use our procedure to construct an explicit form for a solution. After a normalisation we may assume that $P(0, 0) = 0$, and our aim is to construct a solution $f(x)$ with $f(0) = 0$. Let us attempt to find a solution for small $x > 0$; if such a solution exists then for an arbitrary natural number N the function $R_N(x, y) = P(x, y)/x^N$ has 0 as an asymptotic value as $x \to 0$; namely along the curve $y = f(x)$. Of course, our procedure is designed to find all limiting values of R_N as $x \to 0$; we obtain a finite number of algebraic curves $y = \phi_k(x) + tx^{\beta_k}$, which are polynomials in a fractional power of x, each of which reduces $R_N(x, y)$ to a polynomial in a fractional power of x and t; we then put $x = 0$ giving a polynomial in t, whose values for t real are the limiting values; we are concerned with the values of t which yield the limiting value 0, i.e. the zeros of the final polynomial, which is simply taking the procedure to the next stage. Thus 0 will be a limiting value of R_N along a finite number of curves $y = \psi_{k,N}(x)$. It is possible that for one, or more, of the $\psi_{k,N}$ we have $R_N(x, \psi_{k,N}(x)) = 0$; then $\psi_{k,N}(x)$ is a required solution and furthermore we have an explicit formula in powers of x for the solution. On the other hand any actual continuous solution $y = f(x)$ must satisfy for some k

$$f(x) = \phi_k(x) + t(x)x^{\beta_k}, \tag{1.75}$$

where $t(x)$ tends to a zero of the final polynomial as $x \to 0$; in other words one of the $\psi_{k,N}$ is an approximation to f. Since $\psi_{k,N}$ is a partial sum of $\psi_{k,N+r}$, we see that we obtain a series expansion for f by letting $N \to \infty$. At each stage the degree of the leading term does not increase, so eventually becomes a constant ν. Thus for large N the values of p will be integers and each leading term will be a pure νth power of a linear polynomial giving just one multiple zero. This implies that the series expansion for f is a power series in $x^{1/q}$ for some natural number q. Our procedure continued indefinitely by increasing N will yield all the solutions $y = f(x)$ for $x > 0$ near 0 (i.e. all the *branches* of the algebraic function) and each such solution is a convergent power series, and therefore an analytic function, in a fractional power of x.

2

The degree principle and the fundamental theorem of algebra

2.1 The fundamental theorem of algebra

2.1.1

Theorem *Every polynomial*

$$P(z) = \sum_{k=0}^{n} a_k z^k \tag{2.1}$$

of degree n has a unique representation in the form

$$P(z) = c \prod_{k=1}^{n} (z - z_k), \tag{2.2}$$

where c is a non-zero constant and the numbers z_k are the ***zeros*** *of P.*

There are many proofs of this fundamental result and one of the aims of this chapter is to develop a particular method of proof, which will show that the theorem is a consequence of a general topological principle relating the zeros of a continuous function to its rotational properties.

However, before turning to this, it is worth reflecting on the significance of the fundamental theorem. Firstly, we see that every non-constant polynomial has a zero. Indeed, a polynomial of degree n has exactly n zeros, provided that we count each zero 'according to its multiplicity' – the **multiplicity** of a zero ζ of $P(z)$ is the number of values k such that $z_k = \zeta$ in the representation (2.2). Thus, if $\zeta_1, \zeta_2, \ldots, \zeta_m$ denote the distinct zeros of $P(z)$ having multiplicities r_1, $r_2, \ldots, r_m$ respectively, then we can rewrite (2.2) in the form

$$P(z) = c \prod_{k=1}^{m} (z - \zeta_k)^{r_k}. \tag{2.3}$$

Our second observation concerns the duality of the relations (2.1) and (2.2). It is clear that, by 'multiplying out', every expression of the form (2.2) can be written in the form (2.1). We obtain $a_n = c$ and in general, for $1 \leq k \leq n$,

$$a_{n-k} = c(-1)^k \sum z_{i_1} z_{i_2} \ldots z_{i_k}, \tag{2.4}$$

where the numbers $i_1, \ldots, i_k$ run over the combinations of the numbers $1, \ldots, n$ taken k at a time. However, the converse statement, that every expression of the form (2.1) can be written in the form (2.2), is far from trivial, and indeed this assertion, which can be briefly stated as 'every finite linear combination of powers of z has a representation as a finite product of linear factors', constitutes one of the fundamental properties of the field of complex numbers. Many of the deeper properties of functions of a complex variable depend in a crucial way on this duality between sums and products. Moreover, it is worth remarking that, as the complex numbers arise from solutions to quadratic equations, it is a considerable bonus that the resulting field is then sufficiently broad to solve any algebraic equation.

However, the existence of a solution, as provided by the fundamental theorem, is one thing, but the construction of a solution is quite another. Works in algebra (see e.g. Birkhoff and Mac Lane [7]) show how to solve equations up to degree 4. A celebrated result, due to Abel and Galois, shows that such solutions 'in radicals' are, in general, impossible for equations of degree 5 and higher.

2.1.2

If we differentiate logarithmically the relation (2.3), we obtain

$$\frac{P'(z)}{P(z)} = \sum_{k=1}^{m} \frac{r_k}{z - \zeta_k}. \tag{2.5}$$

This equation has numerous important consequences and is fundamental in relating the location of the zeros of P with a variety of algebraic and analytic properties of the polynomial. Notice that the logarithmic derivative P'/P is a rational function with simple poles at the zeros of P, the residue at each pole ζ_k being the multiplicity r_k of the zero ζ_k. Hence, by the residue theorem, if C is a positively oriented, simple closed rectifiable path not passing through any of the ζ_k, we have

$$\frac{1}{2\pi i} \int_C \frac{P'(z)}{P(z)} dz = \sum r_k, \tag{2.6}$$

the sum being taken over all k such that ζ_k is enclosed by C. This equation expresses what is known as the **argument principle**, about which we shall have a great deal to say in the major part of this chapter. Notice that the right-hand side is the sum of the multiplicities of the zeros enclosed by C; this is usually expressed as 'the number of zeros enclosed by C counted according to multiplicity'. However, as we shall see later, the more accurate terminology 'sum of the multiplicities' will emerge as the correct interpretation, permitting a very general version of the argument principle. Let us consider two cases.

2.1.3

If C_k is a small circle with centre ζ_k enclosing no other zeros of P, we have

$$\frac{1}{2\pi i}\int_{C_k}\frac{P'(z)}{P(z)}dz = r_k, \tag{2.7}$$

which gives us an analytic expression for the multiplicity of the zero ζ_k.

2.1.4

If C_R is a large circle of radius R enclosing all the zeros of P, we have

$$\frac{1}{2\pi i}\int_{C_R}\frac{P'(z)}{P(z)}dz = n. \tag{2.8}$$

We have proved this on the basis that the fundamental theorem is true. However, if we can prove it directly, then it follows immediately from Cauchy's theorem, that $P(z)$ has at least one zero enclosed by C_R (as otherwise the integral would evaluate to 0).

2.1.5

As every polynomial $P(z)$ of degree n is an analytic function of z, the mapping $w = P(z)$ is **locally 1–1** near every point except for those points where $P'(z) = 0$. As P' is a polynomial of degree $n-1$, there are $n-1$ such points, counted according to multiplicity. The zeros of P' have a special significance and they will play an important role. They are called the **critical points** of P. The equation (2.5) can be used to deduce a very useful and elegant result, which relates the location of the critical points to the location of the zeros. To state the result, we need to define a convex set. A set E is said to be **convex**, if for each pair of points z_1 and z_2 in E the line segment joining z_1 to z_2 also lies in E. It is clear that the intersection of any family of convex sets is itself convex (though possibly empty). Also the whole plane is trivially convex. Thus given any non-empty set A, there is a unique smallest convex set $K(A)$ containing A, namely the intersection of the family of all convex sets containing A. The set $K(A)$ is called the **convex hull** of A. We leave it as an exercise to show that, if the points $z_1, z_2, \ldots, z_n$ lie in A, then every point of the form

$$w = \sum_{k=1}^{n} \alpha_k z_k, \tag{2.9}$$

where $\alpha_k \geq 0$ and $\sum_{k=1}^{n} \alpha_k = 1$, lies in $K(A)$.

2.1.6

Gauss–Lucas theorem *Let $P(z)$ be a polynomial of degree n with zeros $z_1, \ldots, z_n$. The critical points of $P(z)$ lie in the convex hull of the set $\{z_1, \ldots, z_n\}$.*

Proof Let ζ be a zero of P'. If ζ is also a zero of P, then the result is clear. Otherwise ζ is a zero of P'/P, and hence by (2.5) we have

$$\sum_{k=1}^{n} \frac{1}{\zeta - z_k} = 0. \tag{2.10}$$

Conjugating this equation and using the identity $1/\overline{a} = a/|a|^2$, we obtain

$$\sum_{k=1}^{n} \frac{\zeta - z_k}{|\zeta - z_k|^2} = 0, \tag{2.11}$$

which gives

$$\zeta = \sum_{k=1}^{n} \alpha_k z_k, \tag{2.12}$$

where

$$\alpha_k = \frac{\dfrac{1}{|\zeta - z_k|^2}}{\displaystyle\sum_{j=1}^{n} \frac{1}{|\zeta - z_j|^2}} \tag{2.13}$$

for $1 \leq k \leq n$. We see that each $\alpha_k > 0$ and $\sum_{k=1}^{n} \alpha_k = 1$. Thus ζ lies in the convex hull of the set of zeros z_k, as required.

2.1.7 *Problems*

1. Show that, in order to prove the fundamental theorem of algebra, it is enough to show that every polynomial of degree ≥ 1 has at least one zero. [Hint: if

$P(z)$ has degree n and $P(a) = 0$, then $P(z)/(z-a)$ is a polynomial of degree $n-1$; use induction.]

2. Show that, if $P(z)$ is a non-constant polynomial, then $P(z) \to \infty$ as $z \to \infty$. Deduce that, if $P(z)$ has no zeros, then $1/P(z)$ is a bounded entire function and therefore the fundamental theorem of algebra follows from Liouville's theorem.
3. Show that, if $P(z)$ is a polynomial of degree n, then $zP'(z)/P(z) \to n$ as $z \to \infty$. Hence give a direct proof of (2.8) and deduce the fundamental theorem of algebra.
4. Show that, if $E_n = \{z_1, \dots, z_n\}$ is a set of n distinct points, then the convex hull of E_n is the set of points w of the form $w = \sum_{k=1}^{n} \alpha_k z_k$, where each $\alpha_k \geq 0$ and $\sum_{k=1}^{n} \alpha_k = 1$.
5. Let $c_k > 0$, $|z_k| \leq 1$ $(1 \leq k \leq n)$ and set

$$R(z) = \sum_{k=1}^{n} \frac{c_k}{(z - z_k)^m},$$

where $m \geq 1$ is a natural number. Show that $R(z)$ has all its finite zeros in $\{|z| \leq \frac{1}{2^{1/m}-1}\}$.

2.2 Continuous functions in the plane

2.2.1

Many of the very subtle results for polynomials which we shall obtain in later chapters depend on the fundamental theorem, which is why we are giving it some prominence. Because polynomials are analytic functions, it is natural to appeal to proofs which make use of the Cauchy theory, such as those indicated in Problems 2.1.7.2 and 2.1.7.3. Such proofs have the advantage of being short and making use of results which are part of everybody's mathematical education. Indeed the Cauchy theory is so dominant that there is a tendency to fear any complex function which is not analytic. As an example, let us consider a quite minor variation on the problem of the existence of zeros of a polynomial. If $P(z)$ is a polynomial, does the function $f(z) = \bar{z} + P(z)$ have a zero? (The answer is 'yes' except possibly when $P(z) = az + b$ with $|a| = 1$; e.g. $f(z) = \bar{z} + z + i = 2x + i$ has no zeros.) This problem is a special case of a very natural algebraic problem, namely to solve simultaneously the pair of algebraic equations

$$S(x, y) = u, \qquad T(x, y) = v, \tag{2.14}$$

where $S(x, y)$ and $T(x, y)$ are real polynomials in the real variables x and y and where u and v are given real numbers. Writing $P = S + iT$, $w = u + iv$, this is equivalent to finding a zero of the complex-valued function $f = P - w$. If we make the substitution $x = \frac{1}{2}(z + \overline{z})$, $y = -\frac{1}{2}i(z - \overline{z})$, then f takes the form of a polynomial in the variables z and $\overline{z}$. Functions of this form are called **real analytic polynomials** and we have considered these from an algebraic point of view in chapter 1. Finding zeros, or even proving the existence of zeros for real analytic polynomials, is not a trivial task and can involve the methods of algebraic geometry. Nevertheless, by developing a general approach in the plane to problems of this type, we shall be able to see where some of the major difficulties lie and deal with a proportion of cases.

2.2.2

For analytic functions the number of zeros in a region is determined by the argument principle. In its simplest form this states that the number of zeros of $f(z)$ inside a region bounded by a simple curve is given by the change in argument of the function as the curve is traversed divided by 2π. This quantity is one which, in principle, we could calculate for any function and therefore we may ask: what does it tell us about the zeros of the function inside the curve? For functions analytic except for poles (**meromorphic** functions), the quantity gives us (the number of zeros) – (the number of poles). Therefore, we may say that the function on the boundary 'notices' both the zeros and the poles inside the region. A closer look at the proof of the argument principle (usually obtained as a special case of the residue theorem) reveals that the change in argument on the boundary measures the sum of the local changes in argument near the singularities of the function, where by 'singularity' we mean an isolated point at which the function fails to be both analytic and non-zero. We shall see that this form of the argument principle extends to a general topological principle in which we replace analytic functions by continuous functions. In the development which follows, the underlying themes are the rotational mapping properties of functions on curves, the polar coordinate representation of a complex number and the continuity of the complex logarithm.

2.2.3

Lemma *Let γ be a curve not passing through the origin. Then every parametric representation $z = z(t)$ $(a \leq t \leq b)$ of γ can be written in the form*

$$z = z(t) = e^{\alpha(t)}, \tag{2.15}$$

where $\alpha(t)$ is continuous for $a \leq t \leq b$. Furthermore the quantity $\alpha(b) - \alpha(a)$ depends only on the curve γ and not on the choice of parametric representation.

Proof This is simply an assertion of the existence of a continuous branch of $\log z(t)$. As it is fundamental to our discussion we include the proof. Since $z(t)$ is non-zero, we can write it in polar coordinate form $z(t) = r(t)e^{i\theta(t)}$, where $r(t)$ is continuous and >0, and where $0 \leq \theta(t) < 2\pi$. In general this choice of $\theta(t)$ will not remain continuous on the interval. To obtain continuity we shall need to add to it a suitable multiple of 2π at each point. To determine the construction of this multiple we proceed as follows. Firstly, we note that $\exists\delta > 0$ such that $r(t) \geq \delta$ for $a \leq t \leq b$. Also as $z(t)$ is uniformly continuous on $[a, b]$, $\exists\lambda > 0$ such that $|z(t) - z(s)| < \delta$ if $|t - s| < \lambda$. We choose $a = t_0 < t_1 < \cdots < t_n = b$ so that $t_k - t_{k-1} < \lambda$ for $k = 1, 2, \ldots, n$. We define $\phi(a) = \theta(a)$ and then proceed with the following inductive construction. Suppose that $\phi(t_{k-1})$ has been defined and set $\phi(t) = \theta(t) + 2\nu\pi$ in $(t_{k-1}, t_k]$, where ν is an integer chosen so that $\phi(t_{k-1}) - \pi < \phi(t) \leq \phi(t_{k-1}) + \pi$. If we find that $\phi(t) = \phi(t_{k-1}) + \pi$, then $e^{i\phi(t)} = -e^{i\phi(t_{k-1})}$, and hence $z(t)/|z(t)| = -z(t_{k-1})/|z(t_{k-1})|$. But this implies that $|z(t) - z(t_{k-1})| > |z(t_{k-1})| \geq \delta$, which contradicts $t - t_{k-1} < \lambda$. Thus $\phi(t)$ satisfies

$$|\phi(t) - \phi(t_{k-1})| < \pi \quad (t_{k-1} \leq t \leq t_k). \tag{2.16}$$

Furthermore $\phi(t)$ is continuous in $[t_{k-1}, t_k]$; to see this choose τ in this interval; since $e^{i\phi(t)} = z(t)/|z(t)|$, we have $e^{i(\phi(t)-\phi(\tau))} \to 1$ as $t \to \tau$ and hence $\cos(\phi(t) - \phi(\tau)) \to 1$ as $t \to \tau$. For points t in the interval $[t_{k-1}, t_k]$ the inequality (2.16) implies that the only possible limiting value of $\phi(t) - \phi(\tau)$ as $t \to \tau$ is 0, proving the desired continuity. We can now define $\alpha(t) = \log r(t) + i\phi(t)$. To prove that $\alpha(b) - \alpha(a)$ does not depend on the parametrisation, let $w(s)\,(c \leq s \leq d)$ be another parametrisation of γ with $w(s) = e^{\beta(s)}$, so that $\beta(s) = \log|w(s)| + i\psi(s)$, where $\psi(s)$ is continuous. Now $w(s)$ takes the form $w(s) = z(t(s))$, where $t(s)$ is a continuous strictly increasing function mapping $[c, d]$ onto $[a, b]$. It follows that $\beta(s) - \alpha(t(s))$ is a continuous function on $[c, d]$, which is an integer multiple of $2\pi i$ at each s. This function is therefore constant and in particular takes the same value at $s = c$ and $s = d$. This gives $\beta(d) - \beta(c) = \alpha(t(d)) - \alpha(t(c)) = \alpha(b) - \alpha(a)$, as required.

2.2.4

We see from this lemma that, if γ is a **loop**, i.e. if γ has the same initial and terminal points, then $\alpha(b) - \alpha(a)$ is an integer multiple of $2\pi i$ which depends only on γ. This leads to the following definition. Let γ be a loop and let a be a point not on γ. Then we can write $z(t) - a = e^{\alpha(t)}$, where $z(t)\ (a \leq t \leq b)$ is a parametrisation of γ. We define

$$n(\gamma, a) = \frac{\alpha(b) - \alpha(a)}{2\pi i}, \tag{2.17}$$

and call this the **winding number of** γ **about** a. Note that it represents the continuous change in $\arg(z-a)$ divided by 2π as γ is traversed, and so measures the number of times γ 'winds around' the point a in the anti-clockwise direction.

2.2.5

Consider now a function f continuous and non-zero on a curve γ parametrised by $z = z(t)$ $(a \leq t \leq b)$. The image curve $f \circ \gamma$ has a parametrisation $w = w(t) = f(z(t))$ $(a \leq t \leq b)$ and $0 \notin f \circ \gamma$. Hence by the lemma we can write $f(z(t)) = e^{\alpha(t)}$, where $\alpha(t)$ is continuous and where $\alpha(b) - \alpha(a)$ depends only on f and γ. We define

$$d(f, \gamma) = \frac{1}{2\pi}\Im(\alpha(b) - \alpha(a)), \tag{2.18}$$

and call this the **degree of** f **on** γ. Note that the degree represents the continuous change in $\arg f(z)$ divided by 2π as z traverses γ. If γ is a loop, then $d(f, \gamma)$ is an integer and

$$d(f, \gamma) = n(f \circ \gamma, 0). \tag{2.19}$$

On the other hand, if $a \notin \gamma$, then

$$n(\gamma, a) = d(z - a, \gamma). \tag{2.20}$$

2.2.6 Problem

1. (a) If a curve γ is partitioned into sub-curves γ_i $(1 \leq i \leq n)$, show that $d(f, \gamma) = \sum_{i=1}^{n} d(f, \gamma_i)$ for a function f continuous and non-zero on γ. Show also that $d(f, -\gamma) = -d(f, \gamma)$, where $-\gamma$ represents the curve describing γ in reverse.
   (b) Show that, if f is a non-zero constant on the curve γ, then $d(f, \gamma) = 0$.
   (c) Show that, if f and g are continuous and non-zero on γ, then

$$d(fg, \gamma) = d(f, \gamma) + d(g, \gamma), \qquad d\left(\frac{f}{g}, \gamma\right) = d(f, \gamma) - d(g, \gamma). \tag{2.21}$$

Deduce that for any integer n

$$d(f^n, \gamma) = nd(f, \gamma). \tag{2.22}$$

2.2.7

We shall be mainly concerned with functions defined on open sets. A non-empty, open, connected set is called a **domain**. We remind the reader that a **connected** set is one which cannot be written as the union of two non-empty separated subsets (sets E and F are **separated** if $\overline{E} \cap F = \emptyset = E \cap \overline{F}$). An open set is connected, iff it cannot be written as the union of two non-empty disjoint open subsets. Also an open set U is connected iff any pair of points in U can be joined by a curve lying wholly in U. If $f(z)$ is a function continuous and non-zero in a domain D, a (continuous) **branch** of $\log f(z)$ is defined as a function $\phi(z)$ continuous in D such that $e^{\phi(z)} = f(z)$ for all $z \in D$.

2.2.8

Theorem *If $f(z)$ is continuous and non-zero in a domain D, then a branch of $\log f(z)$ exists in D if, and only if,*

$$d(f, \gamma) = 0 \tag{2.23}$$

for every loop γ in D.

Proof If such a branch $\phi(z)$ exists, then for a loop γ with parametrisation $z = z(t)\,(a \le t \le b)$, we have $f(z(t)) = e^{\phi(z(t))}$ and so $2\pi d(f, \gamma) = \Im(\phi(z(b)) - \phi(z(a))) = 0$, since $z(b) = z(a)$. Conversely, suppose that $d(f, \gamma) = 0$ for every loop γ in D. Choose $\zeta \in D$ and let ω satisfy $e^{\omega} = f(\zeta)$. If $z \in D$, $\exists$ a curve Γ in D joining ζ to z . Set

$$\psi(z) = 2\pi d(f, \Gamma) + \Im(\omega). \tag{2.24}$$

Then $\psi(z)$ depends only on the point z and not on the choice of the curve Γ. Indeed, if $\tilde{\Gamma}$ is another such curve, then $\Gamma - \tilde{\Gamma}$ is a loop and so $d(f, \Gamma - \tilde{\Gamma}) = 0$. Thus $d(f, \Gamma) = d(f, \tilde{\Gamma})$. It is now easily verified that $f(z) = |f(z)|e^{i\psi(z)}$, and therefore it is sufficient to show that $\psi(z)$ is continuous in D. Since D is open, for a suitable $\delta > 0$ the disc $\delta = \{z + h \colon |h| \le \delta\} \subset D$. Let l denote a curve in Δ joining z to $\partial\Delta$, the boundary of Δ, parametrised say by $l = l(t)\,(a \le t \le b)$. Then $f(l(t)) = e^{\beta(t)}$, where $\beta(t)$ is continuous in $[a, b]$. We have

$$\psi(l(s)) - \psi(z) = 2\pi d(f, l_s) = 2\pi\Im(\beta(s) - \beta(a)) \to 0 \text{ as } s \to a, \tag{2.25}$$

where l_s is the curve parametrised by $l(t)\,(a \le t \le s)$. Thus $\psi(z + h) \to \psi(z)$ as $h \to 0$ along any curve in $\{|h| \le \delta\}$, and so ψ is continuous at z.

2.2.9 Problems

1. Show that, if γ is a rectifiable loop, then for each $a \notin \gamma$

$$n(\gamma, a) = \frac{1}{2\pi i}\int_\gamma \frac{dz}{z-a}. \tag{2.26}$$

2. Let $f(z)$ be continuous and non-zero in a domain D. Show that $\exists\phi(z)$ continuous in D and satisfying $\phi^2(z) = f(z)$ for $z \in D$, if, and only if, $d(f, \gamma)$ is an even integer for every loop γ in D; $\phi(z)$ is called a **branch** of $\sqrt{f(z)}$ in D. More generally, show that a branch of $(f(z))^{1/n}$ exists in D if, and only if, $d(f, \gamma)$ is divisible by n for every loop γ in D.
3. Let $D = \mathbb{C} - [-1, 1]$, i.e. the plane cut along the line segment joining -1 to 1. Show that
   (a) there is a continuous branch of $\log \frac{1+z}{1-z}$ in D,
   (b) there is no continuous branch of $\log(1 - z^2)$ in D,
   (c) there is a continuous branch of $\sqrt{1 - z^2}$ in D.
4. Let $\mathbb{U} = \{z \colon |z| < 1\}$, i.e. the unit disc. Show that, if $F(z)$ is continuous and 1–1 in $\mathbb{U}$ and satisfies $F(0) = 0$, then $\exists$ a continuous 1–1 branch $f(z)$ of $\sqrt{F(z^2)}$ in $\mathbb{U}$. More generally, for every natural number n, continuous branches of $(F(z^n))^{1/n}$ exist in $\mathbb{U}$ and are 1–1 in $\mathbb{U}$.

2.3 The degree principle

2.3.1

The question which we shall consider in this section is: for a given function f and a given loop γ what is the value of $d(f, \gamma)$? Our point of view towards this question will be analogous to the question: for an analytic function f and a rectifiable loop γ what is the value of $\int_\gamma f(z)dz$? The degree principle will then be seen to be analogous to Cauchy's theorem and our strategy for building up the proof will be identical to the strategy for proving the general version of Cauchy's theorem (see Ahlfors [2]). We have already seen the first such analogue in theorem 2.2.8. The analogue for analytic functions is that for a function f analytic in a domain D, $\int_\gamma f(z)dz = 0$ for every rectifiable loop γ in D iff f is a primitive in D (i.e. takes the form $f = F'$ for some function F in D). For Cauchy's theorem the usual next step is to prove it for a triangle using the Goursat dissection method. We shall follow the same path. We require firstly a very useful lemma.

2.3.2

Lemma *Suppose that $f(z)$ is continuous on a loop γ and that*

$$f(z)+c \neq 0 \tag{2.27}$$

for all $z \in \gamma$ and for all $c \geq 0$. Then

$$d(f,\gamma)=0. \tag{2.28}$$

Proof Let $z(t)\,(a \leq t \leq b)$ be a parametrisation of γ. As the case $c=0$ gives $f(z(t)) \neq 0$, we can write $f(z(t)) = e^{\alpha(t)}$, where $\alpha(t) = \mu(t) + i\phi(t)$ is continuous on $[a,b]$. The condition (2.27) implies that the real function ϕ is never equal to an odd multiple of π (for at such a point $f(z)$ would take on a negative value). Since $\phi(t)$ is continuous on $[a,b]$, the intermediate value theorem implies that the values of $\phi(t)$ lie in an open interval of length 2π and in particular $|\phi(b)-\phi(a)| < 2\pi$. But $|2\pi d(f,\gamma)| = |\phi(b)-\phi(a)|$ and thus $d(f,\gamma)$ is an integer of absolute value < 1. Hence $d(f,\gamma)=0$.

2.3.3

The degree principle for a triangle *Let $f(z)$ be continuous and non-zero inside and on a triangle Δ with perimeter Γ. Then*

$$d(f,\Gamma)=0. \tag{2.29}$$

Proof Using the Goursat method, we join the midpoints of the sides of Δ, dividing the triangle into four congruent sub-triangles with perimeters Γ_i say. Taking Γ and the Γ_i as being positively oriented, we obtain

$$d(f,\Gamma) = \sum_{i=1}^{4} d(f,\Gamma_i). \tag{2.30}$$

If we assume that $d(f,\Gamma) \neq 0$, then for at least one i, $d(f,\Gamma_i) \neq 0$. We now repeat the process on the sub-triangle with this perimeter Γ_i, and continue in this way indefinitely. We obtain a sequence of triangles $\Delta^{(n)}$ with perimeters $\Gamma^{(n)}$ such that $d(f,\Gamma^{(n)}) \neq 0$ for every n. The sequence $\{\Delta^{(n)}\}$ is nested and clearly converges to a point ζ which belongs to every $\Delta^{(n)}$. Since f is continuous and non-zero at ζ, $\exists\delta > 0$ such that

$$|f(z)-f(\zeta)| < |f(\zeta)| \tag{2.31}$$

for $|z-\zeta|<\delta$ and $z \in \Delta$. Also $\exists N$ such that, if $n>N$ and $z \in \Delta^{(n)}$, then $|z-\zeta|<\delta$. Hence for $n>N$ we have

$$|f(z)-Re^{i\phi}|<R \quad \left(z \in \Gamma^{(n)}\right), \tag{2.32}$$

where $f(\zeta)=Re^{i\phi}$. We deduce that for all $c \geq 0$ and $z \in \Gamma^{(n)}$ we have

$$e^{-i\phi}f(z)+c \neq 0, \tag{2.33}$$

and hence by lemma 2.3.2

$$d\left(f,\Gamma^{(n)}\right)=d\left(e^{-i\phi}f,\Gamma^{(n)}\right)=0, \tag{2.34}$$

contradicting the construction of $\Gamma^{(n)}$.

2.3.4

Our next step is to prove the degree principle for the class of starlike domains. A domain D is said to be **starlike with respect to the point** ζ, if for each point $z \in D$ the line segment $[\zeta, z]$ lies wholly in D. Note that a domain is convex iff it is starlike with respect to each of its points.

2.3.5

The degree principle for a starlike domain *Let $f(z)$ be continuous and non-zero in a starlike domain D. Then for every loop γ in D we have*

$$d(f,\gamma)=0. \tag{2.35}$$

Proof By theorem 2.2.8 we need to show that $\exists\phi(z)$, real and continuous in D such that $f(z)=|f(z)|e^{i\phi(z)}$ for $z \in D$. Let D be starlike with respect to $\zeta \in D$ and let w be real and satisfy $f(\zeta)=|f(\zeta)|e^{iw}$. For $z \in D$ let L_z denote the line segment $[\zeta, z]$ and define $\phi(z)=2\pi d(f,L_z)+w$. This is clearly a value of $\arg f(z)$ and therefore it remains to show that $\phi(z)$ is continuous in D. Now $\exists\delta>0$ such that for $|h|<\delta$ the points $z+h \in D$ and therefore the triangle with vertices $\zeta, z, z+h$ together with its interior lies wholly in D. Hence, if Γ is the perimeter of this triangle, then $d(f,\Gamma)=0$ and this implies

$$\phi(z+h)-\phi(z)=2\pi d(f,[z,z+h]) \to 0 \text{ as } h \to 0 \tag{2.36}$$

along a line segment, as in the proof of theorem 2.2.8. Now, if h is small, the points $f(z+h)$ lie in a small disc near the non-zero point $f(z)$, and so for a suitable real μ we have $e^{-i\mu}f(z+h)+c \neq 0$ for all $c \geq 0$ and small h. Thus by lemma 2.3.2 $d(f, \gamma) = 0$ on any loop consisting of points $z + h$. We deduce that

$$\phi(z + h) - \phi(z) = 2\pi d(f, \{z, z + h\}) \qquad (2.37)$$

where $\{z, z + h\}$ is any curve joining z to $z + h$ and remaining close to z. We obtain, as in the proof of theorem 2.2.8, that $\phi(z+h)-\phi(z) \to 0$ as $h \to 0$ along any curve, and so ϕ is continuous at z.

2.3.6

Corollary: The degree principle for a circle *If $f(z)$ is continuous and non-zero inside and on a circle C, then $d(f, C) = 0$.*

Proof Let $C = \{z = \zeta + \rho e^{i\theta}: 0 \leq \theta \leq 2\pi\}$. If $z = \zeta + Re^{i\theta}$, where $R > \rho$, we define $f(z) = f(\zeta + \rho e^{i\theta})$. This extends f to the whole plane. As f is now continuous and non-zero in the plane, the result follows from the theorem.

2.3.7 *Properties of the winding number*

Lemma *Let γ be a loop. The function $z \mapsto n(\gamma, z)$ is constant on each component of γ^c (the complement of γ). Furthermore, the function is zero on the exterior of γ.*

Proof Let $a \in \gamma^c$. Since γ^c is open, $\exists \delta > 0$ such that $\{z : |z - a| \leq \delta\} \subset \gamma^c$. Choose b satisfying $|b - a| \leq \delta$ and set

$$f(z) = \frac{z - b}{z - a} \quad (|z - a| > \delta). \qquad (2.38)$$

Then it is easily verified that $f(z) + c \neq 0$ for all $c \geq 0$ and $z \in \gamma$. From lemma 2.3.2 we obtain $d(f, \gamma) = 0$ and so $d(z - a, \gamma) = d(z - b, \gamma)$, i.e. $n(\gamma, a) = n(\gamma, b)$. Thus the set of points z, at which $n(\gamma, z) =$ a particular value $n(\gamma, a)$, is open, and so $n(\gamma, z)$ is constant on each component of γ^c.

To prove $n(\gamma, z) = 0$ on the exterior (i.e. unbounded component) of γ^c, we need show only that $n(\gamma, z) = 0$ for $|z| > M$ with M large. Choose M so that $\gamma \subset \{z: |z| < M\}$ and then choose a with $|a| > M$. Then $z - a$ is continuous and non-zero for $|z| < M$ and so $n(\gamma, a) = d(z - a, \gamma) = 0$ by 2.3.5.

2.3.8

Lemma *If C is a positively oriented circle bounding a disc D, then for each $a \in D$,*

$$n(C, a) = 1. \tag{2.39}$$

Proof By 2.3.7 we may assume that a is the centre of the circle, so that, if r is the radius, the function $a + re^{it}$ $(0 \leq t \leq 2\pi)$ is a parametrisation of C. Then, if $f(z) = z - a$, $f \circ C$ has the parametrisation re^{it} $(0 \leq t \leq 2\pi)$. Thus clearly $n(C, a) = d(f, C) = 1$.

2.3.9

At this stage we could assert the degree principle for a simply-connected domain by appealing to the Riemann mapping theorem, since it is evident that the concept of degree will carry over in an obvious way under topological mappings. However, we believe it is altogether more illuminating, and certainly more elementary, to continue with our current development running parallel to the development of Cauchy's theorem. Our next result will enable us to replace the degree of a function on a general curve with the degree of the function on a rectilinear polygon, i.e. a curve consisting of a finite number of line segments each parallel to one of the coordinate axes.

2.3.10

Lemma *Let γ be a curve in a domain D. There exists a rectilinear polygon Γ in D with the same endpoints as γ such that*

$$d(f, \gamma) = d(f, \Gamma) \tag{2.40}$$

for every function f continuous and non-zero in D.

Proof Let $z(t)$ $(a \leq t \leq b)$ be a parametrisation of γ. Then $\exists \delta > 0$ such that

$$|z(t) - w| \geq \delta \quad (a \leq t \leq b, w \notin D). \tag{2.41}$$

Also $\exists \lambda > 0$ such that for $|t - s| < \lambda$ $(a \leq s \leq b, a \leq t \leq b)$, we have $|z(t) - z(s)| < \delta$. Choose a partition $P = \{t_0, t_1, \ldots, t_n\}$ of $[a, b]$ so that $t_k - t_{k-1} < \lambda$ for $1 \leq k \leq n$. Let

$$\gamma_k = \{z(t) \colon t_{k-1} \leq t \leq t_k\}, \tag{2.42}$$

$$\Gamma_k = [z(t_{k-1}), (x(t_k), y(t_{k-1}))] + [(x(t_k), y(t_{k-1})), z(t_k)] \tag{2.43}$$

so that Γ_k joins $z(t_{k-1})$ to $z(t_k)$ by means of a horizontal segment followed by a vertical segment. Let D_k denote the open discs of radius δ and centres $z(t_k)$ $(0 \le k \le n)$. Then $D_k \subset D$. Furthermore, $\gamma_k \subset D_{k-1}$ and in particular the points $z(t_{k-1})$ and $z(t_k)$ lie in D_{k-1} for $1 \le k \le n$. Hence each polygon $\Gamma_k \subset D_{k-1}$. Thus $\gamma_k - \Gamma_k$ is a loop in D_{k-1} and therefore for f continuous and non-zero in D

$$d(f, \gamma_k) = d(f, \Gamma_k) \quad (1 \le k \le n) \tag{2.44}$$

by 2.3.5. Since $\gamma = \sum_{k=1}^{n} \gamma_k$, we have, writing $\Gamma = \sum_{k=1}^{n} \Gamma_k$ and summing over k,

$$d(f, \gamma) = d(f, \Gamma). \tag{2.45}$$

2.3.11

The best and most general form of Cauchy's theorem is obtained by integrating functions on cycles rather than individual curves. This solves at a stroke the geometrical awkwardness in attempting to generalise the theorem from simple curves to more complicated ones. For example, the proof of the residue theorem is greatly simplified if one uses cycles rather than trying to apply Cauchy's theorem on a simple loop, since in the latter case one is artificially joining up disjoint curves, when in fact this is never necessary. The same considerations apply to the degree principle. We recall that a **chain** is defined to be a formal sum $\gamma = \sum_{i=1}^{n} c_i \gamma_i$, where the c_i are integers and the γ_i are curves. We define the **degree of f on γ** by $\sum_{i=1}^{n} c_i d(f, \gamma_i)$ for any function f continuous and non-zero on the γ_i. If γ, γ' are chains in a domain D (i.e. the individual curves making up the chains all lie in D), we say γ **is (degree) homologous to γ' relative to** D if $d(f, \gamma) = d(f, \gamma')$ for every function f continuous and non-zero in D. This is an equivalence relation. If a chain is homologous to an integral linear combination of loops, it is called a **cycle**. The **winding number** of a cycle γ about a point $a \notin \gamma$ is defined by $n(\gamma, a) = d(z - a, \gamma)$. A cycle γ in a domain D is said to be a **zero cycle in** D if $n(\gamma, a) = 0$ for all points $a \notin D$. This last definition is intentionally prejudiced. A more natural definition would be that γ is a zero cycle in D if $d(f, \gamma) = 0$ for all functions f continuous and non-zero in D. Clearly cycles satisfying this condition are zero cycles, since the functions $z - a$ are continuous and non-zero in D, precisely for the points $a \notin D$. The content of the degree principle is that the converse is true: the zero cycles in D are homologous to zero in D, indeed homologous to the trivial zero cycle consisting of a single curve taking on only one non-zero value. Before turning to the proof of the degree principle, it is an interesting observation that the fundamental theorem of algebra implies the degree principle for any rational function whose poles and zeros lie outside D. For by factorising numerator and denominator of the rational function, we need only apply the additivity properties of degree as given in Problem 2.2.6.1 (c). From this point of view it is a remarkable topological fact that the principle holds for any continuous function.

2.3.12

Theorem: The degree principle *Let $f(z)$ be continuous and non-zero in a domain D. Then*

$$d(f, \gamma) = 0 \tag{2.46}$$

for every zero cycle γ in D.

Proof We note firstly that by lemma 2.3.10 we may assume that γ is a rectilinear zero cycle in D. Secondly, it is clear from the additive properties of the degree of a function on a curve that, up to the homology equivalence, we may replace a general rectilinear cycle by one consisting of rectilinear loops each of which consists of a finite number of segments alternating between horizontal and vertical segments. Taking γ to have this form we construct a rectangular grid in the plane by drawing at each corner point of the curves in the cycle the two full lines parallel to the two coordinate axes. In this way we divide the plane into a finite number of regions each of which is either a rectangle or an unbounded region. We label the rectangles R_k with perimeters Γ_k assumed positively oriented ($1 \leq k \leq n$). Let a_k denote a point in the interior of the rectangle R_k. We note that, from the construction, all the points of γ lie on the grid and, in fact, γ consists of sides of the rectangles R_k. We will show that

$$\gamma \text{ is homologous to } \sum_{k=1}^{n} n(\gamma, a_k)\Gamma_k \text{ in } D. \tag{2.47}$$

The method of proof is to show that, if L is a side of a rectangle R_k, then the net number of times L occurs in γ is exactly equal to the net number of times L occurs in the above sum. Indeed, suppose that L occurs in γ μ times with a positive orientation relative to the rectangle R_k and ν times with a negative orientation. We form a new cycle γ' by replacing L each time it occurs by the other three sides of the rectangle R_k, otherwise leaving the remaining segments of γ unchanged. This gives $\gamma = \gamma' + (\mu - \nu)\Gamma_k$. The quantity $\mu - \nu$ is clearly the net number of times that L occurs in γ and so we wish to show that L

occurs $\mu - \nu$ times in the sum in (2.47). To prove this, we note first that L is also a side of a neighbouring region R, which is either another rectangle R_j or an unbounded region. Since L does not meet γ', if a is an interior point of R, then the line segment joining a to a_k lies in the complement of γ', and so in a component of this complement, which implies that $n(\gamma', a) = n(\gamma', a_k)$. Suppose now that $R = R_j$. Then $n(\gamma', a_j) = n(\gamma', a_k)$. We deduce that

$$n(\gamma, a_k) - n(\gamma, a_j) = (\mu - \nu)(n(\Gamma_k, a_k) - n(\Gamma_k, a_j)) = \mu - \nu. \tag{2.48}$$

On the other hand, if R is an unbounded region, then for $a \in R$, $n(\gamma', a) = 0$, and so $n(\gamma', a_k) = 0$. It follows that

$$n(\gamma, a_k) = (\mu - \nu)n(\Gamma_k, a_k) = \mu - \nu. \tag{2.49}$$

In each of these two cases the left-hand term is clearly the net number of times that L occurs in the sum in (2.47) (the orientation of L relative to R_j is the opposite of that relative to R_k). To prove (2.47) it only remains to show that the curves appearing in the sum lie in D. In fact, this is not necessarily the case, but if there is a point $a \in D^c$ which also lies in R_k say, then the line segment joining a_k to a lies in the complement of γ, and so $n(\gamma, a_k) = n(\gamma, a) = 0$. Thus for such k we obtain a zero term in the sum and the term may be removed from the sum. With this understanding the relation (2.47) follows.

The degree principle is now a simple consequence, for if f is continuous and non-zero in D, we obtain

$$d(f, \gamma) = \sum_{k=1}^{n} n(\gamma, a_k) d(f, \Gamma_k) = 0, \tag{2.50}$$

since, for those k for which $n(\gamma, a_k) \neq 0$, f is continuous and non-zero inside and on R_k and so $d(f, \Gamma_k) = 0$ by the degree principle for a rectangle. This completes the proof.

2.3.13

A domain D is said to be **simply-connected** if $D^c \cup \{\infty\}$ is connected in the extended plane $\mathbb{C} \cup \{\infty\}$. We recall that by the process of **stereographic projection** one shows that the extended plane is homeomorphic to the surface of a three-dimensional sphere, and is therefore a compact metric space. The point ∞ is identified with the north pole of the sphere. If γ is a loop in a simply-connected domain D, then $n(\gamma, a) = 0$ for all $a \notin D$, for $a \in$ the connected set $D^c \cup \{\infty\} \subset \gamma^c \cup \{\infty\}$ and so $a \in$ the component of $\gamma^c \cup \{\infty\}$ containing ∞. As γ is bounded, $\gamma^c \cup \{\infty\}$ contains a neighbourhood of ∞ and hence $a \in$ the unbounded component of γ^c. Thus in a simply-connected domain every loop is a zero cycle and hence, by the degree principle, homologous to zero (this statement is characteristic of simply-connected domains: if every loop in a domain D is a zero cycle, then D is simply-connected; see Problem 2.3.16.3). We thus have the next result.

2.3.14 The degree principle for a simply-connected domain

Theorem *If $f(z)$ is continuous and non-zero in a simply-connected domain D, then*

$$d(f, \gamma) = 0 \tag{2.51}$$

for every loop γ in D. In particular a continuous branch of $\log f(z)$ *exists in D.*

2.3.15 Remarks on plane topology

There are several different ways of characterising simply-connected domains in the plane, but it would go beyond the purpose of this chapter for us to develop a complete account of the various equivalent statements. The two main ones are as follows:

(a) a domain D is simply-connected iff D is homeomorphic to $\mathbb{C}$; this is a consequence of the **Riemann mapping theorem** which asserts that every simply-connected domain other than the whole plane is conformally equivalent to the open unit disc; as there is a simple (non-conformal) transformation from the open disc to $\mathbb{C}$, this establishes the result;

(b) a domain D is simply-connected iff every loop γ in D can be shrunk to a point in D, or more formally is homotopic to a constant loop in D; the homotopy version of simple-connectedness is the right one for more general topological spaces and formalises the concept of a region containing no holes; we give the homotopic version of the degree principle in section 2.4.

An important class of simply-connected domains is the class of Jordan domains. A loop γ is called a **Jordan curve** if γ is the homeomorphic image of a circle; equivalently, if $z(t)$ ($a \leq t \leq b$) is a parametrisation of γ, then γ is a Jordan curve if $z(t)$ is 1–1 (injective) on $[a, b)$ (of course $z(a) = z(b)$). The fundamental **Jordan curve theorem** states that a Jordan curve γ divides the plane into exactly two regions; to be precise γ^c has exactly two components, the unbounded component (exterior of γ) and the bounded component (interior of γ). A **Jordan domain** is the interior of a Jordan curve. Thus a Jordan curve is the boundary of a Jordan domain. An extended version of the Riemann mapping theorem shows that the union of a Jordan domain and its boundary curve is homeomorphic to a closed disc with the open disc corresponding to the Jordan domain and the circle bounding the disc corresponding to the boundary Jordan curve.

For a detailed modern exposition with full proofs of all these results and much else on winding numbers and the relationships between plane topology and complex analysis we refer to the book of Beardon [6].

2.3.16 Problems

1. **Brouwer's theorem on plane domains.** Let $\omega(z)$ be a 1–1 continuous mapping defined on a domain D. Show that $\omega(D)$ is a domain homeomorphic to D. [Hint: use the Jordan curve theorem.]
2. Let $\omega(z)$ be a 1–1 continuous function mapping the domain D onto the domain Ω and let γ be a curve in D. Let $\Gamma = \omega \circ \gamma$ be the image curve in Ω. Show that, if f is continuous and non-zero in Ω and if $F(z) = f(\omega(z))$, then

$$d(F, \gamma) = d(f, \Gamma). \tag{2.52}$$

   Deduce that the degree principle for a simply-connected domain follows from 2.3.5 and the Riemann mapping theorem.
3. Let E be a compact set in $\mathbb{C}$ and let $a \in E$. Let F be a closed set disjoint from E. Show that there exists a cycle γ in $(E \cup F)^c$ such that $n(\gamma, a) = 1$. [Hint: form a cycle Γ as follows: for small $d > 0$ form a network of squares R_k with perimeters Γ_k such that a is the centre of one square and set $\Gamma = \sum_k \Gamma_k$, where the sum is taken over all k such that R_k contains a point of E.] Hence show that, if D is a domain such that every cycle in D is a zero cycle, then D is simply-connected.

2.4 The degree principle and homotopy

2.4.1

Let γ_0 and γ_1 be curves in a domain D possessing common endpoints. We say that γ_0 and γ_1 are **homotopic** in D if there exists a function $z(s, t)$ continuous from $R = [0, 1] \times [0, 1] \to D$ such that the curves γ_s with parametrisations $z(s, t)$ $(0 \le t \le 1)$ have the same endpoints as γ_0 and γ_1 and are equal to γ_0 and γ_1 when $s = 0$ and $s = 1$ respectively. Thus, briefly, we have a continuous deformation in D from γ_0 to γ_1. Our main result is given next.

2.4.2

Theorem *Let γ_0 and γ_1 be curves homotopic in a domain D and let $f(z)$ be continuous and non-zero in D. Then*

$$d(f, \gamma_0) = d(f, \gamma_1). \tag{2.53}$$

Proof The function $g(s, t) = f(z(s, t))$ is continuous and non-zero on the square R with perimeter Γ say being positively oriented. Hence by the degree principle for a square, we have

$$d(g, \Gamma) = 0. \tag{2.54}$$

Now $g(0, t) = f(z_0)$, where z_0 is the common initial point of the curves γ_s, and so the degree of g on the lower side of the square is 0. Similarly, the degree of g on the upper side of the square is 0. Hence (2.54) implies that the degree of g on the left-hand side of the square proceeding upwards is equal to the degree of g on the right-hand side of the square proceeding upwards. This is precisely the relation (2.53).

2.4.3

This result tells us that our notion of degree is a **homotopy invariant**. To explain this, let D be a domain and ζ a given point in D. If γ_1 and γ_2 are loops in D with common endpoint ζ, then the relation γ_1 *is homotopic to* γ_2 is an equivalence relation on the class of all loops in D with endpoint ζ. The resulting equivalence classes are called **homotopy classes**. The theorem then tells us that if γ is a homotopy class of loops in D, then for a function f continuous and non-zero in D, we can define $d(f, \gamma)$ as the common value of $d(f, \gamma_1)$ for all loops $\gamma_1 \in \gamma$. The constant loop γ_0 with parametrisation $z(t) = \zeta$ $(0 \le t \le 1)$ gives rise to the **null homotopy class** and the loops in this class (i.e. those homotopic to γ_0) are called **null homotopic**. Clearly we have $d(f, \gamma) = 0$ for all null homotopic loops in D and functions f continuous and non-zero in D. Notice that, if $a \notin D$, then choosing $f(z) = z - a$, we obtain $n(\gamma, a) = 0$ for all null homotopic loops in D. In particular, we see that *if every loop γ in D with endpoint ζ is null homotopic, then every loop in D is a zero cycle*. To see this we must show that the homotopy relation does not, in its essentials, depend on the point ζ. If γ' is a loop in D with endpoint ζ' say, then let l denote a curve in D joining ζ to ζ' and put $\gamma = l\gamma' l^{-1}$, which is a loop with endpoint ζ. Then for f continuous and non-zero in D, we have $d(f, \gamma) = d(f, l) + d(f, \gamma') + d(f, l^{-1}) = d(f, \gamma')$, since $d(f, l^{-1}) = -d(f, l)$. Furthermore we obtain a 1–1 correspondence between the homotopy classes defined relative to ζ and the homotopy classes defined relative to ζ'. In particular γ' is null homotopic relative to ζ' iff γ is null homotopic relative to ζ.

In works on general topology the definition of a simply-connected region is usually given as one for which all loops in the region are null homotopic. Notice that with this definition our theorem immediately gives us the degree principle for a simply-connected domain. In particular, we again obtain that $n(\gamma, a) = 0$ for all points $a \notin D$ and all loops γ in D. Hence, by Problem 2.3.16.3, D is simply-connected according to our definition given in 2.3.13. To prove that our definition implies the homotopy definition is less elementary, but is easily seen from the Riemann mapping theorem whose usual proof uses our definition (see Problem 2.4.6.1).

2.4.4

Both homotopy and homology have traditionally been developed with the purpose of introducing algebraic methods into topology. To give no more than the flavour of this, let us show how groups can be associated with each of the two concepts in a natural way. Let D be a domain. For the **homology group** for D we proceed as follows. If γ_1 and γ_2 are cycles in D the relation γ_1 *is homologous to* γ_2, defined by $\gamma_1 - \gamma_2$ *is a zero cycle*, is an equivalence relation whose equivalence classes we will call homology classes. By the degree principle $d(f, \gamma)$ for a function f continuous and non-zero in D depends only on the homology class of γ. If Γ_1 and Γ_2 are homology classes we define $\Gamma_1 + \Gamma_2$ as the homology class of $\gamma_1 + \gamma_2$, where $\gamma_1 \in \Gamma_1$ and $\gamma_2 \in \Gamma_2$. It is easily checked that this gives a well-defined binary operation and satisfies the axioms of an additive group. A simply-connected domain is then one whose homology group is trivial, i.e. consists only of the zero element. The homology group is a topological invariant in the sense that if D' is a domain in $\mathbb{C}$ homeomorphic to D, then the homology group of D' is isomorphic to the homology group for D.

The **homotopy group**, also known as the **fundamental group**, of D is a group structure defined on the set of homotopy classes of D relative to a point ζ say in D. If Γ_1 and Γ_2 are homotopy classes we define $\Gamma_1\Gamma_2$ as the homotopy class of the loop $\gamma_1\gamma_2$ (i.e. traverse γ_1 followed by γ_2), where $\gamma_1 \in \Gamma_1$ and $\gamma_2 \in \Gamma_2$. Again this gives a well-defined binary operation satisfying the axioms of a group (not necessarily commutative) and furthermore the group is independent of the choice of the point ζ. A simply-connected domain is one whose fundamental group is trivial. Finally, it is easily shown that the fundamental group is a topological invariant.

2.4.5

Theorem: Fundamental group of the punctured plane *Let $D = \mathbb{C} - \{0\}$ denote the punctured plane. If γ_1 and γ_2 are loops in D with a common endpoint, then γ_1 is homotopic to γ_2 in D if, and only if,*

$$n(\gamma_1, 0) = n(\gamma_2, 0). \tag{2.55}$$

It follows that the fundamental group of D is isomorphic to $\mathbb{Z}$, the additive group of integers.

Proof If γ_1 is homotopic to γ_2, then taking $f(z)=z$ the relation (2.55) follows from theorem 2.4.2. Conversely, if (2.55) holds, then taking the common endpoint as $z = 1$, we have $n(\gamma, 0) = 0$, where $\gamma = \gamma_1\gamma_2^{-1}$. It will be enough to show that γ is null homotopic in D. Now γ has a parametrisation in the form $z(t) = e^{\alpha(t)}$, where $\alpha(t)$ is continuous for $0 \leq t \leq 1$ and $\alpha(0) = \alpha(1) = 0$. Then $z(s, t) = e^{s\alpha(t)}$ $(0 \leq s \leq 1, 0 \leq t \leq 1)$ defines the required homotopy between the constant loop and γ. To prove the last assertion of the theorem, let F denote the fundamental group of D and for $\Gamma \in F$ define $\lambda(F) = n(\gamma, 0)$, where $\gamma \in \Gamma$. This is easily seen to define a homomorphism from F into $\mathbb{Z}$. To prove it is an isomorphism we need only show that $\exists$ a loop γ in D with $n(\gamma, 0) = 1$. The positively oriented unit circle serves this purpose.

2.4.6 Problems

1. Let D be a simply-connected domain (i.e. $D^c \cup \{\infty\}$ is connected in the extended plane). With the help of the Riemann mapping theorem show that every loop γ in D is null homotopic. Hence show that a domain is simply-connected iff its fundamental group is trivial.
2. Show that the homology group of the punctured plane is isomorphic to $\mathbb{Z}$.
3. Let D be a domain with homology group H and fundamental group F. Show that there is a homomorphism $\psi\colon F \to H$.

2.5 The topological argument principle

2.5.1

We have seen that the degree principle is analogous to Cauchy's theorem. We now show that the analogue of the residue theorem leads to a topological version of the argument principle. To accomplish this we need first to define the analogue of a residue. Let f be a function continuous and non-zero in a punctured disc $\Delta = \{z\colon 0 < |z - \zeta| < R\}$ and for $0 < r < R$ let C_r be the positively oriented circle of radius r and centre ζ. Then if $0 < r < s < R$, the cycle $C_r - C_s$ is a zero cycle in Δ, and hence by the degree principle $d(f, C_r - C_s) = 0$, i.e. $d(f, C_r) = d(f, C_s)$. Thus $d(f, C_r)$ is constant for $0 < r < R$. We define $m(f, \zeta) = d(f, C_r)$ and call this integer the **multiplicity of f at ζ**.

Note that, if $f(z) \to$ a non-zero limit w as $z \to \zeta$, then by putting $f(\zeta) = w$ we obtain a function continuous and non-zero in the disc of radius R and centre ζ and hence $m(f, \zeta) = 0$. In this case we may call the 'singularity' of f at ζ a **removable singularity**. In our situation a **singularity** corresponds to an isolated point at which f fails to be *both continuous and non-zero*. A second type of singularity occurs when f remains continuous at ζ with $f(\zeta) = 0$, i.e. a **zero** of f. If f is analytic in the neighbourhood of ζ, then it is easily seen that $m(f, \zeta)$ is the usual multiplicity of the zero ζ of f. In particular, for analytic functions f, $m(f, \zeta)$ is always non-negative. This fact is one of the fundamental topological properties of analytic functions. Notice that, if f is **co-analytic**, i.e. has the form $f = \overline{g}$, where g is analytic, then $m(f, \zeta)$ is always non-positive, since $m(f, \zeta) = -m(g, \zeta)$. The third type of singularity is a **pole** of f: ζ is a pole if $f(z) \to \infty$ as $z \to \zeta$. If f is analytic in Δ with a pole at ζ, then $m(f, \zeta) = -$(the order of the pole). Any other type of singularity of f at ζ will be called an **essential singularity**. For example, the function $f(z) = e^{1/z}$ has an essential singularity at $z = 0$. Note that $m(f, 0) = 0$.

2.5.2

Theorem: The argument principle *Let $f(z)$ be continuous and non-zero in a domain Ω except on a set $E \subset \Omega$ consisting of isolated points $\{a_j\}$ having no accumulation point in Ω. If γ is a zero cycle in Ω and $\gamma \subset D = \Omega - E$, then*

$$d(f, \gamma) = \sum_j n(\gamma, a_j) m(f, a_j). \tag{2.56}$$

Proof We note first that $n(\gamma, a_j) = 0$ for all but finitely many a_j, say $1 \le j \le n$. To see this, suppose on the contrary that there is an infinite sequence of a_j such that $n(\gamma, a_j) \neq 0$. Since $n(\gamma, a) = 0$ in the exterior of γ, we may assume that the sequence of a_j converges to a point $w \notin \Omega$. Since γ is a zero cycle $n(\gamma, w) = 0$. Furthermore $\exists a_j$ belonging to the same component of γ^c as w, which implies $n(\gamma, a_j) = 0$, a contradiction.

For each j $(1 \le j \le n)$ let C_j be a positively oriented circle in D with centre a_j, whose interior lies in Ω and contains no a_k for $k \neq j$. We consider the cycle Γ given by

$$\Gamma = \gamma - \sum_{j=1}^{n} n(\gamma, a_j) C_j. \tag{2.57}$$

It is easily seen that Γ is a cycle in D such that $n(\Gamma, a) = 0$ for all $a \notin D$, i.e. Γ is a zero cycle in D. Furthermore, f is continuous and non-zero in D and D is a domain. Hence by the degree principle we have $d(f, \Gamma) = 0$. This gives

$$\begin{aligned} d(f, \gamma) &= \sum_{j=1}^{n} n(\gamma, a_j) d(f, C_j) = \sum_{j=1}^{n} n(\gamma, a_j) m(f, a_j) \\ &= \sum_{j} n(\gamma, a_j) m(f, a_j), \end{aligned} \tag{2.58}$$

the required result.

2.5.3

If f is a non-constant meromorphic function in a domain Ω, then the set $\{a_j\}$ of zeros and poles of f is indeed a set of isolated points, and therefore the theorem contains as a particular case the usual version of the argument principle for meromorphic functions.

2.5.4

Example Let Γ be a positively oriented Jordan curve bounding the Jordan domain Ω and let f be a function continuous on $\overline{\Omega}$. If f is non-zero on Γ and if f has at most isolated zeros a_j in Ω, then

$$d(f, \Gamma) = \sum_{j} m(f, a_j). \tag{2.59}$$

Proof We make use of the Riemann mapping theorem for Jordan domains as stated in 2.3.15. Let D denote the open unit disc and C the positively oriented unit circle. There exists a homeomorphism $F\colon \overline{D} \to \overline{\Omega}$ such that $\Gamma = F(C)$. Let $g = f \circ F$ and let $b_j = F^{-1}(a_j)$. Then g is continuous on $\overline{D}$ with zeros only at the points b_j, which lie in D. Note that $d(f, \Gamma) = d(g, C)$. Furthermore, for each j, $m(f, a_j) = m(g, b_j)$ (see Problem 2.5.6.1). Thus we need to show that

$$d(g, C) = \sum_{j} m(g, b_j). \tag{2.60}$$

By defining $g(Re^{i\theta}) = g(e^{i\theta})$ for $R > 1$, g becomes continuous in $\mathbb{C}$. Since $n(C, b_j) = 1$, the result follows from the argument principle.

2.5.5

This example provides us with the most commonly cited version of the argument principle, but now extended to continuous functions possessing only isolated zeros. We still obtain that the number of zeros in a Jordan domain 'counted according to multiplicity' is equal to $2\pi \times$ change of argument of the function around the boundary, the difference from the analytic case being that multiplicities may be negative or even zero. Thus to count the actual number of zeros would require additional information. Even in the absence of such information we can get a lower bound on the number of zeros (counting a zero of multiplicity 3 say as three zeros) and in particular prove quite generally the existence of zeros, whenever the change of argument is non-zero. For smooth functions detailed information concerning the local behaviour of a function near a zero can be obtained by reference to the Jacobian determinant of the function, about which more in subsection 2.7.8. Notice also that even for functions analytic in the open unit disc and continuous in the closed disc, we obtain the classical argument principle and in particular there is no requirement that the function remain analytic on the boundary. Thus the classical proof using integration is seen not to be the best argument.

2.5.6 Problems

1. If Γ is a Jordan curve bounding a Jordan domain D, then $n(\Gamma, a) = \pm 1$ for all $a \in D$. This is usually included as part of the proof of the Jordan curve theorem. When the orientation is positive, $n(\Gamma, a) = 1$. By the Riemann mapping theorem Γ has a parametrisation $w(t) = F(e^{it})\,(0 \leq t \leq 2\pi)$, where $w = F(z)$ gives a 1–1 conformal mapping of the open unit disc onto D and extends continuously to the unit circle C mapping it homeomorphically onto Γ. Then, if $a = F(b)$, we obtain $n(\Gamma, a) = d(F(z) - F(b), C) = 1$, by the argument principle.

   Now let $\Delta = \{z\colon 0 < |z - \zeta| < R\}$ and let Γ be a positively oriented Jordan curve in Δ such that ζ lies in the Jordan domain bounded by Δ. Let C_r denote the circle of centre ζ and radius r. Show that, for $0 < r < R$, the cycle $\Gamma - C_r$ is a zero cycle in Δ. Hence show that if f is continuous and non-zero in Δ, then $d(f, \Gamma) = m(f, \zeta)$.
2. Let D be a Jordan domain bounded by a Jordan curve Γ and let f be continuous on $D \cup \Gamma$. Show that, if $f(z) \neq w_0$ for $z \in \Gamma$ and if $d(f - w_0, \Gamma) \neq 0$, then $\exists \rho > 0$ such that

$$\{w\colon |w - w_0| < \rho\} \subset f(D \cup \Gamma). \tag{2.61}$$

Hence show that, if $g(z)$ is a continuous function in the neighbourhood of a point ζ, and if g has an isolated zero of non-zero multiplicity at ζ, then g attains all values in a neighbourhood of the origin near ζ.

3. Show that the function $f(z) = \overline{z}+z+iz^2$ has an isolated zero at the origin of multiplicity 0. [Hint: show that in $\{|z| < 1\}$ f attains no values on the positive imaginary axis and apply Problem 2.]
4. Show that the function $f(z) = \overline{z} + z - z^2$ has an isolated zero at the origin of multiplicity 1, but that f is not 1–1 in any neighbourhood of the origin.
5. Let $f(z) = P(\overline{z}) + h(z)$, where $P(z)$ is a polynomial of degree n and where $h(z)$ is analytic in a neighbourhood of the origin. Suppose that $P(0) = h(0) = 0$ and that the zero of f at the origin is isolated. Show that, for small $r > 0$, we have

$$m(f, 0) = -n + \left\{\text{number of zeros in } \{|z| < 1\} \text{ of } P\left(\frac{r}{z}\right) + h(rz)\right\}. \tag{2.62}$$

2.6 The coincidence theorem

2.6.1

One of the most useful applications of the argument principle for analytic functions is Rouché's theorem. This has a topological generalisation, which we give next.

2.6.2

Theorem *Let f and g be functions continuous in $\overline{D}$, where D is a Jordan domain with boundary the Jordan curve Γ. Suppose that*

(i) *$g(z) \neq 0$ on Γ and $d(g, \Gamma) \neq 0$,*
(ii) *$|f(z)| \leq |g(z)|$ for $z \in \Gamma$.*

Then the equation $f(z) = g(z)$ has a solution in $\overline{D}$.

Proof Assume that $f(z) \neq g(z)$ on Γ, so that $f - g \neq 0$ on Γ. Condition (ii) then implies that

$$\Re\left(1 - \frac{f(z)}{g(z)}\right) \geq 0 \quad \text{for all } z \in \Gamma, \tag{2.63}$$

and hence by lemma 2.3.2

$$d\left(1 - \frac{f}{g}, \Gamma\right) = 0. \tag{2.64}$$

We obtain

$$d(g-f,\Gamma) = d(g,\Gamma) + d\left(1-\frac{f}{g},\Gamma\right) = d(g,\Gamma) \neq 0 \text{ by (i).} \quad (2.65)$$

Hence by the degree principle $g - f$ has a zero in D.

2.6.3

Rouché's theorem *Let $f(z)$ and $g(z)$ be analytic in a Jordan domain D and continuous in $D \cup \Gamma$, where Γ is the Jordan curve bounding D. If*

$$|f(z)| < |g(z)| \quad \textit{for all } z \in \Gamma, \quad (2.66)$$

then $g(z) - f(z)$ and $g(z)$ have equally many zeros in D counted according to multiplicity.

Proof As before we obtain $d(g-f,\Gamma) = d(g,\Gamma)$, and the result follows from the argument principle.

2.6.4

The difference between these two results is accounted for simply by the fact that, since g and $g - f$ are analytic, we know that their zeros are isolated and have positive multiplicities. Therefore Rouché's theorem would be valid for continuous functions whenever this additional information was available. There is an interesting and important special case of the coincidence theorem.

2.6.5

Brouwer's fixed point theorem *Let $f(z)$ be continuous for $|z| \leq 1$ and suppose that for $|z| = 1$ we have $|f(z)| \leq 1$. Then f has a* ***fixed point*** *in the closed disc, i.e. $\exists\, z$ in $\{|z| \leq 1\}$ such that $f(z) = z$.*

Proof Putting $g(z) = z$, the result follows from the coincidence theorem.

2.6.6

This is a slightly stronger statement than the usual version of Brouwer's theorem (in $\mathbb{R}^2$). The customary hypothesis is that $|f(z)| \leq 1$ for all z in the closed disc; in other words, if D denotes the closed disc $\{|z| \leq 1\}$, f is a mapping from

D into D. The advantage of this latter statement is that it is topologically invariant, in other words, if K is a set homeomorphic to D and g is a continuous mapping from K into K, then g has a fixed point. The existence of fixed points for mappings from a topological space to itself has proved to be very important in both topology and analysis. Fixed points are also important in iteration theory. If f is a mapping from a space $\Omega \mapsto \Omega$, then starting from a given value $z_0 \in \Omega$, we can form the sequence of iterates defined by the recursive relation $z_{n+1} = f(z_n)\,(n = 0, 1, 2, \ldots)$. Fixed points act as 'attractors' in the sense that, if the sequence of iterates $\{z_n\}$ converges, then, assuming Ω is sequentially compact, the limiting value ζ is a fixed point of f. Among much else, this technique can provide a powerful means of finding approximate or numerical solutions to an equation.

We also remark that Brouwer's theorem extends to n dimensions. If f is a continuous mapping from the closed n-dimensional unit ball in $\mathbb{R}^n$ into itself, then f has a fixed point. This is proved by an extension of the concept of degree into higher dimensions.

Rouché's theorem enables us to give a painless proof of the fundamental theorem of algebra.

2.6.7

Theorem *Let $P(z) = \sum_{k=0}^{n} a_k z^k$ be a polynomial of degree n and let*

$$R = \max\left(1, \sum_{k=0}^{n-1} \frac{|a_k|}{|a_n|}\right). \tag{2.67}$$

Then $P(z)$ has exactly n zeros, counted according to multiplicity, in $\mathbb{C}$ and all the zeros lie in the disc $\{|z| \leq R\}$.

Proof Let $g(z) = a_n z^n$ and $f(z) = -\sum_{k=0}^{n-1} a_k z^k$, so that $P(z) = g(z) - f(z)$. If $|z| = r \geq R$, then

$$|f(z)| \leq \sum_{k=0}^{n-1} |a_k| r^k \leq r^{n-1} \sum_{k=0}^{n-1} |a_k| \leq r^{n-1} R |a_n| \leq |a_n| r^n = |g(z)| \tag{2.68}$$

and the result follows from Rouché's theorem applied to the disc $\{|z| \leq r\}$, since g has a single zero of multiplicity n at the origin. We note also that, although of course f and $g - f$ are analytic, the proof does not depend on this result, since the fact that the zeros of a polynomial have positive multiplicity follows by simple algebraic and topological arguments.

2.6.8 Completion of proof of fundamental theorem of algebra

Let $P(z) = \sum_{k=0}^{n} a_k z^k$ be a polynomial of degree n with distinct zeros ζ_k having multiplicity r_k say. It remains to prove the representation (2.3). Denote by $Q(z)$ the polynomial in (2.3) with $c = a_n$. Then $Q(z) - P(z)$ is either identically zero or a polynomial of degree m satisfying $0 \leq m \leq n - 1$. However, in the latter case P and Q have the same zeros with the same multiplicities, and therefore $Q - P$ has at least n zeros counted according to multiplicity. Since $Q - P$ has at most m zeros, we have a contradiction. Thus $P = Q$.

2.6.9 Harmonic polynomials

We have emphasised the topological nature of our proof of the fundamental theorem with the intention of generalising the argument to other types of polynomial expression. One such type is the class of **harmonic polynomials**. A harmonic polynomial is a function of the form

$$P(re^{i\theta}) = \sum_{k=-m}^{n} a_k r^{|k|} e^{ik\theta} \tag{2.69}$$

where m and n are non-negative integers. The numbers a_k are called the **coefficients** of P and the **degree** of P is defined to be $\max(m, n)$, assuming that each of the coefficients a_{-m} and a_n is non-zero. The case $m = 0$ reduces to our algebraic polynomials. The function $P(z)$ can be written in the form

$$P(z) = \overline{S(z)} + T(z), \tag{2.70}$$

where S and T are algebraic polynomials. In fact

$$S(z) = \sum_{k=1}^{m} \overline{a}_{-k} z^k, \qquad T(z) = \sum_{k=0}^{n} a_k z^k. \tag{2.71}$$

$P(z)$ is clearly a harmonic function, i.e. a complex-valued solution of Laplace's equation. Note also that on each circle $|z| = r$, $P(z) = P(re^{i\theta})$ takes the form of a **trigonometric polynomial** in θ. If P is a harmonic polynomial of degree n, it is often convenient to write P in the form

$$P(re^{i\theta}) = \sum_{k=-n}^{n} a_k r^{|k|} e^{ik\theta}, \tag{2.72}$$

where one of the coefficients a_{-n} or a_n may be zero, but not both.

2.6.10

Theorem *Let $P(z)$ be a harmonic polynomial of degree n written in the form (2.72) and suppose that $|a_{-n}| \neq |a_n|$. Let*

$$R = \max\left(1, \frac{\sum_{k=-n+1}^{n-1} |a_k|}{||a_{-n}| - |a_n||}\right). \tag{2.73}$$

Then every zero of P is isolated and lies in $\{|z| \leq R\}$; furthermore P has at least one zero and in fact, if each zero is counted according to the absolute value of its multiplicity, then P has at least n zeros. Finally, P has at most n^2 zeros.

Proof We may assume that $|a_{-n}| < |a_n|$. Let $g(re^{i\theta}) = (a_{-n}e^{in\theta} + a_n e^{in\theta})r^n$ and let C_r denote the positively oriented circle $\{|z| = r\}$. Then

$$|g(re^{i\theta})| \geq (|a_n| - |a_{-n}|)r^n \tag{2.74}$$

and in particular $g(re^{i\theta}) \neq 0$ on C_r $(r > 0)$. If $r > R$, we obtain

$$\begin{aligned} |g(re^{i\theta}) - P(re^{i\theta})| &\leq \sum_{k=-n+1}^{n-1} |a_k| r^{|k|} \leq \left(\sum_{k=-n+1}^{n-1} |a_k|\right) r^{n-1} \\ &< (|a_n| - |a_{-n}|)r^n \leq |g(re^{i\theta})|. \end{aligned} \tag{2.75}$$

We deduce that $P(re^{i\theta}) \neq 0$ and

$$d(P, C_r) = d(g, C_r) = d(h, C_1) \tag{2.76}$$

where $h(z) = a_n z^n - a_{-n}\bar{z}^n = a_n z^n(1 - \frac{a_{-n}}{a_n}\bar{z}^{2n})$. As the bracketed expression has positive real part on the unit circle, we obtain

$$d(P, C_r) = d(h, C_1) = n. \tag{2.77}$$

Thus every zero of P lies in $\{|z| \leq R\}$ and the number of zeros, by the argument principle, is either infinite or at least n, if counted according to the absolute value of their multiplicities; indeed, we obtain at least n zeros even if we count only those with positive multiplicity.

It remains to prove the estimate n^2 for the number of zeros. The harmonic polynomial

$$f = \frac{\overline{a_n}P - a_{-n}\overline{P}}{|a_n|^2 - |a_{-n}|^2} \tag{2.78}$$

has the form $f(z) = z^n + Q(z)$, where Q has degree at most $n-1$. Furthermore, the zeros of f and the zeros of P are identical. Therefore the desired result follows from theorem 1.2.5 of chapter 1.

This argument is the author's modification of the original proof of Wilmshurst [72], whose idea it was to use Bézout's theorem. Wilmshurst has further constructed an elegant example, which we give next, to show that a harmonic polynomial of degree n can have exactly n^2 distinct zeros.

2.6.11 Wilmshurst's example

Consider the polynomial

$$Q(z) = \Im(e^{-i\pi/4}z^n) + i\Im(e^{i\pi/4}(z-1)^n), \tag{2.79}$$

which is clearly harmonic. The zeros of Q occur precisely at those points z at which both the quantities $e^{-i\pi/4}z^n$ and $e^{i\pi/4}(z-1)^n$ are real. Now $e^{-i\pi/4}z^n$ is real on the n lines through the origin making angles $(4k+1)\pi/4n$ $(k = 0, 1, \ldots, n-1)$ with the positive axis; $e^{i\pi/4}(z-1)^n$ is real on the n lines through the point $(1, 0)$ making angles $(4k+3)\pi/4n$ $(k = 0, 1, \ldots, n-1)$ with the positive axis. No line in the group passing through the origin is parallel to a line in the group passing through 1. Hence each line in the origin group has exactly one meeting point with each of the lines in the 1 group, giving an exact total of n^2 zeros of Q.

In his thesis Wilmshurst [71] has constructed a number of examples of harmonic polynomials $P = \overline{S} + T$ of degree n with 'maximal valence' n^2. In all such examples he found that, if T had degree n, then S had degree at least $n-1$. On the basis of this and other considerations Wilmshurst has made the following *conjecture: if S has degree m and T has degree $n > m$, then P has at most $m(m-1) + 3n - 2$ distinct zeros.*

A particularly interesting case is the case $m = 1$. Wilmshurst's conjecture in this case is equivalent to the following *conjecture: if $T(z)$ is a complex polynomial of degree n, then the conjugate $\overline{T(z)}$ has at most $3n-2$ fixed points (i.e. solutions of $\overline{T(z)} = z$).* He has proved this for $n = 2$ and $n = 3$.

Algebraic methods appear unable to handle this conjecture. In an attempt to apply topological methods Wilmshurst has conjectured that *if D is a component of the level set $\{z\colon |T'(z)| < 1\}$ and r denotes the number of zeros of T'' in D, then $\overline{z} + T$ is $(r+1)$-valent in D.*

2.6.12 Harmonic functions at a non-isolated zero

Theorem *Let $f(z)$ be harmonic in a neighbourhood of the origin and suppose that $\exists$ a sequence $\{z_n\}$ of distinct points converging to 0 such that*

$$f(z_n) = 0 \quad (n = 1, 2, \dots). \tag{2.80}$$

Then $\exists$ a simple analytic arc C passing through 0 such that

$$f(z) = 0 \quad (z \in C). \tag{2.81}$$

Furthermore, either $f(z) \equiv 0$, or there are a finite number k of such arcs containing all the zeros of f sufficiently close to 0. In particular, the set of non-isolated zeros of f contains no isolated points.

Proof If f is not identically zero in a neighbourhood of the origin we can write f in an absolutely convergent series of the form

$$f(z) = \sum_{n=k}^{\infty} a_{-n}\bar{z}^n + \sum_{n=k}^{\infty} a_n z^n \tag{2.82}$$

where at least one of a_k and a_{-k} is non-zero. We then have

$$0 = \lim_{m\to\infty} \frac{f(z_m)}{|z_m^k|} = a_{-k}e^{-ik\theta} + a_k e^{ik\theta} \tag{2.83}$$

for some real θ. It follows that $|a_{-k}| = |a_k| \neq 0$. Thus we can write

$$f(z) = \overline{z^k g(z)} + z^k h(z) \tag{2.84}$$

where g and h are analytic near 0 and $|g(0)| = |h(0)| \neq 0$. Now near 0 we have

$$z^k g(z) = \lambda^k(z), \qquad -z^k h(z) = \mu^k(z) \tag{2.85}$$

where λ and μ are analytic and 1–1 with $|\lambda'(0)| = |\mu'(0)|$. Thus for large n

$$\overline{\lambda^k(z_n)} = \mu^k(z_n) \tag{2.86}$$

and so for infinitely many n

$$\overline{\lambda(z_n)} = \omega\mu(z_n) = \nu(z_n) \tag{2.87}$$

where ω is some kth root of unity. Also we may assume that both λ and ν are 1–1.

Let $w = \lambda(z)$ have inverse $z = \tau(w)$ and let $\kappa(w) = \nu(\tau(w))$ and $w_n = \lambda(z_n)$. Then $\overline{w_n} = \kappa(w_n)$ and $\kappa'(0) = e^{i\alpha}$ say, so that

$$\overline{e^{i\alpha/2}w_n} = e^{-i\alpha/2}\kappa(w_n). \tag{2.88}$$

Writing $W = e^{i\alpha/2}w$, $W_n = e^{i\alpha/2}w_n$, $\sigma(W) = e^{-i\alpha/2}\kappa(e^{-i\alpha/2}W)$, this becomes

$$\overline{W_n} = \sigma(W_n) \tag{2.89}$$

where $\sigma(W) = W + A_2W^2 + \cdots$ near 0. Hence

$$\Re W_n = \frac{1}{2}(W_n + \sigma(W_n)) = W_n + \frac{1}{2}A_2W_n^2 + \cdots. \tag{2.90}$$

Now $\zeta = \frac{1}{2}(W + \sigma(W))$ is 1–1 near 0 with inverse $W = K(\zeta)$ say. Then

$$\zeta_n = \frac{1}{2}(W_n + \sigma(W_n)) = \Re K(\zeta_n). \tag{2.91}$$

Thus the points ζ_n are real and distinct near 0 and the real harmonic function $U(\zeta) = \Re(K(\zeta) - \zeta)$ vanishes at the ζ_n. But on the real axis $U(\zeta)$ is an analytic function of ζ and so we have $U(\zeta) = 0$ for all ζ real near 0. Thus $\Re K(\zeta) = \zeta$ for ζ real and so

$$\overline{W} = \sigma(W) \tag{2.92}$$

at all points W near 0 where $W + \sigma(W)$ is real. This is on a 1–1 analytic image of a segment of the real axis. We obtain $\overline{w} = \kappa(w)$ at all points w near 0 such that $e^{i\alpha/2}w + e^{-i\alpha/2}\kappa(w)$ is real. This in turn gives

$$\overline{\lambda(z)} = \omega\mu(z) \tag{2.93}$$

at points z near 0 for which $e^{i\alpha/2}\lambda(z) + e^{-i\alpha/2}\mu(z)$ is real. Thus for such z, $f(z) = 0$. These zeros lie on a simple analytic arc through 0 and, corresponding to the root of unity ω, are the only zeros near 0. The remaining kth roots of unity may lead to further such zero arcs, but at most k. This completes the proof.

2.6.13

Wilmshurst's theorem *Let P be a harmonic polynomial of degree n satisfying*

$$P(z) \to \infty \text{ as } z \to \infty. \tag{2.94}$$

Then for each $w \in \mathbb{C}$ the equation $P(z) = w$ has at most n^2 roots in $\mathbb{C}$.

Proof Applying theorem 1.2.3 of chapter 1 it is sufficient to show that (2.94) implies that P has at most finitely many zeros in $\mathbb{C}$. If not, then there is at least one non-isolated zero of P. Writing $P = \overline{S} + T$ assume that a is such a zero; then $P = 0$ on an analytic curve passing through a and so the rational function $Q = S/T$ satisfies $|Q(z)| = 1$ on this curve. Thus the zeros of P form a compact subset of the level set $\{z : |Q(z)| = 1\}$, which consists of a finite number of loops. By the previous theorem it is clear that one of these loops is a zero set of P. But such a loop will enclose at least one Jordan domain. In this domain P is harmonic and zero on the boundary of the domain. By the minimum and maximum principles for harmonic functions, we deduce that $P \equiv 0$, a contradiction.

2.6.14 Problems

1. Let $P(z)$ be a harmonic polynomial of degree n with coefficients a_k $(-n \leq k \leq n)$ and suppose that $|a_n| \neq |a_{-n}|$. Show that for each ζ

$$|m(P, \zeta)| \leq n. \tag{2.95}$$

[Hint: use Problem 2.5.6.5.]

2. If $a > e$, prove that the equation

$$e^z = az^n \tag{2.96}$$

has exactly n roots in $\{|z| < 1\}$.

3. Let f be continuous in $\mathbb{C}$ and suppose that
   (i) $f(z) \to \infty$ as $z \to \infty$,
   (ii) f has at least one zero and all the zeros of f are isolated points with positive multiplicity.

   Show that $f: \mathbb{C} \to \mathbb{C}$ is surjective.

4. Let f and g be continuous in $\mathbb{C}$ and suppose that
   (i) $f(z) \sim g(z)$ as $z \to \infty$,
   (ii) $g(z) \to \infty$ as $z \to \infty$,
   (iii) $d(g, C_R) > 0$ for large R, where C_R is the positively oriented circle $\{|z| = R\}$.

   Show that $f: \mathbb{C} \mapsto \mathbb{C}$ is surjective.

5. Let $g(z)$ be continuous for $|z| < 1$ and suppose that

$$(1 - |z|^2)g(z) \to 0 \text{ as } |z| \to 1. \tag{2.97}$$

Show that there exists a point z satisfying $|z| < 1$ such that

$$g(z) = \frac{z}{1 - |z|^2}. \tag{2.98}$$

6. Continuity of zeros. Let $P(z) = \sum_{k=0}^{n} a_k z^k$ be a polynomial of degree n with distinct zeros z_j $(1 \leq j \leq m)$. Show that, for each $\epsilon > 0$, if $|z_j - z_i| \neq \epsilon$ for $1 \leq j \leq m$ and $1 \leq i \leq m$, then $\exists \delta > 0$ such that, if δ_k are numbers satisfying $0 < \delta_k < \delta$ for $0 \leq k \leq n$, the polynomial $P^*(z) = \sum_{k=0}^{n} (a_k + \delta_k) z^k$

has the same number of zeros as P in each of the discs $\{z\colon |z - z_j| < \epsilon\}$, zeros being counted according to their multiplicity. [Hint: apply Rouché's theorem.]

2.7 Locally 1–1 functions

2.7.1

Let f be continuous in a domain D. We will say that f is **locally 1–1** in D if, for each point $z_0 \in D$, f is 1–1 in some neighbourhood of z_0. By Brouwer's plane domain theorem (Problem 2.3.16.1) f will have a **locally defined continuous inverse**, i.e. if $w_0 = f(z_0)$ there exists a function $\phi(w)$ defined and continuous in the neighbourhood of w_0 such that $\phi(w_0) = z_0$ and $\phi(f(z)) = z$ in the neighbourhood of z_0. For example, if f is analytic, the condition for f to be 1–1 near z_0 is $f'(z_0) \neq 0$.

A function f continuous and locally 1–1 in a domain D will have an orientation at each point $z_0 \in D$, i.e. a small circle centre z_0 will be mapped by f onto a Jordan curve which will have either positive or negative orientation; f is said to be **sense preserving** at z_0 if this orientation is positive; otherwise f is **sense reversing** at z_0. The function f is sense preserving at z_0 iff $m(f - f(z_0), z_0) = 1$. For sense reversing, the value is -1. The function f will be either sense preserving at every point of D or sense reversing at every point of D. This is proved by a connectedness argument. It is enough to show that the set of points in D at which f is sense preserving is open. Suppose that f is sense preserving at z_0. Then f is continuous and 1–1 in some disc $\{|z - z_0| < \rho + \epsilon\}$, where $\rho > 0$ and $\epsilon > 0$. Let $C_r(\zeta)$ denote the positively oriented circle of radius r and centre ζ. We have $d(f - f(z_0), C_\rho(z_0)) = 1$. Choose z_1 $(0 < |z_1 - z_0| < \rho)$ and $t > 0$ such that $C_t(z_1) \subset C_\rho(z_0)$. The cycle $\Gamma = C_\rho(z_0) - C_t(z_1)$ is a zero cycle in $\{|z - z_1| < \rho + \epsilon\} - \{z_1\}$ and so, since $f(z) - f(z_1) \neq 0$ in this punctured disc, $d(f - f(z_1), \gamma) = 0$. Thus

$$\begin{aligned} m(f - f(z_1), z_1) &= d(f - f(z_1),\ C_t(z_1)) = d(f - f(z_1), C_\rho(z_0)) \\ &= n(\gamma, f(z_1)) = n(\gamma, f(z_0)) = m(f - f(z_0), z_0) = 1, \end{aligned} \tag{2.99}$$

where $\gamma = f(C_\rho(z_0))$. Hence f is sense preserving at z_1, as required.

As an immediate corollary of this result, we see that if f is locally 1–1 in D and if Ω is a Jordan domain whose closure lies in D and Γ is the positively oriented Jordan curve bounding Ω, then, if $f \neq 0$ on Γ, $|d(f, \Gamma)|$ gives the exact number of distinct zeros of f in Ω by the argument principle.

We also see that f is an **open mapping**, i.e. f maps open sets onto open sets. From this it follows that f satisfies the strong form of the **maximum principle**: if $|f| \leq M$ on Γ, then $|f| < M$ in Ω.

2.7.2

Theorem *Let f be a function continuous and locally 1–1 in $\mathbb{C}$ and suppose that*

$$f(z) \to \infty \text{ as } z \to \infty. \tag{2.100}$$

Then f is a homeomorphism of $\mathbb{C}$ onto $\mathbb{C}$.

Proof We may assume that f is sense preserving. We show first that $\exists$ an integer $n \geq 1$ such that for each $w \in \mathbb{C}$ the equation $f(z) = w$ has exactly n roots in $\mathbb{C}$. In particular f is a surjective mapping. Choose $w \in \mathbb{C}$. The zeros of $f - w$ are isolated with each zero (if any) having multiplicity 1. Also (2.100) implies that the number of such zeros is finite; (2.100) further implies that

$$h(z) = \frac{f(z) - f(0)}{f(z) - w} \to 1 \text{ as } z \to \infty \tag{2.101}$$

and so for large $R > 0$, $d(h, C_R) = 0$, where C_R is the positively oriented circle $\{|z| = R\}$. Thus

$$d(f - f(0), C_R) = d(f - w, C_R) \tag{2.102}$$

for all large R. This implies that the equation $f(z) = w$ has exactly the same number of roots in $\mathbb{C}$ as the equation $f(z) = f(0)$, which is some number $n \geq 1$.

It remains to show that $n = 1$. We will establish this by a quite different method, which is of considerable interest in its own right. The idea is to show that a branch of the inverse function can be continued along any curve. Choose $w_0 \in \mathbb{C}$ and pick one of the n values z_0 for which $f(z_0) = w_0$. In the neighbourhood of w_0 we have a uniquely defined inverse function $g(w)$ continuous and satisfying $g(w_0) = z_0$. Let γ be a curve with initial point w_0 and terminal point w_1. We wish to show that there is a unique continuation of $g(w)$ along γ, i.e. exactly one continuous inverse to f defined on γ and equal to $g(w)$ on the initial part of γ. Let γ have a parametrisation $w = w(t)\,(t_0 \leq t \leq t_1)$. Certainly $g(w)$ is well defined on γ for $t_0 \leq t < s$ say. Furthermore, at $w(s)$ there are n possible branches of the inverse function defined in the neighbourhood of $w(s)$ and exactly one of these coincides with $g(w)$ on the part of γ with $t < s$.

Notice that this argument requires that the curve $g(w(t)) = z(t)$ does not approach ∞, which follows because of (2.100). If the curve approaches a finite value ζ (possibly even only as a limit point) as $t \to s$, then the inverse function defined at $f(\zeta)$ with $f(\zeta)$ mapping to ζ provides the required branch. But this branch defines a continuation on γ for values of $t > s$. Continuing in this way we see by the same argument that there is no value s $(t < s \le t_1)$ at which the continuation cannot be defined. Indeed by the Heine–Borel theorem we can reach t_1 in a finite number of steps.

Thus, since $g(w)$ has a unique continuation along any curve from the given initial branch, we can apply the monodromy theorem to deduce that the terminal branch at w_1 is independent of the choice of curve γ joining w_0 to w_1. In other words we obtain a single-valued inverse function $g(w)$ defined over the whole w-plane. We remark that the monodromy theorem is usually applied when the continuation is analytic continuation. However, an examination of the proof (based on homotopy – see e.g. Ahlfors [2]) shows that all that is required for the argument to work is that the continuation process be uniquely defined.

It is now clear that $n = 1$, for if $f(z_1) = f(z_0)$, then the line segment joining z_0 to z_1 is mapped by f onto a loop γ with endpoint w_0 and, since $g(w)$ is uniquely defined, the terminal branch of the continuation along γ leads back to the initial branch and so $z_1 = g(f(z_1)) = g(f(z_0)) = z_0$. This completes the proof.

2.7.3

There are some interesting corollaries of this result, which lead to quite surprising results even for analytic functions. Let D and Ω be simply-connected domains with respective boundaries ∂D and $\partial\Omega$, which we will assume include the point ∞ for unbounded domains. Let f be a function defined on D. We write $f(z) \to \partial\Omega$ as $z \to \partial D$ if for each sequence $\{z_n\} \subset D$ satisfying $z_n \to \zeta \in \partial D$, all the limit points of the sequence $\{f(z_n)\}$ lie on $\partial\Omega$. The set of such limit points is called the **cluster set of** f.

2.7.4

Corollary *Let f be a continuous locally 1–1 mapping from a simply-connected domain D into a simply-connected domain Ω and suppose that*

$$f(z) \to \partial\Omega \text{ as } z \to \partial D. \tag{2.103}$$

Then f is a homeomorphism of D onto Ω.

Proof Since D and Ω are simply-connected, $\exists$ homeomorphisms $\phi \colon \mathbb{C} \to D$, $\psi \colon \Omega \to \mathbb{C}$. Then $g = \psi \circ f \circ \phi$ is locally 1–1 in $\mathbb{C}$ and $g(z) \to \infty$ as $z \to \infty$ (see Problem 2.7.24.2). By the theorem g is a homeomorphism of $\mathbb{C}$ onto $\mathbb{C}$ and so f is a homeomorphism of D onto Ω.

2.7.5

Corollary *Let D be a simply-connected domain and Ω a Jordan domain bounded by a Jordan curve Γ. Suppose that f is continuous and locally 1–1 in D and*

$$f(z) \to \Gamma \textit{ as } z \to \partial D. \tag{2.104}$$

Then f is a homeomorphism of D onto Ω.

Proof To apply the previous corollary we need to show that $f(D) \subset \Omega$. Let $\mathbb{U} = \{|z| < 1\}$ and $\mathbb{T} = \{|z| = 1\}$. We require the **theorem of Schoenflies**: *if Γ is a Jordan curve bounding a Jordan domain Ω, then $\exists$ a homeomorphism $\phi \colon \mathbb{C} \to \mathbb{C}$ such that $\phi(\gamma) = \mathbb{T}$ and $\phi(\Omega) = \mathbb{U}$.* This can be proved using the strong form of the Riemann mapping theorem plus one extra result: if ω is a homeomorphism of $\mathbb{T}$ onto $\mathbb{T}$, then ω can be extended into $\mathbb{U}$ to give a homeomorphism of $\overline{\mathbb{U}}$ onto $\overline{\mathbb{U}}$ (in fact the Rado–Kneser–Choquet theorem shows that the harmonic extension of ω will suffice).

Let ψ be a homeomorphism of $\mathbb{U}$ onto D. We consider $g = \phi \circ f \circ \psi$. This function is continuous and locally 1–1 in $\mathbb{U}$ and

$$g(z) \to \mathbb{T} \text{ as } z \to \mathbb{T}, \tag{2.105}$$

i.e. $|g(z)| \to 1$ as $|z| \to 1$. To complete the proof we need to show that $|g(z)| < 1$ in $\mathbb{U}$. Now g is an open mapping so satisfies the maximum principle. Thus, if $0 < r < R < 1$,

$$\max_{|z|=r} |g(z)| \leq \max_{|z|=R} |g(z)| \leq 1 + \epsilon(R) \tag{2.106}$$

where $\epsilon(R) \to 0$ as $R \to 1$. Letting $R \to 1$ we obtain $|g(z)| \leq 1$ for $|z| < 1$, and since no maximum value of $|g|$ can be attained in $\mathbb{U}$, $|g(z)| < 1$ in $\mathbb{U}$.

Caution This is a result for finite Jordan domains and is invalid even if Ω is a half-plane with a hypothesis that $f(z) \to$ either the bounding line of Ω or ∞ as $z \to \partial D$.

2.7.6

Example If f is analytic in $\mathbb{U}$ and

$$|f(z)| \to 1 \text{ as } |z| \to 1, \tag{2.107}$$

then either f' has a zero in $\mathbb{U}$ or $\exists$ constants α real and z_0 ($|z_0| < 1$) such that

$$f(z) = e^{i\alpha} \frac{z - z_0}{1 - \overline{z_0} z}. \tag{2.108}$$

For, if $f'(z) \neq 0$, then f is a homeomorphism of $\mathbb{U}$ onto $\mathbb{U}$ and it is a consequence of Schwarz's lemma that the only such analytic functions f have the form (2.108).

For example, every finite **Blaschke product**

$$B(z) = \prod_{k=1}^{n} \frac{z - z_k}{1 - \overline{z_k} z} \tag{2.109}$$

of order $n > 1$ has a critical point in $\mathbb{U}$. In fact it is not difficult to show directly that B' has exactly $n - 1$ zeros in $\mathbb{U}$.

2.7.7

The example is a special case of corollary 2.7.5, when f is a function continuous for $\{|z| \leq 1\}$, locally 1–1 for $\{|z| < 1\}$ and satisfying $|f(z)| = 1$ for $|z| = 1$. Then f is a homeomorphism of $\mathbb{U}$ onto $\mathbb{U}$. Notice that it is not necessary to make any assumptions about the behaviour of f on the unit circle. In particular, we do not need to assume that f is 1–1 on $\mathbb{T}$, although, if it is, then f gives a homeomorphism of $\overline{\mathbb{U}}$ onto $\overline{\mathbb{U}}$. In general, under the given hypotheses, we can write

$$f(e^{it}) = e^{i\phi(t)} \quad (0 \leq t \leq 2\pi), \tag{2.110}$$

where $\phi(t)$ is real and continuous. Since f is a homeomorphism on $\mathbb{U}$, we clearly have $\phi(2\pi) - \phi(0) = 2\pi$, assuming a positive orientation. Furthermore ϕ is an increasing function, though not necessarily strictly increasing. The proof of this is left as a problem.

2.7.8 The Jacobian determinant

Let $f(z) = f(x, y)$ be a function defined in a domain D and suppose that the first partial derivatives $\frac{\partial f}{\partial x}, \frac{\partial f}{\partial y}$ exist in D and are continuous. Writing $f = u + iv$

(u, v real) the **Jacobian matrix** of f is given by

$$M(f) = \begin{bmatrix} \dfrac{\partial u}{\partial x} & \dfrac{\partial u}{\partial y} \\ \dfrac{\partial v}{\partial x} & \dfrac{\partial v}{\partial y} \end{bmatrix}. \tag{2.111}$$

The determinant of this matrix is called the Jacobian determinant, or simply the **Jacobian**, and denoted by $J(f)$. Thus

$$J(f) = \frac{\partial u \partial v}{\partial x \partial y} - \frac{\partial u \partial v}{\partial y \partial x}, \tag{2.112}$$

which is a real-valued, continuous function in D. The non-vanishing of the Jacobian provides an important analytic criterion for a function to be locally 1–1.

2.7.9

Inverse function theorem *Let f be a function possessing continuous first partial derivatives in a domain D and let $z_0 \in D$. If $J(f)(z_0) \neq 0$, then there is a neighbourhood of z_0 in which f is 1–1. In this case the mapping is sense preserving at z_0, and $m(f - f(z_0), z_0) = 1$, iff $J(f)(z_0) > 0$. In particular, if $J(f) \neq 0$ in D, then f is locally 1–1 in D and either $J(f) > 0$ in D and the mapping is sense preserving, or $J(f) < 0$ in D and the mapping is sense reversing.*

Proof Without loss of generality we may assume that $z_0 = 0$ and $f(0) = 0$. Suppose that $J(f)(0) \neq 0$. If f is not 1–1 in some neighbourhood of 0, then $\exists$ two sequences $\{z_n\}$, $\{\zeta_n\}$ each converging to 0 such that $f(z_n) = f(\zeta_n)$ and $z_n \neq \zeta_n$ for every n. Let $z_n = x_n + iy_n, \zeta_n = \xi_n + i\eta_n, f = u + iv$. For sufficiently large n we have

$$\begin{aligned} u(z_n) - u(\zeta_n) &= u(x_n, y_n) - u(\xi_n, y_n) + u(\xi_n, y_n) - u(\xi_n, \eta_n) \\ &= (x_n - \xi_n)\frac{\partial u}{\partial x}(s_n, y_n) + (y_n - \eta_n)\frac{\partial u}{\partial y}(\xi_n, \sigma_n) \end{aligned} \tag{2.113}$$

where s_n lies between x_n and ξ_n and σ_n lies between y_n and η_n, applying the mean value theorem. Similarly

$$v(z_n) - v(\zeta_n) = (x_n - \xi_n)\frac{\partial v}{\partial x}(t_n, y_n) + (y_n - \eta_n)\frac{\partial v}{\partial y}(\xi_n, \tau_n) \tag{2.114}$$

where t_n lies between x_n and ξ_n and τ_n lies between y_n and η_n. We deduce that

$$\begin{aligned}
0 &= \frac{f(z_n) - f(\zeta_n)}{|z_n - \zeta_n|} \\
&= \frac{(x_n - \xi_n)}{|z_n - \zeta_n|}\left(\frac{\partial u}{\partial x}(s_n, y_n) + i\frac{\partial v}{\partial x}(t_n, y_n)\right) \\
&\quad + \frac{(y_n - \eta_n)}{|z_n - \zeta_n|}\left(\frac{\partial u}{\partial y}(\xi_n, \sigma_n) + i\frac{\partial v}{\partial y}(\xi_n, \tau_n)\right) \\
&= \alpha_n\left(\frac{\partial u}{\partial x}(s_n, y_n) + i\frac{\partial v}{\partial x}(t_n, y_n)\right) + \beta_n\left(\frac{\partial u}{\partial y}(\xi_n, \sigma_n) + i\frac{\partial v}{\partial y}(\xi_n, \tau_n)\right)
\end{aligned} \tag{2.115}$$

where α_n and β_n are real and satisfy $\alpha_n^2 + \beta_n^2 = 1$. Letting $n \to \infty$ through a suitable subsequence of integers n, we obtain

$$\alpha\frac{\partial f}{\partial x}(0, 0) + \beta\frac{\partial f}{\partial y}(0, 0) = 0, \tag{2.116}$$

where α and β are real and satisfy $\alpha^2 + \beta^2 = 1$. This is equivalent to a non-trivial solution $A = \binom{\alpha}{\beta}$ of the matrix equation $M(f)(0)A = 0$, and therefore implies that the matrix $M(f)(0)$ is singular, i.e. $J(f)(0) = 0$, a contradiction.

For the next part of the proof we assume that $J(f)(0) > 0$ and consider the function

$$p(z) = f(z) - ax - by \tag{2.117}$$

where $z = x + iy$, $a = \frac{\partial f}{\partial x}(0, 0)$, $b = \frac{\partial f}{\partial y}(0, 0)$. Then p has continuous partial derivatives in D, which satisfy

$$\frac{\partial p}{\partial x}(0, 0) = 0 = \frac{\partial p}{\partial y}(0, 0). \tag{2.118}$$

It follows that $p'(0)$ exists and $= 0$, and hence, since $p(0) = 0$,

$$q(z) = \frac{p(z)}{|z|} \to 0 \text{ as } z \to 0. \tag{2.119}$$

Thus for small $r > 0$ we have

$$\begin{aligned}
\frac{f(re^{it})}{r} &= a\cos t + b\sin t + q(re^{it}) = \frac{1}{2}(a - ib)e^{it} + \frac{1}{2}(a + ib)e^{-it} + q(re^{it}) \\
&= \frac{1}{2}(a - ib)e^{it}\left[1 + \frac{a + ib}{a - ib}e^{-2it} + \frac{2q(re^{it})e^{-it}}{a - ib}\right].
\end{aligned} \tag{2.120}$$

Now the condition $J(f)(0) > 0$ is equivalent to $|a + ib| < |a - ib|$ and so, since $q(re^{it}) \to 0$ uniformly in t as $r \to 0$, we see that for small $r > 0$ the expression in the square brackets has positive real part. Thus this expression has degree 0 on the positively oriented circle $C_r = \{|z| = r\}$. Thus

$$m(f, 0) = d(f, C_r) = d(e^{it}, C_1) = 1. \tag{2.121}$$

Thus f is sense preserving at 0. If $J(f)(0) < 0$, then $J(\overline{f})(0) = -J(f)(0) > 0$. Hence $\overline{f}$ is sense preserving, which implies that f is sense reversing. Finally, if $J(f) \neq 0$ in D, then since $J(f)$ is continuous and D is connected, $J(f)$ has the same sign at all points of D. This completes the proof.

2.7.10

Example The function $f(x, y) = x^3 + iy^3$ is 1–1 in $\mathbb{C}$, but $J(f) = 9x^2y^2$, which is zero on both the real and imaginary axes. Thus a 1–1 function need not have a non-vanishing Jacobian.

2.7.11

The inverse function theorem can be used to give a proof of the implicit function theorem.

Implicit function theorem *Let $u(x, y)$ be a real-valued function possessing continuous first partial derivatives in a neighbourhood of (x_0, y_0) and satisfying $u(x_0, y_0) = 0$, and suppose that*

$$\frac{\partial u}{\partial y}(x_0, y_0) \neq 0. \tag{2.122}$$

Then $\exists$ a unique function $f(x)$ defined and differentiable in a neighbourhood U of x_0 with $f(x_0) = y_0$ and satisfying

$$u(x, f(x)) = 0 \quad (x \in U). \tag{2.123}$$

Proof Consider the mapping $P(x, y) = (x, u(x, y))$ and observe that $J(P) = \partial u/\partial y$. Hence the hypothesis implies from the inverse function theorem that P is 1–1 in a neighbourhood of (x_0, y_0). P therefore maps such a neighbourhood onto a neighbourhood of $(x_0, 0)$. Let $g(a, b)$ denote the inverse function defined in this latter neighbourhood. Then $g(x, 0)$ is defined and differentiable in a neighbourhood U of x_0. Since $g(x, u(x, y)) = (x, y)$, we see that $g(x, 0) = (x, y)$, where $u(x, y) = 0$. Thus $f(x) = \Im g(x, 0)$ gives the required function.

2.7.12

Corollary *If $u(x, y)$ is a function with continuous first partial derivatives in a neighbourhood of (x_0, y_0), if $u(x_0, y_0) = 0$ and if one of the first partial derivatives of u does not vanish at (x_0, y_0), then $\exists$ a path γ, unique in the neighbourhood of (x_0, y_0), such that $u(x, y) = 0$ for $(x, y) \in \gamma$.*

2.7.13

If f is analytic, then $J(f) = |f'(z)|^2$, and in this case f is locally 1–1 in a domain D iff $f' \neq 0$ in D, i.e. iff $J(f) > 0$ in D. This result extends to harmonic functions. First, some notations. We can write $f(x, y) = f(\frac{1}{2}(z + \overline{z}), -\frac{1}{2}i(z - \overline{z}))$ and interpreting the latter expression as a function of the variables z and $\overline{z}$, we obtain

$$\frac{\partial f}{\partial z} = \frac{1}{2}\left(\frac{\partial f}{\partial x} - i\frac{\partial f}{\partial y}\right), \qquad \frac{\partial f}{\partial \overline{z}} = \frac{1}{2}\left(\frac{\partial f}{\partial x} + i\frac{\partial f}{\partial y}\right). \tag{2.124}$$

From this we obtain

$$J(f) = \left|\frac{\partial f}{\partial z}\right|^2 - \left|\frac{\partial f}{\partial \overline{z}}\right|^2 \tag{2.125}$$

and so $J(f) > 0$ iff

$$\left|\frac{\partial f}{\partial \overline{z}}\right| < \left|\frac{\partial f}{\partial z}\right|. \tag{2.126}$$

2.7.14

Lewy's theorem *Let f be a complex-valued function harmonic in a domain D. Then f is locally 1–1 and sense preserving in D iff (2.126) holds in D.*

Proof We assume that f is 1–1 and sense preserving for $|z| < R$ and that $f(0) = 0$. We can write f in the form $f(z) = \overline{g(z)} + h(z)$, where g and h are analytic for $|z| < R$ and $g(0) = h(0) = 0$. Note that

$$\frac{\partial f}{\partial \overline{z}} = \overline{g'(z)}, \qquad \frac{\partial f}{\partial z} = h'(z) \tag{2.127}$$

and so we need to show that $|g'(0)| < |h'(0)|$. Suppose firstly that $g'(0) \neq 0$. Then g is 1–1 near 0 and so has an inverse function $z = G(w)$ defined and analytic in a neighbourhood of $w = 0$. The function

$$F(w) = f(G(w)) = \overline{w} + h(G(w)) = \overline{w} + K(w) \tag{2.128}$$

is 1–1 and harmonic near $w = 0$. It is also sense preserving at $w = 0$. Now for suitable $\rho > 0$, if $|w| \leq \rho$, then

$$F\left(w + \frac{e^{i\phi}}{n}\right) - F(w) \neq 0 \tag{2.129}$$

for ϕ real and n sufficiently large. Hence

$$\frac{K\left(w + \frac{e^{i\phi}}{n}\right) - K(w)}{\frac{e^{i\phi}}{n}} \neq -e^{-2i\phi} \tag{2.130}$$

for large n and $|w| \leq \rho$. Now the functions on the left form a sequence of analytic functions converging uniformly to $K'(w)$ in $\{|w| \leq \rho\}$ and each member of the sequence omits the value on the right. A theorem of Hurwitz states that in this situation the limit function $K'(w)$ either omits the value on the right or is identically equal to this value. In the latter case we obtain

$$e^{i\phi} F(w) = 2i \Im(e^{i\phi}\overline{w}) \tag{2.131}$$

and so F is not 1–1. Thus we have

$$K'(w) \neq -e^{-2i\phi} \tag{2.132}$$

and as this holds for every real ϕ, we deduce that $|K'(w)| \neq 1$ and in particular $|K'(0)| \neq 1$. This gives $|g'(0)| \neq |h'(0)|$ and so $J(f)(0) \neq 0$. Since f is sense preserving, $J(f)(0) > 0$. In a similar manner we obtain the same conclusion if $h'(0) \neq 0$. It remains to consider the case when $g'(0) = h'(0) = 0$. Since f is not constant, the above argument applied to the points z near 0 shows that $|g'(z)| < |h'(z)|$ for $0 < |z| < r$ say. Hence $\exists$ an integer $k \geq 2$ such that

$$g(z) = az^k + \cdots, \qquad h(z) = bz^k + \cdots \tag{2.133}$$

for $|z| < r$, where $|a| < |b|$. From this we easily obtain for small $t > 0$

$$d(f, C_t) = k \tag{2.134}$$

where C_t is the positively oriented circle $\{|z| = t\}$. We deduce that for values w sufficiently close to 0

$$d(f - w, C_t) = k \tag{2.135}$$

and, since f is 1–1 with positive Jacobian in the punctured disc, it follows from the argument principle that f attains the value w exactly k times in $\{|z| < t\}$, which gives $k = 1$. This contradiction completes the proof.

2.7.15 Harmonic polynomials on the critical set

With these results established we can now consider in more detail the geometric and topological behaviour of a harmonic polynomial. Let $P(z) = \overline{S(z)} + T(z)$ be a harmonic polynomial of degree n with coefficients a_k $(-n \le k \le n)$ and satisfying $|a_{-n}| < |a_n|$. As we have seen, the sum of the multiplicities of the zeros of P is exactly n. The multiplicity of a particular zero will be either $+1$ or -1 except for those zeros, if any, lying on the level set $\{z\colon |S'(z)| = |T'(z)|\}$. The points of this set are either isolated points at which $S'(z) = T'(z) = 0$ or points on the level set $\{z\colon |R(z)| = 1\}$, where $R(z) = S'(z)/T'(z)$ is a rational function of degree at most $n-1$. Note that $|R(z)| \to |a_{-n}|/|a_n| < 1$ as $z \to \infty$. Hence the **critical set** on which $J(P) = 0$ is compact and is made up of at most $n-1$ 'level curves' (since for any c the equation $R(z) = c$ has at most $n-1$ solutions). Further, we see that P is locally 1–1 and sense preserving in the neighbourhood of ∞.

It follows that $\exists K$ such that if $|w| \ge K$, then $P(z) - w$ has exactly n distinct zeros each of multiplicity 1. This follows, since $P(z) \to \infty$ as $z \to \infty$ and so for large $|w|$, $P(z) = w$ only at points z in the neighbourhood of ∞ exterior to the critical set.

In order to proceed further, we need to consider the various possibilities of the local behaviour of P near points on the critical set. Our first observation is that the degree of P on any circle C on which P is non-zero satisfies

$$|d(P, C)| \le n. \tag{2.136}$$

To see this there is no loss of generality in assuming that C is the unit circle positively oriented and noting that

$$P(e^{i\theta}) = e^{-in\theta} Q(e^{i\theta}) \tag{2.137}$$

where Q is an analytic polynomial of degree $2n$. Hence by the argument principle

$$d(P, C) = -n + \text{number of zeros of } Q \text{ in } \{|z| < 1\}. \tag{2.138}$$

Since Q has at most $2n$ zeros in the disc, (2.136) follows.

In particular we see that for any zero ζ of P

$$|m(P, \zeta)| \le n. \tag{2.139}$$

Notice also that if D is any open disc contained in a region where $J(P) \ne 0$, then P has at most n distinct zeros in D. Since this assertion applies equally well to $P - w$ for any $w \in \mathbb{C}$, we see that P is n-**valent** in D, i.e. P attains no value w more than n times in D. By considering discs of arbitrarily large radius, we further deduce that, if $J(P) > 0$ in a half-plane H, then P is n-valent in H. One final observation in this vein: P has at most $2n - 1$ zeros on any circle C; this follows from (2.137); Q has at most $2n$ zeros on C and if Q has all its zeros on C, then $|a_{-n}| = |a_n|$, so there is under our hypotheses at least one zero of Q not on C.

We consider now a zero of P on the level set $\{|R(z)| = 1\}$. Without loss of generality we may assume that this zero is the origin. We can then write P in the form

$$P(re^{i\theta}) = e^{-in\theta} r^k \big(a_{-n} r^{n-k} + \cdots + a_{-k} e^{i(n-k)\theta} + a_k e^{i(n+k)\theta} + \cdots + a_n r^{n-k} e^{2in\theta}\big) \tag{2.140}$$

where, since $|R(0)| = 1$, we have $|a_{-k}| = |a_k| \neq 0$. Then for small $r > 0$

$$m(P, 0) = -n + \text{number of zeros in } \{|z| < 1\} \text{ of } Q_r(z) \tag{2.141}$$

where

$$Q_r(z) = a_{-n} r^{n-k} + \cdots + a_{-k} z^{n-k} + a_k z^{n+k} + \cdots + a_n r^{n-k} z^{2n}. \tag{2.142}$$

Letting $r \to 0$, $Q_r(z) \to Q_0(z) = z^{n-k}(a_{-k} + a_k z^{2k})$. This has a zero of order $n - k$ at the origin and $2k$ simple zeros on the unit circle. Thus by the continuity of the zeros, $Q_r(z)$ has ν zeros in the unit disc, where $n - k \leq \nu \leq n + k$. Hence

$$|m(P, 0)| \leq k. \tag{2.143}$$

Consider next the mapping behaviour of P on the level curve Γ passing through 0. Except in the case $R'(0) = 0$, when Γ branches at the origin, Γ will have a parametrisation $z = z(t)$ such that

$$R(z(t)) = e^{it} \tag{2.144}$$

since, by definition, Γ is obtained as a branch of the inverse function R^{-1} continued around the positively oriented unit circle. This gives

$$\frac{d}{dt} P(z(t)) = \overline{z'(t) S'(z(t))} + z'(t) T'(z(t)) = 2e^{-it/2} \Re\big(e^{it/2} z'(t) T'(z(t))\big) \tag{2.145}$$

which is a relation enabling us to determine the behaviour of the tangent to the curve $P(\Gamma)$. In fact on any interval of values t on which

$$\Re\big(e^{it/2} z'(t) T'(z(t))\big) \neq 0, \tag{2.146}$$

the tangent to the curve $P(\Gamma)$ is strictly decreasing and so that part of $P(\Gamma)$ is a convex curve described in the negative (clockwise) direction. At points t where

$z'(t)$ is defined, we have

$$z'(t)R'(z(t)) = ie^{it} = iR(z(t)) \tag{2.147}$$

and so

$$\Re(e^{it/2}z'(t)T'(z(t))) = 0 \tag{2.148}$$

if, and only if,

$$\frac{(S'T')^3}{(S''T' - S'T'')^2}(z(t)) \geq 0. \tag{2.149}$$

Now (2.148) cannot hold on an interval of values of t, since otherwise on this interval $P(z(t))$ is constant, contradicting our theorem 2.6.12 that all zeros of $P - w$ are isolated. On the other hand $z(t)$ is locally an analytic function of t and therefore, by the important principle that two analytic curves either are identical or intersect in at most a finite number of points in any bounded portion of the plane, we deduce that (2.149) can hold at only a finite number of points. An alternative way to see this is as follows. The points where (2.148) holds are zeros of the harmonic function $Q = -\overline{A} + B$, where

$$A = \frac{S'^2T'}{S''T' - S'T''}, \qquad B = \frac{S'T'^2}{S''T' - S'T''}. \tag{2.150}$$

Now the zeros of Q in the plane are either zeros of S' or T', and therefore isolated points, or points on the critical set $\{|R(z)| = 1\}$ and so by theorem 2.6.12, if Q has a non-isolated zero, then Q is identically zero on an analytic curve, which, therefore, must coincide with one of the level curves. It follows that $P(\Gamma)$ consists of a finite number of convex curves whose endpoints correspond to points where either (2.149) holds or $R'(z) = 0$.

At a point where (2.149) holds and $R'(z) \neq 0$ there are two possibilities. These are as follows:

(i) $\Re(e^{it/2}z'(t)T'(z(t)))$ changes sign; in this case $P(\Gamma)$ has a cusp at the point with the tangent angle having a forward jump through an angle π (the jump is forward, since the curve past this point must be convex in the clockwise direction);

(ii) $\Re(e^{it/2}z'(t)T'(z(t)))$ does not change sign; in this case $P(\Gamma)$ remains smooth and convex at the point.

2.7.16

We consider now the detailed structure of P near a point on Γ. Again assume this point is the origin and that S and T have a common zero of multiplicity k

at 0. There will be no loss of generality in taking $a_{-k} = a_k = 1$. Then R is a rational function of degree $n - k$, which near the origin has the form

$$R(z) = 1 - az^m + \cdots \tag{2.151}$$

where $a \neq 0$ and $m \geq 1$. If $m > 1$, then R' has a zero at the origin of multiplicity $m - 1$. After a few calculations we find that

$$T(z) - S(z) = \frac{ak}{m+k} z^{m+k} + \cdots \tag{2.152}$$

and

$$a = \frac{m+k}{k}(a_{m+k} - \overline{a_{-m-k}}). \tag{2.153}$$

Thus P takes the form

$$P(z) = 2\Re S(z) + T(z) - S(z) = 2\Re S(z) + \frac{ak}{m+k} z^{m+k} + \cdots \tag{2.154}$$

and so can be regarded as a shift in the direction of the real axis of the polynomial $T - S$, which has a zero of multiplicity $m + k$ at the origin. In addition we may write P as a shift in any other direction as follows. We have

$$\begin{aligned} P(z) &= T(z) - e^{i\phi} S(z) + 2e^{i\phi/2}\Re\big(e^{i\phi/2} S(z)\big) \\ &= 2e^{i\phi/2}\Re(e^{i\phi/2} S(z)) + (1 - e^{i\phi})z^k + \cdots \end{aligned} \tag{2.155}$$

for any fixed ϕ satisfying $0 < \phi < 2\pi$, which expresses P as a shift in the direction of the lines parallel to the ray through 0 and $e^{i\phi/2}$ of the polynomial $T - e^{i\phi} S$, which has a zero of multiplicity k at the origin. To consider the effect of a shift in a specific direction we require our next lemma.

2.7.17

Shift lemma *Let D be a domain* **convex in the direction of the real axis**, *i.e. the intersection with D of every line parallel to the real axis is either empty or connected. Suppose that $u(w)$ is a continuous real-valued function in D such that*

$$W(w) = w + u(w) \quad (w \in D) \tag{2.156}$$

is locally 1–1 in D. Then $W(w)$ is globally 1–1 in D and $W(D)$ is a domain convex in the direction of the real axis.

Proof If $W(a + ib) = W(c + id)$ $(a, b, c, d$ real), then $b = d$ and

$$a + u(a + ib) = c + u(c + ib). \tag{2.157}$$

Now the continuous real-valued function $x \mapsto W(x+ib)-ib = x+u(x+ib)$, which is defined on an interval (p, q) containing the points a and c, is locally 1–1 by the hypothesis, and so is strictly monotonic. Hence (2.157) implies that $a = c$, as required. Finally, if L is the intersection with D of a line parallel to the real axis, then L is a connected segment and $W(L)$ is also a connected segment of the same line. Hence $W(D)$ is convex in the real direction.

2.7.18

We consider now the local behaviour of P in the neighbourhood of the origin in the case $m = k = 1$, i.e. the common zero of S and T is simple and R does not have a critical point at 0. This, in fact, is the usual case and will be the case for the polynomials $P - w$ for all except at most a finite number of points w. Indeed, for many polynomials it will be the only case to arise. We then have

$$S(z) = z + \overline{a_{-2}}z^2 + \cdots + \overline{a_{-n}}z^n, \quad T(z) = z + a_2z^2 + \cdots + a_nz^n, \tag{2.158}$$

$$R(z) = 1 - az + \cdots, \quad T(z) - S(z) = \frac{1}{2}az^2 + \cdots, \tag{2.159}$$

where $a \neq 0$. If we refer back to (2.145)–(2.150) we see that, if a is not real, $P(z)$ maps that part of the critical curve near the origin onto a convex curve K, whose normal direction at the origin is the imaginary axis. It follows that the lines parallel to the imaginary axis will meet this small portion of K in at most one point. Since

$$S(z) + T(z) = P(z) - 2i\Im(S(z)), \tag{2.160}$$

it follows that the analytic function

$$S(z) + T(z) = 2z + \cdots \tag{2.161}$$

maps the part of the critical curve near the origin onto a curve L, which is convex in the direction of the imaginary axis, i.e. the lines parallel to the imaginary axis meet this curve in at most one point. Next we note that the critical curve $z = z(t)$ is an analytic curve, which near the origin is obtained as the 1–1 function R^{-1} acting on a small arc of the unit circle containing the point $w = 1$. Therefore, for small $r > 0$, the curve will divide the disc $\{|z| < r\}$ into exactly two domains D^+ and D^-, each bounded by the curve and an arc of $\{|z| = r\}$. The Jacobian of P will be positive in D^+ and negative in D^-. P is therefore locally 1–1 in each of these two domains. In addition, for small $r > 0$, the function $S + T$ maps the disc $\{|z| < r\}$ onto a convex domain with $\{|z| = r\}$ being mapped onto the convex curve C bounding this domain. It follows that this domain is divided by L into two subdomains E^+ and E^- bounded by L

and complementary arcs of C. Both E^+ and E^- are convex in the direction of the imaginary axis; for any line parallel to the imaginary axis meets C at most twice and L at most once and so meets the Jordan curve J^+ bounding E^+ at most three times; since, if E^+ is not convex in the imaginary direction, there is at least one such line with at least four points of intersection with J^+, the assertion follows for E^+, and similarly for E^-.

Applying lemma 2.7.17 we deduce from (2.160) that P maps D^+ and D^- 1–1 onto domains convex in the imaginary direction. Furthermore, values w on K are attained exactly once near 0 and other values w at most twice. P is therefore 2-valent in the neighbourhood of the origin. In fact the function $w = P(z)$ maps a neighbourhood of the origin onto a two-sheeted region with a fold along the curve K. To see this, we recall that from (2.143) we have the multiplicity $m(P, 0) = -1, 0$ or 1. On the other hand, if $r > 0$ is small, $m(P, 0) =$ sum of multiplicities of $P - w$ in $\{|z| < r\}$ for all values w sufficiently close to 0. If we choose $w = P(z)$, where $z \in D^+$, then $m(P - w, z) = 1$. The value w is attained at most once more as $w = P(z')$, where $z' \in D^-$, and then $m(P - w, z') = -1$. Hence $m(P, 0) = 0$ or 1. Similarly, $m(P, 0) = 0$ or -1. Hence

$$m(P, 0) = 0. \tag{2.162}$$

Thus P omits values on one side of K and attains values twice on the other side of K, once with multiplicity 1 and once with multiplicity -1.

2.7.19

Next we must consider the case where P maps the critical curve near 0 onto a curve K with a cusp at 0. In this case a is real and K consists of two convex curves meeting at 0 each being tangential to the real axis. Furthermore, as we travel along K the orientation along the curves is clockwise and thus the two curves lie on opposite sides of the real axis. K is therefore convex in the direction of the real axis. It follows that

$$T(z) - S(z) = P(z) - 2\Re(S(z)) = \frac{1}{2}az^2 + \cdots \tag{2.163}$$

maps the critical curve near 0 onto a curve L convex in the direction of the real axis. L consists of two curves meeting at 0 and lying on opposite sides of the real axis forming a cusp with the real axis as tangent. Furthermore, the critical curve itself is tangential at 0 to the imaginary axis, and as before divides a small disc $\{|z| < r\}$ into two domains D^+ and D^-. The circle $\{|z| = r\}$ is divided into two arcs. Now, assuming r is sufficiently small, $T - S$ has an increasing tangent and an increasing argument on the circle $\{|z| = r\}$ with total rotation 4π and so maps one of the two circular arcs (say the one which forms, with the critical curve, the boundary of D^+) onto a curve C, with increasing tangent, whose total rotation about 0 is $<2\pi$. The endpoints of C are the endpoints of L and $C \cup L$ forms a Jordan curve bounding a domain Ω which is convex in the direction of the real axis. $T - S$ is 1–1 on the boundary of D^+ and so 1–1 in D^+ and maps D^+ onto Ω. It follows from lemma 2.7.17 that P is 1–1 in D^+. Also values w on K are not attained in D^+. From this we see that $m(P, 0) = 0$ or -1; for suppose $m(P, 0) = 1$; if w is the image under P of a point z in D^-, then $P - w$ has multiplicity -1 at z and so w must be attained at least twice in D^+, a contradiction. If we assume that $m(P, 0) = 0$, then as before P is 2-valent near 0; indeed P would then be 1–1 in D^-, and it is at least intuitively clear from the geometry that this is not possible. If $m(P, 0) = -1$, then P attains every value w near 0 in one of the following ways: (i) once in D^+ and twice in D^-; (ii) once on K and once in D^-; (iii) once in D^-. Thus P is 3-valent at the origin.

We thus see that a harmonic polynomial $P = \overline{S} + T$ such that the level set $\{|R(z)| = 1\}$ contains no critical points of R and no common zeros of S' and T' is finitely valent in the neighbourhood of every point including the point at ∞. We have proved the next theorem.

2.7.20

Theorem *Let $P(z) = \overline{S(z)} + T(z)$ be a non-constant harmonic polynomial of degree n with coefficients a_k $(-n \le k \le n)$ such that $|a_{-n}| \ne |a_n|$. Let*

$$R(z) = \frac{S'(z)}{T'(z)} \tag{2.164}$$

and suppose that on the level set $\{z \colon |R(z)| = 1\}$ there are no common zeros of S' and T' and no zeros of R'. Then P is n-valent in the neighbourhood of ∞, at most 3-valent in the neighbourhood of any point on the level set and at most 1-valent (univalent) in the neighbourhood of any point off the level set, except at a common zero of S' and T'. If S' and T' have a common zero of order k, then in the neighbourhood of this point P is $(k + 1)$-valent.

It has been shown by A. Lyzzaik [28] that even if we allow common zeros of S' and T' or zeros of R' on the level set $\{|R(z)| = 1\}$, there is still a bound on the valence of P in the neighbourhood of such points depending only on n. However, the proof is long and complicated and so omitted here.

2.7.21 The case n = 1

A mapping of the form

$$w = a\bar{z} + b + cz, \tag{2.165}$$

where $|a| \neq |c|$, is called an **affine mapping**. The equation has the unique solution

$$z = \frac{a}{|a|^2 - |c|^2}\bar{w} + \frac{a\bar{b} - b\bar{c}}{|c|^2 - |a|^2} + \frac{\bar{c}}{|c|^2 - |a|^2}w \tag{2.166}$$

and so the mapping is a homeomorphism of $\mathbb{C}$ onto $\mathbb{C}$. Note that the inverse function is also an affine mapping. Writing $z = x + iy$ the mapping has the form

$$w = Ax + B + Cy \tag{2.167}$$

where $\Im(A\overline{C}) \neq 0$. Conversely, a mapping of this form is affine. The composition of two affine mappings is affine and affine mappings map lines onto lines and circles onto ellipses; indeed ellipses onto ellipses. Every ellipse is the affine image of a circle.

The affine mappings are the only harmonic polynomials which are 1–1 in $\mathbb{C}$. For if $P = \overline{S} + T$ is 1–1 and sense preserving in $\mathbb{C}$, then $|S'(z)| < |T'(z)|$ in $\mathbb{C}$. Therefore $T'(z) \neq 0$ in $\mathbb{C}$ and so by the fundamental theorem of algebra T' is a non-zero constant and T has degree 1. Also S' is bounded and so constant by the fundamental theorem of algebra. Hence S has degree at most 1. Since $|S'(0)| < |T'(0)|$, the mapping is affine.

This argument uses only the fundamental theorem of algebra and the positivity of the Jacobian. A more general argument shows that the only harmonic functions which are 1–1 in $\mathbb{C}$ are affine mappings. For if $f = \overline{g} + h$ is 1–1 and sense preserving in $\mathbb{C}$, then $|g'| < |h'|$ in $\mathbb{C}$ and so by Liouville's theorem $g' = ch'$, where c is a constant with $|c| < 1$. It follows that f is the composition of an affine mapping with h and therefore h is an analytic function which is 1–1 in $\mathbb{C}$. It is a well-known deduction from the theory of normal families that the only such h are first degree polynomials.

2.7.22 The case n = 2

We will prove the following result.

Theorem *Let $P(z)$ be a harmonic polynomial of degree 2. The following statements are equivalent.*

(i) *P is 4-valent in $\mathbb{C}$;*
(ii) *for each $w \in \mathbb{C}$, $P(z) - w$ has at most isolated zeros;*
(iii) *$P(z) \to \infty$ as $z \to \infty$.*

Proof With our earlier notations we note that $R(z)$ is either constant or a rational function of degree 1. In the former case we can express P as the composition of an affine mapping with a quadratic polynomial and so P is 2-valent. Otherwise the critical set $\{|R(z)| = 1\}$ is a circle described once, in the negative direction if we assume that $|a_{-2}| < |a_2|$. By means of a transformation of the form $z \mapsto \rho z + \sigma$, we may assume that the centre of this circle is the origin and the radius of the circle is 1. Then $R(z)$ gives a 1–1 sense reversing transformation of the unit circle onto itself and hence $R(z)$ has the form

$$R(z) = e^{i\phi}\frac{1+\bar{a}z}{z+a} \tag{2.168}$$

where $|a| < 1$. This gives

$$S'(z) = c(1+\bar{a}z), \quad T'(z) = d(z+a) \tag{2.169}$$

where $|c| = |d| \neq 0$ and so

$$P(z) = \frac{1}{2}\bar{c}a\bar{z}^2 + \overline{cz} + a_0 + daz + \frac{1}{2}dz^2. \tag{2.170}$$

Clearly we may take $a_0 = 0$ and $c = 1$. Now it is easily verified that we may choose $A(|A| < 1)$ such that the sense preserving affine transformation $A\overline{P} + P$ has the coefficient of z equal to 0. The critical set is still the unit circle and we find that our polynomial reduces to the form

$$P(z) = e^{i\alpha}\bar{z} + \frac{1}{2}z^2. \tag{2.171}$$

A final rotation of P and z reduces to the form

$$P(z) = \bar{z} + \frac{1}{2}z^2. \tag{2.172}$$

In $\{|z| < 1\}$ P is locally 1–1 and sense reversing. P is also 1–1 on the unit circle, for if z, ζ are points on the circle such that $P(z) = P(\zeta)$, then

$$\frac{1}{z} - \frac{1}{\zeta} + \frac{1}{2}(z^2 - \zeta^2) = 0 \tag{2.173}$$

and so either $z = \zeta$ or $z + \zeta = -2\overline{z\zeta}$, which implies $|z + \zeta| = 2$ and so $z = \zeta$. It follows that P is 1–1 in the closed unit disc and maps the disc onto a closed Jordan domain $J \cup D$, where J is the Jordan curve negatively oriented obtained as the image of the circle. For $w \in D$, $P(z) - w$ has exactly one zero of multiplicity -1 in the open unit disc and all its other zeros lie in $\{|z| > 1\}$ which therefore have multiplicity 1. Since the sum of the multiplicities is exactly 2, $P - w$ has exactly three zeros in $\{|z| > 1\}$ and hence exactly four zeros in $\mathbb{C}$. For $w \notin J \cup D$, $P - w$ has zeros only in $\{|z| > 1\}$. As these have multiplicity 1, $P - w$ has exactly two zeros in $\{|z| > 1\}$ and no other zeros in $\mathbb{C}$. Finally, if $w \in J$, $P - w$ has one zero on $\{|z| = 1\}$ of multiplicity 0, 1 or -1 and no zeros in $\{|z| < 1\}$ and hence $P - w$ has either two, one or three zeros in $\{|z| > 1\}$ and hence at most four zeros in $\mathbb{C}$.

To confirm our earlier analysis and look in greater detail at the shape of J and the number of zeros of $P - w$ for $w \in J$, we first find the cusp points on the unit circle. By 2.7.15 these occur at the zeros on the circle of the harmonic function $\bar{z} - z^2$. These are exactly the three cube roots of unity. We find that

$$z = 1 \mapsto w = \frac{3}{2}, z = e^{2\pi i/3} \mapsto w = \frac{3}{2}e^{-2\pi i/3}, z = e^{-2\pi i/3} \mapsto w = \frac{3}{2}e^{2\pi i/3}. \tag{2.174}$$

Note also that

$$e^{2\pi i/3}P(ze^{2\pi i/3}) = P(z) = e^{-2\pi i/3}P(ze^{-2\pi i/3}) \tag{2.175}$$

from which we see that J is a curvilinear triangle consisting of three congruent convex curves described anti-clockwise (as the unit circle is described anti-clockwise) with cusps at the three points $\frac{3}{2}\times$ cube roots of unity which appear in clockwise order (so J is described negatively). If w is a non-cusp point of J, then w corresponds to a zero of $P - w$ of multiplicity 0 and so is attained exactly three times in $\mathbb{C}$. If w is one of the three cusp points of J, then according to our earlier analysis $P - w$ should have a zero of multiplicity 1 at the relevant cube root of unity. By symmetry the multiplicity is clearly the same at all three points. To verify the multiplicity we consider the point $z = 1$. Writing $\zeta = z - 1$, we require the multiplicity of the zero at the origin of

$$Q(\zeta) = \bar{\zeta} + \zeta + \frac{1}{2}\zeta^2. \tag{2.176}$$

This is given by

$$m(Q, 0) = -1 + (\text{number of zeros in } \{|z| < 1\} \text{ of } 1 + z^2 + \frac{1}{2}\rho z^3) \tag{2.177}$$

for small $\rho > 0$. Since $1 + z^2 + \frac{1}{2}\rho z^3$ has one large negative zero and two conjugate zeros, we have $m(Q, 0) = -1$ or $+1$. It is clear from the local nature of P near $z = 1$ that $m(Q, 0) \neq -1$ and indeed a simple calculation gives $m(Q, 0) = 1$. Thus the three cusp values are attained exactly twice in $\mathbb{C}$.

2.7.23

In order to complete our discussion of harmonic polynomials of degree 2 we now consider the case that $|a_{-2}| = |a_2|$. Assuming that $J(P) \not\equiv 0$, which holds iff P takes all its values on a line (so gives the form $P(z) = au(z) + b$, where a, b are constants and $u(z)$ is a real harmonic polynomial) and is therefore infinitely valent, the critical curve $\{|R(z)| = 1\}$ is a straight line; indeed, if $S'(z)$ has its zero at p and $T'(z)$ has its zero at q, then the critical line is the perpendicular bisector of the segment $[p, q]$. By means of a transformation of the form $z \mapsto \rho z + \sigma$ followed by an affine transformation $A\overline{P} + P$, we can bring S and T into the form

$$S(z) = z^2 + 2z, \quad T(z) = e^{i\phi}(z^2 - 2z) \tag{2.178}$$

where ϕ is real. The critical curve is then the imaginary axis. There are two cases.

(i) $e^{i\phi} = -1$. Writing $z = x + iy$, we obtain

$$P(z) = 4x(1 - iy). \tag{2.179}$$

This gives $P(z) = 0$ when $x = 0$, i.e. P vanishes identically on the critical curve. The equation

$$P(z) = w = u + iv \tag{2.180}$$

has the unique solution

$$z = \frac{u}{4} - i\frac{v}{u} \tag{2.181}$$

providing that $u \neq 0$. For $u = 0$, there is no solution to the equation unless also $v = 0$, when every point on $x = 0$ satisfies the equation. Thus P maps the plane onto the plane cut along both the positive imaginary axis and the negative imaginary axis, taking only the value 0 on the imaginary axis. P is a 1–1 mapping of the right half-plane onto itself and a 1–1 mapping of the left half-plane onto itself.

(ii) $e^{i\phi} \neq -1$. We first show that

$$P(z) \to \infty \text{ as } z \to \infty. \tag{2.182}$$

We have

$$P(z) = 2e^{i\phi/2}[\Re(e^{i\phi/2}z^2) - 2i\Im(e^{i\phi/2}z)]. \tag{2.183}$$

Suppose that there is a sequence of points $z = re^{i\theta}$ such that $r \to \infty$ and

$|P(re^{i\theta})|$ remains bounded. Then $\exists M$ such that

$$r^2 \left|\cos\left(2\theta + \frac{1}{2}\phi\right)\right| \le M, \quad r\left|\sin\left(\theta + \frac{1}{2}\phi\right)\right| \le M \tag{2.184}$$

and hence

$$\cos\left(2\theta + \frac{1}{2}\phi\right) \to 0, \qquad \sin\left(\theta + \frac{1}{2}\phi\right) \to 0. \tag{2.185}$$

These two relations holding simultaneously imply that $e^{i\phi} = -1$, a contradiction.

Let C_r denote the positively oriented circle $\{|z| = r\}$. Then (2.182) implies that

$$d(P - w, C_r) = d(P, C_r) \tag{2.186}$$

for large $r > 0$ and each $w \in \mathbb{C}$. Hence the sum of the multiplicities of the zeros in $\mathbb{C}$ of $P - w$ is identical to the sum of the multiplicities of the zeros in $\mathbb{C}$ of P. However, it is easily seen from (2.183) that $P(z)$ has just one zero at $z = 0$. Also

$$m(P, 0) = -2 + \text{number of zeros in } \{|z| < 1\} \text{ of } Q_r(z) \tag{2.187}$$

for any $r > 0$, where

$$Q_r(z) = r + 2z + e^{i\phi}(-2z^3 + rz^4). \tag{2.188}$$

To find the number of such zeros, we note that $Q_r(z) = Q_r(e^{i\phi}; z)$ has no zeros on $\{|z| = 1\}$ for any $r > 0$ and any $e^{i\phi} \neq -1$ on the unit circle. It follows that the number of zeros inside the disc is the same for all such polynomials. In particular we may take $e^{i\phi} = 1$. Then Q_r is a real polynomial, so the zeros either are real or occur in conjugate pairs. For a conjugate pair, either both lie in the disc or both lie outside the disc. Also the product of the zeros is 1. As $r \to \infty$, $Q_r(z)/r \to 1 + z^4$, which has zeros on the circle forming two pairs of conjugate zeros. Hence for large r, Q_r has two pairs of conjugate zeros $\alpha, \overline{\alpha}, \beta$ and $\overline{\beta}$ say. We have $|\alpha|^2|\beta|^2 = 1$, and so $|\alpha| < 1$ and $|\beta| > 1$ or *vice versa*. Thus Q_r has two zeros in the disc. We deduce that

$$m(P, 0) = 0. \tag{2.189}$$

It follows that the net sum of the multiplicities of the zeros of $P - w$ for any $w \in \mathbb{C}$ is zero.

Next, we note that on the imaginary axis we have

$$P(iy) = -(1 + e^{i\phi})(y^2 + 2iy) \tag{2.190}$$

and we see that this is a 1–1 mapping of the imaginary axis onto a parabola. Also P is continuous at ∞ with $P(\infty) = \infty$ and P is locally 1–1 in each of the two half-planes $\{x > 0\}$, $\{x < 0\}$, although having opposite orientations in the two half-planes. It follows that P is 1–1 in the two half-planes and furthermore maps both half-planes onto the same side of the parabola, the other side consisting of omitted values. Thus P is 2-valent in $\mathbb{C}$; P takes every value on the parabola exactly once, every value on one side of the parabola exactly twice and every value on the other side not at all. To determine which is the omitted side we consider the case $e^{i\phi} = 1$. Then we see that the parabola lies in the left half-plane symmetric about the real axis with the negative axis lying in the 'smaller' parabolic region. P is clearly positive on the real axis and so the omitted values are those in the smaller parabolic region. The image of P is a sheet folded along the parabola and twice covering the larger parabolic region.

2.7.24 Problems

1. If z_k are points satisfying $|z_k| < 1\,(1 \le k \le n)$, show that the Blaschke product

$$B(z) = \prod_{k=1}^{n} \frac{z - z_k}{1 - \overline{z_k} z} \tag{2.191}$$

has exactly $n - 1$ critical points in $\{|z| < 1\}$.

2. Let f be a homeomorphism from a domain $D \to \Omega$. Show that

$$f(z) \to \partial\Omega \text{ as } z \to \partial D. \tag{2.192}$$

Conversely, if $f(z) \to \partial\Omega$ then $z \to \partial D$. (The point at ∞ is assumed to be included as a boundary point of an unbounded domain.)

3. Let f be continuous for $\{|z| \le 1\}$ and suppose that f is a sense preserving homeomorphism of $\{|z| < 1\}$ onto $\{|w| < 1\}$. Show that we can write

$$f(e^{it}) = e^{i\phi(t)} \quad (0 \le t \le 2\pi), \tag{2.193}$$

where $\phi(t)$ is real, continuous and increasing and satisfies $\phi(2\pi) - \phi(0) = 2\pi$.

4. Let $\{f_n\}$ be a sequence of functions continuous and 1–1 in a domain D and suppose that $f_n \to f$ locally uniformly in D (i.e. for each $z_0 \in D$ there is a neighbourhood of z_0 in which the convergence is uniform). Show that, if f is locally 1–1 in D, then f is 1–1 in D.

5. Let $g(z)$ be analytic in a convex domain D and suppose that

$$|g'(z)| < 1 \quad (z \in D). \tag{2.194}$$

Show that the harmonic function

$$f(z) = \overline{g(z)} + z \tag{2.195}$$

is 1–1 in D.

6. Determine the valence in $\mathbb{C}$ of the harmonic polynomial

$$P(z) = \overline{z}^k + z^n, \tag{2.196}$$

where $1 \leq k < n$.

7. Show that, if P is a harmonic polynomial satisfying

$$\Re P(z) \geq 0 \quad (z \in \mathbb{C}), \tag{2.197}$$

then $\Re P(z)$ is constant in $\mathbb{C}$ and $J(P) \equiv 0$. Hence show that, if Q is a harmonic polynomial such that $J(Q) \not\equiv 0$, then, for each half-plane H, $Q(z)$ attains at least one value w in H.

2.8 The Borsuk–Ulam theorem

2.8.1

We conclude the chapter with one final application of the degree principle.

Theorem *Let $\mathbb{S}^2$ denote the unit sphere*

$$\mathbb{S}^2 = \left\{X = (x_1, x_2, x_3) \colon x_1^2 + x_2^2 + x_3^2 = 1\right\}. \tag{2.198}$$

Let $f\colon \mathbb{S}^2 \mapsto \mathbb{C}$ be a continuous mapping. Then $\exists X \in \mathbb{S}^2$ such that

$$f(X) = f(-X). \tag{2.199}$$

Proof Let $z = x + iy$ satisfy $|z| \leq 1$, so that $x^2 + y^2 \leq 1$. We define

$$\phi(z) = f(x, y, \sqrt{(1 - x^2 - y^2)}) - f(-x, -y, -\sqrt{(1 - x^2 - y^2)}). \tag{2.200}$$

It is clear that $\phi(z)$ is continuous for $\{|z| \leq 1\}$ and

$$\phi(-z) = -\phi(z) \quad (|z| = 1). \tag{2.201}$$

We are required to prove that ϕ has a zero in $\{|z| \leq 1\}$. Assume on the contrary that ϕ is non-zero in the closed disc. Then if C denotes the positively oriented unit circle, we have by the degree principle

$$d(\phi, C) = 0. \tag{2.202}$$

Let

$$C_1 = \{e^{it} \colon -\pi \leq t \leq 0\}, \qquad C_2 = \{e^{it} \colon 0 \leq t \leq \pi\}. \tag{2.203}$$

Then we see from (2.201) that

$$d(\phi, C_1) = d(\phi, C_2) \tag{2.204}$$

and hence

$$0 = d(\phi, C) = d(\phi, C_1) + d(\phi, C_2) = 2d(\phi, C_2) \tag{2.205}$$

and hence $d(\phi, C_2) = 0$. This implies that, modulo 2π, $\phi(1)$ and $\phi(-1)$ are equal in argument. Since from (2.201) they are equal in absolute value, we deduce that

$$\phi(-1) = \phi(1). \tag{2.206}$$

But (2.201) gives $\phi(-1) = -\phi(1)$, so $\phi(1) = 0$, a contradiction. Hence result.

2.8.2

The above result provides a striking topological explanation of the impossibility of representing the globe on a flat surface without producing a gross distortion; a fact which is immediately verified by glancing at any atlas of the world. All the maps in the atlas look reasonable, even conformal (preserving angles locally), except for the maps (usually in the early pages) of the whole world.

$\mathbb{S}^2$ is of course topologically equivalent to the extended complex plane $\mathbb{C} \cup \{\infty\}$, which is proved by the method of **stereographic projection**, usually with the north pole corresponding to ∞. (Place the sphere with its south pole resting on the origin of the plane, and map each point, other than the north pole, onto the plane by extending the line from the north pole through the point till it cuts the plane, this latter point being the image. The mapping is clearly a homeomorphism from the sphere minus the north pole onto the plane. Points near the north pole correspond to far away points in the plane. Alternatively, the sphere may be placed so that its equator coincides with the unit circle in the plane; then the lower hemisphere maps to the unit disc and the upper hemisphere to the outside of this disc.)

2.8.3 Problem

1. Use the Borsuk–Ulam theorem to show that every 1–1 continuous map $\mathbb{S}^2 \to \mathbb{S}^2$ is a homeomorphism from $\mathbb{S}^2$ onto $\mathbb{S}^2$. Use the monodromy theorem (see 2.7.2) to show that this result is even true for every locally 1–1 continuous map $\mathbb{S}^2 \mapsto \mathbb{S}^2$.

3
The Jacobian problem

3.1 The Jacobian conjecture

3.1.1

A real analytic polynomial

$$P(x, y) = \sum_{j=0}^{m} \sum_{k=0}^{n} a_{j,k} x^j y^k \tag{3.1}$$

with complex coefficients $a_{j,k}$ may be considered as a mapping from $\mathbb{R}^2$ into $\mathbb{R}^2$. Writing $P(x, y) = S(x, y) + iT(x, y)$, where S and T are real analytic polynomials with real coefficients, we have seen earlier that a natural algebraic question is the following. For given real numbers u and v, can we find a simultaneous solution of the pair of equations

$$S(x, y) = u, \qquad T(x, y) = v? \tag{3.2}$$

Writing $w = u + iv$, this is equivalent to asking for a solution to the equation

$$P(x, y) = w. \tag{3.3}$$

In other words we would like to know whether the mapping $w = P(z)$ ($z = (x, y)$) is surjective. If in addition we would like the solution to be unique, then we are asking for a condition on P which guarantees that the mapping $w = P(z)$ gives a homeomorphism from $\mathbb{C}$ onto $\mathbb{C}$. The Jacobian conjecture in its strongest form has been the assertion that

$$J(P)(z) \neq 0 \tag{3.4}$$

for all $z \in \mathbb{C}$ provides a sufficient condition for the mapping $w = P(z)$ to be a bijective mapping on $\mathbb{C}$. Surprisingly this 'simple', schoolchild problem has only recently been solved. S. Pinchuk [38] has constructed a polynomial P

(of degree 25) for which $J(P)(z) > 0\,(z \in \mathbb{C})$, but P is not a homeomorphism of $\mathbb{C}$. We give the details of his construction, together with an exploration of the geometry and topology of the surface onto which P maps, in section 2.

Nevertheless, given additional information it is often possible to show that a polynomial with positive Jacobian is a homeomorphism of $\mathbb{C}$. The main way of using the positivity is the deduction that the functions are locally 1–1, and much of our discussion will use only this fact. Note that the example $f(z) = e^z$, which has a positive Jacobian determinant on $\mathbb{C}$, shows that a locally 1–1 transcendental function may be a long way from a homeomorphism.

If P is a harmonic polynomial, then it has already been shown in chapter 2, (2.7.21) that condition (3.4) implies that P is an affine mapping, and therefore the conjecture is certainly true in this case. However, P then has degree 1 and therefore harmonic polynomials do not cover any of the interesting cases of real analytic polynomials satisfying the condition (3.4). We shall see shortly that there are explicit examples of polynomials satisfying the condition for every $n > 1$.

3.1.2

By means of the substitutions

$$x = \tfrac{1}{2}(z + \bar{z}), \quad y = \tfrac{1}{2i}(z - \bar{z}) \tag{3.5}$$

a real analytic polynomial of the form (3.1) can be written in the form

$$P(z) = \sum_{j=0}^{M} \sum_{k=0}^{M} b_{j,k} \bar{z}^j z^k \tag{3.6}$$

where $M \leq m + n$. On the other hand the substitution $z = x + iy$ reduces a polynomial of the form (3.6) back to the form (3.1). We may also write P in polar coordinate form

$$P(re^{i\theta}) = \sum_{\nu=0}^{N} T_\nu(\theta) r^\nu \tag{3.7}$$

where $N \leq 2M$ and also $N \leq m + n$ and where each $T_\nu(\theta)$ is a trigonometric polynomial of degree at most ν. Indeed, we find that

$$T_\nu(\theta) = \sum_{j+k=\nu} b_{j,k} e^{i(k-j)\theta} = e^{-i\nu\theta} Q_\nu(e^{2i\theta}) \tag{3.8}$$

where each Q_ν is an analytic polynomial of degree at most ν. On the other hand, a polynomial of the form (3.7), where $T_\nu(\theta)$ has the form (3.8), reduces back to the form (3.6) and so is real analytic. Providing that $T_N(\theta) \not\equiv 0$, the number N is the degree of P. We may also write a real analytic polynomial P in the form

$$P(re^{i\theta}) = \sum_{k=-N}^{N} S_k(r)e^{ik\theta} \tag{3.9}$$

where each S_k is an analytic polynomial of degree at most N. Indeed, we find that

$$S_k(r) = r^{|k|}q_k(r^2) \tag{3.10}$$

where q_k is a polynomial of degree at most $\frac{1}{2}(N - |k|)$. On the other hand, since $r^2 = \bar{z}z$, a polynomial of the form (3.9), where $S_k(r)$ is of the form (3.10), is real analytic.

If $P(z)$ is a locally 1–1 function in $\mathbb{C}$, then by theorem 2.7.2 of chapter 2 the condition for P to be a homeomorphism of $\mathbb{C}$ onto $\mathbb{C}$ is that $P(z) \to \infty$ as $z \to \infty$. Thus the Jacobian problem reduces to considering the continuity of P at ∞. Our first result establishes the finite valence of P.

3.1.3

Theorem *Let $P(z)$ be a real analytic polynomial of degree N locally 1–1 in $\mathbb{C}$. Then P is N-valent in $\mathbb{C}$. Furthermore, if $z(t)$ is a complex-valued non-constant polynomial in the real variable t, then*

$$P(z(t)) \to \infty \text{ as } t \to \infty. \tag{3.11}$$

In particular

$$P(z) \to \infty \text{ as } z \to \infty \text{ along any line in } \mathbb{C}. \tag{3.12}$$

Proof If $z(t)$ is a non-constant polynomial, then $z(t) \to \infty$ as $t \to \infty$. Furthermore, $P(z(t))$ is a polynomial in t and so either (3.11) holds or $P(z(t))$ is constant, i.e. P is constant on a curve and this contradicts P being locally 1–1 in $\mathbb{C}$. Since every line has a parametrisation in the form $z(t) = At + b$, we deduce (3.12).

To prove that P is N-valent, we may assume that P is sense preserving so that, for each $w \in \mathbb{C}$, every zero of $P(z) - w$ is isolated and has multiplicity 1. We choose R large so that $P(z) \neq w$ on $C_R = \{z = Re^{i\theta}; 0 \leq \theta \leq 2\pi\}$. Applying (3.9) we can write

$$P(Re^{i\theta}) - w = \sum_{k=-N}^{N} S_k(R)e^{ik\theta} = e^{-iN\theta}Q(R, e^{i\theta}) \tag{3.13}$$

where for fixed R, $\zeta \mapsto Q(R, \zeta)$ is a polynomial in ζ of degree at most $2N$. Hence Q has at most $2N$ zeros in $\{|\zeta| < 1\}$ and therefore the degree of Q on the unit circle is a number between 0 and $2N$. We deduce that

$$-N \leq d(P - w, C_R) \leq N \tag{3.14}$$

and so by the argument principle $P - w$ has at most N zeros in $\{|z| < R\}$. Since R is arbitrarily large, $P - w$ has at most N zeros in $\mathbb{C}$. Hence result.

Note that this double use of the argument principle gives the N-valence for a much larger class of polynomial-type mappings than real analytic polynomials. Also the N-valence is an improvement on the N^2-valence, which can be deduced from Beźout's theorem.

3.1.4

Example Consider the polynomial

$$P(x, y) = (1 - xy)(x + iy). \tag{3.15}$$

Observe firstly that on the curve $y = 1/x$, $P(x, y) = 0$, so certainly it is not true that $P(z) \to \infty$ as $z \to \infty$. On the other hand, if $z(t) = x(t) + iy(t)$ is a non-constant polynomial in t, say $x(t)$ has degree m and $y(t)$ has degree n so $m + n \geq 1$, then $P(x(t), y(t))$ has degree ≥ 1 and so $P(x(t), y(t)) \to \infty$ as $t \to \infty$. Thus although $P \to \infty$ along every 'polynomial curve' including lines, nevertheless P is constant on an infinite curve.

3.1.5

Let $P(x, y)$ be a real analytic polynomial such that $J(P) \neq 0$. The previous example is intended as a caution against jumping to conclusions on the basis of theorem 3.1.3. However, if $z(t)$ is a regular parametrisation of any smooth curve, i.e. $z'(t)$ is continuous and non-zero, then $w(t) = P(z(t))$ is regular, so $w'(t) \neq 0$. For

$$w'(t) = x'(t)\frac{\partial P}{\partial x} + y'(t)\frac{\partial P}{\partial y} = 0 \tag{3.16}$$

implies that $x'(t) = y'(t) = 0$ from the non-vanishing of the Jacobian determinant. Alternatively, we can write

$$w'(t) = \overline{z'(t)}\frac{\partial P}{\partial \bar{z}} + z'(t)\frac{\partial P}{\partial z} \tag{3.17}$$

and therefore, assuming that $J(P) > 0$, we have

$$|w'(t)| \geq \left[\left|\frac{\partial P}{\partial z}\right| - \left|\frac{\partial P}{\partial \bar{z}}\right|\right]|z'(t)| > 0. \tag{3.18}$$

Let us assume now that our polynomial P satisfies the strengthened condition

$$\left|\frac{\partial P}{\partial z}\right| - \left|\frac{\partial P}{\partial \bar{z}}\right| \geq \delta > 0 \tag{3.19}$$

for some $\delta > 0$ and all $z \in \mathbb{C}$. We will show that in this case the Jacobian conjecture for P is correct, i.e. P is indeed a homeomorphism from $\mathbb{C}$ onto $\mathbb{C}$. Suppose on the contrary that P is not a homeomorphism. Let $w_0 = P(0)$ and let $Q(w)$ denote the well defined local inverse of P in the neighbourhood of w_0. We apply the method of continuation of Q described in chapter 2, 2.7.2. If Q can be continued along any curve with initial point w_0, then applying the monodromy theorem will lead us to the conclusion that P is a homeomorphism on $\mathbb{C}$. Therefore there exists a curve Γ with initial point w_0 along which Q cannot be continued. There will then be a first point $w_1 \in \Gamma$ before which Q can be continued and past which it cannot. Let $w(t)\,(a \leq t \leq b)$ denote the parametrisation of this curve with endpoints w_0 and w_1: in fact a refinement of the continuation argument shows that we can even choose Γ to be a line segment. In particular we can assume that Γ has finite length. Consider the curve γ with parametrisation $z(t) = Q(w(t))\ (a \leq t \leq b)$, so $w(t) = P(z(t))$. If $z(t)$ has any finite limiting value z_1 as $t \to b$, then $P(z_1) = w_1$ and, as P is 1–1 in the neighbourhood of z_1, P has a well-defined inverse $\widetilde{Q}$ defined near w_1. We have

$$P(\widetilde{Q}(w(t))) = w(t) = P(z(t)) = P(Q(w(t))) \tag{3.20}$$

for t near b and so $\widetilde{Q} = Q$ on Γ. It is easily seen that $\widetilde{Q}$ provides a continuation of the inverse beyond w_1. It follows that $z(t) \to \infty$ as $t \to b$ and in particular that γ is a curve of infinite length. We have

$$P(z) \to w_1 \text{ as } z \to \infty \text{ along } \gamma. \tag{3.21}$$

Thus w_1 is an asymptotic value of P. Now applying (3.19) we obtain

$$\text{length of } \Gamma = \int_a^b |w'(t)|dt \geq \delta \int_a^b |z'(t)|dt = +\infty, \tag{3.22}$$

giving a contradiction.

3.1.6 *Proof of the conjecture in the case* $n = 2$

Let P be a real analytic polynomial of degree 2 satisfying $J(P) > 0$. We may assume that $P(0) = 0$. Further, we note that the first degree term is a sense preserving affine mapping and therefore with an appropriate substitution we may take this term $= z$. Then P has the form

$$P(z) = z + a\bar{z}^2 + b\bar{z}z + cz^2. \tag{3.23}$$

The condition $J(P) > 0$ can be expressed in the form

$$\frac{\partial P}{\partial z} + \zeta \frac{\partial P}{\partial \bar{z}} \neq 0 \quad (|\zeta| \leq 1, z \in \mathbb{C}). \tag{3.24}$$

We obtain

$$1 + (b + 2a\zeta)\bar{z} + (2c + b\zeta)z \neq 0 \tag{3.25}$$

and so this first degree polynomial is not surjective and this implies

$$|b + 2a\zeta| = |2c + b\zeta| \quad (|\zeta| \leq 1). \tag{3.26}$$

We easily deduce that

$$a = \tfrac{1}{2}be^{i\phi}, \quad c = \tfrac{1}{2}be^{-i\phi} \tag{3.27}$$

where ϕ is real. Substituting into (3.25) we obtain

$$e^{i\phi/2} + 2b(1 + \zeta e^{i\phi})\Re(ze^{-i\phi/2}) \neq 0 \tag{3.28}$$

which implies that

$$\frac{e^{i\phi/2}}{2b(1 + \zeta e^{i\phi})} \tag{3.29}$$

is non-real, if $|\zeta| \leq 1$, $\zeta \neq e^{-i\phi}$. Hence in this case

$$\Im(be^{-i\phi/2}(1 + \zeta e^{i\phi})) \neq 0. \tag{3.30}$$

Let $b = \rho e^{i\beta}$. We deduce that

$$\sin\left(\beta - \tfrac{1}{2}\phi\right) + \sin\left(\theta + \tfrac{1}{2}\phi\right) \tag{3.31}$$

does not change sign for all real θ. Hence $|\sin(\beta - \frac{1}{2}\phi)| = 1$ and so $e^{i\beta} = \pm ie^{i\phi/2}$. We obtain

$$P(z) = z \pm \tfrac{1}{2}i\rho e^{3i\phi/2}\bar{z}^2 \pm i\rho e^{i\phi/2}\bar{z}z \pm \tfrac{1}{2}i\rho e^{-i\phi/2}z^2 \tag{3.32}$$

from which we easily obtain

$$\begin{aligned} e^{-i\phi/2}P(ze^{i\phi/2}) &= z + i\lambda(|z|^2 + \Re(z^2)) \\ &= x + iy + 2i\lambda x^2, \end{aligned} \tag{3.33}$$

where λ is real. This is clearly a homeomorphism of $\mathbb{C}$ onto $\mathbb{C}$.

Thus for real analytic polynomials of degree 2, the condition $J(P) > 0$ implies that, by no more than affine transformations on either the function or the variable, we can reduce P to a quite explicit form. Further we find that in this form we obtain $J(P) = 1$, which implies that the condition $J(P) \neq 0$ can only hold for polynomials of degree 2 when $J(P)$ is a non-zero constant. Also the original problem of solving a pair of simultaneous equations can be achieved explicitly:

$$x + iy + 2i\lambda x^2 = u + iv \tag{3.34}$$

has the unique solution

$$x = u,\ y = v - 2\lambda u^2 \tag{3.35}$$

and we see that the inverse function is itself a real analytic polynomial of degree 2. In short P is **polynomially invertible**.

This fact corresponds to the classical statement of the **Keller Jacobian conjecture** which, in its simplest form, is for polynomials in two complex variables, rather than real variables: we consider

$$P(z, \zeta) = (S(z, \zeta), T(z, \zeta)) \tag{3.36}$$

where S, T are polynomials in the two complex variables z, ζ so that P is a mapping from $\mathbb{C}^2 \to \mathbb{C}^2$. The condition $J(P) \neq 0$ then implies that $J(P)$ is a non-zero constant, since $J(P)$ is itself a polynomial in z, ζ so by the fundamental theorem of algebra will certainly have zeros unless it reduces to a constant. After a normalisation one can then take $J(P) = 1$. Under these circumstances the conjecture then is that P **is polynomially invertible**. The conjecture in this form has been proved for polynomials of degree ≤ 100 by Moh [35], but is in general unresolved. We give an account of this case in section 3.3. There are many examples in $\mathbb{R}^2$, where $J(P) \neq 0$, but $J(P)$ is not constant.

We note that in the case $n = 2$ we have reduced P to a polynomial in the form

$$P(x, y) = q(x) + ir(x, y) \tag{3.37}$$

where $q(x)$ is strictly monotonic and $y \mapsto r(x, y)$ is strictly monotonic for each fixed x . Since q and r are polynomials, it is easily seen that such a function is a homeomorphism from $\mathbb{C}$ onto $\mathbb{C}$. The example

$$P(x, y) = x + iy + ix^n \tag{3.38}$$

shows that P may have arbitrary degree. Note also that

$$J(P) = q'(x)\frac{\partial r}{\partial y} \tag{3.39}$$

which is non-zero iff both q' and $\partial r/\partial y$ are non-zero. There is no difficulty in seeing that $J(P)$ need not be constant, e.g. take $q'(x) = 1 + x^2$.

3.1.7 The leading term

Since proving the Jacobian conjecture for a polynomial P amounts to showing that $P(z) \to \infty$ as $z \to \infty$, it is important to consider the leading term of P, namely the homogeneous polynomial in x, y

$$\sum_{j+k=N} a_{j,k} x^j y^k \tag{3.40}$$

where N is the degree of P. If we write P in its polar form (3.7), this is the term $T_N(\theta)r^N$, where $T_N(\theta)$ is a trigonometric polynomial of degree at most N. This term will dominate $|P(re^{i\theta})|$ on any ray $\theta = \theta_0$ for which $T_N(\theta_0) \neq 0$. For such a ray we obtain

$$|P(re^{i\theta_0})| \to \infty \text{ as } r \to \infty. \tag{3.41}$$

Indeed, we have

$$|P(re^{i\theta})| \to \infty \text{ as } r \to \infty \tag{3.42}$$

uniformly on any interval $[\theta_1 + \delta, \theta_2 - \delta]$, where $\delta > 0$ is small and θ_1 and θ_2 are successive zeros in $[0, 2\pi)$ of $T_N(\theta)$. In particular, if $T_N(\theta)$ has no such zeros, then $P(z) \to \infty$ as $z \to \infty$, and we obtain a proof of the conjecture in this case. The example

$$P(re^{i\theta}) = re^{i\theta}(1 + r^{N-1}) \tag{3.43}$$

shows that this case can occur. Note that, for this function to be real analytic, N must be odd.

On the other hand, in our example (3.38) the leading term vanishes on the line $x = 0$, so for particular polynomials the conjecture may hold despite zeros of $T_N(\theta)$.

A trigonometric polynomial T_N of degree at most N has at most $2N$ zeros in a periodic interval, and so there are at most $2N$ directions in which P does not tend to ∞. Note that for a real analytic polynomial P, $T_N(\theta_0) = 0$ implies $T_N(\theta_0 \pm \pi) = 0$ applying (3.8). Thus the zeros of T_N occur in pairs.

In polar coordinates the condition $J(P) > 0$ takes the form

$$\frac{1}{r}\Im(\overline{Q_r}\, Q_\theta) > 0, \tag{3.44}$$

where $Q(r, \theta) = P(r\cos\theta, r\sin\theta)$, the subscripts denoting partial differentiation; thus $J(P) > 0$ iff

$$\Im\left(\frac{\sum_{k=0}^{N} T_k'(\theta)r^k}{\sum_{k=1}^{N} kT_k(\theta)r^k}\right) > 0. \tag{3.45}$$

Letting $r \to \infty$ gives

$$\Im\left(\frac{T_N'(\theta)}{T_N(\theta)}\right) \geq 0, \tag{3.46}$$

at all points θ for which $T_N(\theta) \neq 0$. Thus between successive zeros the function $\theta \mapsto \arg T_N(\theta)$ is increasing. In some important cases, this function turns out to be constant on successive intervals.

Note also that $J(P) > 0$ implies that, for each $r > 0$, the tangent vector to the curve $\{Q(r, \theta) \colon 0 \leq \theta \leq 2\pi\}$ has total net rotation 2π.

3.1.8 Topology of the mapping

The most important general fact about a locally 1–1 mapping of the plane is the following.

Let P be a locally 1–1 mapping of $\mathbb{C}$. Then P acts as a homeomorphism of $\mathbb{C}$ onto a simply-connected surface, which projects into $\mathbb{C}$.

For $z_0 \in \mathbb{C}$, the mapping P gives not only an image point $w_0 = P(z_0)$, but in addition a locally defined inverse function $\phi(w)$ mapping a neighbourhood of w_0 onto a neighbourhood of z_0. The mapping $z_0 \mapsto (w_0, \phi(w))$ is easily seen to define the essential structure of the homeomorphism. Of course, for precision we need to consider equivalence classes of the different ϕ and also define a topology for such pairs. This is easily achieved; it is the standard method of constructing the Riemann surface of a function and an excellent account of the process is given in Ahlfors [2].

The above mapping of points of $\mathbb{C}$ onto pairs is clearly bijective and bicontinuous, and so a homeomorphism. The surface is therefore simply-connected, which means that any loop on the surface can be shrunk to a point on the surface. All this is true for the logarithmic surface arising out of the mapping e^z. However, our mappings are finitely valent, and this has a profound effect on the nature of the surface, which is quite dissimilar to surfaces of logarithmic type. The asymptotic values at ∞ provide borders to the surface at its different levels, and because of the monodromy theorem the surface flows up to these borders, its projection onto the plane covering any simply-connected region containing no border points. For real analytic polynomials we have given a precise account in chapter 1 of the construction and nature of asymptotic values at ∞, and we shall shortly apply this information. We regard the **Jacobian problem** as being the classification of the possible surface structures which can arise and their associated algebraic structure.

However, before turning to this, we include some topological considerations, valid for any mapping P that is an N-valent function locally 1–1 in $\mathbb{C}$. Consider a Jordan loop Γ in the $w(= P(z))$-plane and assume that at least one point on Γ is attained by P. From this point continue a suitably chosen branch of the inverse function around the loop. There are two possibilities. Either we meet an asymptotic value of P or the continuation will continue indefinitely as we traverse the loop. In the latter case, since P is N-valent, after a finite number ν of circuits, where $1 \leq \nu \leq N$, we return to the initial branch. The loop Γ described ν times is then the image under P of a Jordan curve γ in the z-plane described once. Let D be the Jordan domain enclosed by γ. Then P is locally 1–1 in D and $P(z) \to \Gamma$ as $z \to \gamma$. It follows from corollary 2.7.5 of chapter 2 that P gives a homeomorphism from D onto Ω, where Ω is the Jordan domain enclosed by Γ. From the argument principle we obtain $\nu = 1$. In other words, if the chosen branch of the inverse function does not lead to an asymptotic value of P on Γ, then the branch returns to the initial branch after one circuit. Furthermore, the branch has a unique continuation into Ω, defining a single-valued inverse function in Ω, so in particular the branch does not lead to an asymptotic value of P in Ω.

Now, if P does have an asymptotic value w in Ω, then there is a curve Λ in Ω with one endpoint w along which a branch of the inverse function is defined leading to the asymptotic value w. At a point w_1 on Λ this branch will map w_1 onto a point z_1 in the z-plane. If the mapping is of polynomial type (e.g. has a polar representation as a finite sum of trigonometric polynomials with coefficients functions of r), then on a line from z_1 to ∞ we know that $P \to \infty$ and so the branch at w_1 can be continued along the image of this line, which tends to ∞ and in particular meets the curve Γ. Thus we obtain a branch on Γ whose continuation into Ω leads to an asymptotic value w. It follows from the above that this same branch leads to an asymptotic value on Γ.

Thus, *if P has an asymptotic value w, then P has an asymptotic value on every loop enclosing w. Moreover, a branch of the inverse function exists on a path joining these two asymptotic values and remaining on or inside the loop. In particular, the set of asymptotic values is infinite, unbounded and contains no isolated points.*

Before discussing further the algebra and topology of locally 1–1 polynomial mappings, we give a detailed account of Pinchuk's counter-example to the $\mathbb{R}^2$ Jacobian conjecture.

3.2 Pinchuk's example

3.2.1

S. Pinchuk [38] has constructed the following example of a real analytic polynomial (p, q) satisfying $J(p, q) > 0$, but which is not a homeomorphism of the plane. Pinchuk defines

$$t = xy - 1, \qquad s = 1 + xt, \qquad h = ts, \qquad f = s^2(t^2 + y) \tag{3.47}$$

and then sets

$$p = h + f, \qquad q = -t^2 - 6th(h+1) - u(f, h), \tag{3.48}$$

where u is a polynomial to be determined. We then have

$$J(p, q) = -J(p, t^2) - 6J(p, th) - 6J(p, th^2) - J(p, u), \tag{3.49}$$

where we have made use of the algebraic properties of the Jacobian operator:

$$J(a, b + c) = J(a, b) + J(a, c), \qquad J(a, bc) = bJ(a, c) + cJ(a, b) \tag{3.50}$$

valid for any polynomials a, b, c. From this we obtain

$$\begin{gathered} J(p, t^2) = 2tJ(p, t), \qquad J(p, th) = tJ(p, h) + hJ(p, t), \\ J(p, th^2) = thJ(p, h) + hJ(p, th) = 2thJ(p, h) + h^2 J(p, t) \end{gathered} \tag{3.51}$$

and therefore

$$J(p, q) = -(2t + 6h + 6h^2)J(p, t) - (6t + 12th)J(p, h) - J(p, u). \tag{3.52}$$

Now

$$J(h, t) = J(ts, t) = tJ(s, t) + sJ(t, t) = tJ(s, t) = t(s - 1) = h - t \tag{3.53}$$

and

$$\begin{aligned} J(f, t) &= J(s^2 t^2, t) + J(s^2 y, t) = 2stJ(st, t) + yJ(s^2, t) + s^2 J(y, t) \\ &= 2st^2(s - 1) + 2ys(s - 1) - s^2 y = -f + 3h^2 - 2s(t^2 - y(s - 1)) \\ &= -f + 3h^2 + 2h \end{aligned} \tag{3.54}$$

where we have made use of the relation

$$y(s - 1) = t(t + 1). \tag{3.55}$$

This gives

$$J(p, t) = -f + 3h^2 + 3h - t. \tag{3.56}$$

Next we have

$$\begin{aligned}J(f,h) &= J(s^2t^2+s^2y,st) = J(s^2y,st) = yJ(s^2,st)+s^2J(y,st)\\ &= ysJ(s^2,t)+s^3J(y,t)+s^2tJ(y,s)\\ &= 2ys^2(s-1)-s^3y-s^2t(2t+1)\\ &= s^2(2t(t+1)-sy-t(2t+1))\\ &= s^2(t-sy) = s^2(t-y-t^2-t) = -f. \end{aligned} \tag{3.57}$$

This gives

$$J(p,h) = -f. \tag{3.58}$$

Thus

$$\begin{aligned}J(p,q) &= (2t+6h+6h^2)(f+t-3h-3h^2)\\ &\quad +(6t+12th)f - J(p,u)\\ &= f(8t+6h+6h^2+12th)\\ &\quad +(2t+6h+6h^2)(t-3h-3h^2) - J(p,u) && (3.59)\\ &= f(8t+6h+6h^2+12th)+2t^2\\ &\quad -18h^2(h+1)(h+1) - J(p,u) && (3.60)\\ &= f(8t+6h+6h^2+12th-18(h-t)(h+1))\\ &\quad +2t^2 - J(p,u) && (3.61)\\ &= f(26t-12h-12h^2+30th)+2t^2-J(p,u) && (3.62)\\ &= t^2+f^2+(t+f(13+15h))^2-f^2-f^2(13+15h)^2\\ &\quad -12h(h+1)f-J(p,u). && (3.63)\end{aligned}$$

We obtain

$$J(p,q) = t^2+f^2+(t+f(13+15h))^2 \tag{3.64}$$

provided that we can find u such that

$$J(p,u) = -fv, \tag{3.65}$$

where

$$v = f+f(13+15h)^2+12h(h+1). \tag{3.66}$$

Now, if $u = u(f, h)$, then by the chain rule

$$J(p, u) = J(f + h, u(f, h)) = J(f, h)\left(\frac{\partial u}{\partial h} - \frac{\partial u}{\partial f}\right) = -f\left(\frac{\partial u}{\partial h} - \frac{\partial u}{\partial f}\right), \tag{3.67}$$

and so we require u to satisfy

$$v(f, h) = \frac{\partial u}{\partial h} - \frac{\partial u}{\partial f}. \tag{3.68}$$

We construct u in the form

$$u = A(h)f + B(h) \tag{3.69}$$

from which we require

$$v = A'f + B' - A, \tag{3.70}$$

and this is possible if

$$A' = 1 + (13 + 15h)^2, \qquad B' = 12h(h + 1) + A. \tag{3.71}$$

We may thus choose

$$A = h + \frac{1}{45}(13 + 15h)^3, \ B = 4h^3 + 6h^2 + \frac{1}{2}h^2 + \frac{1}{2700}(13 + 15h)^4. \tag{3.72}$$

Note that, since h has degree 5 and f has degree 10, therefore p has degree 10 and q has degree 25. Finally, it is clear that $J(p, q) > 0$, provided that $J(p, q) \neq 0$. But $J(p, q) = 0$, only if $f = t = 0$. But, if $t = 0$, then $s = 1$ and $xy = 1$, so $y \neq 0$. As then $f = s^2(t^2 + y) = y$, we see that $J(p, q) \neq 0$. The mapping (p, q) is not a homeomorphism of the plane, since when $s = 0$, then $p = 0$. But $s = 0$ on the curve $y = \frac{x-1}{x^2}$. This is in fact two curves, one when $x > 0$ and one when $x < 0$, each curve tending to infinity in the (x, y)-plane at either end. Such a pair of curves cannot be the inverse image or part of the inverse image of the imaginary axis in the (p, q)-plane for a homeomorphic mapping. Thus Pinchuk's result is established.

3.2.2 The Pinchuk surface

In order to ascertain the topological and geometric structure of Pinchuk's function, we need to calculate its asymptotic behaviour at infinity. We will first use the constructive method of chapter 1 to find the asymptotic values and tracts of the function p. Observe firstly that

$$\begin{aligned} h(x,y) = st = (1 + x(xy-1))(xy-1) &= x(xy-1)^2 + xy - 1 \\ &= x^3y^2 - 2x^2y + xy + x - 1 \end{aligned} \tag{3.73}$$

and

$$\begin{aligned} f(x,y) &= (1 + x(xy-1))^2(y + (xy-1)^2) \\ &= (1 + 2x(xy-1) + x^2(xy-1)^2)(y + x^2y^2 - 2xy + 1) \\ &= (1 - 2x + x^2 + 2x^2y - 2x^3y + x^4y^2)(1 + y - 2xy + x^2y^2) \\ &= 1 - 2x + y + x^2 - 4xy + 7x^2y - 4x^3y + 3x^2y^2 - 8x^3y^2 \\ &\quad + 6x^4y^2 + 3x^4y^3 - 4x^5y^3 + x^6y^4. \end{aligned} \tag{3.74}$$

This gives

$$\begin{aligned} p(x,y) = -x + y + x^2 - 3xy + 5x^2y - 4x^3y + 3x^2y^2 - 7x^3y^2 \\ + 6x^4y^2 + 3x^4y^3 - 4x^5y^3 + x^6y^4. \end{aligned} \tag{3.75}$$

The leading term is x^6y^4, which vanishes when $x = 0$ or $y = 0$. Therefore the asymptotic tracts of p lie only in these directions. To find the tracts corresponding to $y = 0$, we make the substitution $x \to \frac{1}{x}$, $y \to \frac{y}{x}$. We obtain

$$\begin{aligned} p\left(\frac{1}{x}, \frac{y}{x}\right) = \frac{1}{x^{10}}(y^4 - 4y^3x^2 + 3y^3x^3 + 6y^2x^4 - 7y^2x^5 + (3y^2 - 4y)x^6 \\ + 5yx^7 + (1 - 3y)x^8 + (y-1)x^9). \end{aligned} \tag{3.76}$$

The multiplicities of the zero at the origin are given by

$$r_0 = 4, r_1 = \infty, r_2 = 3, r_3 = 3, r_4 = 2, r_5 = 2, r_6 = 1, r_7 = 1, r_8 = r_9 = 0. \tag{3.77}$$

We require p as the minimum positive value of $\frac{j}{r_0 - r_j}$. This gives $p = 2$. We therefore make the substitution $y = \alpha x^2$. We obtain

$$\begin{aligned} &\frac{1}{x^2}(\alpha^4 - 4\alpha^3 + 3\alpha^3x + 6\alpha^2 - 7\alpha^2x + 3\alpha^2x^2 - 4\alpha + 5\alpha x \\ &\quad + 1 - 3\alpha x^2 + \alpha x^3 - x) = \frac{1}{x^2}((\alpha-1)^4 + (3\alpha^3 - 7\alpha^2 + 5\alpha - 1)x \\ &\qquad + (3\alpha^2 - 3\alpha)x^2 + \alpha x^3). \end{aligned} \tag{3.78}$$

Writing $\alpha - 1 = \beta$, this becomes

$$\frac{1}{x^2}(\beta^4 + \beta^2(3\beta + 2)x + 3\beta(\beta+1)x^2 + (\beta+1)x^3). \tag{3.79}$$

The substitution $\beta = \gamma x^{\frac{1}{2}}$ gives

$$\gamma^4 + 2\gamma^2 + o(1) \text{ as } x \to 0. \tag{3.80}$$

We thus obtain the asymptotic values $\gamma^4 + 2\gamma^2$ in the tract defined by the equation

$$y = \frac{1}{x} + \frac{\gamma}{x^{3/2}} \tag{3.81}$$

with $x \to +\infty$. Note that in this tract, writing $xt^2 = w$,

$$t = \frac{\gamma}{x^{1/2}} \to 0 \text{ as } x \to +\infty, \tag{3.82}$$

$$h = t + w \to \gamma^2 \text{ as } x \to +\infty \tag{3.83}$$

and

$$f = w^2 + (3t+1)w + 3t^2 + 2t + y \to \gamma^4 + \gamma^2 \text{ as } x \to +\infty. \tag{3.84}$$

In this notation we have

$$p = w^2 + (3t+2)w + 3t^2 + 3t + y; \tag{3.85}$$

w, t, and y satisfy the relation

$$wy = t^2(t+1). \tag{3.86}$$

We also observe that $q = -u(f, h) + o(1)$ in the tract, and so q has asymptotic values

$$\begin{aligned} -u(\gamma^4 + \gamma^2, \gamma^2) &= -A(\gamma^2)(\gamma^4 + \gamma^2) - B(\gamma^2) \\ &= -\left(\gamma^2 + \frac{1}{45}(13 + 15\gamma^2)^3\right)(\gamma^4 + \gamma^2) - 4\gamma^6 - 6\gamma^4 - \frac{1}{2}\gamma^4 \\ &\quad - \frac{1}{2700}(13 + 15\gamma^2)^4. \end{aligned} \tag{3.87}$$

Next we find the asymptotic values of p in the direction $x \to -\infty$. We will still find that $t \to 0$ and $y \to 0$, so it is enough to find the asymptotic values of w in this direction. We have $w = x(xy-1)^2$, and changing x to $-x$ this becomes $-x(xy+1)^2$, where we will now consider $x \to +\infty$. Omitting the details we obtain the asymptotic values $-\gamma^2$ for w in the tract given by

$$y = \frac{1}{x} + \frac{\gamma}{(-x)^{3/2}} \text{ as } x \to -\infty. \tag{3.88}$$

Thus h has asymptotic values $-\gamma^2$, f has asymptotic values $\gamma^4 - \gamma^2$ and p has asymptotic values $\gamma^4 - 2\gamma^2$. The asymptotic values of q are given by

$$\begin{aligned}-u(\gamma^4-\gamma^2,-\gamma^2) &= -A(-\gamma^2)(\gamma^4-\gamma^2)-B(-\gamma^2)\\ &= -\left(-\gamma^2+\frac{1}{45}(13-15\gamma^2)^3\right)(\gamma^4-\gamma^2)+4\gamma^6-6\gamma^4\\ &\quad -\frac{1}{2}\gamma^4-\frac{1}{2700}(13-15\gamma^2)^4. \end{aligned}\tag{3.89}$$

Our next task is to find the asymptotic values of p in the imaginary direction. We consider

$$\begin{aligned}p\left(\frac{x}{y},\frac{1}{y}\right) &= \frac{1}{y^{10}}(x^6-4x^5y^2+3x^4y^3+6x^4y^4-7x^3y^5+3x^2y^6-4x^3y^6\\ &\quad +5x^2y^7-3xy^8+x^2y^8+y^9-xy^9).\end{aligned}\tag{3.90}$$

The multiplicities of the zeros at the origin are given by

$$\begin{aligned}r_0=6, r_1=\infty, r_2=5, r_3=4, r_4=4, r_5=3,\\ r_6=2, r_7=2, r_8=1, r_9=0.\end{aligned}\tag{3.91}$$

The choice of p is $\frac{3}{2}$ so we make the substitution $x = \alpha y^{3/2}$. Our expression becomes

$$\begin{aligned}&\frac{1}{y}(\alpha^6-4\alpha^5y^{1/2}+3\alpha^4+6\alpha^4y-7\alpha^3y^{1/2}+3\alpha^2-4\alpha^3y^{3/2}\\ &\quad +5\alpha^2y-3\alpha y^{1/2}+\alpha^2y^2+1-\alpha y^{3/2}).\end{aligned}\tag{3.92}$$

The leading term is $(\alpha^2+1)^3$, which has no zeros, and hence $p \to +\infty$ as $y \to 0$ through positive values. Thus in the original plane p has no asymptotic values in the imaginary direction with $y \to +\infty$.

To find the asymptotic values with $y \to -\infty$ we consider

$$\begin{aligned}p\left(-\frac{x}{y},-\frac{1}{y}\right) &= \frac{1}{y^{10}}(x^6-4x^5y^2-3x^4y^3+6x^4y^4+7x^3y^5\\ &\quad +3x^2y^6-4x^3y^6-5x^2y^7-3xy^8+x^2y^8-y^9+xy^9).\end{aligned}\tag{3.93}$$

Again we make the substitution $x = \alpha y^{3/2}$. Our expression becomes

$$\begin{aligned}&\frac{1}{y}(\alpha^6-4\alpha^5y^{1/2}-3\alpha^4+6\alpha^4y+7\alpha^3y^{1/2}+3\alpha^2-4\alpha^3y^{3/2}\\ &\quad -5\alpha^2y-3\alpha y^{1/2}+\alpha^2y^2-1+\alpha y^{3/2}).\end{aligned}\tag{3.94}$$

This has the form

$$\frac{1}{y}((\alpha^2-1)^3-(4\alpha^5-7\alpha^3+3\alpha)y^{1/2}+(6\alpha^4-5\alpha^2)y)+o(1) \text{ as } y\to 0. \quad (3.95)$$

Writing $Y = y^{1/2}$ this becomes

$$\frac{1}{Y^2}((\alpha^2-1)^3-\alpha(\alpha^2-1)(4\alpha^2-3)Y+\alpha^2(6\alpha^2-5)Y^2)+o(1). \quad (3.96)$$

The leading term has zeros at $\alpha = 1$ and $\alpha = -1$. Firstly put $\beta = \alpha - 1$. We obtain

$$\frac{1}{Y^2}(\beta^3(\beta+2)^3-\beta(\beta+1)(\beta+2)(4\beta^2+8\beta+1)Y$$
$$+(\beta+1)^2(6\beta^2+12\beta+1)Y^2). \quad (3.97)$$

We now put $\beta = \gamma Y^{1/2}$ obtaining

$$\frac{1}{Y^{1/2}}(\gamma^3(2+\gamma Y^{1/2})^3$$
$$-\gamma(1+\gamma Y^{1/2})(2+\gamma Y^{1/2})(1+8\gamma Y^{1/2}+4\gamma^2 Y)+Y^{1/2})+o(1)$$
$$=\frac{1}{Y^{1/2}}(8\gamma^3-2\gamma+(12\gamma^4-19\gamma^2+1)Y^{1/2})+o(1). \quad (3.98)$$

The leading term $8\gamma^3 - 2\gamma$ has simple zeros at $\gamma = 0, \frac{1}{2}, -\frac{1}{2}$ each leading to a simply described real axis of asymptotic values.

For example, for the zero $\gamma = 0$ we make the substitution $\gamma = \lambda Y^{1/2}$ obtaining the asymptotic values $-2\lambda + 1$ with the tract

$$x = \alpha y^{3/2} = (\beta+1)y^{3/2} = (1+\gamma y^{1/4})y^{3/2} = y^{3/2}+\lambda y^2, \quad (3.99)$$

which becomes in the original (x, y)-plane

$$-\frac{x}{-y} = \frac{1}{(-y)^{3/2}}+\frac{\lambda}{y^2}, \quad (3.100)$$

i.e.

$$x = \frac{-y}{(-y)^{3/2}}-\frac{\lambda}{y}. \quad (3.101)$$

This gives $y + (xy + \lambda)^2 = 0$, or $(t + 1 + \lambda)^2 = -y$.

For the zero $\gamma = \frac{1}{2}$ we make the substitution $\gamma = \frac{1}{2} + \lambda Y^{1/2}$ giving the asymptotic values $4\lambda - 3$ with the tract

$$x = y^{3/2}\left(1+\frac{1}{2}y^{1/4}+\lambda y^{1/2}\right) = y^{3/2}+\frac{1}{2}y^{7/4}+\lambda y^2, \quad (3.102)$$

which becomes

$$x = \frac{1}{(-y)^{1/2}} + \frac{\frac{1}{2}}{(-y)^{3/4}} - \frac{\lambda}{y}. \tag{3.103}$$

For the zero $\gamma = -\frac{1}{2}$ we make the substitution $\gamma = -\frac{1}{2} + \lambda Y^{1/2}$ giving the asymptotic values $4\lambda - 3$ with the tract

$$x = y^{3/2}\left(1 - \frac{1}{2}y^{1/4} + \lambda y^{1/2}\right) = y^{3/2} - \frac{1}{2}y^{7/4} + \lambda y^2, \tag{3.104}$$

which becomes

$$x = \frac{1}{(-y)^{1/2}} - \frac{\frac{1}{2}}{(-y)^{3/4}} - \frac{\lambda}{y}. \tag{3.105}$$

Next we put $\beta = \alpha + 1$. We obtain

$$\frac{1}{Y^2}(\beta^3(\beta-2)^3 - \beta(\beta-1)(\beta-2)(4\beta^2 - 8\beta + 1)Y + (\beta-1)^2(6\beta^2 - 12\beta + 1)Y^2). \tag{3.106}$$

We now put $\beta = \gamma Y^{1/2}$ obtaining

$$\begin{aligned} &\frac{1}{Y^{1/2}}(\gamma^3(\gamma Y^{1/2} - 2)^3 \\ &\quad - \gamma(\gamma Y^{1/2} - 1)(\gamma Y^{1/2} - 2)(1 - 8\gamma Y^{1/2} + 4\gamma^2 Y) + Y^{1/2}) + o(1) \\ &= \frac{1}{Y^{1/2}}(-8\gamma^3 - 2\gamma + (-12\gamma^4 + 19\gamma^2 + 1)Y^{1/2}) + o(1). \end{aligned} \tag{3.107}$$

The leading term $-8\gamma^3 - 2\gamma$ has one simple zero at $\gamma = 0$ leading to one simply described real axis of asymptotic values. We make the substitution $\gamma = \lambda Y^{1/2}$ obtaining the asymptotic values $-2\lambda + 1$ with the tract

$$x = \alpha y^{3/2} = (\beta - 1)y^{3/2} = (-1 + \gamma y^{1/4})y^{3/2} = -y^{3/2} + \lambda y^2, \tag{3.108}$$

which becomes in the original (x, y)-plane

$$-\frac{x}{-y} = \frac{-1}{(-y)^{3/2}} + \frac{\lambda}{y^2}, \tag{3.109}$$

i.e.

$$x = \frac{y}{(-y)^{3/2}} - \frac{\lambda}{y}. \tag{3.110}$$

This gives $y + (xy + \lambda)^2 = 0$, or $(t + 1 + \lambda)^2 = -y$.

Our next task is to consider in detail the actual curve of finite asymptotic values constructed for the mapping (p, q). The real part of the curve is obtained from the parametrisations $\gamma^4 + 2\gamma^2$ in one tract and $\gamma^4 - 2\gamma^2$ in the other tract. This amounts to $\lambda^2 + 2\lambda$ traversed for λ going from $+\infty$ to 0 followed by 0 to $+\infty$ in the first tract, and traversed for λ going from $-\infty$ to 0 followed by 0 to $-\infty$ in the second tract. We have a similar situation for the imaginary part. Here the basic curve is given by

$$-\left(\lambda + \frac{1}{45}(13 + 15\lambda)^3\right)(\lambda^2 + \lambda) - 4\lambda^3 - \frac{13}{2}\lambda^2 - \frac{1}{2700}(13 + 15\lambda)^4, \tag{3.111}$$

with the traversal as above. To simplify this we make the substitution $\mu = \lambda + 1$. We obtain

$$\begin{aligned} -&\left(\mu - 1 + \frac{1}{45}(15\mu - 2)^3\right)(\mu^2 - \mu) - 4(\mu - 1)^3 \\ &- \frac{13}{2}(\mu - 1)^2 - \frac{1}{45 \times 60}(15\mu - 2)^4 \\ =& -75\mu^5 + (30 + 75 - 75/4)\mu^4 - (5 + 30 + 4 - 10)\mu^3 \\ &+ \left(\frac{15}{2} + \frac{8}{45} + 4 - 2\right)\mu^2 - \frac{683}{1350} \\ =& -75\mu^5 + \frac{345}{4}\mu^4 - 29\mu^3 + \frac{871}{90}\mu^2 - \frac{683}{1350}. \end{aligned} \tag{3.112}$$

Note then the real part is $\mu^2 - 1$ and so the curve is given by the parametric representation

$$w(\mu) = \left(\mu^2 - 1, -75\mu^5 + \frac{345}{4}\mu^4 - 29\mu^3 + \frac{871}{90}\mu^2 - \frac{683}{1350}\right) \tag{3.113}$$

with μ real, and this has a cusp at $\mu = 0$. Furthermore the curve is simple: for if the curve meets itself, then for some μ we have $w(\mu) = w(-\mu)$. We then obtain

$$-75\mu^5 - 29\mu^3 = 0, \tag{3.114}$$

which is possible only when $\mu = 0$. We denote by Γ this curve with parametric representation $w(\mu)$. (Note that, if we stipulate that $w'(0) = 0$ and $w(\mu)$ is simple, then $w(\mu)$ must have degree ≥ 3.) We write $w_0 = w(0)$ and $w_1 = w(1)$. The curve corresponding to $\mu \geq 1$ is denoted by Γ_1; the curve corresponding to $\mu \leq 0$ is denoted by Γ_0; the curve corresponding to $0 \leq \mu \leq 1$ is denoted by Γ_{01}. In the positive x direction the mapping at ∞ describes the curve Γ_1 from ∞ to w_1 and back again. In the negative x direction the mapping at ∞ describes $\Gamma_0 \cup \Gamma_{01}$ from ∞ to w_1 and back again. On the basis that the mapping has no other finite asymptotic values we obtain the following description of the simply-connected surface onto which the plane is mapped by (p, q). As Γ is simple it divides the plane into two simply-connected regions, A to the left of Γ and B to the right of Γ. Let us denote by $A1$ the domain A bordered by $\Gamma_1 \cup w_0$. Γ is thus divided into two arcs corresponding to $0 < \mu < 1$ and $\mu < 0$, and we regard the locally defined inverse mapping of (p, q) as starting in A and as able to cross either of these arcs. On crossing Γ_{01} we move into $B1$, which is defined as B bordered by $\Gamma_1 \cup \Gamma_0$. Thus from B1 we can cross only that part of Γ corresponding to $0 < \mu < 1$. On crossing Γ_0 from $A1$ we move into $B2$, which is defined as B bordered by Γ_{01}. Thus from $B2$ we can cross that part of Γ corresponding to $\mu < 0$ and $\mu > 1$. On crossing Γ_1 from $B2$ we move into $A2$ which is defined as A bordered by $\Gamma_{01} \cup \Gamma_0$. These four regions define the Pinchuk surface onto which (p, q) maps. We thus see that (p, q) *is a 2-valent mapping of* $\mathbb{C}$*, attaining all values in either A or B exactly twice. All points on* Γ *are attained in* $\mathbb{C}$ *exactly once except for the two points* w_0 *and* w_1*, which are not attained at all. Thus the mapping fails to be surjective*, as it omits these two values.

3.2.3 General observations for the $\mathbb{R}^2$ problem

Pinchuk's example is very instructive for the geometric structure revealed and the manner in which it arises. Although the set of asymptotic values is formed from a single simple curve, it is the manner in which the curve is described that opens up the simply-connected surface. Firstly, the repetition of asymptotic values occurs by the third method described in chapter 1, subsection 1.3.3; i.e. in one tract half the curve Γ is described (from ∞ to w_1) and then reversed; in another tract the other half of the curve is described and then reversed. However, this on its own is not sufficient to generate a simply-connected surface. To be precise, we wish to describe the nature of the mapping for values of $z = (x, y)$ in the tract near ∞. To do this we may consider the image of a large circle $|z| = r$ as z crosses the tract. The values on the image curve C_r will follow very closely the asymptotic curve as it is described by the parametric representation $A(0, t)$ (see chapter 1). Because the mapping is locally 1–1 and orientation preserving, the domain to the left of C_r as $|z| = r$ crosses the tract in the positive direction is the image of $|z| < r$. This behaviour extends to the asymptotic curve itself: the image of the tract lies to the left of the curve as it is described. However, there is an exceptional point, namely the point w_0. C_r loops negatively around w_0 crossing Γ from B to A and back again, proceeding then down to w_1. The effect of this is to open up Γ so that the surface flows across that part of Γ below w_0. In essence we have an instantaneous double reversal of direction in Γ giving a zero net effect in the describing of Γ itself, but having a massive effect on the nearby mapping. Of course the behaviour of the mapping is completely given in the full function $A(s, t)$. But clearly it is most convenient if we can understand the behaviour from knowledge of the function $A(0, t)$. The behaviour is in fact signalled by the fact that $A(0, t)$ has vanishing derivative at the point corresponding to w_0. Whether or not in general a vanishing derivative is necessary to produce this double reversal behaviour is unclear. What we can say is that the transformed mapping $A(s, t)$ must branch at the point $(0, t_0)$, so that the values $A(0, t)$ are attained not only on $s = 0$, but also on another curve passing through $(0, t_0)$.

An interesting observation is that the instantaneous reversal at w_0 is a necessary feature in order to obtain a simply-connected surface from a single simple curve. One might ask, could one replace the instantaneous reversal with a finite reversal from say w_0 to w_2 back up the curve, followed by another reversal at w_2 coming back down the curve? This would certainly produce a surface enabling one to loop around the segment $[w_0, w_2]$ of the curve. However, the rules are that repetition of the asymptotic curve must occur by one of the three methods: repetition, reversal or the third splitting method. In all three cases the finite reversal gets repeated, and we get too much opening up for the surface to remain simply-connected. As the repetition rules apply only to the asymptotic curve itself (and not to nearby curves), the instantaneous reversal does not have to be repeated. Indeed in Pinchuk's case, when we pass w_0 on the way back up, no instantaneous reversal occurs.

The general nature of a surface arising from a locally 1–1 polynomial mapping may be considered in the form of a graph. If we label by R_k each region on the surface, whose projection into the plane is a simply-connected region bounded by parts of asymptotic value curves, and then join R_k to R_j if there is a continuation across a boundary between the two (the open boundary being an opening in part of an asymptotic value curve), we obtain a graph. In general given any R_j and R_k there is a unique route through the graph from one to the other. Consequently the graph takes the form of a tree. The method of labelling is not unique, but depends on where one starts.

One important conclusion, which can be drawn from these topological considerations and the repetition property, is that, *if P is 1–1, then P is a homeomorphism of $\mathbb{C}$*. In other words, P cannot be an injective mapping onto a subdomain of $\mathbb{C}$. For assuming that P is injective onto a domain D with boundary Γ, then Γ consists of a finite number of asymptotic value curves, and each boundary value has to be repeated. If a curve is repeated in the same direction, then clearly nearby points of D are attained twice. If a curve is repeated by the splitting method, then points near the disjunction are attained twice. It follows that all curves are repeated by reversal, and so both sides of every curve are attained values. Therefore no such curve can divide the plane. An asymptotic value curve, therefore, is a curve proceeding from ∞ to a point, and then reversing back from the point to ∞. Thus in a single tract both sides of the curve are attained, and this behaviour is repeated (though in reverse) in a second tract, so all the nearby values are again attained. Thus in all cases repetition contradicts the injectivity.

3.2.4 Estimations of degree

Pinchuk's example has degree 25. It is clearly of interest to estimate the smallest possible degree of a counter-example to the real Jacobian conjecture. Since such an example must have at least one curve of asymptotic values, we certainly have one means for giving a lower bound for the degree of the polynomial. If P has degree N, then the leading term of P has degree N and the asymptotic values arise out of the zeros of this leading term. (However, in two ways, since we must consider both $x > 0$ and $x < 0$.) A zero of multiplicity m can only lead to an asymptotic value curve of degree $\leq m$. Thus, allowing for repetition of curves, the sum of the degrees of the asymptotic value curves cannot exceed the degree of P.

For a Pinchuk surface, which is the simplest possible, we require a basic curve, which is described by the splitting method. This doubles the degree of the curve. Also we must allow for the instantaneous double reversal. If we assume that this phenomenon arises from a zero of the derivative (though we have not proved this), then the minimum basic curve will have degree ≥ 3. This gets us up to degree 6. However, the real and imaginary parts of P must also reflect the amount of movement of P on large circles, and this means there must be additional asymptotic tracts for each of these. We require at least two simply described straight lines for each, getting us up to degree 8. Thus the real Jacobian conjecture should be true for polynomials of degree ≤ 7.

To give some justification for the hypothesis that the double instantaneous reversal is signalled by the function $A(0, t)$ having a vanishing derivative, we consider a mapping of the form $A = U + iV$, where

$$U(s, t) = u(t) + s^{\rho} K(s, t), \qquad V(s, t) = s^{\sigma} L(s, t). \tag{3.115}$$

For the double reversal we would require that $V(s, t) = 0$ on two curves emanating from 0 into the region $s > 0$. For the Jacobian we obtain

$$\begin{aligned} J(s, t) = s^{\sigma-1}\{&-\sigma u'(t)L - su'(t)L_s + s^{\rho}(\rho K L_t - \sigma K_t L) \\ &+ s^{\rho+1}(K_s L_t - K_t L_s)\}. \end{aligned} \tag{3.116}$$

We are assuming that $K(s,t)$ and $L(s,t)$ are not divisible by s. Since by hypothesis $J \neq 0$ for $s \neq 0$, we see that $u'(t)L(0,t)$ has constant sign (though it may have some zeros). Now if $u'(t) \neq 0$ for t near 0 and $L(0,0) = 0$, then $L_t(0,0) = 0$. Also

$$\frac{J(s,t)}{s^\sigma} = -\sigma u'(t)\frac{L}{s} - u'(t)L_s + o(1)s^{\rho-1}, \tag{3.117}$$

and so, when $t = 0$, the right-hand side tends to $-(\sigma+1)u'(0)L_s(0,0)$. On the other hand $J(s,t)/s^{\sigma-1}$ has constant sign, and so the left-hand side must tend to 0. It follows that $L_s(0,0) = 0$. Thus both partial derivatives of L vanish at the origin. This gives some support for the desired conclusion that L has constant sign near 0.

This is a somewhat rough-and-ready argument and we shall now give a different approach, which appears to get nearer to the heart of the problem. An asymptotic value curve of P is associated with an asymptotic tract, which is expressed by an algebraic transformation

$$x = \frac{1}{s^q}, \qquad y = \frac{\phi(s) + ts^\beta}{s^q}. \tag{3.118}$$

This tract serves as an asymptotic tract for a whole family of polynomials, and in particular there is a polynomial of smallest degree, which then generates the family together with certain other associated polynomials. In other words, we can write

$$P(x,y) = F(\alpha_1(x,y), \alpha_2(x,y), \ldots, \alpha_m(x,y)), \tag{3.119}$$

where F is a polynomial and where α_m has exactly the tract, say T_m, for its asymptotic values; the remaining α_k are associated with tracts T_k such that $j < k \Rightarrow T_j \supset T_k$ and α_j has a constant limit in T_k.

To obtain a tract with the splitting property, we need q even, $\phi(s)$ a polynomial in s^2 and β odd. Furthermore, in order that the polynomial not be constant on some curve, we require $q < \deg \phi < \beta$. Without loss of generality we will choose a tract with $y \to 0$ and construct a polynomial of smallest degree satisfying these criteria. We take $q = 2$, $\phi(s) = s^4$ and $\beta = 5$. This leads to a tract of the form

$$y = \frac{c}{x} + \frac{t}{x^{3/2}}, \tag{3.120}$$

which gives the basic polynomial $\gamma = x(xy - c)^2$. Since $c \neq 0$, we shall take $c = 1$. The associated subtracts are then $y = t/x$ and $y = t$, from which we obtain the polynomials xy and y. Thus our candidate for a polynomial of smallest degree providing a counter-example to the Jacobian conjecture takes the form

$$P(x, y) = F(y, xy - 1, x(xy - 1)^2). \tag{3.121}$$

Pinchuk's example has precisely this form. The polynomial $F(\alpha, \beta, \gamma)$ will have at least one term in a power of γ alone. Thus P has degree ≥ 5. We will show that P has degree ≥ 10. The alternative is that F is linear in γ. Now the Jacobian of P is given by

$$J(P) = y\Im\overline{F_2}F_1 - x(xy-1)^2\Im\overline{F_2}F_3 + (xy-1)(3xy-1)\Im\overline{F_3}F_1, \tag{3.122}$$

where the subscripts denote partial differentiation. In the tract, $x(xy-1)^2 = t^2$ and then $J(P) \to -t^2\Im\overline{F_2}F_3(0, 0, t^2)$. Now F is linear in the third variable and so after an affine transformation, if we consider the polynomial $A(s, t)$, which arises from P out of the transformation, we obtain

$$A(s, t) = P\left(\frac{1}{s^2}, s^2 + ts^3\right) = F(s^2 + ts^3, ts, t^2), \tag{3.123}$$

where

$$F(s^2 + ts^3, ts, t^2) = t^2 + s^\rho K(s, t) + is^\sigma L(s, t) \tag{3.124}$$

and we obtain for the Jacobian of A

$$J(A) = s^{\sigma-1}(-2t\sigma L + \text{ terms involving powers of } s), \tag{3.125}$$

and as we saw above $L(0, t)$ has constant sign. Now from this form we see that the term s^σ must arise from a product of the form $(s^2 + ts^3)^m(ts)^n$, where $2m + n = \sigma$, and then we see that $L(0, t) = t^n(at^2 + b)$, and this cannot vanish for $t \neq 0$ and still have constant sign. This eliminates the possibility of a Pinchuk surface arising, as for this to occur L must vanish on curves emanating from some point $(0, t_0)$, where $t_0 \neq 0$. It follows that F must contain a term in γ of degree higher than 1, and so the original polynomial has degree ≥ 10. Furthermore, as we saw in studying Pinchuk's example, in order for the basic curve to be simple, it needs to contain terms of odd degree >1. Thus we would expect F to contain at least a cubic term in γ and so it seems highly unlikely that a counter-example to the Jacobian conjecture can have degree <15; in other words the conjecture is probably true up to degree 14. We have by no means carried out all the algebra necessary to prove this. Our argument is certain up to degree 5, almost certain up to degree 10. However, the algebra of this basic polynomial needs to be completed; also the small number of other possible basic polynomials requires consideration.

One interesting observation from this algebra is that the basic generating polynomial $x(xy-1)^2$ cannot give a counter-example to the following weaker version of the conjecture: *if* $J(P) \geq \delta > 0$, *then* P *is a homeomorphism.* For, whatever the function F, $J(P) \to 0$ along a path. In particular, it would require a higher degree generating polynomial to provide a counter-example to Keller's conjecture.

3.2.5

Using the variables α, β and γ we can give an interesting geometric interpretation of the mapping as the restriction of a polynomial mapping $H: \mathbb{R}^3 \mapsto \mathbb{R}^3$ to a surface in $\mathbb{R}^3$ mapping to a plane in $\mathbb{R}^3$. With $\alpha = y, \beta = xy - 1, \gamma = x(xy-1)^2$ we note that $\alpha\gamma = \beta^2(\beta+1)$. The set S of points (α, β, γ) satisfying this equation defines a surface in $\mathbb{R}^3$. Indeed the function $(y, xy - 1, x(xy - 1)^2)$ is a 1–1 mapping of $\mathbb{R}^2$ into S and so the image I of the mapping is a subset of S homeomorphic to $\mathbb{R}^2$. On the other hand, given $(\alpha, \beta, \gamma) \in S$ we can define $y = \alpha$ and $x = \frac{\beta+1}{\alpha}$ or $\frac{\gamma}{\beta^2}$ provided that we do not have both $\alpha = 0$ and $\beta = 0$. Thus $I = S$ minus the line $\{(0, 0, \gamma)\}$. Now S is symmetric in α and γ, as $(\alpha, \beta, \gamma) \in S \Leftrightarrow (\gamma, \beta, \alpha) \in S$; hence $I' = S$ minus the line $\{(\alpha, 0, 0)\}$ is homeomorphic to $\mathbb{R}^2$. It follows that $S - \{(0, 0, 0)\}$ is a surface. The origin (0,0,0) is a singular point of S and S is not a surface. Indeed, the intersection of a small ball centred at the origin with $S - \{(0, 0, 0)\}$ is disconnected and the origin divides the neighbourhood on S into two disjoint parts. The mapping H is defined as $H(\alpha, \beta, \gamma) = (F(\alpha, \beta, \gamma), \alpha\gamma - \beta^2(\beta + 1))$.

3.3 Polynomials with a constant Jacobian

3.3.1

We consider the case of the Jacobian conjecture in which the Jacobian determinant is constant. Thus, if our mapping is given by $P = (u, v)$, then we assume that $J(u, v) = 1$. The full form of the conjecture states that in this case the inverse mapping exists and is itself a polynomial; i.e. P is **polynomially invertible**.

Consider such a mapping P, where u has degree m and v has degree n and where without loss of generality we may assume $m \leq n$ (otherwise consider the mapping (v, u)). We may also assume that the expansion of u contains a term ay^m and the expansion of v a term by^n. We then obtain

$$\begin{aligned} u\left(\frac{1}{x}, \frac{y}{x}\right) &= \frac{u_0(y) + u_1(y)x + \cdots + u_m(y)x^m}{x^m}, \\ v\left(\frac{1}{x}, \frac{y}{x}\right) &= \frac{v_0(y) + v_1(y)x + \cdots + v_n(y)x^n}{x^n}. \end{aligned} \tag{3.126}$$

Now

$$J\left(u\left(\frac{1}{x},\frac{y}{x}\right),v\left(\frac{1}{x},\frac{y}{x}\right)\right)=-\frac{1}{x^3}, \tag{3.127}$$

but on the other hand the above expansion gives

$$J\left(u\left(\frac{1}{x},\frac{y}{x}\right),v\left(\frac{1}{x},\frac{y}{x}\right)\right)=(-mu_0(y)v_0'(y)+nu_0'(y)v_0(y))x^{-(m+n+1)}$$
$$+\text{terms in higher powers of } x. \tag{3.128}$$

If $m=0$, then u is constant, when clearly $J(u,v)=0$. Thus unless $m=n=1$, we deduce that

$$-mu_0(y)v_0'(y)+nu_0'(y)v_0(y)=0. \tag{3.129}$$

3.3.2

Lemma *Let $u(z)$ and $v(z)$ be polynomials of degrees m and n respectively, where $m\le n$ and suppose that*

$$nu'(z)v(z)-mu(z)v'(z)=0. \tag{3.130}$$

If m and n have highest common factor ω, we can write

$$u(z)=\phi^\mu(z),\ v(z)=\alpha\phi^\nu(z), \tag{3.131}$$

where ϕ is a polynomial of degree ω, with $m=\mu\omega$ and $n=\nu\omega$, and where α is a constant.

The proof is left as a simple exercise. Returning to our discussion of the Jacobian, if $m=n$, we deduce from the lemma that $u_0(y)=cv_0(y)$ for some constant c. However, if we now make the affine transformation $(u,v)\mapsto(u-cv,v)$, then the new polynomial still has Jacobian $=1$, but its real part has degree smaller than n. Thus we may assume that $m<n$. The above relation then implies from the lemma that there is a polynomial ϕ of degree ω say such that

$$u_0=a\phi^\mu \text{ and } v_0=b\phi^\nu, \tag{3.132}$$

where a and b are constants, ω is the greatest common divisor of m and n and where μ and ν are relatively prime. We see from this that u_0 and v_0 have the same zeros, namely those of ϕ; let α be a zero of ϕ of multiplicity ρ. Then u_0 has a zero at α of multiplicity $r_0=\mu\rho$, and v_0 has a zero at α of multiplicity $s_0=\nu\rho$. Consider the transformation

$$(x,y)\mapsto(x,\alpha+yx^p). \tag{3.133}$$

Let r_j and s_k denote respectively the multiplicities of the zeros at α of u_j and v_k $(0 \leq j \leq m, 0 \leq k \leq n)$. Observe that u_j has degree at most $m - j$ and v_k has degree at most $n - k$. With the above transformation we obtain the new expansions

$$\frac{\sum_{j=0}^{m} y^{r_j} U_j(\alpha + yx^p)x^{p(r_j - r_0)+j}}{x^{m-pr_0}},$$
$$\frac{\sum_{k=0}^{n} y^{s_k} V_k(\alpha + yx^p)x^{p(s_k - s_0)+k}}{x^{n-ps_0}}, \tag{3.134}$$

where $U_j(\alpha)$ and $V_k(\alpha)$ are $\neq 0$. If $0 < p \leq$ the smallest positive value of $j/(r_0 - r_j)$, then the leading term in the first expression becomes

$$A(y)x^{-m+pr_0} \text{ where } A(y) = \sum_j U_j(\alpha)y^{r_j}, \tag{3.135}$$

and where the sum is taken over those terms for which $p(r_j - r_0) + j = 0$. Similarly, if $0 < p \leq$ the smallest positive value of $k/(s_0 - s_k)$, then the leading term in the second expression becomes

$$B(y)x^{-n+ps_0} \text{ where } B(y) = \sum_k V_k(\alpha)y^{s_k}, \tag{3.136}$$

and where the sum is taken over those terms for which $p(s_k - s_0) + k = 0$. As the transformation from the original (x, y)-plane is given by

$$(x, y) \mapsto (x^{-1}, \alpha x^{-1} + yx^{p-1}), \tag{3.137}$$

the Jacobian of the transformed function is $-x^{p-3}$. The leading term for the Jacobian then becomes

$$-((m - pr_0)A(y)B'(y) - (n - ps_0)A'(y)B(y))x^{-m-n-1+pr_0+ps_0}. \tag{3.138}$$

This term is zero as long as $m + n > p(r_0 + s_0 - 1) + 2$; assuming $m + n \geq 3$, this is certainly true for small $p > 0$. Furthermore, if $0 < p <$ the smallest positive value of $j/(r_0 - r_j)$, then $A(y) = U_0(\alpha)y^{r_0}$; if $0 < p <$ the smallest positive value of $k/(s_0 - s_k)$, then $B(y) = V_0(\alpha)y^{s_0}$. Thus, if p satisfies both inequalities, then it is easily calculated that the above expression is 0, and hence $m + n > p(r_0 + s_0 - 1) + 2$.

Suppose now that $p =$ the smallest positive value of the numbers $j/(r_0 - r_j)$ and $k/(s_0 - s_k)$. We will show that, except in one circumstance, p is the minimum of the values $j/(r_0 - r_j)$ and also the minimum of the values $k/(s_0 - s_k)$; i.e. that the two minima coincide. If the above expression is zero, then by the lemma

A and B have common zeros; however, if the two minima do not coincide, then one of A and B has only one distinct zero at the origin; but then so does the other, and this contradicts the definition of p. If the expression is non-zero, then the zeros of A and B are distinct and simple. Hence, if the two minima do not coincide, then $r_0 = 1$ and so $\mu = \rho = 1$. This situation will be considered below. Thus apart from this case we have

$$p = \frac{j}{r_0 - r_j} = \frac{k}{s_0 - s_k} \tag{3.139}$$

for some j and k. This implies that

$$js_0 - kr_0 = js_k - kr_j, \tag{3.140}$$

i.e.

$$\rho(j\nu - k\mu) = js_k - kr_j. \tag{3.141}$$

Now, since $m = \omega\mu, n = \omega\nu, r_0 = \rho\mu, s_0 = \rho\nu$, our leading coefficient takes the form

$$(\omega - p\rho)(\mu AB' - \nu A'B) \tag{3.142}$$

and therefore

$$\mu A(y)B'(y) - \nu A'(y)B(y) = \text{constant}. \tag{3.143}$$

Let us consider the relations implied if the constant is $\neq 0$. We then have $m + n = p(r_0 + s_0 - 1) + 2$, i.e.

$$p = \frac{m + n - 2}{r_0 + s_0 - 1} = \frac{(\mu + \nu)\omega - 2}{(\mu + \nu)\rho - 1}. \tag{3.144}$$

Now, since $p\rho < \omega$, we obtain $(\mu+\nu)\omega\rho - 2\rho < (\mu+\nu)\omega\rho - \omega$, i.e. $\omega < 2\rho$. From this it follows that there is at most one zero α for which the constructed polynomials A and B can satisfy (3.143) with the constant $\neq 0$.

Note further that in the case $\rho = 1$ we obtain $\omega = 1$, which gives

$$p = \frac{\mu + \nu - 2}{\mu + \nu - 1}. \tag{3.145}$$

We will show that this implies that $\mu = 1$; indeed, as A and B have no common zeros, at least one of the numbers $A(0)$ and $B(0)$ is $\neq 0$. If $A(0) \neq 0$, then at least one of the relevant multiplicities r_j is zero, and therefore $p = j/r_0 = j/\mu$ for some j. We thus have

$$p = 1 - \frac{1}{\mu + \nu - 1} = \frac{j}{\mu} \tag{3.146}$$

which gives

$$\frac{1}{\mu + \nu - 1} = \frac{\mu - j}{\mu} \tag{3.147}$$

and therefore

$$\mu = (\mu - j)(\mu + \nu - 1) \geq \mu + \nu - 1. \tag{3.148}$$

Hence $\nu \leq 1 \leq \mu$, a contradiction. It follows that $B(0) \neq 0$ and a similar argument gives $\mu \leq 1$ and hence $\mu = 1$, as asserted.

Furthermore the mapping has no asymptotic values at ∞ and so the Jacobian conjecture for this mapping is correct. We can in fact go further than this and find the actual form which the mapping takes.

3.3.3

Let $u = ax + by + c$. Then we obtain

$$av_y - bv_x = 1. \tag{3.149}$$

Clearly, not both a and b are zero. If $a = 0$, then v_x is constant and so $v = dx + \psi(y)$, where d is a non-zero constant and ψ is any polynomial in y. If $a \neq 0$, then writing

$$V(u, y) = v\left(\frac{1}{a}(u - by), y\right), \tag{3.150}$$

we obtain $aV_y = 1$, and so $V = \frac{y}{a} + \psi(u)$, where ψ is any polynomial in u. Thus in general v takes the form

$$v = \psi(u) + w(x, y) \tag{3.151}$$

where ψ is an arbitrary polynomial and w is a polynomial of first degree satisfying $w_y - w_x = 1$. Note particularly the cases given by the mappings

$$u = x, \qquad v = y + x^n. \tag{3.152}$$

These functions are called **elementary mappings**.

3.3.4 The case $\omega = 1$

We will show that, if m and n are relatively prime, i.e. $\omega = 1$, then the constant in (3.143) is necessarily $\neq 0$, and therefore $m = 1$ and the mapping has the above form. In particular the conjecture is true for all polynomials of prime degree, which is a result due to Magnus [29].

If on the contrary the constant is zero, then, as A and B have respective degrees m and n, it follows from the lemma that A and B have exactly one zero, say $\beta \neq 0$, and $A = a(x-\beta)^m$, $B = b(x-\beta)^n$; $\beta \neq 0$ because of the definition of p. It follows easily that $p = 1 = m/r_0 = n/s_0$, and we have asymptotic values. This implies that the mapping is constant on a line and therefore the Jacobian vanishes on that line, a contradiction. Thus the constant is $\neq 0$ and the result asserted follows.

As remarked above, if n is prime, the conjecture follows immediately. The conjecture can also be proved, if m is prime. For in this case n is a multiple of m, i.e. $n = \omega m$. If we now make use of the elementary transformation $u \mapsto u, v \mapsto v + au^{\omega}$, for a suitable constant a, then we see that the new function still has constant Jacobian, but its degree for suitable a is smaller than n (this is because of the relation between the homogeneous terms of highest degree in u and v). Repeating this process a finite number of times reduces the degree of the polynomial to m.

3.3.5 The case $\omega = 2$

We remark that this argument can be applied whenever m divides n. For example, consider a polynomial mapping of degree 2ν, where ν is prime. Then $n = 2\nu$ and so if $m \neq 1$, then $\omega = 2$ or ν. If $\omega = \nu$, then $m = \nu$ is prime, so the conjecture follows. Hence we may assume that $\omega = 2$ and $m = 2\mu$, where $1 < \mu < \nu$. Therefore ϕ has degree 2 and so has either two distinct simple zeros or one double zero; i.e. $\rho = 1$ or $\rho = 2$. If $\rho = 1$, then A has degree μ and B has degree ν and we have

$$\nu A'B - \mu AB' = 0. \tag{3.153}$$

A has a single zero $\neq 0$ and of order μ and B has the same zero of order ν. We have

$$p = \frac{j}{\mu} = \frac{k}{\nu} \tag{3.154}$$

for some j and k, from which it follows that p is an integer, since μ and ν are relatively prime. Also A contains $\mu + 1$ non-vanishing terms, i.e. the r_j powers appearing in A run through the integers from 0 to μ with

$$p = \frac{j}{\mu - r_j} \tag{3.155}$$

for suitable j lying between 1 and $2\mu - 1$. Hence $p < 2$ so $p = 1$. We now find that we are back in the identical situation which applied when we considered the case that m and n are relatively prime. The denominator power of x has been reduced by μ and the numerator powers of x do not exceed μ. Because the non-vanishing of the Jacobian implies that the mapping cannot be constant on a curve, we find as before that $\mu = 1$, i.e. $m = 2$ is prime and the conjecture follows. This leaves the case $\rho = 2$. Then A has degree 2μ and B has degree 2ν and

$$\nu A'B - \mu AB' = 0. \tag{3.156}$$

Hence $A = \psi^\mu$ and $B = c\psi^\nu$, where ψ has degree 2. If $\psi(0) \neq 0$, then

$$p = \frac{j}{2\mu} = \frac{k}{2\nu} \tag{3.157}$$

for some j $(1 \leq j \leq 2\mu - 1)$ and k $(1 \leq k \leq 2\nu - 1)$. Since μ and ν are relatively prime, this implies that $p = \frac{1}{2}$. Thus for the relevant powers r_j occurring in A we have

$$\frac{j}{2\mu - r_j} = \frac{1}{2}, \tag{3.158}$$

which gives $r_j = 2\mu - 2j$. Thus each r_j is even so A is a polynomial in y^2. It follows that ψ is a polynomial of degree 1 in y^2. As $p = \frac{1}{2}$, the denominators have been reduced to x^μ and x^ν and the numerator is a polynomial in $x^{1/2}$. The substitution $x \mapsto x^{1/2}$ will bring the denominators back to $x^{2\mu}$ and $x^{2\nu}$ and the numerators to polynomials in x of degrees $\leq 2\mu$ and 2ν respectively. As ψ has two simple zeros, we are back to the situation just considered above (though at the next stage). As there we find that $\mu = 1$ and so $m = 2$. It remains to consider the case that $\psi(0) = 0$. In this case A has a zero of order μ at the origin and a zero of order μ elsewhere. It follows that for some j, $r_j = \mu$ and so

$$p = \frac{j}{\mu} = \frac{k}{\nu}, \tag{3.159}$$

from which we deduce that $p = 1$. But

$$p < \frac{\mu + \nu - 2}{\mu + \nu - 1} < 1, \tag{3.160}$$

since $\omega = \rho = 2$. The contradiction shows that this case cannot occur. Thus the conjecture is proved for polynomials of degree 2ν, where ν is prime. Indeed, more generally, we have proved the conjecture for $\omega = 2$, i.e when m and n have greatest common divisor $= 2$. Note that in this case it turns out that $m = 2$, and so m divides n. These results are due to Nakai and Baba [36].

3.3.6 The closing gap problem

We see from the above arguments that as we proceed through the stages of the resolution process the leading terms A and B arising from the two components of the mapping satisfy the equation

$$\nu A'B - \mu AB' = c, \tag{3.161}$$

where c is a constant, which is zero until such time as the power of the leading term of the Jacobian matches the actual power corresponding to the Jacobian. This leads to some inequalities, which we now explain. Firstly, we will take the variables to be complex, so that the mapping P is a polynomial mapping from $\mathbb{C}^2$ to $\mathbb{C}^2$. The non-vanishing of the Jacobian of such a mapping implies, by the fundamental theorem of algebra, that the Jacobian is a non-zero constant. The resolution process generalises straightforwardly to this case, and is indeed simpler in that the leading terms of a component always have the full number of possible zeros according to their degree, and so all tracts constructed lead to asymptotic values for the component. Also, as the variables are complex, the separate consideration of $x > 0$ and $x < 0$ is irrelevant.

To keep the notations simple, we will maintain the transformations in fractional form in the x variable (i.e. we refrain from making the transformations of the form $x = s^q$). At the rth stage of the resolution process the power of x for the leading term of the Jacobian is

$$\sum_{i=0}^{r-1} p_i\left(r_0^{(i)} + s_0^{(i)}\right) - m - n - 1, \tag{3.162}$$

and the power of x for the Jacobian itself is

$$\sum_{i=0}^{r-1} p_i - 3, \tag{3.163}$$

where we take $p_0 = 0$; p_i is the p chosen by our minimum process at stage i (though in that formula the js will in general be fractional); the $r_0^{(i)}$ and the $s_0^{(i)}$ are the multiplicities for each of the components at stage i of the zero chosen. As long as the inequality

$$\sum_{i=0}^{r-1} p_i\left(r_0^{(i)} + s_0^{(i)}\right) - m - n - 1 < \sum_{i=0}^{r-1} p_i - 3 \tag{3.164}$$

holds, then the constant c is zero and leading terms satisfy our above equation. Therefore the tracts for the two components proceed in tandem. Indeed, writing the two leading terms as A_i and B_i, we have

$$A_i = \phi_i^{\mu} \text{ and } B_i = \phi_i^{\nu} \tag{3.165}$$

so that $r_0^{(i)} = \mu\rho_i$ and $s_0^{(i)} = \nu\rho_i$, where ρ_i is the multiplicity of the chosen zero of ϕ_i. The above inequality can therefore be written as

$$(\mu+\nu)\left(\sum_{i=0}^{r-1} p_i\rho_i - \omega\right) < \sum_{i=0}^{r-1} p_i - 2. \tag{3.166}$$

The conjecture is equivalent to the assertion that along any tract constructed through the resolution process the gap between these two quantities will close, i.e. the two quantities become equal. For it is only then that the constant c becomes non-zero and therefore the tracts for the two components separate.

Note that, as we reach the rth stage, the left expression increases by $(\mu+\nu)p_r\rho_r$ while the right side increases by p_r. Thus the gap is closing as we proceed through the stages. However, during the intermediate stages we have $\sum p_i\rho_i < \omega$, and so the left expression remains negative. Therefore for the possibility to remain open that the gap eventually closes, the right expression must remain negative, i.e. we must have $\sum p_i < 2$. We can sharpen this as follows. Choose the smallest natural number q such that each $p_i q$ is an integer; then $\omega q - \sum p_i q\rho_i \geq 1$, and hence the left expression is $\leq -(\mu+\nu)/q$. Therefore for the gap to be closable we require that

$$\sum_{i=0}^{r-1} p_i < 2 - \frac{\mu+\nu}{q}. \tag{3.167}$$

In particular this inequality must hold if the conjecture is true. This gives us some interesting information concerning the minimum size of q. Since each $p_i q$ is an integer, we obtain the inequality

$$q \geq \frac{1}{2}(\mu+\nu+r), \tag{3.168}$$

where r denotes the number of stages we have reached. In particular $q > 1$, even at the very first step. In other words, if we found an example where the first step gave p_1 an integer (e.g. $p_1 = 1$), then we would have a counter-example to the conjecture.

Now at stage r of the reduction process we pick out a zero of ϕ_r of multiplicity ρ_r and use the minimum formula to calculate p_r. There are two important quantities to consider. The first is

$$\alpha_r = \frac{\omega - \sum_{i=0}^{r-1} p_i\rho_i}{\rho_r}. \tag{3.169}$$

If $p_r < \alpha_r$, then we do not obtain asymptotic values and the process continues to stage $r+1$. The second quantity is

$$\beta_r = \frac{(\mu+\nu)\omega - 2 - \sum_{i=0}^{r-1} p_i((\mu+\nu)\rho_i - 1)}{(\mu+\nu)\rho_r - 1}. \tag{3.170}$$

If $p_r < \beta_r$, then at stage $r+1$ the gap between the two Jacobian powers remains positive and the tracts for the two components do not separate. If the conjecture is true, then at some stage r we obtain $p_r = \beta_r < \alpha_r$. Now

$$\alpha_r - \beta_r = \frac{2\rho_r - \omega + \sum_{i=0}^{r-1} p_i(\rho_i - \rho_r)}{\rho_r((\mu+\nu)\rho_r - 1)}. \tag{3.171}$$

Denoting by γ_r the numerator of this expression, at some stage we require $\gamma_r > 0$. On the other hand, if $\gamma_r > 0$, then we cannot obtain $p_r = \alpha_r$ (for the smaller choice $p_r = \beta_r$ will give the constant $c \neq 0$ in the differential equation for the components and this can only occur with a proper minimum choice of p); therefore, at the very least, the process will continue to the next stage. We note that

$$\gamma_{r+1} - \gamma_r = (\rho_{r+1} - \rho_r)\left(2 - \sum_{i=0}^{r} p_i\right). \tag{3.172}$$

Since $\rho_{r+1} \leq \rho_r$, we see that the sign of $\gamma_{r+1} - \gamma_r$ is the same as the sign of $\sum_{i=0}^{r} p_i - 2$, except when $\rho_{r+1} = \rho_r$ giving $\gamma_{r+1} = \gamma_r$.

It follows from our previous reasoning that, if the conjecture is true, then the sequence $\{\gamma_r\}$ is decreasing and so $\gamma_r \leq \gamma_1$. Since for some r we have $\gamma_r > 0$, we obtain $\gamma_1 > 0$, i.e. $2\rho_1 - \omega > 0$ or $\rho_1 > \frac{1}{2}\omega$. Thus every zero of ϕ_1 has multiplicity at least half the degree of ϕ_1. Clearly this is only possible if ϕ_1 has exactly one zero of multiplicity ω, i.e. $\rho_1 = \omega$.

We then obtain $\alpha_1 = 1$, $\beta_1 = (m+n-2)/(m+n-1)$ and $\gamma_1 = \omega$. Also, as we showed in subsection 3.3.2, at least one $r_j^{(0)} \neq 0$ and so p_1 has the form $p_1 = j/\omega$ for some integer $j < \omega$. Let us assume that j and ω are relatively prime (true, for example, if ω is prime). Then $\phi_2(t)$ is a polynomial of degree ω and is also a polynomial in t^ω (see chapter 1, subsection 1.3.1). Also ϕ_2 has at least one zero $\neq 0$. It follows that $\phi_2(t) = b(t^\omega + a)$, where a and b are non-zero constants. Thus at the second stage we get ω simple zeros and every $\rho_2 = 1$. This gives

$$\gamma_2 = \omega - (\omega - 1)(2 - p_1), \tag{3.173}$$

and (assuming $\omega > 1$) we easily calculate that this expression is >0, only if

$j = \omega - 1$, giving

$$\alpha_2 = 1, \qquad \beta_2 = \frac{\mu + \nu - 1 - 1/\omega}{\mu + \nu - 1}, \qquad \gamma_2 = 1/\omega. \tag{3.174}$$

Now, if the constant is 0 at stages $3, \ldots,$ then $\rho_r = 1$ and so $\gamma_r = \gamma_2$ for $r \geq 3$. However, $\rho_2 = \rho_r$ implies that p_{r-1} is an integer multiple of $1/\omega$ (as the powers of the x variable are multiples of $1/\omega$). Since the sum of the multiples of $1/\omega$ cannot exceed 1, at some stage r the constant becomes $\neq 0$ and we obtain

$$p_r = \beta_r = \frac{(\omega - \lambda)(\mu + \nu - 1) - 1}{\omega(\mu + \nu - 1)} \tag{3.175}$$

for some integer λ satisfying $1 \leq \lambda < \omega$. However, this implies that at least one of the components at this stage is a polynomial in $t^{\mu+\nu-1}$. As the two components have respective degrees μ and ν, this is possible only if $\mu = 1$. Therefore we have proved a particular case (the case ω is a prime) of a theorem of Wright [73] that *if* $P\colon \mathbb{C}^2 \to \mathbb{C}^2$ *is a homeomorphic polynomial mapping with components* U *of degree* m *and* V *of degree* n*, where* $m < n$*, then* m *divides* n. Because of the degree reduction process outlined in subsection 3.3.4, this is equivalent to the result that *all homeomorphic polynomial mappings of* $\mathbb{C}^2$ *are formed by a finite number of compositions of affine mappings and elementary mappings.*

3.3.7 The case $\omega = 3$

The results of the previous subsection bring into the foreground the stringent conditions required by the conjecture: there can be only one zero of the initial leading term and m divides n. Thus almost all conceivable configurations would be counter-examples, if they could be shown to occur. If the conjecture is true, then number-theoretic properties seem to play a large role. However, even in relatively simple situations, it is difficult to find a decisive algebraic argument, as the following case $\omega = 3$ shows.

Assume we have a polynomial counter-example with $\omega = 3$. Using the notations of the previous subsection, we have

$$\sum p_i \rho_i = \omega = 3, \qquad \sum p_i \geq 3, \tag{3.176}$$

where we assume a full tract to an asymptotic value curve. This gives

$$\sum p_i(\rho_i - 1) \leq 0, \tag{3.177}$$

which implies that all $\rho_i = 1$ and $\sum p_i = 3$. Since all the ρ_i are equal, every

p_i is an integer. We obtain an asymptotic tract of the form

$$y = c_1 x + c_2 + \frac{c_3}{x} + \frac{t}{x^2}. \tag{3.178}$$

If we make an affine transformation $(x, y) \mapsto (x, y - c_1 x - c_2)$, then the Jacobian is unchanged and the tract takes the above form with $c_1 = c_2 = 0$. We then see that $c_3 \neq 0$, since otherwise the mapping would be constant on a curve. With another affine transformation $(x, y) \mapsto (c_3 x, y)$, we reduce to a tract with $c_3 = 1$. Thus without loss of generality, we may take the asymptotic tract to have the form

$$y = \frac{1}{x} + \frac{t}{x^2}. \tag{3.179}$$

This is an invertible tract and the mapping P is transformed to

$$A(s, t) = P\left(\tfrac{1}{s}, s + ts^2\right), \tag{3.180}$$

which is a polynomial mapping in (s, t), whose Jacobian is also a non-zero constant. As explained in subsection 3.2.4, this implies that the mappings P and A can be written in the form

$$P(x, y) = F(y, xy, x(xy - 1)), \qquad A(s, t) = F(s + ts^2, 1 + st, t), \tag{3.181}$$

where $F(\alpha, \beta, \gamma)$ is a polynomial mapping from $\mathbb{C}^3 \to \mathbb{C}^2$. Notice that both mappings are represented by $F(\alpha, \beta, \gamma)$ confined to the complex algebraic space $\alpha\gamma = \beta^2 - \beta$. Writing $F = (U, V)$ and denoting partial derivatives by subscripts, we can calculate the Jacobian of P obtaining

$$\begin{aligned} J(P) &= y(U_\beta V_\alpha - U_\alpha V_\beta) + (x^2 y - x)(U_\gamma V_\beta - U_\beta V_\gamma) \\ &\quad + (2xy - 1)(U_\gamma V_\alpha - U_\alpha V_\gamma) \\ &= \alpha(U_\beta V_\alpha - U_\alpha V_\beta) + \gamma(U_\gamma V_\beta - U_\beta V_\gamma) + (2\beta - 1)(U_\gamma V_\alpha - U_\alpha V_\gamma). \end{aligned} \tag{3.182}$$

Expressed in terms of α, β and γ we obtain the same expression for $-J(A)$. Consider now the mapping $H = (U, V, W)\colon \mathbb{C}^3 \to \mathbb{C}^3$, where $W = \beta^2 - \beta - \alpha\gamma$. We have $W_\alpha = -\gamma$, $W_\beta = 2\beta - 1$ and $W_\gamma = -\alpha$. Calculating the Jacobian determinant of H gives $J(H) = J(P)$. Thus we have the result that the $\mathbb{C}^3$ mapping H has constant non-vanishing Jacobian on the space $\Omega = \{W = 0\}$.

H is therefore locally 1–1 in the neighbourhood of Ω and maps Ω into a space homeomorphic to $\mathbb{C}^2$($\mathbb{C}^3$ with the third coordinate 0). Now the mapping $(x, y) \mapsto (y, xy, x(xy - 1))$ is easily seen to be a 1–1 mapping from $\mathbb{C}^2$ into Ω; the image is therefore a subspace of Ω homeomorphic to $\mathbb{C}^2$. However, the image excludes the points $(0, 1, \gamma) \in \Omega$, so the mapping is not surjective. Nevertheless, any point of Ω off this complex line is attained: given $(\alpha, \beta, \gamma) \in \Omega$ define $y = \alpha$ and $x = \frac{\beta}{\alpha}$ for $\alpha \neq 0$ or $x = \frac{\gamma}{\beta - 1}$ for $\beta \neq 1$. Thus we obtain a value for x unless both $\alpha = 0$ and $\beta = 1$. By considering both the mappings $P(x, y)$ and $A(s, t)$, we see that Ω is a two-dimensional complex manifold embedded in $\mathbb{C}^3$ (and restricting the variables to being real Ω_R is a surface embedded in $\mathbb{R}^3$). Ω is a space homeomorphic to $\mathbb{C}^2+$ a complex line (a plane). Similarly Ω_R is $\mathbb{R}^2 +$ a line. Indeed Ω_R is a quadric surface, whose overall shape can be seen as follows. We change coordinates by the transformations $X = \alpha - \gamma$, $Y = 2\beta - 1$, $Z = \alpha + \gamma$. Then $X^2 + Y^2 - Z^2 = -4\alpha\gamma + 4\beta^2 - 4\beta + 1 = 1$ on Ω_R. Thus the plane $Z = c$ cuts the transformed surface in the circle with radius $\sqrt{1 + c^2}$ and centre $(0, 0, c)$. The transformed surface is the union of these circles (as c varies over $\mathbb{R}$), of which the smallest has radius 1. The conical figure which we get is homeomorphic to an infinite cylinder, as can be seen from the transformation

$$F(X, Y, Z) = \left(\frac{X}{\sqrt{1 + Z^2}}, \frac{Y}{\sqrt{1 + Z^2}}, Z \right). \tag{3.183}$$

The cylindrical surface is not homeomorphic to $\mathbb{R}^2$, since none of the generating circles can be shrunk to a point on the surface. According to our hypothesis, H is a locally 1–1 mapping of Ω into $\mathbb{C}^2$. If P had no other asymptotic values, then the mapping H would be surjective, as the inverse function could be continued along all paths in $\mathbb{C}^2$. The monodromy theorem would imply that the mapping H was a homeomorphism of Ω onto $\mathbb{C}^2$, which is impossible, as the two spaces are not homeomorphic. If there are other asymptotic values, then these correspond to at most two extra lines 'at infinity'. The lines do not interfere with one another, as they correspond to distinct tracts, and otherwise have the same local properties as above. We can then continue the inverse function over the whole of $\mathbb{C}^2$, deducing that the original mapping is a homeomorphism of the extended space. In particular the mapping P is 1–1 on $\mathbb{C}^2$, which means there are no asymptotic values. These contradictions establish the conjecture in the case $\omega = 3$.

3.3.8 Problem

1. The following construction provides an example of a simply-connected surface, for which the repetition property is satisfied. Let L_1 and L_2 be two straight lines each parallel to the imaginary axis with L_1 to the right of L_2, which are to be described as follows. Proceed along L_1 from the top to a point b, passing a point a on the way. Reverse direction at b and proceed to a; reverse direction at a and proceed to the bottom of L_1. Next, L_2 is described in an identical manner with two points c and d. Third, L_1 is described in

the reverse manner: start at the bottom, proceed to a, reverse direction and proceed to b, reverse direction and proceed to the top. Fourth, L_2 is described in the same manner as at step 2. Show that with this boundary behaviour we can obtain a 4-valent locally 1–1 mapping of the plane, and calculate for each point w how often it is attained as the image of a finite point.

3.4 A topological approach

3.4.1

The previous arguments deal only with algebraically simple cases; they indicate the potential for considerable algebraic complexity in more general cases. Furthermore, in the case $\omega = 3$ we made direct appeal to topological considerations. Algebraic difficulties can sometimes be resolved by adopting a geometric or topological point of view. We have already seen that, although the real Jacobian conjecture is not true, nevertheless there are severe topological restrictions on the possible counter-examples. Assuming the Jacobian is constant does not give one any obvious additional geometric information in the plane. However, this condition does imply that the corresponding $\mathbb{C}^2$ mapping, obtained by allowing the variables to be complex, is locally 1–1, and conversely a locally 1–1 polynomial mapping from $\mathbb{C}^2$ to $\mathbb{C}^2$ necessarily has a constant non-vanishing Jacobian. Thus it is entirely natural to ask whether the topological restrictions imposed by the polynomial nature of the mappings force a locally 1–1 mapping to be globally 1–1; alternatively, if this is not the case, can we nevertheless so restrict the topological situation as to begin the process of constructing a counter-example. We saw that this was indeed possible in the real case, when Pinchuk's example began to emerge from a combination of geometric and algebraic considerations. Geometric examples constructed by Vitushkin [67] and more recently Orevkov [37] indicate that a purely topological resolution of the conjecture is not possible. Therefore we should take seriously Pinchuk's suggestion [38] that a counter-example be sought.

3.4.2

Assume that we have a locally 1–1 polynomial mapping $P : \mathbb{C}^2 \to \mathbb{C}^2$ and that P is not a global homeomorphism and so possesses asymptotic values at ∞. As the algebraic construction of asymptotic values is essentially identical to the method given in chapter 1, we find that the set of asymptotic values is a finite union of polynomial images of $\mathbb{C}$ (often called **complex curves**), which have the form $\{(A(t), B(t)) : t \in \mathbb{C}\}$, where $A(t)$ and $B(t)$ are polynomials, not both constant.

Let $\mathcal{A}$ denote this finite union of complex curves and $\mathcal{A}^c$ its complement.

3.4.3

Lemma *The set $\mathcal{A}$ is a closed set which is **non-separating** in $\mathbb{C}^2$*; by this we mean that, for every domain $W \subset \mathbb{C}^2$, the set $W - \mathcal{A}$ is a domain.

Proof $\mathcal{A}$ is clearly closed, and so $W - \mathcal{A}$ is open. W contains a neighbourhood of the form $C \times D$, where C and D are open discs of $\mathbb{C}$ and $\times$ denotes the cartesian product. Sets of this form are a basis for the topology of $\mathbb{C}^2$. If $c \in C$ and $d \in D$, then (c, d) lies in the plane $\{(u, v): u = c\}$. This plane either lies entirely in $\mathcal{A}$ or intersects $\mathcal{A}$ in a finite set of points: for among the polynomial pairs $(A(t), B(t))$, either $A(t) = c$ identically or the equation $A(t) = c$ has a finite number of solutions t. Each such solution gives just one value $B(t)$. The number of complex numbers c for which one of the $A(t) = c$ identically is finite; therefore picking a c for which this is not the case, the plane $u = c$ contains points $(c, v) \notin \mathcal{A}$ but in $C \times D$. Thus $W - \mathcal{A}$ is not empty.

To establish the connectedness, consider first a set W of the form $W = C \times D$, where C and D are open discs, and choose $(c, d) \in W - \mathcal{A}$. Consider the plane $\Pi_a = \{(u, v): v = a(u - c) + d\}$; this meets $\mathcal{A}$ in a finite set of points and therefore the intersection with $W - \mathcal{A}$ is connected via (c, d). A similar remark applies to the plane $\Pi_\infty = \{u = c\}$. Thus $\bigcup_a (\Pi_a \cap (W - \mathcal{A}))$ is a union of connected sets containing a common point, and so is connected. However, this set is $W - \mathcal{A}$, for if $(p, q) \in W - \mathcal{A}$, then with the choice $a = (q - d)/(p - c)$ (which includes ∞ if $p = c$), we get $(p, q) \in \Pi_a$. Thus $W - \mathcal{A}$ is connected.

For the general case, let W be a domain and let γ denote a curve in W joining two points in $W - \mathcal{A}$. Then γ can be covered with a finite number n of basis sets $W_k = C_k \times D_k$ $(1 \le k \le n)$ and from the above each $W_k - \mathcal{A}$ is connected. We may assume that $W_j \cap W_{j+1} \neq \emptyset$ for $1 \le j \le n - 1$ and observe that $(W_j - \mathcal{A}) \cup (W_{j+1} - \mathcal{A})$ is connected, as $(W_j \cap W_{j+1}) - \mathcal{A} \neq \emptyset$. The union of the $W_k - \mathcal{A}$ is connected, since it can be expressed in the form $\bigcup_k Q_k$, where $Q_k = \bigcup_{j=1}^{k} (W_j - \mathcal{A})$, which is a union of connected sets containing a common point. It follows that the endpoints of γ can be joined in $W - \mathcal{A}$ and the proof is complete.

3.4.4

We note next that the mapping P is finitely valent on $\mathbb{C}^2$; for, if $P = (u, v)$ and $w_1 = (u_1, v_1)$ is a given point, then $u - u_1$ and $v - v_1$ are relatively prime polynomials, as any common factor of this pair must also be a common factor of the Jacobian $J(P)$, which by hypothesis is a non-zero constant. It follows from Beźout's theorem that the equation $P = w_1$ has at most mn solutions, where m is the degree of u and n is the degree of v.

Let $\mathcal{B} = P(\mathbb{C}^2)$ be the set of attained values of the mapping. As P is an open mapping, $\mathcal{B}$ is a domain and so $\mathcal{B} - \mathcal{A}$ is a domain. All values in $\mathcal{B} - \mathcal{A}$ are attained equally often. To see this choose w_0 and w_1 in $\mathcal{B} - \mathcal{A}$ and suppose that the equation $P = w_0$ has ν distinct solutions. Let γ denote a path joining w_0 and w_1 in $\mathcal{B} - \mathcal{A}$. Each of the ν branches of the inverse function at w_0 can be continued along γ giving ν branches at w_1. These latter branches are necessarily distinct, for if two coincided, we would obtain just one branch at w_0 by continuing back along γ. Thus the equation $P = w_1$ also has ν distinct solutions in $\mathcal{B} - \mathcal{A}$.

The set $\mathcal{B} \supset \mathcal{A}^c$, the complement of $\mathcal{A}$; for suppose that $w_1 \in \mathcal{A}^c$; since $\mathcal{B} - \mathcal{A}$ is non-empty, $\exists w_0 \in \mathcal{B} - \mathcal{A}$; both w_0 and w_1 lie in the domain $\mathcal{A}^c$ and so can be joined by a path $\gamma \subset \mathcal{A}^c$; since $w_0 \in \mathcal{B}$, a branch of the inverse function can be continued along γ and therefore $w_1 \in \mathcal{B}$. Thus we see that

$$\mathcal{B} - \mathcal{A} = \mathcal{A}^c. \tag{3.184}$$

Also $\mathcal{A}^c$ is dense in $\mathbb{C}^2$; for if $w \in \mathcal{A}$ and W is a neighbourhood of w, then as $W - \mathcal{A}$ is non-empty, it contains points of $\mathcal{A}^c$, so w lies in the closure of $\mathcal{A}^c$. Thus $\mathcal{A}^c \cup \mathcal{A} = \overline{\mathcal{A}^c}$.

If ν as above is the cardinality of the covering of $\mathcal{A}^c$, then no value $w \in \mathcal{B}$ can be attained more than ν times. For, if w is attained μ times, then the inverse image of a small neighbourhood W of w will give μ disjoint neighbourhoods of points in $\mathbb{C}^2$ mapping to W and, since $W - \mathcal{A}$ is non-empty, we can find a value in each of the μ neighbourhoods which maps to the same point in $W - \mathcal{A}$. Thus $\mu \leq \nu$. In particular, we see that, if P is not 1–1, then P attains values in $\mathcal{A}^c$ more than once. We can then perform the following construction: Let p and q be distinct points such that $P(p) = P(q) = w \in \mathcal{A}^c$. Since P is locally 1–1, we may cover the line segment $[p, q]$ with a finite number of open neighbourhoods in each of which P is 1–1. P maps the segment onto a loop with endpoint w and this loop is covered by the images under P of these open neighbourhoods, which are themselves open. In each such image neighbourhood W_k we have defined a branch of the inverse function mapping back to the neighbourhood of $[p, q]$. Each set $W_k - \mathcal{A}$ is open and connected and by traversing each $W_k - \mathcal{A}$ in turn, we can construct a simple loop $\Gamma \subset \mathcal{A}^c$ with endpoint w, where the continuation of the inverse function maps to a simple curve in the domain space joining p to q. Now this inverse function can be continued indefinitely around Γ and must therefore return to the initial branch after a finite number μ of circuits ($\mu \leq \nu$). In this way we obtain a simple loop γ with endpoint p which is mapped by P onto Γ described μ times.

We now employ a theorem of knot theory, which states that there are no knotted loops in $\mathbb{R}^4$ and higher dimensions; in other words, given a simple loop $\gamma \subset \mathbb{R}^4$ there is a homeomorphism of the space which maps γ onto a circle. Since a circle lies in a plane which contains a disc, whose boundary is the circle, the loop γ lies on a planar surface and bounds a topological disc on this surface. P maps this region onto some kind of pseudo-surface in $\mathbb{R}^4$ – in fact, a finite number of pieces of surface stuck together. We will show that the image under P of the closed topological disc cannot be an orientable surface with or without boundary, if $\mu \geq 2$. This is a consequence of the following result, which generalises the topological arguments of subsection 3.1.8.

3.4.5

The loop lemma *Let f be a locally 1–1 and continuous mapping of $\mathbb{R}^2$ into an orientable surface Σ and let γ be a simple loop in $\mathbb{R}^2$. Suppose that f maps γ into a simple loop $\Gamma \subset \Sigma$, i.e. $f(\gamma) \subset \Gamma$. Then f is a homeomorphism of γ onto Γ.*

Proof Without loss of generality we may assume that γ is the unit circle C bounding the open unit disc D. We will show that f is an injective mapping of D mapping C injectively onto Γ; f is finitely valent on $\overline{D} = D \cup C$; for f is injective in a neighbourhood of each point of $\overline{D}$ and therefore we can cover $\overline{D}$ with a finite number, say N, of such neighbourhoods. Then in $\overline{D}$, f can attain no value more than N times.

Suppose that a point of D is mapped by f onto Γ; then the branch of the inverse function ϕ corresponding to this point can be continued around Γ. The values attained will lie in D, as otherwise the resulting curve would travel out to C and at the meeting point with C we would contradict the local injectivity of f. Thus the branch can be continued around Γ a finite number of times until returning to the initial branch. This gives a simple loop in D which is mapped by f onto Γ described a finite number of times. The number of such loops in D is finite and all such loops are disjoint. Let Δ denote the (in general) multiply-connected domain bounded by C and these loops. Thus we are ignoring those loops which lie inside others. $f(\Delta) = S$ is an orientable surface. Furthermore, our construction of Δ ensures that no point of Γ lies in S. Now either (a) $\overline{S} = S \cup \Gamma$ is a compact bordered surface with the one border contour Γ or (b) S lies on either side of Γ and so $\overline{S}$ is a compact surface. These assertions are easily verified. Our terminology and results for surface topology are taken from Ahlfors and Sario [3], chapter 1.

Assume (a). The finite valence implies the finite character of $\overline{S}$, which is therefore a polyhedron. According to the classical classification theorem, such a surface is determined topologically by the characteristic, which is determined by the homology group. Thus topologically $\overline{S}$ is a hemispherical surface with handles, the hemisphere being bordered by Γ. We employ a classical device to construct a simply-connected domain contained in Δ by cutting Δ along a finite number of simple disjoint arcs α_k joining a fixed point ζ of C to each of the simple interior loops: one curve for each loop l_k, which we take as being described with negative orientation. Also we choose the endpoint of α_k on l_k to be a point whose image under f is the same point on Γ as $f(\zeta)$. The boundary of this domain is a curve Λ described in the form $\alpha_1 l_1 \alpha_1^{-1} \ldots \alpha_m l_m \alpha_m^{-1} C$, where C has positive orientation. Because the domain bounded by Λ can be mapped homeomorphically onto a rectangle, we see that the mapping f provides a homotopy through $\overline{S}$ of $f(\Lambda)$ to the constant element. However, $f(\Lambda)$ has the form $\beta_1(\Gamma)^{\nu_1}\beta_1^{-1} \ldots \beta_m(\Gamma)^{\nu_m}\beta_m^{-1}(\Gamma)^{\mu}$, where the β_k are loops and the ν_k and μ are integers. These integers all have the same sign: for as f is locally 1–1, the orientation is preserved as Λ is described; on the other hand the points of S lie only on one side of the border Γ, and so each time we meet Γ in traversing Λ, Γ is traversed in the same direction. The element $f(\Lambda)$ reduces to the identity element in the fundamental group of $\overline{S}$. If we abelianise it we obtain $\nu\Gamma$ as the 0 element in the homology group for a non-zero integer ν. Since $\overline{S}$ has no torsion elements and therefore no elements of finite order, it follows that Γ is homologous to zero. Now, if the homology group is non-trivial, each border contour is a basis element for the group, and a basis element cannot vanish. It follows that the homology group is trivial; in other words the hemispherical surface has no handles, and is therefore topologically a closed disc. From the construction of S, it follows that the appropriate branch of the inverse function can be continued along all paths in S, since the branch can only map into Δ. Since we have now shown that S is simply-connected, we deduce from the monodromy theorem that the branch of the inverse function is single-valued and injective in S. From this it follows that there are no inner loops and $\Delta = D$; f is 1–1 in D and the curve Γ is described exactly once.

Next we eliminate case (b). If we have an inner loop in D (a boundary loop of Δ), then we obtain another possibly multiply-connected domain of the same type as Δ lying inside this loop and being mapped by f onto S. If we continue this process through D we eventually reach an innermost loop mapping onto Γ with simply-connected interior mapping onto S. Thus topologically we have a closed disc being mapped by a locally 1–1 function f onto the compact surface $\overline{S}$ with the open disc U mapping onto S. Furthermore, the points of Γ do not lie in S. Therefore the inverse function can be continued along all paths in S. Thus f maps $\partial U \nu$ times onto Γ and therefore provides a homotopy of Γ^{ν} to the trivial element of $\overline{S}$. Γ is thus homologous to zero in the homology group of $\overline{S}$. Now assume that ω is a non-trivial element in the homology group of S. Then continuing the inverse function around the loop ω a finite number μ times, we obtain a loop in U. Again f provides a homotopy of ω^{μ} to the trivial element in the fundamental group of S. But this implies that ω is trivial in the homology group of S, which is a contradiction. Hence the homology group of S is trivial, and so S is simply-connected. From the monodromy theorem f is a homeomorphism of U onto S, and so $\overline{S}$ is topologically a closed disc and therefore is not a compact surface. The orientability of $\overline{S}$ is important here: the non-orientable cross-cap surface (equivalent to the projective plane) is precisely an open hemisphere + the bounding circle described twice.

3.4.6

We thus see that our counter-example P will map any topological disc in $\mathbb{R}^4$ bounded by the loop γ onto a pseudo-surface, which has **intrinsic** self-intersections; that is, small movements of the topological disc (keeping γ fixed) will not avoid the self-intersections of the image. The self-intersections will be linked to the fixed curve Γ, so that any point on Γ will border several pieces of surface pointing into the space in different directions, and it is not possible, by manipulating the topological disc in the domain space, to flatten the structure on Γ so that we obtain an orientable surface.

For example, suppose that we took a topological disc D bounded by Γ and considered the component of the inverse image $P^{-1}(D)$ which contains γ. We would obtain an orientable surface Σ in the domain space, which is necessarily unbounded; for otherwise, the inverse function could be continued along all paths in D, and so as D is simply-connected, the inverse function would be single-valued in D, the mapping would be 1–1 and Γ would be described once. Furthermore, Σ cannot contain any topological disc bounded by γ, though of course it contains γ itself.

Thus our counter-example leads to the strange situation that, although Γ, described a finite number of times, can be shrunk to a point through the pseudo-surface, which lies in $\mathcal{B}$, Γ itself cannot be shrunk to a point in $\mathcal{A}^c$. This phenomenon occurs in $\mathbb{R}^4$. Let us consider an analogous situation in the visualisable space $\mathbb{R}^3$. Assume we have a locally 1–1, finitely valent mapping $f: \mathbb{R}^3 \to \mathbb{R}^3$, whose set of asymptotic values is the real axis $\{(u, v, w): v = 0, w = 0\}$. Let Γ denote the circle $\{u = 0, v^2 + w^2 = 1\}$, which encircles the real axis. Any component of the inverse image of Γ is a simple loop γ in the domain space, which is mapped by f onto Γ described a finite number of times. In $\mathbb{R}^3$ γ may be knotted, which is a complication not occurring in higher dimensions. If, for simplicity, we assume it is not, then we have the same situation as above; γ bounds a topological disc, which maps onto a pseudo-surface containing the circle Γ. Thus Γ described finitely often can be shrunk to a point through this pseudo-surface, and this cannot be done without passing through the real axis (seemingly at least as many times as the number of circuits). Thus some of the asymptotic values are also attained values. However, Γ described once cannot be shrunk to a point without passing through genuine asymptotic values. The author has been unable to construct a geometric example of this strange phenomenon, and wonders whether it is possible in $\mathbb{R}^3$. Put simply, is the loop lemma true, if the image space of the mapping is $\mathbb{R}^3$?

3.5 The resultant and the Jacobian

We conclude the chapter with some remarks related to the question of actually solving the equations $P_i = w_i$ and indicating the possibility of a purely algebraic resolution of the conjecture. We will confine our remarks to the $\mathbb{R}^2$ case.

3.5.1

Consider a polynomial mapping of $\mathbb{R}^2$ expressed in complex form $P(\overline{z}, z)$. To solve the equation $P(\overline{z}, z) = w$, we may use the Sylvester method of elimination by calculating the resultant of the pair of polynomials $P(\overline{z}, z) - w$ and $\overline{P(\overline{z}, z)} - \overline{w}$ obtained by eliminating the variable $\overline{z}$. This gives us a polynomial of the form $D(z, \overline{w}, w)$, whose zeros in z will contain all roots of the equation. If P has degree n, it is easily shown that, as a real analytic polynomial in w, D has degree at most n. Suppose now that P is polynomially invertible, so that we have an inverse polynomial $\phi(\overline{w}, w)$. Then $z - \phi$ is clearly a factor of D. Using this, it is easily shown, using an induction argument, that ϕ has exact degree n. Therefore D has no other factor involving w. On the basis that there is no pure factor in z, this would mean that D had the form $D = c(z - \phi)$, where c is a non-zero constant. From this we can easily calculate ϕ, i.e. we can solve the equation. Furthermore, we see that D has degree 1 in the variable z. On the other hand, if D has degree 1 in z, then the equation has a unique solution.

Thus we may rephrase the conjecture as: *P has constant non-vanishing Jacobian, if, and only if, the resultant D has exact degree* 1. This is an algebraic statement, which suggests an underlying general relationship between the Jacobian and the resultant.

3.5.2 Problem

1. Prove that the Jacobian of P vanishes identically if, and only if, the resultant D is independent of the variable z.

4

Analytic and harmonic functions in the unit disc

4.1 Series representations

4.1.1

A function analytic in the neighbourhood of a point z_0 has a unique representation as a convergent power series about z_0. Unless the function is entire, the power series will have a finite radius of convergence R. After the transformation $z = z_0 + R\zeta$, we obtain a power series in ζ convergent in the unit disc. Thus the study of power series converging in the disc is equivalent to the local study of general analytic functions. On the other hand, power series in the unit disc have many quite specific properties which can be exploited to draw out remarkable and deep interrelationships.

Consider a function

$$f(z) = \sum_{n=0}^{\infty} a_n z^n \tag{4.1}$$

analytic in $\mathbb{U} = \{|z| < 1\}$. Let $\mathbb{T} = \{|z| = 1\}$ and consider the series on $\mathbb{T}$, with $z = e^{i\theta}$,

$$\sum_{n=0}^{\infty} a_n e^{in\theta}. \tag{4.2}$$

This, of course, may be far from convergent. However, we recognise it as a 'right-sided' trigonometric series. In complex form the general trigonometric series takes the form

$$\sum_{n=-\infty}^{\infty} a_n e^{in\theta} \tag{4.3}$$

and so from an analytic function we obtain such a series with coefficients $a_n = 0$

for $n < 0$. The analogue of this series in $\mathbb{U}$ is the series

$$\sum_{n=-\infty}^{\infty} a_n r^{|n|} e^{in\theta} \tag{4.4}$$

where $z = re^{i\theta}$. In other words we interpret $e^{-in\theta}$ as $\overline{e^{in\theta}}$, obtaining for the left-sided part a power series in $\overline{z}$. The expression (4.4) is the general form of the series representation of a harmonic function in $\mathbb{U}$. Note that we do not choose the more obvious analogue of (4.3) as the Laurent-type expansion

$$\sum_{n=-\infty}^{\infty} a_n z^n \tag{4.5}$$

as this will only converge in $\mathbb{U}$ if the left-sided terms vanish. The Laurent expansion is relevant to functions analytic in an annulus containing $\mathbb{T}$.

Even if one's interest is mainly in analytic functions, it is nevertheless helpful to bear in mind the harmonic point of view. There are a number of properties of analytic functions which are quite mysterious in the analytic context, but which become readily understandable if one thinks in terms of two-sided series. In particular we shall show how some general results on polynomials, which would otherwise be quite difficult to prove, can be established by these means.

In chapter 2 our point of view was topological and centred on zeros. In this chapter we develop what might be termed the 'additive' point of view. Analytic polynomials are finite power series and trigonometric polynomials are the restriction to the unit circle of finitely represented harmonic functions. (Also of a particular type of rational function.) Additive or linear ideas together with the finiteness can be exploited in a number of ways. In chapter 5 we will combine both methods to develop the 'multiplicative' theory.

4.1.2 Convolutions

Let

$$f(re^{i\theta}) = \sum_{n=-\infty}^{\infty} a_n r^{|n|} e^{in\theta}, \qquad g(re^{i\theta}) = \sum_{n=-\infty}^{\infty} b_n r^{|n|} e^{in\theta} \tag{4.6}$$

be functions harmonic in $\mathbb{U}$. We define the **convolution** of the two functions by the formula

$$(f * g)(re^{i\theta}) = \sum_{n=-\infty}^{\infty} a_n b_n r^{|n|} e^{in\theta}. \tag{4.7}$$

This series is absolutely convergent in $\mathbb{U}$ and we have the following formula.

$$(f * g)(re^{i\theta}) = \frac{1}{2\pi}\int_0^{2\pi} f\left(\frac{r}{R}e^{i(\theta-t)}\right)g(Re^{it})dt, \tag{4.8}$$

valid for $0 \leq r < R < 1$ and θ real. If g is harmonic in $\{|z| \leq 1\}$ (so the series for g has radius of convergence > 1), then the formula is valid when $R = 1$. In this case no additional assumption on f is required and the convolution series (4.7) is absolutely convergent in $\mathbb{U}$.

The proof of this is obtained by verifying directly that the integral on the right in (4.8) evaluates to the sum on the right in (4.7). We obtain for the integral

$$\begin{aligned}&\frac{1}{2\pi}\int_0^{2\pi}\sum_{m=-\infty}^{\infty} a_m\left(\frac{r}{R}\right)^{|m|} e^{im(\theta-t)}\sum_{n=-\infty}^{\infty} b_n R^{|n|}e^{int}dt\\ &= \sum_{m=-\infty}^{\infty}\sum_{n=-\infty}^{\infty} a_m\left(\frac{r}{R}\right)^{|m|} e^{im\theta}b_n R^{|n|}\frac{1}{2\pi}\int_0^{2\pi} e^{i(n-m)t}dt\end{aligned} \tag{4.9}$$

where we have re-arranged the sums and applied term-by-term integration, making use of the fact that the series are absolutely and uniformly convergent. Since

$$\frac{1}{2\pi}\int_0^{2\pi} e^{i(n-m)t}dt = \begin{cases}0 & \text{if } m \neq n,\\ 1 & \text{if } m = n,\end{cases} \tag{4.10}$$

the final expression in (4.9) reduces to the sum in (4.7).

As an immediate corollary we deduce

$$M(r, f * g) \leq M\left(\frac{r}{R}, f\right)I(R, g) \quad (0 < r < R < 1) \tag{4.11}$$

where

$$M(\rho, F) = \max_{|z|=\rho}|F(z)|, \qquad I(\rho, F) = \frac{1}{2\pi}\int_0^{2\pi}|F(\rho e^{it})|dt. \tag{4.12}$$

Let $\mathcal{H}$ denote the space of functions harmonic in $\mathbb{U}$ and $\mathcal{A}$ the space of functions analytic in $\mathbb{U}$. Together with normal addition and scalar multiplication of functions, the convolution operation gives each of $\mathcal{H}$ and $\mathcal{A}$ the structure of a complex linear algebra. Each of the two spaces has its own unit element. From the definition of convolution it is clear that these are the functions within the spaces whose coefficients are all equal to 1. Thus for $\mathcal{H}$ we obtain the unit element as

$$\mathcal{P}(re^{i\theta}) = \sum_{n=-\infty}^{\infty} r^{|n|}e^{in\theta} = \frac{1-r^2}{|1-re^{i\theta}|^2} = \Re\left(\frac{1+re^{i\theta}}{1-re^{i\theta}}\right). \tag{4.13}$$

Applying the convolution formula (4.8) we obtain the important **Poisson's formula**

$$f(re^{i\theta}) = (\mathcal{P} * f)(re^{i\theta}) = \frac{1}{2\pi}\int_0^{2\pi} \frac{R^2 - r^2}{|Re^{it} - re^{i\theta}|^2} f(Re^{it})dt \quad (4.14)$$

valid for any $f \in \mathcal{H}$ and R $(r < R < 1)$. We may take $R = 1$, if f remains harmonic on $\mathbb{T}$. Similarly for $\mathcal{A}$ the unit element is given by

$$\sum_{n=0}^{\infty} z^n = \frac{1}{1-z} \quad (4.15)$$

and we obtain for every function $f \in \mathcal{A}$

$$f(z) = \frac{1}{2\pi}\int_0^{2\pi} \frac{Re^{it}}{Re^{it} - z} f(Re^{it})dt = \frac{1}{2\pi i}\int_{|\zeta|=R} \frac{f(\zeta)}{\zeta - z} d\zeta \quad (4.16)$$

which is, of course, Cauchy's integral formula for f. An important consequence of these representations is the following. If $\{f_n(z)\}$ is a sequence of functions in $\mathcal{H}(\mathcal{A})$ converging locally uniformly in $\mathbb{U}$ to $f(z)$, then $f \in \mathcal{H}(\mathcal{A})$. Thus local uniform convergence provides us with a concept of convergence for elements of $\mathcal{H}(\mathcal{A})$, and so with a topology on these spaces, called the **compact open topology**. Henceforth we shall assume that the two spaces are endowed with this topology. There is a fundamental convergence theorem in this topology, which is expressed in terms of the concept of a **normal family**. A family $\mathcal{F}$ of functions in $\mathcal{H}$ is said to be **normal** if every sequence of functions in $\mathcal{F}$ possesses a convergent subsequence (the Bolzano–Weierstrass property). The theorem states that *$\mathcal{F}$ is normal if, and only if, $\mathcal{F}$ is* ***locally bounded****, i.e. for each compact subset $E \subset \mathbb{U}$ $\exists$ a constant $M(E)$ such that $|f(z)| \le M(E)$ for $z \in E$ and all $f \in \mathcal{F}$.*

4.1.3

It is easy to see how the above ideas can be extended. If S is any set of integers, we can define $\mathcal{H}_S$ as the set of $f \in \mathcal{H}$, whose coefficients $a_n(f)$ satisfy

$$a_n(f) = 0 \text{ for } n \notin S. \quad (4.17)$$

Then $\mathcal{H}_S$ is a complex linear algebra under convolution with unit element

$$e_S(re^{i\theta}) = \sum_{n \in S} r^{|n|} e^{in\theta}. \quad (4.18)$$

Notice also that for any $f \in \mathcal{H}$

$$f * e_S \in \mathcal{H}_S. \quad (4.19)$$

If we take $S = \{0, 1, \ldots, n\}$, then $\mathcal{H}_S$ is the class of analytic polynomials of degree at most n. The unit element is the polynomial

$$e_n(z) = 1 + z + \cdots + z^n = \frac{1 - z^{n+1}}{1 - z}. \tag{4.20}$$

If we take $S = \{-n, -n+1, \ldots, 0, 1, \ldots, n\}$, then $\mathcal{H}_S$ is the class of harmonic polynomials of degree at most n. The unit element is the polynomial

$$d_n(z) = \bar{z}^n + \bar{z}^{n-1} + \cdots + \bar{z} + 1 + z + \cdots + z^n. \tag{4.21}$$

Finally, if we take $S = \{n\}$ for an arbitrary integer n, then the unit element is $r^{|n|}e^{in\theta}$ and the convolution formula gives, for $0 < r < 1$,

$$a_n = \frac{1}{2\pi r^{|n|}} \int_0^{2\pi} f(re^{-it})e^{int}dt = \frac{1}{2\pi r^{|n|}} \int_0^{2\pi} f(re^{it})e^{-int}dt \tag{4.22}$$

for any $f \in \mathcal{H}$ of the form (4.4). These are some of the more obvious examples of forming the convolution of a general harmonic function with a specific polynomial. The connections with the theory of Fourier series of functions on $\mathbb{T}$ will be apparent to the reader. If $f(t)$ is a Lebesgue integrable function on $[0, 2\pi]$, then the Poisson integral

$$F(re^{i\theta}) = \frac{1}{2\pi} \int_0^{2\pi} \frac{1 - r^2}{\left|1 - re^{i(\theta - t)}\right|^2} f(t)dt \tag{4.23}$$

gives the 'harmonic extension' of f into $\mathbb{U}$. The coefficients of F are exactly the Fourier coefficients of f. If we take f to be periodic on $\mathbb{R}$ with period 2π, then if f is continuous at θ,

$$F(z) \to f(\theta) \text{ as } z \to e^{i\theta} \text{ from inside } \mathbb{U}. \tag{4.24}$$

Less stringently,

$$F(re^{i\theta}) \to \tfrac{1}{2}(f(\theta+) + f(\theta-)) \text{ as } r \to 1 \tag{4.25}$$

whenever the right-hand expression exists. Thus in a genuine sense, F extends f into $\mathbb{U}$ and f is the boundary function of F.

The 'annihilation' properties of convolution will provide us with a simple, but powerful, technique. We illustrate this by proving two well-known inequalities.

4.1.4 Inequalities of Fejér and Riesz, and of Hilbert

Lemma *Let $f(z)$ be analytic for $|z| \leq 1$. Then*

$$\int_{-1}^{1} |f(x)|dx \leq \frac{1}{2} \int_0^{2\pi} |f(e^{it})|dt. \tag{4.26}$$

This is the Fejér–Riesz inequality. It is easily extended to all f belonging to the Hardy space $\mathbb{H}^1$.

Proof We use the formula

$$zf(z) = zf(z) * \left(\frac{z}{1-z} - \frac{\bar{z}}{1-\bar{z}}\right) = zf(z) * 2i\,\Im\left(\frac{z}{1-z}\right) \tag{4.27}$$

and therefore, if $-1 < x < 1$, we obtain applying (4.8)

$$f(x) = \frac{i}{\pi}\int_0^{2\pi} \Im\left(\frac{e^{-it}}{1-xe^{-it}}\right) e^{it} f(e^{it}) dt. \tag{4.28}$$

Applying Fubini's theorem we deduce that

$$\int_{-1}^{1} |f(x)| dx \le A \int_0^{2\pi} |f(e^{it})| dt \tag{4.29}$$

where

$$A = \sup_{0\le t\le 2\pi} \frac{1}{\pi}\int_{-1}^{1}\left|\Im\left(\frac{e^{-it}}{1-xe^{-it}}\right)\right| dx = \sup_{0\le t\le 2\pi}\left|\frac{1}{\pi}\int_{-1}^{1}\Im\left(\frac{e^{-it}}{1-xe^{-it}}\right)dx\right|$$
$$= \sup_{0<t<\pi}\left|\frac{1}{\pi}\int_{-1}^{1}\frac{d}{dx}\arg\left(\frac{1}{1-xe^{-it}}\right)dx\right| = \sup_{0<t<\pi}\frac{1}{\pi}\left|\arg\left(\frac{1-e^{-it}}{1+e^{-it}}\right)\right| = \frac{1}{2}. \tag{4.30}$$

From this we easily deduce the following result.

Theorem: Hilbert's inequality *Let $\{a_n\}$ be a sequence of complex numbers such that $\sum_0^\infty |a_n|^2$ is convergent. Then*

$$\sum_{m=0}^{\infty}\sum_{n=0}^{\infty}\frac{|a_m||a_n|}{m+n+1} \le \pi \sum_{n=0}^{\infty}|a_n|^2. \tag{4.31}$$

Proof Let

$$f(z) = \sum_0^\infty |a_n| z^n. \tag{4.32}$$

We apply the Fejér–Riesz inequality to the function $f^2(z)$, which belongs to $\mathbb{H}^1$. We obtain

$$\sum_{m=0}^{\infty}\sum_{n=0}^{\infty}\frac{|a_m||a_n|}{m+n+1} = \int_0^1 f^2(x)dx \le \frac{1}{2}\int_0^{2\pi}|f(e^{it})|^2 dt = \pi\sum_{n=0}^{\infty}|a_n|^2. \tag{4.33}$$

The last equality expresses **Parseval's formula**. In general for $f \in \mathbb{L}^2$

$$\frac{1}{2\pi}\int_0^{2\pi} |f(e^{it})|^2 dt = \sum_{n=-\infty}^{\infty} |a_n|^2 \tag{4.34}$$

where a_n are the Fourier coefficients of f. This may be proved by considering the convolution $f(z) * \overline{f(\overline{z})}$.

4.1.5 Linear functionals

A linear functional on $\mathcal{H}$ is a continuous linear mapping $\Lambda: \mathcal{H} \to \mathbb{C}$. Thus for $f, g \in \mathcal{H}$ and complex constants α and β, we have

$$\Lambda(\alpha f + \beta g) = \alpha \Lambda f + \beta \Lambda g. \tag{4.35}$$

It is a result of Toeplitz that all such functionals have a unique representation, which we shall obtain as follows. *For each integer n let*

$$\lambda_n = \Lambda\left(r^{|n|}e^{in\theta}\right). \tag{4.36}$$

Then the function

$$L(re^{i\theta}) = \sum_{n=-\infty}^{\infty} \lambda_n r^{|n|} e^{in\theta} \tag{4.37}$$

is harmonic for $|z| \leq 1$. If $f \in \mathcal{H}$ has coefficients a_n, then

$$\Lambda f = \sum_{n=-\infty}^{\infty} \lambda_n a_n = (L * f)(1) = \lim_{r\to 1} \frac{1}{2\pi}\int_0^{2\pi} f(re^{-it})L(e^{it})dt. \tag{4.38}$$

The first equality follows immediately from the linearity and continuity of Λ. In particular the series

$$\sum_{n=-\infty}^{\infty} \lambda_n a_n \tag{4.39}$$

is convergent for every $f \in \mathcal{H}$. Since every function, whose coefficients a_n satisfy $|a_n| = 1$, is in $\mathcal{H}$, it follows that

$$\sum_{n=-\infty}^{\infty} |\lambda_n| < +\infty \tag{4.40}$$

and so the radius of convergence of the series for L is at least 1. We need to show that it is greater than 1. Suppose, on the contrary, that the radius of convergence $= 1$. Then $\exists$ a subsequence $\{\lambda_{n_k}\}$ such that

$$\left|\lambda_{n_k}\right|^{1/|n_k|} \to 1 \text{ as } k \to \infty. \tag{4.41}$$

We consider the function g whose coefficients b_n are given by

$$b_n = \begin{cases} 1/|\lambda_{n_k}| & \text{if } n = n_k, \\ 0 & \text{otherwise.} \end{cases} \tag{4.42}$$

Then $g \in \mathcal{H}$ and

$$\Lambda g = \sum_k \frac{\lambda_{n_k}}{|\lambda_{n_k}|} \tag{4.43}$$

which is not convergent, a contradiction. The second equality in (4.38) follows immediately and the third follows from Abel summation.

We note also the converse statement, that any L harmonic in $\{|z| \le 1\}$ represents a continuous linear functional Λ on $\mathcal{H}$ given by

$$\Lambda f = (L * f)(1) \quad (f \in \mathcal{H}). \tag{4.44}$$

4.1.6 Linear operators

Let F be a fixed function in $\mathcal{H}$. The operator defined by

$$F * f \quad (f \in \mathcal{H}) \tag{4.45}$$

defines a continuous linear mapping $\mathcal{H} \to \mathcal{H}$. Thus $\Lambda = F*$ is an example of a **linear operator** on $\mathcal{H}$ and we call this case a **convolution operator**. Linear functionals are also cases of linear operators, where the image functions are constants. The general theory of such operators is properly part of functional analysis and is not our main concern. However, as there are other types of operator which we wish to introduce shortly, we shall find it useful to generalise Toeplitz's theorem in order to represent a linear operator. We require first the following notation. For each integer n let

$$\sigma_n(re^{i\theta}) = r^{|n|}e^{in\theta}. \tag{4.46}$$

Theorem *Let Λ be a linear operator on $\mathcal{H}$. There exists a unique kernel $H(z, \zeta)$ with the following properties.*

(a) *For each r $(0 < r < 1)$ $\exists\delta(r) > 0$ such that*

$$H(z, \zeta) = \sum_{j=-\infty}^{\infty} \sum_{k=-\infty}^{\infty} c_{j,k}\sigma_j(z)\sigma_k(\zeta), \tag{4.47}$$

the series converging absolutely and uniformly for $|z| \le r$, $|\zeta| < 1 + \delta(r)$. In particular the series is absolutely convergent for $|z| < 1$, $|\zeta| \le 1$.

(b) *Denoting by $H_z(\zeta)$ the function $\zeta \mapsto H(z,\zeta)$, we have for each $f \in \mathcal{H}$*

$$(\Lambda f)(z) = (H_z * f)(1) \quad (|z| < 1). \tag{4.48}$$

Conversely, any function H satisfying (a) represents a linear operator Λ on $\mathcal{H}$ given by (4.48).

Proof Let $\mathcal{P}(z)$ denote the Poisson function given by (4.13) and write $\mathcal{P}_\zeta(z)$ for the function $z \mapsto \mathcal{P}(z\zeta)$. The kernel H is obtained from

$$H(z,\zeta) = \Lambda(\mathcal{P}_\zeta(z)) = \sum_{k=-\infty}^{\infty} \tau_k(z)\sigma_k(\zeta) \tag{4.49}$$

for $|z| < 1$, $|\zeta| \le 1$, where

$$\tau_k(z) = \Lambda(\sigma_k(z)). \tag{4.50}$$

The linearity and continuity of Λ imply that, if $f = \sum_{-\infty}^{\infty} a_k\sigma_k \in \mathcal{H}$, then

$$(\Lambda f)(z) = \sum_{k=-\infty}^{\infty} a_k\tau_k(z), \tag{4.51}$$

the series converging locally uniformly in $\mathbb{U}$. Writing f_ζ for the function $z \mapsto f(\zeta z)$, we obtain for $|\zeta| < 1$

$$(\Lambda f_\zeta)(z) = \sum_{-\infty}^{\infty} a_k\tau_k(z)\sigma_k(\zeta) = f(\zeta) * H_z(\zeta). \tag{4.52}$$

The first equality remains valid when $|\zeta| = 1$ and, once we have established (a), the second equality will also hold for $|\zeta| = 1$.

Let τ_k have coefficients $\tau_j^{(k)}$ and set

$$A(r,\tau_k) = \sum_{j=-\infty}^{\infty} |\tau_j^{(k)}| r^{|j|}, \qquad M(r,\tau_k) = \max_{|z|=r} |\tau_k(z)|. \tag{4.53}$$

To prove (a) and (b) it will be enough to show that

$$\limsup_{|k|\to\infty} A(r,\tau_k)^{1/|k|} < 1 \quad (0 < r < 1). \tag{4.54}$$

Let $f = \sum_{-\infty}^{\infty} a_k\sigma_k \in \mathcal{H}$ and let $\mathcal{F}(f)$ denote the family of functions $g = \sum_{-\infty}^{\infty} b_k\sigma_k$ for which $|b_k| \le |a_k|$ for all integers k. $\mathcal{F}(f)$ is locally bounded and therefore a normal family of functions. The continuity of Λ implies that the family $\Lambda(\mathcal{F}(f))$ is normal and therefore locally bounded. Hence $\exists K_r$ such that for each $g \in \mathcal{F}$ we have

$$\left| \sum_{k=-\infty}^{\infty} b_k \tau_k(z) \right| \leq K_r \quad (|z| \leq r). \tag{4.55}$$

For a given z' $(|z'| \leq r)$ choose λ_k $(|\lambda_k| = 1)$ such that $\lambda_k \tau_k(z') = |\tau_k(z')|$. Then $\sum_{-\infty}^{\infty} |a_k| \lambda_k \sigma_k \in \mathcal{F}(f)$ and so applying (4.55) at $z = z'$ we obtain

$$\sum_{k=-\infty}^{\infty} |a_k||\tau_k(z)| \leq K_r \quad (|z| \leq r). \tag{4.56}$$

Now $A(r, \tau_k) = (\tau_k * \mu_k)(r)$, where μ_k has coefficients of absolute value 1. Hence, if $0 < r < \rho < R < 1$

$$A(r, \tau_k) \leq \frac{\rho + r}{\rho - r} M(\rho, \tau_k) \leq \frac{\rho + r}{\rho - r} \frac{R + \rho}{R - \rho} \frac{1}{2\pi} \int_0^{2\pi} |\tau_k(Re^{i\theta})| d\theta \tag{4.57}$$

where we have performed obvious estimates using the convolution formula and Poisson's formula. We deduce that

$$\sum_{k=-\infty}^{\infty} |a_k| A(r, \tau_k) \leq \frac{\rho + r}{\rho - r} \frac{R + \rho}{R - \rho} K_R. \tag{4.58}$$

As this holds for all $f \in \mathcal{H}$, we deduce (4.54) using the argument of 4.1.5. The converse is clear from the uniform convergence and the convolution formula (4.8).

We note that an operator Λ is analytic, i.e. Λ maps $\mathcal{A}$ to $\mathcal{A}$, iff each $\tau_k \in \mathcal{A}$ for $k \geq 0$. Restricting the operator to $\mathcal{A}$ we see that the kernel takes the form

$$H(z, \zeta) = \sum_{j=0}^{\infty} \sum_{k=0}^{\infty} c_{j,k} z^j \zeta^k \tag{4.59}$$

with the convergence property (a).

The kernel of the convolution operator $g*$ takes the simple form

$$H(z, \zeta) = g(z\zeta). \tag{4.60}$$

4.1.7 Harmonic multiplication operators

If $g \in \mathcal{A}$, then the operator Λ defined by $\Lambda f = gf$ is clearly a linear operator on $\mathcal{A}$. However, if $f \in \mathcal{H}$, then in general $gf \notin \mathcal{H}$ and so Λ is not a linear operator on $\mathcal{H}$. Our aim is to extend the concept of pointwise multiplication on $\mathcal{A}$ in such a way as to define a multiplication operation on $\mathcal{H}$ which preserves a good portion of the algebraic structure. Let us consider a simple example. The effect of multiplication by z on $\mathcal{A}$ is to shift all the coefficients of a function one place to the right, and this leaves a zero coefficient in the a_0 spot. Thus multiplication by z is a shift operator and this interpretation extends to the whole of $\mathcal{H}$. Hence, if $f = \sum_{-\infty}^{\infty} a_k \sigma_k \in \mathcal{H}$, then we define

$$z \bullet f = \sum_{k=-\infty}^{\infty} a_k \sigma_{k+1} = \sum_{k=-\infty}^{\infty} a_{k-1} \sigma_k. \tag{4.61}$$

More generally, we define for each integer n

$$\sigma_n \bullet f = \sum_{k=-\infty}^{\infty} a_{k-n} \sigma_k \tag{4.62}$$

so that $\sigma_n \bullet$ shifts the coefficients $|n|$ places, to the right if $n > 0$, to the left if $n < 0$. This leads to the following definition of multiplication. Let $g = \sum_{-\infty}^{\infty} b_k \sigma_k \in \mathcal{H}$. We define

$$g \bullet f = \sum_{k=-\infty}^{\infty} b_k (\sigma_k \bullet f) \tag{4.63}$$

when the series on the right is convergent in $\mathbb{U}$.

4.1.8

Theorem *The operator $g \bullet$ is a continuous linear operator on $\mathcal{H}$ if, and only if, g is harmonic in $\{|z| \le 1\}$, i.e. the series has radius of convergence >1. We then have*

$$(g \bullet f)(z) = \lim_{R \to 1} \frac{1}{2\pi} \int_0^{2\pi} \mathcal{P}(ze^{-it}) f(Re^{it}) g(e^{it}) dt, \tag{4.64}$$

the limit converging locally uniformly in $\mathbb{U}$.

Proof If $g \bullet$ is an operator on $\mathcal{H}$, then the kernel

$$H(z, \zeta) = g \bullet \mathcal{P}_{\zeta}(z) = \sum_{-\infty}^{\infty} \sigma_k(\zeta)(g \bullet \sigma_k)(z) \tag{4.65}$$

expands into a double series absolutely convergent for $|z| \le r < 1$, $|\zeta| < 1 + \delta(r)$. In particular $H(0, \zeta) = g(\overline{\zeta})$ has radius of convergence >1, i.e. g is harmonic in the closed disc. Conversely, assume that g is harmonic for $|z| \le 1$. Then we can write

$$g(z) = \sum_{-\infty}^{\infty} c_k \rho^{|k|} \sigma_k(z) \tag{4.66}$$

where $0 < \rho < 1$ and $\sum_{-\infty}^{\infty} c_k\sigma_k \in \mathcal{H}$. We need to show that the series in (4.65) expands into a double series absolutely convergent in the stated manner. We have for $|z| \leq r < 1$,

$$\begin{aligned} |H(z,\zeta)| &= \left| \sum_{k=-\infty}^{\infty} \sigma_k(\zeta) \sum_{j=-\infty}^{\infty} c_j \rho^{|j|} \sigma_{k+j}(z) \right| \\ &\leq \sum_{k=-\infty}^{\infty} \sum_{j=-\infty}^{\infty} |c_j| \rho^{|j|} |\zeta|^{|k|} r^{|k+j|} \\ &\leq \sum_{k=-\infty}^{\infty} \sum_{j=-\infty}^{\infty} |c_j| |\zeta|^{|k|} R^{|j|+|k+j|} \end{aligned} \tag{4.67}$$

where $r < R < 1$, $\rho < R < 1$. Let

$$C_m = \max_{|j|\leq m} |c_j|, \qquad D_m = \max_{|j|\leq m} |\zeta|^{|j|}. \tag{4.68}$$

Then we obtain

$$\begin{aligned} |H(z,\zeta)| &\leq \sum_{m=0}^{\infty} \left(\sum_{|j|+|k+j|=m} |c_j||\zeta|^{|k|} \right) R^m \\ &\leq \sum_{m=0}^{\infty} (4m+1) C_m D_m R^m. \end{aligned} \tag{4.69}$$

It is easily seen that the series $\sum_0^{\infty} C_m R^m$ is convergent, and so it remains to consider the series $\sum_1^{\infty} D_m R^m$. If $|\zeta| \leq 1$, then $D_m \leq 1$, and if $|\zeta| > 1$, $D_m = |\zeta|^m$. Thus the series converges for $|\zeta| < 1/R$. It follows that $H(z,\zeta)$ has the required convergence property.

This property clearly implies that, if $f = \sum_{-\infty}^{\infty} a_k\sigma_k \in \mathcal{H}$ and $g = \sum_{-\infty}^{\infty} b_k\sigma_k$ is harmonic in $\{|z| \leq 1\}$, then

$$(g \bullet f)(z) = \sum_{j=-\infty}^{\infty} \sum_{k=-\infty}^{\infty} a_j b_k \sigma_{j+k}(z), \tag{4.70}$$

the series converging absolutely and locally uniformly in $\mathbb{U}$. On the other hand, if f is harmonic in $\{|z| \leq 1\}$, then

$$\frac{1}{2\pi} \int_0^{2\pi} \mathcal{P}(ze^{-it}) f(e^{it}) g(e^{it}) dt = \sum_{j=-\infty}^{\infty} \sum_{k=-\infty}^{\infty} a_j b_k \sigma_{j+k}(z). \tag{4.71}$$

Thus writing $f_R(z) = f(Rz)$ $(0 < R < 1)$, we see that for each $f \in \mathcal{H}$

$$(g \bullet f_R)(z) = \frac{1}{2\pi} \int_0^{2\pi} \mathcal{P}(ze^{-it}) f(Re^{it}) g(e^{it}) dt. \tag{4.72}$$

Now $f_R \to f$ locally uniformly in $\mathbb{U}$ as $R \to 1$, and so $g \bullet f_R \to g \bullet f$ locally uniformly in $\mathbb{U}$.

4.1.9

We denote by $\mathcal{M}$ the class of functions g harmonic in $\{|z| \leq 1\}$ and so representing a harmonic multiplication operator on $\mathcal{H}$. $\mathcal{M}$ has a number of interesting properties which we mention without proof. $\mathcal{M}$ is a complex linear algebra under $\bullet$ multiplication with unit element $g = 1$. An element g is invertible iff g is non-zero on $\mathbb{T}$. Furthermore, every $g \in \mathcal{M}$, not identically zero, can be written in the form

$$g = P \bullet h \tag{4.73}$$

where P is an analytic polynomial with zeros only on $\mathbb{T}$ and h is an invertible element of $\mathcal{M}$.

If $g \in \mathcal{M}$ is invertible, $g \bullet \overline{g}$ is positive and invertible. Furthermore, the harmonic extensions into $\mathbb{U}$ of the boundary functions $\log|g(e^{i\theta})|$ and $|g(e^{i\theta})|^p$ (p real) are elements of $\mathcal{M}$. In particular the harmonic function with boundary values $|g(e^{i\theta})|$ is in $\mathcal{M}$. This result is not true in general. The harmonic function with boundary values $|1 - e^{i\theta}|$ is not in $\mathcal{M}$.

If $g \in \mathcal{M}$ is invertible, then every function $f \in \mathcal{H}$ has a unique series representation in the form

$$f(z) = \sum_{k=-\infty}^{\infty} c_k (g \bullet \sigma_k)(z), \tag{4.74}$$

the series converging absolutely and locally uniformly in $\mathbb{U}$. Conversely, if such a series converges pointwise in $\mathbb{U}$, then it converges absolutely and locally uniformly to a function f harmonic in $\mathbb{U}$. The necessary and sufficient condition for the convergence of the series is

$$\limsup_{|k|\to\infty} |c_k|^{1/|k|} \leq 1. \tag{4.75}$$

The action of the general continuous linear operator $\Lambda\colon \mathcal{H} \to \mathcal{H}$ can be expressed in terms of $\bullet$ multiplication as follows. If Λ has kernel $H(z, \zeta) = H_z(\zeta)$, then

$$(\Lambda f)(z) = (H_z \bullet \hat{f})(0) \tag{4.76}$$

for every $f \in \mathcal{H}$, where $\hat{f}$ denotes the function $\zeta \to f(\overline{\zeta})$.

Finally, we note the identity

$$(g \bullet \mathcal{P}_\zeta)(z) = g(\overline{\zeta})\mathcal{P}(\zeta z) \quad (|\zeta| = 1,\ |z| < 1), \tag{4.77}$$

valid for every $g \in \mathcal{M}$. In particular, if g is not invertible, then for a suitable $\zeta \in \mathbb{T}$, $g \bullet \mathcal{P}_\zeta \equiv 0$.

4.1.10 Problems

1. Let $f \in \mathcal{H}$ and suppose that $f(0) = 0$ and $|f(z)| \leq 1$ for $z \in \mathbb{U}$. Show that

$$|f(z)| \leq \frac{2}{\pi} \arg\left(\frac{1+ir}{1-ir}\right) = \frac{2}{\pi} \arctan\left(\frac{2r}{1-r^2}\right) = \frac{4}{\pi} \arctan r \leq \frac{4}{\pi} r \tag{4.78}$$

for $|z| \leq r < 1$. [Hint: $f \in \mathcal{H}_S$, where S is the set of non-zero integers; hence

$$|f(re^{i\theta})| \leq \frac{1}{2\pi} \int_0^{2\pi} |\mathcal{P}(re^{it}) - \alpha(r)| dt \tag{4.79}$$

where $\alpha(r)$ is any function of r; by choosing $\alpha(r) = (1-r^2)/(1+r^2)$, show that

$$|f(re^{i\theta})| \leq \frac{1}{\pi} \int_{-\pi/2}^{\pi/2} \mathcal{P}(re^{it}) dt - 1 = \frac{2}{\pi} \int_0^{\pi/2} \mathcal{P}(re^{it}) dt - 1.] \tag{4.80}$$

2. If $f \in \mathcal{H}$, show that for $|a| < 1$,

$$\lim_{z \to a} (f(z) - f(a)) \bullet (z-a)^{-1} = h'(a), \tag{4.81}$$

where h is the analytic part of f (the superscript -1 denotes the harmonic multiplication inverse).

4.2 Positive and bounded operators

4.2.1

A linear operator Λ on $\mathcal{H}$ is said to be **positive** if for $u \in \mathcal{H}$ satisfying $u(z) \geq 0$ in $\mathbb{U}$, we have $(\Lambda u)(z) \geq 0$ in $\mathbb{U}$. If $u > 0$ implies that $\Lambda u > 0$, then Λ is called **strictly positive**.

Because the Poisson function $\mathcal{P}(z) > 0$ in $\mathbb{U}$, it is easily verified that an operator Λ is positive (strictly positive) if, and only if, the kernel H satisfies

$$H(z, e^{i\theta}) \geq 0 \quad (> 0) \tag{4.82}$$

for $z \in \mathbb{U}$ and θ real.

For convolution operators, we deduce that the operator $f*$ is positive iff $f(z) \geq 0$ in $\mathbb{U}$. Furthermore, either $f \equiv 0$ or $f > 0$. Hence apart from the zero operator, a positive convolution operator is strictly positive.

For $\bullet$ multiplication operators, we obtain that for $g \in \mathcal{M}$ the operator $g\bullet$ is positive (strictly positive) iff $g(e^{i\theta}) \geq 0$ (> 0) for θ real.

4.2.2

Positive operators are important for their structure preserving properties and many of the results in this book will depend on these – although sometimes the connection is quite deep and well hidden. The first question to ask is, if Λ preserves the property of positivity, then what is its effect on other non-positive functions in $\mathcal{H}$? Clearly the condition (4.82) implies that if u is real, then Λu is real. Is it true that, if in addition u is bounded above, say $u \leq M$, then Λu is bounded above? We have $M - u \geq 0$ and so $\Lambda(M - u) \geq 0$, i.e. $\Lambda u \leq \Lambda M = M\Lambda 1$. However, $\Lambda 1 = H(z, 0)$ may be any positive harmonic function (e.g. $\mathcal{P}(z)$) and not necessarily bounded above. To maintain the bounded above property we require that $\Lambda 1$ be bounded above. Similarly, to maintain the property of bounded below, we require that $\Lambda 1$ be bounded below. Thus, if Λ is a positive operator such that $\Lambda 1$ is bounded, then for every bounded function $f \in \mathcal{H}$, Λf is bounded; for if $|f(z)| \leq M$, then

$$
\begin{aligned}
|(\Lambda f)(z)| &= \left| \lim_{r \to 1} \frac{1}{2\pi} \int_0^{2\pi} f(re^{-it}) H(z, e^{it}) dt \right| \\
&\leq \frac{M}{2\pi} \int_0^{2\pi} |H(z, e^{it})| dt = \frac{M}{2\pi} \int_0^{2\pi} H(z, e^{it}) dt = M\Lambda 1. \qquad (4.83)
\end{aligned}
$$

An operator which preserves the property of boundedness is called a **bounded operator**. We see from the above that a *sufficient* condition for an operator Λ with kernel $H(z, \zeta)$ to be bounded is that the function

$$
z \mapsto \frac{1}{2\pi} \int_0^{2\pi} |H(z, e^{it})| dt \qquad (4.84)
$$

be bounded in $\mathbb{U}$. This condition is also *necessary*, which can be seen as follows. Let Λ be a bounded operator on $\mathcal{H}$. Then Λ is a continuous linear mapping from $h^\infty \to h^\infty$, where h^∞ denotes the space of bounded functions in $\mathcal{H}$ given the supremum norm; h^∞ is a Banach space and therefore the uniform boundedness principle implies that $\exists K$ such that $\|f\| \leq M \Rightarrow \|\Lambda f\| \leq KM$. Let H be the kernel of Λ. We show that this implies that the function (4.84) is bounded above by K. Fix $z \in \mathbb{U}$. Either (a) $H(z, e^{it}) = 0$ for $0 \leq t \leq 2\pi$, or (b) $H(z, e^{it})$ has at most finitely many zeros in $[0, 2\pi]$. In case (a) $(\Lambda f)(z) = 0$ for any $f \in \mathcal{H}$. In case (b) we consider the harmonic function $f(\zeta)$ whose boundary values are

$$
\frac{|H(z, e^{-it})|}{H(z, e^{-it})} \qquad (4.85)
$$

on the unit circle, except at a finite number of points. We see immediately that for this function

$$(\Lambda f)(z) = \frac{1}{2\pi}\int_0^{2\pi} |H(z, e^{it})| dt. \tag{4.86}$$

On the other hand we have $|f(\zeta)| \leq 1$ in $\mathbb{U}$ and hence $|(\Lambda f)(z)| \leq K$. Since $z \in \mathbb{U}$ is arbitrary, the desired conclusion follows.

We define

$$\|\Lambda\| = \sup_{z\in\mathbb{U}} \left(\frac{1}{2\pi}\int_0^{2\pi} |H(z, e^{it})| dt \right) \tag{4.87}$$

and call this the **norm** of Λ. Note that $\|\Lambda\|$ is the smallest constant K such that, for $f \in \mathcal{H}$,

$$|f(z)| \leq M \ (z \in \mathbb{U}) \Rightarrow |(\Lambda f)(z)| \leq K\, M \ (z \in \mathbb{U}). \tag{4.88}$$

4.2.3

Summarising, we have established the following result.

Theorem *Let Λ be a linear operator on $\mathcal{H}$. Λ is bounded if, and only if, $\|\Lambda\|$ defined by (4.87) is finite. We then have, for each bounded $f \in \mathcal{H}$,*

$$\sup_{z\in\mathbb{U}} |(\Lambda f)(z)| \leq \|\Lambda\| \sup_{z\in\mathbb{U}} |f(z)|. \tag{4.89}$$

The case which will be of most concern to us is when $\|\Lambda\| \leq 1$. Such operators preserve (or reduce) the bounds on a function and we call them **bound preserving** operators. Returning to positive operators we have the next result.

4.2.4

Theorem *Let Λ be a positive operator satisfying $\Lambda 1 = 1$. Then Λ is a bounded operator with $\|\Lambda\| = 1$. Furthermore, if $f \in \mathcal{H}$ and, for $z \in \mathbb{U}$, $f(z)$ takes all its values in a convex set E, then $(\Lambda f)(z)$ takes all its values in $\overline{E}$.*

Conversely, if Λ is a bounded operator with $\|\Lambda\| \leq 1$ and if $\Lambda 1 = 1$, then Λ is a positive operator.

Proof The first assertion is clear. To prove the second, let w be a point not in the closure of E. Since $\overline{E}$ is a closed convex set, $\exists\delta > 0$ and ϕ real such that $\Re(e^{-i\phi}(f(z)-w)) \geq \delta$ for all $z \in \mathbb{U}$, which is to say that $f(z)$ takes all its values in a closed half-plane not containing the point w. It follows that

$$\Lambda(\Re(e^{-i\phi}(f(z) - w))) \geq \delta \tag{4.90}$$

which gives

$$\Re(e^{-i\phi}((\Lambda f)(z) - w)) \geq \delta \tag{4.91}$$

and so $\Lambda f \neq w$. To prove the final assertion we note that

$$1 = \frac{1}{2\pi}\int_0^{2\pi} H(z, e^{it})dt \leq \frac{1}{2\pi}\int_0^{2\pi} |H(z, e^{it})|dt = 1 \tag{4.92}$$

from which it easily follows that $H(z, e^{it}) = |H(z, e^{it})| \geq 0$.

Note that we can improve the 'convexity preserving' part of this theorem as follows. Either Λf takes all its values in E, or Λf takes all its values on a straight line portion of the boundary of E. In particular, if E is a convex set with a finite smooth boundary (e.g. a circle) then the possibility must be allowed that Λf is constant in $\mathbb{U}$. This can only occur if for some $\zeta \in \mathbb{T}$, $\Lambda(\mathcal{P}_\zeta)(z) = 0$ for all $z \in \mathbb{U}$. It is easily checked that this possibility cannot occur for convolution operators, and therefore for such operators we can replace $\overline{E}$ by E in the theorem.

4.2.5 Convolution operators and the extension property

Let $F \in \mathcal{H}_S$ and suppose that the operator $F*: \mathcal{H}_S \mapsto \mathcal{H}_S$ is bounded, i.e. $\exists K$ such that for each bounded function $f \in \mathcal{H}_S$,

$$|(F * f)(z)| \leq K \sup|f(z)|. \tag{4.93}$$

It is not necessarily the case that this inequality extends to the whole of $\mathcal{H}$, e.g. take $S = \{0, 1\}$ and $F(z) = 1 + z$. Then for $f \in \mathcal{H}_S$, $F * f = f$, so the inequality holds with $K = 1$. However, were the inequality true for all bounded $f \in \mathcal{H}$ with coefficients a_k say, we would obtain $|a_0| + |a_1| \leq 1$ if $|f(z)| \leq 1$ $(z \in \mathbb{U})$. This is contradicted by the function

$$f(z) = \frac{z + a}{1 + \bar{a}z} = a + (1 - |a|^2)z + \cdots \tag{4.94}$$

for $0 < |a| < 1$. However, it will be observed that the operator $\mathcal{P}* = F*$ on the space $\mathcal{H}_S$ and, as $\mathcal{P} * f = f \; \forall f \in \mathcal{H}$, $\mathcal{P}*$ is an extension of $F*$ to the whole of $\mathcal{H}$ which does maintain the boundedness of the operator. We wish to show that such an extension can always be achieved for any convolution operator bounded on a subspace $\mathcal{H}_S$.

4.2.6

Theorem *Let $F \in \mathcal{H}$. The operator $F*$ is a bounded operator on $\mathcal{H}$ with norm ≤ 1 if, and only if,*

$$\frac{1}{2\pi}\int_0^{2\pi} |F(re^{it})|dt \leq 1 \quad (0 \leq r < 1). \tag{4.95}$$

If this condition holds, we have for all $f \in \mathcal{H}$

$$M(r, F * f) \leq M(r, f) \quad (0 \leq r < 1), \tag{4.96}$$

$$\frac{1}{2\pi}\int_0^{2\pi} |(F * f)(re^{it})|^p dt \leq \frac{1}{2\pi}\int_0^{2\pi} |f(re^{it})|^p dt \quad (0 \leq r < 1,\ p \geq 1). \tag{4.97}$$

If F is a positive harmonic function satisfying $F(0) \leq 1$, then the condition holds. Conversely, if the condition holds and $F(0) = 1$, then F is positive.

4.2.7

Theorem *Let $G \in \mathcal{H}$. The operator $G*$ is a bounded operator $\mathcal{H}_S \to \mathcal{H}$ of norm ≤ 1 if, and only if, $\exists F \in \mathcal{H}$ satisfying (4.95) such that*

$$G * \mathcal{P}_S = F * \mathcal{P}_S. \tag{4.98}$$

Proofs Theorem 4.2.6 is straightforward. For theorem 4.2.7, the sufficiency of the condition (4.98) is clear. To prove its necessity, suppose that $G*$ is a bounded operator $\mathcal{H}_S \to \mathcal{H}$ of norm ≤ 1. Let

$$(G * \mathcal{P}_S)(z) = \sum_{k\in S} c_k\sigma_k(z) \tag{4.99}$$

and define for $f = \sum_{k=-\infty}^{\infty} a_k\sigma_k(z)$ harmonic in $\{|z| \leq 1\}$ the linear functional Λ by

$$\Lambda f = \sum_{k\in S} c_k a_k. \tag{4.100}$$

If $f \in \mathcal{H}_S$, then we can write $f(z) = g(rz)$, where $g \in \mathcal{H}$ and $0 < r < 1$. Then g has coefficients $b_k = a_k/r^{|k|}$ and

$$\Lambda f = (G * \mathcal{P}_S * g)(r). \tag{4.101}$$

Hence

$$|\Lambda f| \leq \max_{|z|=r} |g(z)| = \max_{|z|=1} |f(z)|. \tag{4.102}$$

Thus Λ is a bounded linear functional of norm ≤ 1 on the linear space $\mathcal{H}_S$ intersected with the class of functions harmonic in the closed disc given the supremum norm. We apply the Hahn–Banach theorem to deduce that Λ can be extended to a bounded linear functional acting on the normed linear space h^∞ of bounded harmonic functions without increasing the norm of Λ. Now for $k \in S$ we have $\Lambda\sigma_k = c_k$ and $|c_k| \leq 1$. For the extended functional Λ we can define $\Lambda\sigma_k = c_k$ for all integers k and we still have $|c_k| \leq 1$. Let

$$F(z) = \sum_{k=-\infty}^{\infty} c_k \sigma_k(z). \tag{4.103}$$

Then $F \in \mathcal{H}$ satisfies (4.98). Furthermore, if $f \in h^\infty$ and $z \in \mathbb{U}$, then

$$|(F * f)(z)| = |\Lambda f_z| \leq \|f_z\|_\infty \leq \|f\|_\infty \tag{4.104}$$

where f_z denotes the function $\zeta \mapsto f(\zeta z)$. Thus $F*$ is a bounded linear operator on $\mathcal{H}$ of norm ≤ 1, as required.

4.2.8

To illustrate the importance of this extension theorem let us consider the case that S is the set of non-negative integers, so that $\mathcal{H}_S = \mathcal{A}$. Let $F \in \mathcal{A}$. The operator $F*\colon \mathcal{A} \to \mathcal{A}$ is a bounded operator on $\mathcal{A}$ iff $\exists G \in \mathcal{A}$ satisfying $G(0) = 0$ such that

$$\sup_{0<r<1} \frac{1}{2\pi} \int_0^{2\pi} |\overline{G}(re^{it}) + F(re^{it})| dt < +\infty. \tag{4.105}$$

The norm of the operator is then the *minimum over all such G* of this quantity. In other words the analytic functions F representing bounded linear operators on $\mathcal{A}$ are exactly those functions which have the form

$$F(z) = H(z) * \frac{1}{1-z} \tag{4.106}$$

where $H \in h^1$, the class of harmonic functions with bounded integral means. The simplest case occurs when $F(0) = 1$ and $F*$ is a bounded operator of norm 1. In this case the extension theorem tells us that $H*$ has norm 1 and $H(0) = 1$, and hence H is positive. Applying the convolution formula to (4.106) we deduce that

$$\Re F(z) > \tfrac{1}{2} \quad (z \in \mathbb{U}) \tag{4.107}$$

and

$$H(z) = 2\Re F(z) - 1. \tag{4.108}$$

Conversely, any function F analytic in $\mathbb{U}$ and satisfying $F(0) = 1$ and (4.107) has the extension to a positive harmonic function given by (4.108) and so $F*$ is a bounded operator of norm 1 on $\mathcal{A}$. Furthermore, $F*$ has the convexity preserving property on $\mathcal{A}$: if $f \in \mathcal{A}$ maps $\mathbb{U}$ into a convex domain K, then $F * f$ maps $\mathbb{U}$ into K.

4.2.9 Problem

1. If $r \geq 0$ and $j \geq 1$ are integers, show that the operator $F*$, where

$$F(z) = z^r + z^{r+j} + z^{r+2j} + \cdots, \tag{4.109}$$

is a bounded operator of norm 1 on $\mathcal{A}$. Deduce the following: if $f(z) = \sum_{k=0}^{\infty} a_k z^k \in \mathcal{A}$ and satisfies $|f(z)| \leq 1$ $(z \in \mathbb{U})$, then

$$\left| \sum_{k=0}^{\infty} a_{r+kj} z^{kj} \right| \leq 1 \quad (z \in \mathbb{U}). \tag{4.110}$$

Hence show that, if $P(z) = \sum_{k=0}^{n} a_k z^k$ is a polynomial of degree n, then

$$|a_i| + |a_j| \leq \max_{|z|=1} |P(z)| \tag{4.111}$$

provided that $0 \leq i < j \leq n < 2j - i$.

4.3 Positive trigonometric polynomials

4.3.1

We recall that a trigonometric polynomial is an expression of the form

$$T_n(\theta) = \sum_{k=-n}^{n} c_k e^{ik\theta} \tag{4.112}$$

which we may interpret as the restriction to the unit circle of

$$\text{either } P_n(z) = \sum_{k=-n}^{n} c_k \sigma_k(z) \quad \text{or } Q_n(z) = \sum_{k=-n}^{n} c_k z^k, \tag{4.113}$$

i.e. the restriction to $\mathbb{T}$ of either a harmonic polynomial or a rational function whose only pole is at the origin; n is the **degree** of T_n provided that c_n and c_{-n} are not both zero. In the harmonic interpretation we have

$$T_n(\theta) = P_n(e^{i\theta}). \tag{4.114}$$

If $f(z) = \sum_{-\infty}^{\infty} a_k \sigma_k(z) \in \mathcal{H}$, then

$$P_n(z) * f(z) = \sum_{k=-n}^{n} c_k a_k \sigma_k(z) \tag{4.115}$$

is a harmonic polynomial of degree at most n, whose boundary values are given by the trigonometric polynomial

$$(T_n f)(e^{i\theta}) = \sum_{k=-n}^{n} c_k a_k e^{ik\theta}. \tag{4.116}$$

We may call this expression **the T_n mean of f**. Note the formula

$$(T_n f)(e^{i\theta}) = \lim_{r \to 1} \frac{1}{2\pi} \int_0^{2\pi} f(re^{i(\theta - t)}) T_n(t) dt. \tag{4.117}$$

Trigonometric means of a function play an important role in **summability theory**. The main idea is to construct a sequence $\{T_n\}$ of trigonometric polynomials of degrees $n = 1, 2, \ldots$ such that each mean $T_n f$ maintains a certain aspect of the structure of f and further, assuming that, for example, f extends to a continuous function on $\mathbb{T}$, $T_n f \to f$ as $n \to \infty$ uniformly on $\mathbb{T}$. This latter property requires that

$$P_n(z) \to \mathcal{P}(z) \text{ as } n \to \infty \text{ locally uniformly in } \mathbb{U}. \tag{4.118}$$

To see this, note that, writing $\mathcal{P}_r(z) = \mathcal{P}(rz)$,

$$P_n(re^{i\theta}) = T_n \mathcal{P}_r(e^{i\theta}) \to \mathcal{P}(re^{i\theta}) \text{ as } n \to \infty \tag{4.119}$$

uniformly in $[0, 2\pi]$ and so (4.118) holds uniformly on each circle $|z| = r < 1$; (4.118) follows easily. On the other hand, if (4.118) holds then for every $f \in \mathcal{H}$

$$(P_n * f)(z) \to f(z) \text{ as } n \to \infty \text{ locally uniformly in } \mathbb{U} \tag{4.120}$$

because of the continuity of the operator $f*$. However, it is not necessarily true that $T_n f \to f$ on $\mathbb{T}$ even when we assume that f is continuous on $\mathbb{T}$. In order to draw this conclusion, we need further conditions on the polynomials T_n.

Suppose we assume that each $T_n \geq 0$ on $[0, 2\pi]$ and further that (4.118) holds. Since $T_n(0) = P_n * 1 \to 1$ as $n \to \infty$, there is no loss of generality in assuming that $T_n(0) = 1$ for every n. Consider

$$\begin{aligned} |(T_n f)(e^{i\theta}) - f(e^{i\theta})| &= \left| \frac{1}{2\pi} \int_{-\pi}^{\pi} \left(f\left(e^{i(\theta - t)}\right) - f(e^{i\theta}) \right) T_n(t) dt \right| \\ &\leq \frac{1}{2\pi} \int_{-\pi}^{\pi} \left| f\left(e^{i(\theta - t)}\right) - f(e^{i\theta}) \right| T_n(t) dt. \end{aligned} \tag{4.121}$$

If f is continuous at θ, then given $\epsilon > 0$ $\exists\delta > 0$ such that for $|t| < \delta$ we have

$$\left|f(e^{i(\theta-t)}) - f(e^{i\theta})\right| < \epsilon. \tag{4.122}$$

Taking $\theta \mapsto f(e^{i\theta})$ to be periodic on $\mathbb{R}$ with period 2π we divide the range of integration in (4.121) into two, namely $\{|t| < \delta\}$ and $E = \{|t| \geq \delta\} \cap [-\pi, \pi]$. We obtain

$$|(T_n f)(e^{i\theta}) - f(e^{i\theta})| \leq \epsilon + \frac{1}{2\pi}\int_E \left|f\left(e^{i(\theta-t)}\right) - f(e^{i\theta})\right| T_n(t)dt. \tag{4.123}$$

If $|f| \leq M$ on $\mathbb{T}$, we obtain

$$|(T_n f)(e^{i\theta}) - f(e^{i\theta})| \leq \epsilon + \frac{M}{\pi}\int_E T_n(t)dt. \tag{4.124}$$

Thus in order to show that

$$(T_n f)(e^{i\theta}) \to f(e^{i\theta}) \text{ as } n \to \infty \tag{4.125}$$

it is sufficient to show that

$$\int_E T_n(t)dt \to 0 \text{ as } n \to \infty \tag{4.126}$$

for each set $E = E_\delta = \{|t| \geq \delta\} \cap [-\pi, \pi]$, where $0 < \delta < \pi$. Consider the function

$$\mu_n(x) = \frac{1}{2\pi}\int_0^x T_n(t)dt; \tag{4.127}$$

$\mu_n(x)$ is an increasing function on $[0, 2\pi]$ with $\mu_n(0) = 0$, $\mu_n(2\pi) = 1$. From the Helly selection theorem it follows that $\exists$ an increasing function $\mu(x)$ such that for a suitable subsequence $\{\mu_{n_k}\}$

$$\mu_{n_k}(x) \to \mu(x) \text{ as } k \to \infty \quad (0 \leq x \leq 2\pi) \tag{4.128}$$

and for every continuous function $\phi(t)$ on $[0, 2\pi]$

$$\int_0^{2\pi} \phi(t)d\mu_{n_k}(t) \to \int_0^{2\pi} \phi(t)d\mu(t) \text{ as } k \to \infty. \tag{4.129}$$

Now $d\mu_{n_k}(t) = T_{n_k}(t)dt$ and we see applying (4.118) that for $z \in \mathbb{U}$

$$\begin{aligned}\mathcal{P}(z) = \lim_{k\to\infty} \mathcal{P}_{n_k}(z) &= \lim_{k\to\infty} \frac{1}{2\pi}\int_0^{2\pi} \mathcal{P}(ze^{-it})T_{n_k}(t)dt \\ &= \int_0^{2\pi} \mathcal{P}(ze^{-it})d\mu(t),\end{aligned} \tag{4.130}$$

from which we deduce that for every integer ν

$$\int_0^{2\pi} e^{i\nu t} d\mu(t) = 1. \tag{4.131}$$

Since $\mu(t)$ is an increasing function satisfying $\mu(0) = 0$, $\mu(2\pi) = 1$, we easily deduce from this last relation that $\mu(x) = 0$ $(0 \leq x < 2\pi)$, so that μ is a step function with a single jump at the multiples of 2π. The relation (4.126) follows. We have established the next result.

4.3.2

Theorem *Let $\{T_n(t)\}$ be a sequence of non-negative trigonometric polynomials satisfying*

$$\frac{1}{2\pi} \int_0^{2\pi} T_n(t)dt = 1 \quad (n = 1, 2, \dots) \tag{4.132}$$

and such that for the sequence of harmonic extensions $\mathcal{P}_n(z)$ we have

$$\mathcal{P}_n(z) = \frac{1}{2\pi} \int_0^{2\pi} \mathcal{P}(ze^{-it})T_n(t)dt \to \mathcal{P}(z) \text{ as } n \to \infty \text{ locally uniformly in } \mathbb{U}. \tag{4.133}$$

Let $f(e^{i\theta})$ be a function bounded and integrable on $\mathbb{T}$ and suppose that f is continuous on a closed arc $C \subset \mathbb{T}$. Then the sequence of T_n means of f satisfies

$$(T_n f)(e^{i\theta}) = \frac{1}{2\pi} \int_0^{2\pi} f\left(e^{i(\theta - t)}\right)T_n(t)dt \to f(e^{i\theta}) \text{ as } n \to \infty \text{ uniformly on } C. \tag{4.134}$$

The uniformity follows easily from our estimates and the fact that f is uniformly continuous on C.

Depending on the sequence $\{T_n\}$ we thus have a variety of ways of obtaining a function as a limit of trigonometric polynomials. The sequence $\{T_n\}$ is known as an **approximate identity**. By far the best-known example comes from the **Fejér means**, the generating sequence being defined by

$$\begin{aligned} F_n(\theta) &= \frac{1}{n}\left|1 + e^{i\theta} + e^{2i\theta} + \cdots + e^{i(n-1)\theta}\right|^2 \\ &= \frac{1}{n}\left(\frac{\sin\frac{n\theta}{2}}{\sin\frac{\theta}{2}}\right)^2 = \sum_{k=-n+1}^{n-1} \frac{n - |k|}{n} e^{ik\theta}. \end{aligned} \tag{4.135}$$

The third of these expressions shows that $\{F_n\}$ forms an approximate identity, since

$$\frac{1}{n}\left(\frac{\sin\frac{n\theta}{2}}{\sin\frac{\theta}{2}}\right)^2 \le \frac{1}{n}\left(\frac{1}{\sin\frac{\delta}{2}}\right)^2 \quad (\delta \le \theta \le 2\pi - \delta) \tag{4.136}$$

for $0 < \delta < \pi$ and (4.126) is immediately verified. Since we may not always have such a simple inequality available for the verification, we note also that the conclusion can be drawn from the fourth of the expressions in (4.135), since for $n > \nu$ the coefficient of $e^{i\nu\theta}$ is given by

$$c_\nu^{(n)} = \frac{n - |\nu|}{n} \to 1 \text{ as } n \to \infty. \tag{4.137}$$

It follows that the relation (4.131) holds and this is sufficient for the desired conclusion.

Thus as an addendum to the theorem we have: *if the coefficients* $c_\nu^{(n)}$ *of* T_n *satisfy*

$$c_\nu^{(n)} \to 1 \textit{ as } n \to \infty \textit{ for each fixed } \nu, \tag{4.138}$$

then $\{T_n\}$ *is an approximate identity*. Of course, we also require that $T_n \ge 0$. This condition is a replacement for (4.133).

4.3.3

A non-negative trigonometric polynomial $T(\theta)$ satisfying

$$\frac{1}{2\pi}\int_0^{2\pi} T(\theta)d\theta = 1 \tag{4.139}$$

clearly generates a convexity preserving operator $P*$, where $P(z)$ is the harmonic extension of T into $\mathbb{U}$. In addition, if f is an integrable function on $\mathbb{T}$ and if $f(\mathbb{T}) \subset K$, where K is a closed convex set, then $(Tf)(\mathbb{T}) \subset K$. Tf is of course a trigonometric polynomial of degree no greater than that of T. In particular, if $f \ge 0$, then $Tf \ge 0$ and if $|f| \le M$, then $|Tf| \le M$. For example, taking $f(e^{i\theta}) = e^{i\nu\theta}$, we see that each coefficient c_ν of T satisfies

$$|c_\nu| \le 1. \tag{4.140}$$

Much stronger structural preservation properties are possible, as we shall see later. However, even this amount is enough to generate a considerable quantity of useful information.

4.3.4

If $P(z)$ is an analytic polynomial of degree n, then it is clear that

$$T(\theta) = |P(e^{i\theta})|^2 = P(e^{i\theta})\overline{P(e^{i\theta})} \tag{4.141}$$

is a non-negative trigonometric polynomial of degree n. We shall now show that all non-negative trigonometric polynomials $T(\theta)$ can be written in this form. If T has degree n, then we can write

$$T(\theta) = e^{-in\theta} Q(e^{i\theta}) \tag{4.142}$$

where Q is an analytic polynomial of degree $2n$. Since T is real, $T = \overline{T}$, and so

$$e^{in\theta}\overline{Q(e^{i\theta})} = e^{-in\theta} Q(e^{i\theta}) \tag{4.143}$$

and hence the polynomial Q satisfies

$$z^{2n}\overline{Q\left(\frac{1}{\overline{z}}\right)} = Q(z) \tag{4.144}$$

at every point on $\mathbb{T}$. Since both sides of this equation are polynomials of degree $2n$, it follows that this equation holds identically in $\mathbb{C}$; the polynomial Q is **self-inversive**. By the fundamental theorem of algebra we can write

$$Q(z) = c_n \prod_{k=1}^{2n} (z - z_k) = \overline{c_n} \prod_{k=1}^{2n} (1 - \overline{z_k} z) \tag{4.145}$$

where the z_k are non-zero. If z_k does not lie on $\mathbb{T}$, then $1/\overline{z_k}$ is a distinct zero of Q and we see that

$$(e^{i\theta} - z_k)(e^{i\theta} - 1/\overline{z_k}) = -\frac{e^{i\theta}}{\overline{z_k}}|e^{i\theta} - z_k|^2. \tag{4.146}$$

Thus with a suitable labelling of the zeros we can pair off the $2m$ zeros not on $\mathbb{T}$ and obtain

$$Q(e^{i\theta}) = c_n(-1)^m e^{im\theta} \prod_{k=1}^{m}\left(\frac{1}{\overline{z_k}}\right) \prod_{k=1}^{m} |e^{i\theta} - z_k|^2 \prod_{j=1}^{2n-2m} (e^{i\theta} - \zeta_j) \tag{4.147}$$

and so

$$T(\theta) = |H(e^{i\theta})|^2 R(\theta) \tag{4.148}$$

where $H(z)$ is a polynomial of degree m and R is a non-negative trigonometric polynomial of degree $n - m$ with $2n - 2m$ zeros on $\mathbb{T}$. In other words it remains to prove the result in the case that Q has all its zeros on $\mathbb{T}$. Now (4.142) expresses T as the rational function $Q(z)/z^n$ restricted to $\mathbb{T}$. By hypothesis this function maps $\mathbb{T}$ into the non-negative real axis and hence, by the conformal properties of an analytic function, every zero of the function is a multiple zero of even multiplicity. Thus Q has the form

$$Q(z) = P^2(z) \tag{4.149}$$

where P has degree n. We obtain immediately

$$T(\theta) = |T(\theta)| = |P(e^{i\theta})|^2 \tag{4.150}$$

as required.

Note further that we can choose P so that

$$P(z) \neq 0 \quad (z \in \mathbb{U}) \tag{4.151}$$

for, if z_k is a zero with $|z_k| < 1$, we can replace the factor $(z - z_k)$ by the factor $(1 - \overline{z_k}z)$.

4.3.5

Theorem *Let $T(\theta)$ be a non-negative trigonometric polynomial of degree n with coefficients c_k $(-n \leq k \leq n)$. Then $\exists$ an analytic polynomial*

$$P(z) = \sum_{j=0}^{n} a_j z^j \tag{4.152}$$

of degree n with no zeros in $\mathbb{U}$ such that

$$T(\theta) = |P(e^{i\theta})|^2 = \sum_{k=-n}^{n} \left(\sum_{j-i=k} \overline{a_i} a_j \right) e^{ik\theta}. \tag{4.153}$$

We have

$$c_0 = \sum_{k=0}^{n} |a_k|^2, \tag{4.154}$$

$$c_{-k} = \overline{c_k} \quad (1 \leq k \leq n), \tag{4.155}$$

$$c_k = \sum_{i=0}^{n-k} \overline{a_i} a_{i+k} \quad (1 \leq k \leq n). \tag{4.156}$$

4.3.6 Problems

1. Show that the polynomial P in theorem 4.3.5 is unique up to a multiplicative constant of absolute value 1.

2. Let $P(z)$ be a polynomial of degree n. Show that the coefficient sequence taken from the left of the non-negative trigonometric polynomial $|P(e^{i\theta})|^2$ is identical to the coefficient sequence of the polynomial $P(z)P^*(z)$, where $P^*(z) = z^n \overline{P(1/\overline{z})}$. If

$$P(z) = \sum_{k=0}^{n} a_k z^k = c \prod_{k=1}^{n} (z - z_k), \tag{4.157}$$

then

$$P^*(z) = \sum_{k=0}^{n} \overline{a}_{n-k}\, z^k = \overline{c} \prod_{k=1}^{n} (1 - \overline{z_k} z). \tag{4.158}$$

If P has all its zeros on the unit circle, then $|P(e^{i\theta})|^2 = a P^2(e^{i\theta})$, where a is constant, $|a| = 1$.

4.4 Some inequalities for analytic and trigonometric polynomials

4.4.1

We shall now put together some of the ideas and results of this chapter to prove some general inequalities for polynomials. Let

$$T(z) = \sum_{k=-n+1}^{n-1} c_k \sigma_k(z) \tag{4.159}$$

be a harmonic polynomial of degree at most $n - 1$ non-negative in $\mathbb{U}$, with $c_0 = T(0) = 1$. Thus $T(e^{i\theta})$ is a non-negative trigonometric polynomial, and $T(z)$ is its harmonic extension. For fixed ζ satisfying $|\zeta| \leq 1$, the function

$$T(z) \bullet \mathcal{P}(\zeta z^n) = \sum_{k=-\infty}^{\infty} ((T \bullet \sigma_{nk})(z))\sigma_k(\zeta) \tag{4.160}$$

is a non-negative harmonic function in $\mathbb{U}$ which takes the value 1 at $z = 0$. Let

$$P(z) = \sum_{k=0}^{n} a_k z^k \tag{4.161}$$

be a polynomial of degree at most n. We then have, for $z \in \mathbb{U}$,

$$|P(z) * (T(z) \bullet \mathcal{P}(\zeta z^n))| \leq \max_{|z|=1} |P(z)| = \|P\|_\infty. \tag{4.162}$$

The explicit form of this inequality gives

$$\left| \sum_{k=0}^{n-1} a_k c_k z^k + \zeta \sum_{k=1}^{n} a_k \overline{c}_{n-k} z^k \right| \leq \|P\|_\infty \tag{4.163}$$

and as this holds for every ζ on $\mathbb{T}$, we deduce that

$$\left|\sum_{k=0}^{n-1} a_k c_k z^k\right| + \left|\sum_{k=1}^{n} a_k \overline{c}_{n-k} z^k\right| \leq \|P\|_\infty \quad (|z| \leq 1). \tag{4.164}$$

To re-iterate: this inequality is valid *for every polynomial P, with coefficients a_k, of degree not exceeding n and every non-negative trigonometric polynomial T, with coefficients c_k and $c_0 = 1$, of degree not exceeding $n - 1$*. We also remark that, although this is a result for polynomials, the proof makes essential use of the harmonic extension and is not a proof taking place on the unit circle; for the function (4.160) is not defined on the circle.

The simplest case is $T(z) = 1$. We obtain

$$|a_0| + |a_n| \leq \|P\|_\infty. \tag{4.165}$$

If we take $T(e^{i\theta}) = F_n(\theta)$, the Fejér polynomial of degree $n - 1$, we have $c_k = \frac{n-k}{n}$ for $0 \leq k \leq n - 1$, and we deduce the next, important, result.

4.4.2

Theorem *Let $P(z)$ be a polynomial of degree at most n. Then*

$$|nP(z) - zP'(z)| + |zP'(z)| \leq n\|P\|_\infty \quad (|z| \leq 1). \tag{4.166}$$

Equality occurs at each point on $\mathbb{T}$ at which $|P(z)|$ attains its maximum value.

The last assertion follows from the inequality

$$|nP(z) - zP'(z)| + |zP'(z)| \geq n|P(z)|. \tag{4.167}$$

Note that, in particular, we have from (4.166)

$$\|P'\|_\infty \leq n\|P\|_\infty \tag{4.168}$$

which is a result due to Bernstein. This inequality is sharp for $P(z) = z^n$. Nevertheless the inequality can be improved in a number of ways.

Consider first the case of a self-inversive polynomial. A polynomial $P(z)$ of degree n is said to be **self-inversive** if

$$z^n \overline{P\left(\frac{1}{\overline{z}}\right)} \equiv P(z). \tag{4.169}$$

If P has coefficients a_k $(0 \leq k \leq n)$ this is equivalent to the relations

$$\overline{a}_{n-k} = a_k \quad (0 \leq k \leq n), \tag{4.170}$$

for the **conjugate polynomial to P** is given by

$$P^*(z) = z^n \overline{P\left(\frac{1}{\bar{z}}\right)} = \sum_{k=0}^{n} \bar{a}_{n-k} z^k. \tag{4.171}$$

For example, suppose that P is a polynomial which has all of its n zeros on $\mathbb{T}$. Then it is clear that P^* and P have the same zeros with the same multiplicities, and hence the rational function

$$R(z) = \frac{P(z)}{P^*(z)} \tag{4.172}$$

has no zeros or poles in the plane and is therefore a constant, which clearly has absolute value $=1$. Thus for a suitable real constant α, $e^{i\alpha}P$ is a self-inversive polynomial.

Let P be self-inversive. Then we have, applying (4.170),

$$nP(z) - zP'(z) = \sum_{k=0}^{n}(n-k)a_k z^k = \sum_{k=0}^{n} k\overline{a_k}\, z^{n-k} = (zP')^*. \tag{4.173}$$

Now on the unit circle a polynomial and its conjugate have equal absolute values, and hence, if P is self-inversive, we have for $z \in \mathbb{T}$

$$|nP(z) - zP'(z)| = |P'(z)|. \tag{4.174}$$

Applying the theorem we obtain the next result.

4.4.3

Theorem *Let $P(z)$ be a self-inversive polynomial of degree n. Then*

$$\|P'\|_\infty = \frac{n}{2}\|P\|_\infty. \tag{4.175}$$

Furthermore, at each point on $\mathbb{T}$, where $|P|$ attains its maximum value, $|P'|$ attains its maximum value. In particular, this is true if P has all its zeros on $\mathbb{T}$.

The example $P(z) = 1 + z^n$ shows that $|P'|$ may have a maximum value at points where $|P|$ does not.

4.4.4

As a second application of theorem 4.2.2 we establish the following result, due to P. Lax.

Theorem *Let $P(z)$ be a polynomial of degree n such that*

$$P(z) \neq 0 \quad (z \in \mathbb{U}). \tag{4.176}$$

Then

$$\|P'\|_\infty \le \frac{n}{2}\|P\|_\infty. \tag{4.177}$$

Proof By the fundamental theorem of algebra we can write P in the form

$$P(z) = c\prod_{k=1}^{n}(1 - z_k z) \tag{4.178}$$

where c is a constant and $|z_k| \le 1$ $(1 \le k \le n)$. Then for $z \in \mathbb{U}$

$$\Re\left(\frac{zP'(z)}{nP(z)}\right) = \frac{1}{n}\sum_{k=1}^{n}\Re\left(\frac{-z_k z}{1 - z_k z}\right) < \frac{1}{2} \tag{4.179}$$

from which we deduce that

$$|zP'(z)| < |nP(z) - zP'(z)| \quad (z \in \mathbb{U}). \tag{4.180}$$

Combining this with (4.166) we obtain

$$|zP'(z)| < \frac{n}{2}\|P\|_\infty \quad (z \in \mathbb{U}) \tag{4.181}$$

and the result follows.

4.4.5

Our method of combining the harmonic product with convolution to obtain the general inequality (4.164) was designed specifically for analytic polynomials. If we wish to obtain inequalities for trigonometric polynomials, the following method is particularly effective.

As before let $T(z)$ be a harmonic polynomial of degree at most $n - 1$ non-negative in $\mathbb{U}$ with coefficients c_k and $c_0 = 1$. For ζ satisfying $|\zeta| \le 1$ the harmonic function $S_\zeta(z)$ given by

$$S_\zeta(z) = T(z) \bullet \mathcal{P}(\zeta z^{2n}) \tag{4.182}$$

is non-negative in $\mathbb{U}$ with $S_\zeta(0) = 1$, and hence the operator $(z^n \bullet S_\zeta)*$ is a bounded operator of norm 1 on $\mathcal{H}$. We apply this operator to a general harmonic polynomial

$$P(z) = \sum_{k=-n}^{n} a_k \sigma_k(z) \tag{4.183}$$

of degree not exceeding n. We obtain

$$\left|\sum_{k=1}^{n}\overline{c}_{n-k}a_k z^k\right| + \left|\sum_{k=1}^{n}c_{n-k}a_{-k}\overline{z}^k\right| \le \|P\|_\infty \quad (|z| \le 1). \tag{4.184}$$

This inequality is valid *for every harmonic polynomial P, with coefficients a_k, of degree not exceeding n and every non-negative trigonometric polynomial T, with coefficients c_k and $c_0 = 1$, of degree not exceeding $n-1$.*

4.4.6

Note that, if P is a real harmonic polynomial, then $a_{-k} = \overline{a_k}$. Since we may replace $T(z)$ by $\overline{T(\overline{z})}$, we deduce the following result.

Theorem *Let*

$$P(z) = \sum_{k=0}^{n} a_k z^k \tag{4.185}$$

be a polynomial of degree at most n and let

$$T(\theta) = \sum_{k=-n+1}^{n-1} c_k e^{ik\theta} \tag{4.186}$$

be a non-negative trigonometric polynomial of degree at most $n-1$ with $c_0 = 1$. Then

$$\left| \sum_{k=1}^{n} c_{n-k} a_k z^k \right| \leq \|\Re P\|_\infty = \max_{|z|=1} |\Re(P(z))| \quad (|z| \leq 1). \tag{4.187}$$

In particular

$$\|P'\|_\infty \leq n \|\Re P\|_\infty. \tag{4.188}$$

The final inequality comes from choosing $T(\theta) = F_n(\theta)$ and is an improvement of Bernstein's inequality due to Szegö. We also see that applying (4.184) with $F_n(\theta)$ we can deduce Bernstein's original result, in the next subsection.

4.4.7

Theorem *Let $P(\theta)$ be a trigonometric polynomial of degree n. Then*

$$|P'(\theta)| \leq n \max |P(\theta)| \quad (\theta \text{ real}). \tag{4.189}$$

4.4.8 Problems

1. Let $P(z)$ be a polynomial of degree n all of whose zeros lie in $\{|z| \leq 1\}$. Show that

$$\|P'\|_\infty \geq \frac{n}{2} \|P\|_\infty. \tag{4.190}$$

2. Let $P(z) = \sum_0^n a_k z^k$ be a polynomial of degree n. Show that

$$|a_n| \leq \|\Re P\|_\infty. \tag{4.191}$$

3. Let $T(\theta) = \sum_1^n a_k \sin k\theta$ be a sine polynomial of degree n (the a_k are complex). Suppose that

$$|T(\theta)| \leq M \quad (\theta \text{ real}). \tag{4.192}$$

Show that

$$|T(\theta)| \leq Mn|\sin\theta| \quad (\theta \text{ real}). \tag{4.193}$$

[Hint: $2iT(\theta) = P(e^{i\theta}) - P(e^{-i\theta})$, where $P(z) = \sum_1^n a_k z^k$; hence from (4.184)

$$|P'(z)| + |P'(\overline{z})| \leq 2nM \quad (z \in \overline{\mathbb{U}}); \tag{4.194}$$

also

$$\begin{aligned} 2iT(\theta) &= \int_{e^{-i\theta}}^{e^{i\theta}} P'(w)dw \\ &= (e^{i\theta} - e^{-i\theta}) \int_0^1 P'((1-t)e^{-i\theta} + te^{i\theta})dt \\ &= (e^{i\theta} - e^{-i\theta}) \int_0^1 P'((1-t)e^{i\theta} + te^{-i\theta})dt.] \end{aligned} \tag{4.195}$$

4. Let $P(z)$ be a polynomial of degree n satisfying $|P(x)| \leq M$ for $-1 \leq x \leq 1$. Show that

$$|P'(x)| \leq \frac{Mn}{\sqrt{1-x^2}} \quad (-1 < x < 1). \tag{4.196}$$

Hence show that

$$|P'(x)| \leq Mn^2 \quad (-1 \leq x \leq 1). \tag{4.197}$$

[Hint: consider $P(\cos\theta)$.]

4.5 Cesàro means

4.5.1

For $\alpha > -1$ we define

$$C_n(\alpha;\theta) = \sum_{k=-n}^{n} c_k^{(n)}(\alpha) e^{ik\theta} \tag{4.198}$$

where $c_0^{(n)}(\alpha) = 1$, $c_{-k}^{(n)}(\alpha) = c_k^{(n)}(\alpha)$ for $k \geq 1$ and

$$c_k^{(n)}(\alpha) = \prod_{j=1}^{k} \left(1 + \frac{\alpha}{n-k+j}\right)^{-1} = \frac{A_{n-k}^{(\alpha)}}{A_n^{(\alpha)}} \tag{4.199}$$

for $1 \leq k \leq n$, where $A_0^{(\alpha)} = 1$ and for $n \geq 1$

$$A_n^{(\alpha)} = \frac{(\alpha+1)(\alpha+2)\ldots(\alpha+n)}{n!}. \tag{4.200}$$

We have the identity

$$(1-z)^{-\alpha-1} = \sum_{n=0}^{\infty} A_n^{(\alpha)} z^n \quad (|z| < 1). \tag{4.201}$$

Note that $C_n(1;\theta) = F_{n+1}(\theta)$. Also note that

$$c_k^{(n)}(\alpha) \to 1 \text{ as } n \to \infty. \tag{4.202}$$

For a given α, the $C(\alpha)$ Cesàro means of a function $f(e^{i\theta})$ integrable on $\mathbb{T}$ are defined to be the trigonometric polynomials $C_n(\alpha;\theta) * f(e^{i\theta})$. Thus for fixed α, the sequence $\{C_n(\alpha;\theta)\}$ will form an approximate identity, if we can prove that

$$C_n(\alpha;\theta) \geq 0 \quad (\theta \text{ real}, \ n \geq 1). \tag{4.203}$$

We will prove this for $\alpha \geq 1$. Consider the sequence $\{b_k\}$ defined by

$$b_k = \begin{cases} A_{n-k}(\alpha) & \text{for } 0 \leq k \leq n, \\ 0 & \text{for } k > n. \end{cases} \tag{4.204}$$

We leave it to the reader to verify that, for $\alpha \geq 1$, the sequence $\{b_k\}$ is non-negative, non-increasing and convex; i.e. we have

$$b_k \geq 0, \quad b_{k+1} - b_k \leq 0, \quad b_{k+2} - 2b_{k+1} + b_k \geq 0 \quad (k = 0, 1, 2, \ldots). \tag{4.205}$$

That these conditions are sufficient to imply the desired result is a consequence of a useful criterion for positivity, which we give next.

4.5.2

Rogosinski's lemma *Let $u(z) = \sum_{-\infty}^{\infty} b_k \sigma_k(z)$ be a real harmonic function in $\mathbb{U}$, not identically zero, whose coefficients b_k satisfy the conditions (4.205). Then*

$$u(z) > 0 \quad (z \in \mathbb{U}). \tag{4.206}$$

Proof Since u is real, we have $b_{-k} = \overline{b_k} = b_k$. As u is not identically zero, the conditions imply that $b_0 > 0$. Thus $u(r) > 0$ for $0 \leq r < 1$. Fix r $(0 < r < 1)$ and consider

$$\begin{aligned}|1 - e^{i\theta}|^2 u(re^{i\theta}) &= (2 - e^{i\theta} - e^{-i\theta}) \sum_{k=-\infty}^{\infty} b_k r^{|k|} e^{ik\theta} \\ &= \sum_{k=-\infty}^{\infty} \left(2b_k r^{|k|} - b_{k-1} r^{|k-1|} - b_{k+1} r^{|k+1|}\right) e^{ik\theta} \\ &= 2(b_0 - b_1 r) - 2\sum_{k=1}^{\infty} (b_{k-1} - 2b_k r + b_{k+1} r^2) r^{k-1} \cos k\theta.\end{aligned} \tag{4.207}$$

Now applying the conditions (4.205) we obtain

$$\begin{aligned}b_{k+1} r^2 - 2b_k r + b_{k-1} &= b_{k+1} - 2b_k + b_{k-1} + 2b_k(1 - r) - b_{k+1}(1 - r^2) \\ &\geq (1 - r)(2b_k - (1 + r)b_{k+1}) \\ &\geq (1 - r^2)(b_k - b_{k+1}) \geq 0\end{aligned} \tag{4.208}$$

and hence, since $\cos k\theta \leq 1$,

$$|1 - e^{i\theta}|^2 u(re^{i\theta}) \geq 2(b_0 - b_1 r) - 2\sum_{k=1}^{\infty} (b_{k-1} - 2b_k r + b_{k+1} r^2) r^{k-1} = 0. \tag{4.209}$$

Thus $u(z) \geq 0$ in $\mathbb{U}$, and by the maximum principle, $u(z) > 0$ in $\mathbb{U}$, as required.

4.5.3

The trigonometric polynomials $C_n(\alpha; \theta)$ are not positive, if $\alpha < 1$. It is left as an exercise to show that, for $0 < \alpha < 1$

$$C_n(\alpha; \pi) < 0 \quad (n = 1, 2, \ldots). \tag{4.210}$$

For $\alpha < 0$, we have

$$c_1^{(n)}(\alpha) > c_0^{(n)} = 1 \tag{4.211}$$

and so $C_n(\alpha; \theta)$ is not non-negative.

For $\alpha = 0$,

$$C_n(\alpha; \theta) = D_n(\theta) = \sum_{k=-n}^{n} e^{ik\theta} = \frac{\sin\left(n + \frac{1}{2}\right)\theta}{\sin\frac{1}{2}\theta} \tag{4.212}$$

which clearly takes negative values. For f integrable on $\mathbb{T}$, the means $D_n * f$ are the important **Dirichlet means**, whose convergence is equivalent to the convergence of the real form of the Fourier series of f; i.e. the convergence of the series

$$a_0 + \sum_{n=1}^{\infty} (a_{-n} e^{-in\theta} + a_n e^{in\theta}), \tag{4.213}$$

where a_n are the Fourier coefficients of f. It is well known that the Dirichlet means converge, when f is a function of bounded variation on $\mathbb{T}$. Indeed in this case we have

$$\lim_{n\to\infty} (D_n * f)(e^{i\theta}) = \tfrac{1}{2}(f(e^{i\theta+}) + f(e^{i\theta-})) \tag{4.214}$$

for every point on $\mathbb{T}$ (we are taking f to be defined on $\mathbb{T}$, or equivalently, to be periodic with period 2π for θ real).

This result remains valid for the Cesàro means $C_n(\alpha) * f$, when $\alpha \geq 0$; and, as we have seen, when $\alpha \geq 1$, we can weaken considerably the bounded variation condition on f. For more information on the subject of summability of Fourier series we refer the reader to Zygmund[74].

Note that, applying theorem 4.4.6, we have for any polynomial $P(z) = \sum_0^n a_k z^k$ of degree $\leq n$

$$\left| \sum_{k=0}^{n-1} A_k(\alpha) a_{k+1} z^k \right| \leq A_{n-1}(\alpha) \|\Re P\|_\infty \quad (z \in \mathbb{U}) \tag{4.215}$$

when $\alpha \geq 1$.

4.5.4 Problems

1. Suppose that the finite sequence $\{0, c_1, \ldots, c_n\}$ is increasing and convex; i.e.

$$0 < c_1 \leq c_2 - c_1 \leq \cdots \leq c_n - c_{n-1}. \tag{4.216}$$

Let $P(z) = \sum_0^n a_k z^k$ be a polynomial of degree at most n. Prove that

$$\left| \sum_{k=1}^{n} c_k a_k z^k \right| \leq c_n \|\Re P\|_\infty \quad (z \in \mathbb{U}). \tag{4.217}$$

2. Let $P(z) = \sum_0^n a_k z^k$ be a polynomial of degree n satisfying

$$|\Re P(z)| \leq 1 \quad (z \in \mathbb{U}). \tag{4.218}$$

Show that, for all complex numbers $\lambda_0, \lambda_1, \ldots, \lambda_{n-1}$,

$$\left| \sum_{m=1}^{n} \sum_{k=0}^{m-1} \overline{\lambda}_k \lambda_{n-m+k} a_m \right| \leq \sum_{k=0}^{n-1} |\lambda_k|^2. \tag{4.219}$$

Deduce that, for each integer r $(0 \le r \le n-1)$,

$$\left|\sum_{k=1}^{n-r} k a_{k+r}\right| \le n - r. \tag{4.220}$$

3. Show that, if $\alpha > 1$, then

$$C_n(\alpha;\theta) > 0 \quad (\theta \text{ real}, \ n = 1, 2, \ldots). \tag{4.221}$$

4. Show that

$$\sum_{n=0}^{\infty}\sum_{k=0}^{n} c_k^{(n)}(\alpha) z^k w^n = \frac{1}{1-wz}\frac{1}{(1-w)^{\alpha+1}}, \tag{4.222}$$

and deduce that the $C(\alpha)$ Cesàro means of a function $f(z)$ analytic in $\mathbb{U}$ are the coefficients of the power series expansion in w of $f(wz)/(1-w)^{\alpha+1}$.

4.6 De la Vallée Poussin means

4.6.1

It seems likely that the best structure preserving means arising from a sequence of non-negative trigonometric polynomials $T_n(\theta)$ will lie among those satisfying the condition

$$T_n(\theta) = |P_n(e^{i\theta})|^2, \tag{4.223}$$

where each polynomial P_n has all its zeros on $\mathbb{T}$. This is certainly the case for the Fejér means. With this in mind the most obvious candidates are those generated by the polynomials $(1+z)^n$. (The significance of these particular polynomials will emerge in chapter 5). Since the middle coefficient is the binomial factor $\binom{2n}{n}$, we are led to the trigonometric polynomials

$$V_n(\theta) = \frac{1}{\binom{2n}{n}}|1+e^{i\theta}|^{2n} = \left(2\cos\frac{\theta}{2}\right)^{2n}\prod_{k=1}^{n}\left(\frac{k}{n+k}\right) \tag{4.224}$$

$$= 1 + \sum_{k=1}^{n}\left(\prod_{j=1}^{k}\frac{n+j-k}{n+j}\right)(e^{ik\theta}+e^{-ik\theta}).$$

These are the de la Vallée Poussin polynomials. The coefficients of $e^{ik\theta}$ for fixed k clearly tend to 1 as $n \to \infty$, and so the sequence $\{V_n\}$ is an approximate identity. For a function f integrable on $\mathbb{T}$, the means $V_n * f$ are called the **de la Vallée Poussin means of** f.

In a remarkable paper [40] G. Pólya and I.J. Schoenberg showed that the operator V_n* is **variation diminishing**. To explain the meaning of this, consider first a finite sequence of real numbers $\{a_1, a_2, \dots, a_n, a_1\}$ taken to be cyclic, i.e. starting and finishing at the same value a_1, which we assume is non-zero. Define $v(\{a_k\})$ to be the number of changes in sign as we traverse these numbers in order (we do not count a zero as a change in sign, so effectively we ignore the zero terms in the sequence). Because of the cyclic property, v is clearly an even integer ≥ 0. Also we may, if we wish, start at any $a_i \neq 0$, and count the sign changes in the sequence $\{a_i, a_{i+1}, \dots, a_n, a_1, \dots, a_i\}$; for the value v will be the same. If all the a_k are zero we define $v(\{a_k\}) = 0$.

Next, consider a real-valued function $f(z)$ on $\mathbb{T}$ and define

$$v(f) = \sup v(\{f(z_k)\}), \tag{4.225}$$

the supremum being taken over all finite sequences $\{f(z_1), \dots, f(z_n), f(z_1)\}$ such that $\{z_1, \dots, z_n, z_1\}$ forms a cyclic partition of $\mathbb{T}$: so if $z_k = e^{it_k}$, then $t_1 < t_2 < \cdots < t_n < t_1 + 2\pi$. If $v(f)$ is finite, it is an even integer which counts the number of sign changes of f as $\mathbb{T}$ is traversed. We may call this quantity the **sign variation of** f. Pólya and Schoenberg's result is next.

4.6.2

Theorem *Let f be a real-valued function integrable on $\mathbb{T}$ and assume that the set of points at which $f \neq 0$ has positive measure. Then*

$$v(V_n * f) \leq Z(V_n * f) \leq v(f) \quad (n = 0, 1, 2, \dots), \tag{4.226}$$

where $Z(g)$ denotes the number of zeros of g on $\mathbb{T}$.

The first inequality is clear, since the V_n means are trigonometric polynomials, and so continuous. Therefore the inequality follows from the intermediate value theorem. Thus the main content of their result lies in the second of the two inequalities: the number of zeros of a V_n mean does not exceed the sign variation of f.

As $V_n * f$ is not identically zero, it has at most $2n$ zeros, so we may assume that $v(f) = 2\nu$ for some integer ν $(0 \leq \nu < n)$. If $\nu = 0$, then f has no sign changes, so we may assume $f \geq 0$ on $\mathbb{T}$. Then by hypothesis $f > 0$ on a set of positive measure, and so $V_n * f > 0$. Therefore the result follows in this case. Hence we may assume that $1 \leq \nu < n$. Then the sup in (4.225) is an attained maximum and so $\mathbb{T}$ can be partitioned into 2ν arcs $I_1, \ldots, I_{2\nu}$ pairwise adjoining with

$$(-1)^{k-1} f(t) \geq 0 \quad (t \in I_k, 1 \leq k \leq 2\nu). \tag{4.227}$$

Furthermore, f will be non-zero on at least one point of each arc. Now without altering any integrals, and so without altering $V_n * f$, we may change f on a set of measure zero. On each arc I_k we make f identically zero if the set of points on the arc at which it is non-zero has zero measure. This has the effect of amalgamating the arc with the two adjoining arcs and thus reducing ν by 1. Thus we see that we can assume that f is non-zero on a set of positive measure on each arc I_k and zero at the endpoints. We thus have

$$(-1)^{k-1} \int_{I_k} f(t)dt > 0 \quad (1 \leq k \leq 2\nu). \tag{4.228}$$

Now we have

$$\begin{aligned}(V_n * f)(e^{i\theta}) &= \frac{1}{2\pi} \int_0^{2\pi} V_n(\theta - t) f(t)dt \\ &= \sum_{k=1}^{2\nu} (-1)^{k-1} \frac{1}{2\pi} \int_{I_k} V_n(\theta - t)|f(t)|dt \\ &= \sum_{k=1}^{2\nu} (-1)^{k-1} \int_{I_k} V_n(\theta - t) d\mu(t)\end{aligned} \tag{4.229}$$

where μ is a positive Borel measure for which $\mu(I_k) > 0\,(1 \leq k \leq 2\nu)$. We are required to show that, whatever the measure, this quantity has at most 2ν zeros in $[0, 2\pi)$. We shall apply the next lemma. The proof given was pointed out to the author by Peter Borwein and is simpler than an earlier proof worked out by the author.

4.6.3

Carathéodory's lemma *Let X be a set in $\mathbb{R}^n$ and let x belong to the convex hull of X. Then x is a convex combination of $n + 1$ points of X; i.e. $\exists x_k \in X$ $(1 \leq k \leq n + 1)$ and non-negative numbers α_k satisfying $\sum_1^{n+1} \alpha_k = 1$ such that*

$$x = \sum_{k=1}^{n+1} \alpha_k x_k. \tag{4.230}$$

Proof The convex hull of X is easily seen to be the set of all finite convex combinations of points of X (since this set is clearly itself convex and must lie in any convex set containing X). Therefore it will be enough to show that every convex combination of $N \geq n+2$ given points of X can be written as a convex combination of $n+1$ of those same points. Furthermore, a simple induction argument shows that it will be enough to prove this in the case $N = n+2$. We consider then $n+2$ points x_k of X. Let x be a point in the convex hull of the x_k, so that

$$x = \sum_{k=1}^{n+2} \alpha_k x_k, \tag{4.231}$$

where we may assume that $\alpha_k > 0\,(1 \leq k \leq n+2)$ and $\sum_1^{n+2} \alpha_k = 1$. We then have

$$\sum_{k=1}^{n+2} \alpha_k (x_k - x) = 0. \tag{4.232}$$

Now the points $x_1 - x, \ldots, x_{n+1} - x$ are $n+1$ points in $\mathbb{R}^n$ and so are linearly dependent. Thus $\exists \beta_k$ real $(1 \leq k \leq n+1)$ and not all zero such that

$$\sum_{k=1}^{n+1} \beta_k (x_k - x) = 0. \tag{4.233}$$

Choosing $\beta_{n+2} = 0$ the last two relations show that for arbitrary real t we have

$$\sum_{k=1}^{n+2} (\alpha_k + t\beta_k)(x_k - x) = 0. \tag{4.234}$$

We can choose t to satisfy (i) $\alpha_k + t\beta_k \geq 0$ for $1 \leq k \leq n+1$ and (ii) $\exists i$ such that $\alpha_i + t\beta_i = 0$, where $1 \leq i \leq n+1$ (proof for the reader). Then $\alpha_{n+2} + t\beta_{n+2} = \alpha_{n+2} > 0$, and so we obtain

$$x = \sum_{\substack{k=1 \\ k \neq i}}^{n+2} \gamma_k x_k, \tag{4.235}$$

where

$$\gamma_k = \frac{\alpha_k + t\beta_k}{\sum_k (\alpha_k + t\beta_k)}, \tag{4.236}$$

noting that $\gamma_k \geq 0$ and $\sum_k \gamma_k = 1$. Thus we have represented x as a convex combination of $n+1$ of the points x_k, as required.

4.6.4

We apply this result with the sets $X_k = \{(e^{it}, \dots, e^{int}) \colon t \in I_k\} \subset \mathbb{C}^n$ and the points $x^{(k)}$ defined by

$$x^{(k)} = \left(\int_{I_k} e^{it} d\mu(t), \dots, \int_{I_k} e^{int} d\mu(t) \right) \Big/ \mu(I_k), \tag{4.237}$$

which belongs to the convex hull of X_k. We can write therefore

$$x^{(k)} = \sum_{j=1}^{2n+1} \alpha_j^{(k)} \left(e^{i\tau_j^{(k)}}, \dots, e^{in\tau_j^{(k)}} \right), \tag{4.238}$$

where $\tau_j^{(k)} \in I_k$ $(1 \le j \le 2n+1)$, $\alpha_j^{(k)} \ge 0$ and $\sum_{j=1}^{2n+1} \alpha_j^{(k)} = 1$. Denoting the coefficients of V_n by A_r $(-n \le r \le n)$, it follows that

$$\begin{aligned}
(V_n * f)(e^{i\theta}) &= \sum_{k=1}^{2\nu} (-1)^{k-1} \int_{I_k} \sum_{r=-n}^{n} A_r e^{ir(\theta - t)} d\mu(t) \\
&= \sum_{k=1}^{2\nu} (-1)^{k-1} \sum_{r=-n}^{n} A_r e^{ir\theta} \int_{I_k} e^{-irt} d\mu(t) \\
&= \sum_{k=1}^{2\nu} (-1)^{k-1} \sum_{r=-n}^{n} A_r e^{ir\theta} \mu(I_k) \sum_{j=1}^{2n+1} \alpha_j^{(k)} e^{-ir\tau_j^{(k)}} \\
&= \sum_{k=1}^{2\nu} (-1)^{k-1} \mu(I_k) \sum_{j=1}^{2n+1} \alpha_j^{(k)} V_n\left(\theta - \tau_j^{(k)}\right) = \sum_{j=1}^{m} c_m V_n(\theta - \tau_m),
\end{aligned} \tag{4.239}$$

where $\tau_1 < \tau_2 < \cdots < \tau_m < \tau_1 + 2\pi$ and the number of sign changes in the sequence $(c_1, \dots, c_m, c_1)$ is 2ν. Now

$$\begin{aligned}
V_n(\theta - \pi - \tau) &= \left(2 \sin \frac{\theta - \tau}{2} \right)^{2n} \prod_{k=1}^{n} \left(\frac{k}{n+k} \right) \\
&= \left(2 \cos \frac{\theta}{2} \cos \frac{\tau}{2} \right)^{2n} \left(\tan \frac{\theta}{2} - \tan \frac{\tau}{2} \right)^{2n} \prod_{k=1}^{n} \left(\frac{k}{n+k} \right) \\
&= \frac{(x - \xi)^{2n}}{(1 + x^2)^n (1 + \xi^2)^n} \prod_{k=1}^{n} \left(\frac{4k}{n+k} \right),
\end{aligned} \tag{4.240}$$

where $x = \tan \frac{\theta}{2}$ and $\xi = \tan \frac{\tau}{2}$. We thus see that the number of zeros on $[0, 2\pi)$ of $V_n * f$ is the same as the number of zeros on $(-\infty, +\infty)$ of the polynomial

$$\sum_{j=1}^{m} d_m (x - \xi_m)^{2n}, \tag{4.241}$$

where $\xi_m = \tan \frac{\tau_m}{2}$, $\xi_1 < \xi_2 < \cdots < \xi_m$ and $d_m = c_m/(1+\xi_m^2)$. As the number of sign changes in the sequence $\{d_m\}$ is the same as that for $\{c_m\}$, Pólya and Schoenberg's theorem follows from a result of Sylvester, which we give next.

4.6.5

Sylvester's theorem *Let $P(x)$ be a real polynomial, not identically zero, which can be written in the form*

$$P(x) = \sum_{k=1}^{n} a_k(x-\lambda_k)^m, \tag{4.242}$$

where $m \geq 1$, $n \geq 2$, $a_k \neq 0\,(1 \leq k \leq n)$ and $\lambda_k < \lambda_{k+1}$ $(1 \leq k \leq n-1)$. Let Z denote the number of real zeros of P and C the number of sign changes in the sequence

$$a_1, a_2, \ldots, a_n, (-1)^m a_1. \tag{4.243}$$

Then

$$Z \leq C. \tag{4.244}$$

Proof If the a_k have the same sign and m is even, then $C = Z = 0$. If the a_k have the same sign and m is odd, then $C = 1$. Also $P'(x)$ has no zeros, and hence by Rolle's theorem, $Z \leq 1$.

Assume that $C \geq 1$ and that the result is true for polynomials of this form which have at most $C-1$ changes of sign. We can choose successive terms a_j, a_{j+1} which have different signs, where $1 \leq j < n$. Let

$$P(x) = \sum_{k=0}^{m} b_k x^k \tag{4.245}$$

and choose λ to satisfy

$$\lambda_j < \lambda < \lambda_{j+1}, \lambda \neq -\frac{b_{m-1}}{mb_m} \quad \text{if } b_m \neq 0 \text{ and } P(\lambda) \neq 0. \tag{4.246}$$

Let

$$F(x) = \frac{P(x)}{(x-\lambda)^m}. \tag{4.247}$$

We then have

$$\begin{aligned}(x-\lambda)^{m+1}F'(x) &= (x-\lambda)P'(x) - mP(x) = \sum_{k=1}^{m} ma_k(\lambda_k - \lambda)(x-\lambda_k)^{m-1} \\ &= -\sum_{k=1}^{m} c_k(x-\lambda_k)^{m-1}.\end{aligned} \tag{4.248}$$

Now $\lambda_k < \lambda$ for $1 \le k \le j$ and $\lambda_k > \lambda$ for $j+1 \le k \le n$. Therefore c_k has the same sign as a_k for $1 \le k \le j$, and the opposite sign to a_k for $j+1 \le k \le n$. It follows that the number of sign changes in the sequence $c_1, \ldots, c_n, (-1)^{m-1}c_1$ is exactly $C-1$. Applying the induction hypothesis, the number of real zeros of $(x-\lambda)^{m+1}F'(x)$ does not exceed $C-1$, and so $F'(x)$ has at most $C-1$ real zeros.

We show that this implies that F, and therefore P, has at most C real zeros, the required result. To see this, we let

$$Q(x) = x^m P\left(\frac{1}{x}\right) = \sum_{k=0}^{m} b_{m-k}x^k \tag{4.249}$$

which gives

$$F\left(\frac{1}{x}\right) = \frac{Q(x)}{(1-\lambda x)^m}. \tag{4.250}$$

Differentiating logarithmically we obtain

$$\frac{-\dfrac{1}{x^2}F'\left(\dfrac{1}{x}\right)}{F\left(\dfrac{1}{x}\right)} = \frac{Q'(x)}{Q(x)} + \frac{m\lambda}{1-\lambda x} \to \frac{b_{m-1}}{b_m} + m\lambda \text{ as } x \to 0. \tag{4.251}$$

We deduce that, if $b_m \ne 0$, then

$$\frac{x^2F'(x)}{F(x)} \to -\frac{b_{m-1}+m\lambda b_m}{b_m} \text{ as } x \to \infty. \tag{4.252}$$

By the choice of λ, this limit is non-zero. It follows that F' and F have either the same sign near $-\infty$ (and $+\infty$) or opposite sign near $+\infty$ (and $-\infty$). F' has in the first case a zero preceding all the zeros of F, and in the second case a zero succeeding all the zeros of F. In either case, by Rolle's theorem, F has at most C real zeros.

It remains to consider the case when

$$b_k = 0\ (s+1 \le k \le m);\ b_s \ne 0. \tag{4.253}$$

We put

$$R(x) = x^s P\left(\frac{1}{x}\right) = b_s + b_{s-1}x + \cdots + b_0x^s \tag{4.254}$$

and note that

$$F\left(\frac{1}{x}\right) = \frac{x^{m-s}R(x)}{(1-\lambda x)^m}. \tag{4.255}$$

We obtain

$$\frac{-\frac{1}{x^2}F'\left(\frac{1}{x}\right)}{F\left(\frac{1}{x}\right)} = \frac{m-s}{x} + \frac{m\lambda}{1-\lambda x} + \frac{R'(x)}{R(x)} \tag{4.256}$$

and therefore

$$\frac{xF'(x)}{F(x)} \to s - m \text{ as } x \to \pm\infty. \tag{4.257}$$

As this limit is negative, it follows that F' and F have the same sign near $-\infty$ and opposite sign near $+\infty$. Therefore in fact F' has a zero preceding all the zeros of F and a zero succeeding all the zeros of F. Hence, by Rolle's theorem, F has at most $C - 1$ real zeros. The result follows by induction.

4.7 Integral representations

4.7.1

In theorem 4.2.7 we showed that a bounded convolution operator on $\mathcal{H}_S$ could be extended to a bounded operator on $\mathcal{H}$. It is then represented by a function in h^1. There is an important representation theorem for functions in h^1 in terms of complex Borel measures. The advantage of such a representation is that it enables us to obtain immediately a representation for any bounded convolution operator on $\mathcal{H}_S$.

4.7.2

Theorem *Let $f \in \mathcal{H}$ and suppose that*

$$\frac{1}{2\pi}\int_0^{2\pi} |f(re^{i\theta})|d\theta \le 1 \quad (0 \le r < 1). \tag{4.258}$$

Then there exists a complex Borel measure μ on $\mathbb{T}$ satisfying

$$\int_{\mathbb{T}} |d\mu| \le 1 \tag{4.259}$$

such that

$$f(z) = \int_{\mathbb{T}} \mathcal{P}(ze^{-it})d\mu(t) \quad (z \in \mathbb{U}). \tag{4.260}$$

In particular, if $f > 0$ on $\mathbb{U}$ with $f(0) = 1$, then μ is a probability measure (i.e. a positive measure of total mass 1).

The proof of this result can be found in many books (e.g. P. Duren [16]), so we omit it. In essence it is a generalisation of the Poisson formula to the boundary $\mathbb{T}$.

4.7.3

We deduce immediately from theorem 4.2.7.

Theorem *Let $f*$ be a bounded convolution operator on $\mathcal{H}_S$ of norm ≤ 1. Then $\exists$ a complex Borel measure μ on $\mathbb{T}$ satisfying (4.259) such that*

$$(f * \mathcal{P}_S)(z) = \int_{\mathbb{T}} \mathcal{P}_S(ze^{-it})d\mu(t) \quad (z \in \mathbb{U}). \tag{4.261}$$

In particular, if $f \in \mathcal{A}$ and $f*$ is a bounded operator on $\mathcal{A}$ of norm ≤ 1, then

$$f(z) = \int_{\mathbb{T}} \frac{d\mu(t)}{1 - ze^{-it}}. \tag{4.262}$$

If, in addition, $f(0) = 1$, then the measure is a probability measure.

Note that the converse results also hold; if f has such a representation, then $f*$ is a bounded operator on the relevant subspace; for if $g \in \mathcal{H}_S$ and f is given by (4.261) then

$$(f * g)(z) = \int_{\mathbb{T}} g(ze^{-it})d\mu(t). \tag{4.263}$$

4.8 Generalised convolution operators

4.8.1

We conclude the chapter with a brief account of a class of operators, which arise naturally in a number of situations. Some examples will be given in chapter 8. Although they can be defined on the class $\mathcal{H}$, most applications are for the analytic case, so we shall confine our attention to operators on the class $\mathcal{A}$.

A continuous linear operator Λ: $\mathcal{A} \to \mathcal{A}$ is said to be a **generalised convolution operator** (abbreviated as **g.c. operator**) if, for each integer $n \geq 0$,

$$\Lambda z^n = z^n \tau_n(z), \tag{4.264}$$

where $\tau_n \in \mathcal{A}$. Simple convolution is the case when the τ_n are constants. Two other examples are as follows:

(1) multiplication by a fixed function $g \in \mathcal{A}$; $\Lambda f = gf$ for every $f \in \mathcal{A}$;
(2) subordination or composition with a Schwarz function $\omega \in \mathcal{A}$; $\Lambda f = f \circ \omega$, where $\omega(0) = 0$ and $|\omega(z)| < 1$ for $z \in \mathbb{U}$. Schwarz functions satisfy **Schwarz's lemma**: that under the two conditions $|\omega(z)| \leq |z|$ for $z \in \mathbb{U}$.

The kernel of the operator is given by

$$H(z, \zeta) = \Lambda\left(\frac{1}{1 - z\zeta}\right) = \sum_{n=0}^{\infty} \tau_n(z)\zeta^n z^n. \tag{4.265}$$

As this is analytic for $|z| < 1$ and $|\zeta| \leq 1$, it follows that the function

$$K(z, \zeta) = \sum_{n=0}^{\infty} \tau_n(\zeta) z^n \tag{4.266}$$

is analytic in the open polydisc $\{|\zeta| < 1, |z| < 1\}$ and the action of the operator is given by

$$(\Lambda f)(z) = (K(z, \zeta) * f(z))_{\zeta = z}. \tag{4.267}$$

This formula defines a 1–1 correspondence between g.c. operators Λ and functions K analytic in the open polydisc $\{|\zeta| < 1, |z| < 1\}$. We call K the **g.c. kernel** of Λ.

A second characterisation of g.c. operators is as follows: Λ is g.c. iff

$$\Lambda\left(\sum_{n=0}^{\infty} a_n z^n\right) = \sum_{n=0}^{\infty} b_n z^n, \tag{4.268}$$

where each coefficient b_n is a linear combination (defined by Λ) of the coefficients $a_0, a_1, \ldots, a_n$.

4.8.2

Lemma *Let Λ be a g.c. operator p-**mean preserving** on $\mathcal{A}$, where $p \geq 1$.* (By this we mean that for each function f belonging to the Hardy space H^p, we have $\|\Lambda f\|_p \leq \|f\|_p$.) *Then for each $f \in \mathcal{A}$ we have*

$$\int_0^{2\pi} |\Lambda f(re^{i\theta})|^p d\theta \leq \int_0^{2\pi} |f(re^{i\theta})|^p d\theta \quad (0 \leq r < 1). \tag{4.269}$$

Proof Let $f = \sum_0^\infty a_n z^n \in \mathcal{A}$ and for $\zeta = re^{i\theta}$ $(0 < r < 1, 0 \leq \theta \leq 2\pi)$ define f_ζ as the function $z \mapsto f(\zeta z)$. Let $K(z, \zeta) = \sum_0^\infty \tau_n(\zeta) z^n$ be the g.c. kernel of Λ. Then for $0 < \rho < 1$

$$\begin{aligned}
\int_0^{2\pi} |\Lambda f_\zeta(\rho e^{i\phi})|^p d\phi &= \int_0^{2\pi} \left| \sum_0^\infty a_n \tau_n(\rho e^{i\phi}) \zeta^n \rho^n e^{in\phi} \right|^p d\phi \\
&= \int_0^{2\pi} \left| \sum_0^\infty a_n \tau_n\left(\frac{R}{r} e^{i\psi} e^{-i\theta}\right) R^n e^{in\psi} \right|^p d\psi,
\end{aligned} \tag{4.270}$$

where $R = \rho r$. The first term increases with ρ and so by hypothesis the third term is

$$\leq \lim_{\rho \to 1} \int_0^{2\pi} |f(\zeta \rho e^{i\phi})|^p d\phi = \int_0^{2\pi} |f(re^{i\phi})|^p d\phi. \qquad (4.271)$$

This holds for all $e^{-i\theta}$ and the third term is subharmonic in x for $|x| \leq 1$, where x has replaced $e^{-i\theta}$. Hence by the maximum principle we may replace $e^{-i\theta}$ by r, and then letting $R \to r$ we obtain the result.

4.8.3

In particular we may let $p \to \infty$ obtaining

Lemma *If Λ is a bound preserving g.c. operator on $\mathcal{A}$, then for each $f \in \mathcal{A}$ we have*

$$M(r, \Lambda f) \leq M(r, f) \quad (0 < r < 1). \qquad (4.272)$$

4.8.4

The main result for convexity preserving g.c. operators is the following; the proof is left as an exercise.

Theorem *Let Λ be a g.c. operator with g.c. kernel $K(z, \zeta)$. Λ is convexity preserving on $\mathcal{A}$ if, and only if,*

$$\Re K(z, \zeta) > \frac{1}{2} \text{ and } K(0, z) = 1, \qquad (4.273)$$

for $|z| < 1$ and $|\zeta| < 1$. Every such operator is strictly convexity preserving and p-mean preserving on $\mathcal{A}$ for $p \geq 1$.

4.8.5

Rogosinski's coefficient theorem *Let Λ be a 2-mean preserving g.c. operator on $\mathcal{A}$. If $f = \sum_0^\infty a_n z^n \in \mathcal{A}$ and $g = \Lambda f = \sum_0^\infty b_n z^n$, then for each $n = 0, 1, 2, \ldots$ we have*

$$\sum_{k=0}^{n} |b_k|^2 \leq \sum_{k=0}^{n} |a_k|^2. \qquad (4.274)$$

In particular, this inequality holds for every convexity preserving g.c. operator.

Proof From the algebra of g.c. operators we have $e_n * \Lambda f = e_n * \Lambda(e_n * f)$. Therefore for $0 < r < 1$ we have

$$\begin{aligned}\sum_0^n |b_k|^2 r^{2k} &= \frac{1}{2\pi}\int_0^{2\pi} |(e_n * \Lambda f)(re^{i\theta})|^2 d\theta \\ &= \frac{1}{2\pi}\int_0^{2\pi} |(e_n * \Lambda(e_n * f)(re^{i\theta})|^2 d\theta \\ &\leq \frac{1}{2\pi}\int_0^{2\pi} |\Lambda(e_n * f)(re^{i\theta})|^2 d\theta \\ &\leq \frac{1}{2\pi}\int_0^{2\pi} |(e_n * f)(re^{i\theta})|^2 d\theta = \sum_0^n |a_k|^2 r^{2k}. \qquad (4.275)\end{aligned}$$

Letting $r \to 1$ the result follows.

Rogosinski proved this result in the case that g is subordinate to f with essentially the above argument. A second case, that of majorisation, was proved by Clunie: *if* $|g| \leq |f|$ *in* $\mathbb{U}$, *then the coefficient inequality holds*. For $h = g/f$ is analytic and satisfies $|h| \leq 1$ in $\mathbb{U}$. The operator $\Lambda F = hF$ $(F \in \mathcal{A})$ is g.c. and trivially 2-mean preserving. Since $\Lambda f = g$, the conclusion follows.

5

Circular regions and Grace's theorem

5.1 Convolutions and duality

5.1.1

We saw in chapter 2 that the fundamental theorem of algebra leads to the interesting result that the critical points of an analytic polynomial lie in the convex hull of the set of zeros of the polynomial. This result has a remarkable generalisation to a theorem of J. H. Grace [18], which correlates the distribution of the zeros of certain families of polynomials. There are a number of equivalent forms of the result, all of which have considerable interest in their own right, and we shall develop the theory by adopting the linear approach due to G. Szegö [63]. This gives a very useful form of the theorem for applications. Furthermore it can be proved in a simple and elegant manner and is open to further generalisations.

In order to explain the main idea and to set the tone for these generalisations we introduce a general concept of duality within the space $\mathcal{A}$ of functions analytic in the open unit disc $\mathbb{U}$. Let $\mathcal{M}$ denote a fixed non-empty subset of $\mathcal{A}$. If $\mathcal{F}$ is a subset of $\mathcal{A}$, we define

$$\mathcal{F}^*(\mathcal{M}) = \{\phi \in \mathcal{A}: \phi * f \in \mathcal{M}\ \forall f \in \mathcal{F}\} = \{\phi \in \mathcal{A}: (\phi*)(\mathcal{F}) \subset \mathcal{M}\}. \tag{5.1}$$

We call this set the $\mathcal{M}$-**dual of** $\mathcal{F}$. It may of course be empty. Assuming it is not, we can then consider the **second** $\mathcal{M}$**-dual of** $\mathcal{F}$, defined as the $\mathcal{M}$-dual of $\mathcal{F}^*(\mathcal{M})$, and denote this by $\mathcal{F}^{**}(\mathcal{M})$. We then make the quite trivial observation that

$$\mathcal{F} \subset \mathcal{F}^{**}(\mathcal{M}). \tag{5.2}$$

The logical construction involved in the definition of the second dual can be conveniently expressed in the following manner: *a function g lies in the second*

$\mathcal{M}$-dual of $\mathcal{F}$ iff for all $\phi \in \mathcal{A}$

$$\phi * f \in \mathcal{M} \, \forall f \in \mathcal{F} \Rightarrow \phi * g \in \mathcal{M}. \tag{5.3}$$

The reader will recognise that this is a generalisation of a familiar concept of duality in functional analysis. Indeed in chapter 4 we made some use of the idea in considering 'bound preserving' and 'convexity preserving' operators, which correspond to taking $\mathcal{M}$ respectively as the class of functions bounded by 1 and the class of functions of positive real part normalised at the origin by $f(0) = 1$. The main purpose of introducing the second dual is to exploit the containment relation (5.2). It may happen that a relatively small class $\mathcal{F}$ generates a substantial second dual space. When this is the case, the interpretation of the second dual in terms of the implication relation (5.6) establishes a powerful theorem.

Since our main interest is in the distribution of zeros, it is important (as we saw in chapter 2) that we have information about functions with no zeros. With this in mind we take $\mathcal{M} = \mathcal{V} =$ the class of functions $f \in \mathcal{A}$ satisfying $f(z) \neq 0$ $(z \in \mathbb{U})$ and (for the purposes of normalisation) $f(0) = 1$. By Hurwitz's theorem the class $\mathcal{V}$ is closed (in the topology of local uniform convergence). However, it is certainly not compact.

5.1.2

We can now state Grace's theorem in the form due to Szegö [63] as a particularly beautiful example of the above ideas.

Grace–Szegö theorem *Let $f \in \mathcal{A}$ satisfy*

$$f(z) * (1 + z)^n \neq 0 \quad (z \in \mathbb{U}), \tag{5.4}$$

where n is a natural number. Then for each polynomial $P(z)$ of degree not exceeding n satisfying

$$P(z) \neq 0 \quad (z \in \mathbb{U}), \tag{5.5}$$

we have

$$f(z) * P(z) \neq 0 \quad (z \in \mathbb{U}). \tag{5.6}$$

In our duality terminology the theorem states that the second $\mathcal{V}$-dual generated by the single function $(1 + z)^n$ is the class of all polynomials of degree $\leq n$ whose zeros lie outside $\mathbb{U}$.

There are a number of alternative ways of expressing the theorem, most of which we shall be exploring later in the chapter. However, one of these ways is immediate and will serve to introduce another theme which will occur later in the book. *If P and Q are functions each satisfying the condition* (5.4), *so that we have*

$$P(z) * (1+z)^n \neq 0 \text{ and } Q(z) * (1+z)^n \neq 0 \quad (z \in \mathbb{U}), \tag{5.7}$$

then

$$P(z) * Q(z) * (1+z)^n \neq 0 \quad (z \in \mathbb{U}). \tag{5.8}$$

In other words the ***Grace class*** *$\mathcal{G}_n$ of functions P satisfying (5.4) is closed under convolution.* There is no essential loss in assuming that the functions in $\mathcal{G}_n$ are polynomials of degree $\leq n$ normalised at the origin by $P(0) = 1$. $\mathcal{G}_n$ is then obtained from the class $\mathcal{V}_n$ of non-vanishing polynomials of degree $\leq n$ by convolving each member of $\mathcal{V}_n$ with the polynomial

$$I_n(z) = \sum_{k=0}^{n} \frac{1}{\binom{n}{k}} z^k = \sum_{k=0}^{n} \frac{k!(n-k)!}{n!} z^k, \tag{5.9}$$

which is the convolution inverse in the class $\mathcal{A}_n$ of polynomials of degree $\leq n$ of the function $(1+z)^n$. Note that the convolution unit element in $\mathcal{A}_n$

$$e_n(z) = 1 + z + \cdots + z^n \tag{5.10}$$

is a member of $\mathcal{G}_n$.

Thus we see that Grace's theorem asserts that, considered as convolution operators, the members of the Grace class are structure preserving elements in $\mathcal{A}_n$. It is usually the case that a class which preserves one type of structure also preserves other structures. This is certainly the case with $\mathcal{G}_n$. For example *the functions in $\mathcal{G}_n$ are convexity preserving on $\mathcal{A}_n$*. To prove this, consider a polynomial P of degree $\leq n$ satisfying

$$\Re P(z) > 0 \quad (z \in \mathbb{U}). \tag{5.11}$$

Then for each w satisfying $\Re w \leq 0$ we have $P(z) - w \neq 0$ $(z \in \mathbb{U})$. Hence if $f \in \mathcal{G}_n$

$$f(z) * (P(z) - w) \neq 0 \quad (z \in \mathbb{U}), \tag{5.12}$$

and so $f(z) * P(z) \neq f(z) * w = w$ (since $f(0) = 1$), and therefore

$$\Re(f(z) * P(z)) > 0 \quad (z \in \mathbb{U}), \tag{5.13}$$

so the assertion follows.

We deduce from the results of chapter 4 that each function $f \in \mathcal{G}_n$ can be written in the form

$$f(z) = \int_{\mathbb{T}} e_n(ze^{-it})d\mu(t) \tag{5.14}$$

where μ is a probability measure on $\mathbb{T}$. It follows that, *if $P(z)$ is a polynomial of degree $\le n$ with no zeros in $\mathbb{U}$ and satisfying $P(0) = 1$, then we can write P in the form*

$$P(z) = \int_{\mathbb{T}} (1 + ze^{-it})^n d\mu(t), \tag{5.15}$$

where μ is a probability measure on $\mathbb{T}$.

Our proof of Grace's theorem will be based on a duality lemma for finite products, which also will have later applications.

5.1.3

Finite product duality lemma *Let $\alpha_1, \dots, \alpha_n$ be given complex numbers and denote by $\Pi(\alpha_1, \dots, \alpha_n)$ the family of functions of the form*

$$\prod_{k=1}^{n}(1 + x_k z)^{\alpha_k} \quad (|x_k| \le 1, 1 \le k \le n, z \in \mathbb{U}). \tag{5.16}$$

Suppose that for certain complex numbers β_j $(1 \le j \le m)$ the family of functions $\Pi(\beta_1, \dots, \beta_m)$ lies in the second $\mathcal{V}$-dual of $\Pi(\alpha_1, \dots, \alpha_n)$. Then, if

$$\lambda = \sum_{k=1}^{n} \alpha_k - \sum_{j=1}^{m} \beta_j, \tag{5.17}$$

the family $\Pi(\lambda, \beta_1, \dots, \beta_m)$ lies in the second $\mathcal{V}$-dual of $\Pi(\alpha_1, \dots, \alpha_n)$.

Proof For fixed $x \in \mathbb{U}$ we set

$$a(\tau) = \frac{\tau + x}{1 + \bar{x}\tau}, \qquad b(z) = \frac{z + \bar{x}}{1 + xz} \tag{5.18}$$

and note the identity

$$1 + a(\tau)z = \frac{1 + xz}{1 + \bar{x}\tau}(1 + b(z)\tau). \tag{5.19}$$

Hence, if $|\zeta_k| \le 1$ $(1 \le k \le n)$, $|w| \le 1$ and $\alpha = \sum_1^n \alpha_k$, we have

$$\prod_{k=1}^{n}(1 + a(\zeta_k w)z)^{\alpha_k} = (1 + xz)^{\alpha} \prod_{k=1}^{n} \left(\frac{1 + b(z)\zeta_k w}{1 + \bar{x}\zeta_k w} \right)^{\alpha_k}. \tag{5.20}$$

Now $\tau \mapsto a(\tau)$ gives a 1–1 mapping of $\overline{\mathbb{U}}$ onto $\overline{\mathbb{U}}$, and therefore, if $\phi \in \Pi^*(\alpha_1, \dots, \alpha_n)$, we obtain for $|\zeta_k| \le 1$ $(1 \le k \le n)$ and $|w| \le 1$

$$\phi(z) * (1 + xz)^\alpha \prod_{k=1}^{n} \left(\frac{1 + b(z)\zeta_k w}{1 + \overline{x}\zeta_k w} \right)^{\alpha_k} \ne 0 \quad (z \in \mathbb{U}). \tag{5.21}$$

We now observe that for fixed $z \in \mathbb{U}$ the function

$$w \mapsto \prod_{k=1}^{n} (1 + b(z)\zeta_k w)^{\alpha_k} = \prod_{k=1}^{n} (1 + \zeta_k w)^{\alpha_k} *_w \frac{1}{1 - b(z)w} \tag{5.22}$$

and hence (5.21) implies

$$\prod_{k=1}^{n} (1 + \zeta_k w)^{\alpha_k} *_w \left\{ \phi(z) *_z (1 + xz)^\alpha \frac{1}{1 - b(z)w} \right\} \ne 0, \tag{5.23}$$

which implies that for each $z \in \mathbb{U}$ the function

$$w \mapsto \phi(z) *_z (1 + xz)^\alpha \frac{1}{1 - b(z)w} \quad (w \in \mathbb{U}) \tag{5.24}$$

is in $\Pi^*(\alpha_1, \dots, \alpha_n)$. By the hypothesis this function is also in $\Pi^*(\beta_1, \dots, \beta_m)$ and so we obtain

$$\prod_{j=1}^{m} (1 + \zeta_j w)^{\beta_j} *_w \left\{ \phi(z) *_z (1 + xz)^\alpha \frac{1}{1 - b(z)w} \right\} \ne 0 \tag{5.25}$$

for $|\zeta_j| \le 1$ $(1 \le j \le m)$, $|w| < 1$, $|z| < 1$. This gives

$$\phi(z) *_z (1 + xz)^\lambda \prod_{j=1}^{m} (1 + za(\zeta_j w))^{\beta_j} \ne 0 \tag{5.26}$$

and the result follows allowing $|w| \to 1$, $|x| \to 1$ and applying Hurwitz's theorem.

5.1.4 Proof of the Grace–Szegö theorem

We prove the result by induction. The theorem is trivial when $n = 0$ and immediately verified when $n = 1$. Assume that $n \ge 2$ and the result is verified up to $n - 1$. If f satisfies (5.4), then by the fundamental theorem of algebra we can write

$$f(z) * (1 + z)^n = c \prod_{k=1}^{n} (1 + z_k z), \tag{5.27}$$

where c is a non-zero constant and $|z_k| \leq 1$ $(1 \leq k \leq n)$. Then

$$f(z) * (1 + xz) = c\left(1 + \frac{1}{n}\sum_{k=1}^{n} z_k xz\right) \neq 0 \tag{5.28}$$

if $|x| \leq 1$ and $z \in \mathbb{U}$. Thus using the notations of the duality lemma we have that $\Pi(1)$ lies in the second dual of $\Pi(n)$, and so applying the lemma, $\Pi(n-1, 1)$ does so as well; i.e. we have

$$f(z) * (1 + xz)^{n-1}(1 + yz) \neq 0 \quad (|x| \leq 1, |y| \leq 1, |z| < 1). \tag{5.29}$$

In particular, taking $y = 0$, $x = 1$ we deduce that

$$f(z) * (1 + z)^{n-1} \neq 0 \quad (z \in \mathbb{U}). \tag{5.30}$$

By the induction hypothesis this means that

$$f(z) * \prod_{k=1}^{n-1}(1 + x_k z) \neq 0 \quad (|x_k| \leq 1, 1 \leq k \leq n-1, z \in \mathbb{U}) \tag{5.31}$$

and so $\Pi(\alpha_1, \ldots, \alpha_{n-1})$ lies in the second dual of $\Pi(n)$, where $\alpha_1 = \cdots = \alpha_{n-1} = 1$. Applying the duality lemma again we deduce that $\Pi(\alpha_1, \ldots, \alpha_n)$ lies in the second dual of $\Pi(n)$, where $\alpha_1 = \cdots = \alpha_n = 1$. The result follows from the fundamental theorem of algebra.

5.1.5

Our next task is to modify the statement of the Grace–Szegö theorem to permit generalisation to regions other than $\mathbb{U}$. To this end we shall interpret the result in terms of linear functionals. Let Λ be a continuous linear functional on $\mathcal{A}$ and suppose that

$$\Lambda(1 + xz)^n \neq 0 \quad (|x| \leq 1). \tag{5.32}$$

By Toeplitz's theorem $\exists f$ analytic in $\overline{\mathbb{U}}$ such that $\forall g \in \mathcal{A}$

$$\Lambda g = (f * g)(1). \tag{5.33}$$

Therefore we have

$$f(x) * (1 + x)^n \neq 0 \quad (|x| \leq 1). \tag{5.34}$$

Since $f(x) * (1+x)^n$ is a polynomial in x, its zeros must lie in the open exterior of the unit disc and therefore $\exists R > 1$ such that

$$f(Rz) * (1 + z)^n \neq 0 \quad (z \in \mathbb{U}). \tag{5.35}$$

By the Grace–Szegö theorem

$$f(Rz) * P(z) \neq 0 \quad (z \in \mathbb{U}) \tag{5.36}$$

for every $P \in \mathcal{V}_n$. Putting $z = 1/R$ we obtain

$$\Lambda P \neq 0 \quad (P \in \mathcal{V}_n). \tag{5.37}$$

Thus a continuous linear functional which is non-vanishing on the set of functions $\{(1 + xz)^n : |x| \leq 1\}$ remains non-vanishing on the whole of $\mathcal{V}_n$.

Similar reasoning shows that, if Λ is a linear functional such that

$$\Lambda(1 + xz)^n \neq 0 \quad (|x| < 1), \tag{5.38}$$

then

$$\Lambda P \neq 0 \tag{5.39}$$

for every polynomial P of degree $\leq n$ satisfying

$$P(z) \neq 0 \quad (|z| \leq 1). \tag{5.40}$$

5.2 Circular regions

5.2.1

A **circular region** is the image of either the open unit disc or the closed unit disc under a non-singular mapping of the form

$$w = \frac{az + b}{cz + d} \tag{5.41}$$

where a, b, c, d are constants. Without loss of generality the non-singularity condition may be taken to be

$$ad - bc = 1. \tag{5.42}$$

These mappings are 1–1 mappings of the extended plane onto itself, which have the property of mapping every circle onto either a circle or a line, and every line onto either a circle or a line (a line may be regarded as a circle passing through ∞). A circular region is therefore one of the following: an open disc, a closed disc, an open half-plane, a closed half-plane including ∞, the open exterior of a circle including ∞ or the closed exterior of a circle including ∞.

Let Λ be a linear functional defined on the class of polynomials of degree $\leq n$ and let R be a circular region such that (5.41) maps the unit disc (open or closed) to R. Consider a polynomial $P(w)$ $(w \in R)$ of degree $\leq n$ and let

$$Q(z) = (cz + d)^n P\left(\frac{az + b}{cz + d}\right). \tag{5.43}$$

Then Q is a polynomial of degree $\leq n$ and

$$P(w) = (a - cw)^n Q\left(\frac{dw - b}{a - cw}\right). \tag{5.44}$$

Conversely, if Q is a polynomial of degree $\leq n$, then P defined by this relationship is a polynomial of degree $\leq n$. Therefore we can define a new linear functional Λ' on the class of polynomials of degree $\leq n$ by the relation

$$\Lambda'(Q) = \Lambda(P). \tag{5.45}$$

Suppose now that P is a polynomial of degree n such that

$$P(w) \neq 0 \ (w \in R) \text{ and } \Lambda P = 0. \tag{5.46}$$

Then $Q(z)$ is a polynomial of degree $\leq n$ such that

$$Q(z) \neq 0 \ (z \in \mathbb{U} \text{ or } \overline{\mathbb{U}}) \text{ and } \Lambda' Q = 0 \tag{5.47}$$

($\mathbb{U}$ if R is open, $\overline{\mathbb{U}}$ if R is closed). Hence $\exists x$ ($|x| \leq 1$ or $|x| < 1$) such that

$$\Lambda'(1 + xz)^n = 0. \tag{5.48}$$

This gives

$$\Lambda(a - cw + x(dw - b))^n = 0. \tag{5.49}$$

Now if $\infty \in R$, then $dx - c \neq 0$, and we obtain

$$\Lambda(w - \zeta)^n = 0, \tag{5.50}$$

for some $\zeta \notin R$. On the other hand, if $\infty \notin R$, then we deduce that

$$\text{either } \Lambda 1 = 0 \text{ or } \Lambda(w - \zeta)^n = 0, \tag{5.51}$$

for some $\zeta \notin R$. Furthermore, in the case that $\infty \notin R$, the reasoning remains valid for polynomials P of degree $\leq n$, for it is then still true that $P(w) \neq 0 \Rightarrow Q(z) \neq 0$. This latter implication is false, when $\infty \in R$ and P has degree $< n$, for then $Q(z) = 0$ at the point $z = -d/c$, which lies in $\mathbb{U}$ or $\overline{\mathbb{U}}$.

5.2.2

Theorem *Let R be a circular region, Λ a linear functional defined on the class of polynomials of degree n and P a polynomial of degree $\leq n$. Suppose that*

$$P(w) \neq 0 \ (w \in R) \text{ and } \Lambda P = 0. \tag{5.52}$$

Then (i) if $\infty \in R$ and P has degree m, $\exists \zeta \notin R$ such that

$$\Lambda(w - \zeta)^m = 0; \tag{5.53}$$

and (ii) if $\infty \notin R$, either $\Lambda 1 = 0$ or $\exists \zeta \notin R$ such that

$$\Lambda(w-\zeta)^n = 0. \tag{5.54}$$

Since the complement of a circular region is a circular region, the result may be re-phrased as follows.

Let R be a circular region, Λ a linear functional and P a polynomial of degree n with all its zeros in R. Then, if $\Lambda P = 0$, either $\exists \zeta \in R$ such that $\Lambda(w-\zeta)^n = 0$, or $\infty \in R$ and $\Lambda 1 = 0$.

5.2.3

Theorem *Let R be a circular region, Λ_1 and Λ_2 linear functionals and P a polynomial of degree n with all its zeros in R. Suppose that*

$$\Lambda_2(w-\zeta)^n \neq 0 \;\; \forall \zeta \in R, \tag{5.55}$$

and further, if $\infty \in R, \Lambda_2 1 \neq 0$. Then either $\exists \zeta \in R$ such that

$$\frac{\Lambda_1 P}{\Lambda_2 P} = \frac{\Lambda_1(w-\zeta)^n}{\Lambda_2(w-\zeta)^n} \tag{5.56}$$

or $\infty \in R$ and

$$\frac{\Lambda_1 P}{\Lambda_2 P} = \frac{\Lambda_1 1}{\Lambda_2 1}. \tag{5.57}$$

For certainly, by the previous theorem, $\Lambda_2 P \neq 0$, and so if $\Lambda_1 P/\Lambda_2 P = \alpha$, we can apply this theorem to the linear functional $\Lambda_1 - \alpha\Lambda_2$.

5.2.4 Apolar polynomials

Grace's theorem has an elegant formulation in terms of apolar polynomials. Let

$$P(z) = \sum_{k=0}^{n} A_k z^k, \; Q(z) = \sum_{k=0}^{n} B_k z^k \tag{5.58}$$

be polynomials of degree n. We say that P and Q are **apolar** if

$$\sum_{k=0}^{n} \frac{A_k B_{n-k}}{\binom{n}{k}}(-1)^k = 0. \tag{5.59}$$

Note that this relation is symmetric.

5.2.5

Grace's apolarity theorem *Let P and Q be apolar polynomials of degree n and suppose that P has all its zeros in a circular region R. Then Q has at least one zero in R.*

Proof If $T(z) = \sum_{k=0}^{n} C_k z^k$ is a polynomial of degree $\leq n$, we define the linear functional Λ by

$$\Lambda T = \sum_{k=0}^{n} \frac{A_k C_{n-k}}{\binom{n}{k}} (-1)^k. \tag{5.60}$$

Then $\Lambda Q = 0$ and therefore, if Q has no zeros in R, Q has all its zeros in the complement of R, which is a circular region. Since $\Lambda 1 = A_n(-1)^n \neq 0$, it follows that $\exists \zeta \notin R$ such that $\Lambda(z - \zeta)^n = 0$. This gives $P(\zeta) = 0$, contradicting the hypothesis.

5.2.6 Symmetric linear forms

A **linear form** $L(z_1, \ldots, z_n)$ is a polynomial in the complex variables $z_1, \ldots, z_n$ which is linear in each of the variables, i.e. if the variables z_k $(k \neq j)$ are kept fixed, the resulting polynomial in z_j is of the first degree. The form is **symmetric** if L is independent of any permutation of the n variables. We denote by $L_k(z_1, \ldots, z_n)$ the symmetric linear form consisting of the sum of all products of combinations of $z_1, \ldots, z_n$ taken k at a time $(1 \leq k \leq n)$; also set $L_0(z_1, \ldots, z_n) = 1$. We then note that

$$\prod_{k=1}^{n} (z + z_k) = \sum_{k=0}^{n} L_{n-k}(z_1, \ldots, z_n) z^k. \tag{5.61}$$

It is then clear that every symmetric linear form of degree n can be written in the form

$$L(z_1, \ldots, z_n) = \sum_{k=0}^{n} A_k L_k(z_1, \ldots, z_n) \tag{5.62}$$

for suitable coefficients A_k. We then have

$$L(z, \ldots, z) = \sum_{k=0}^{n} \binom{n}{k} A_k z^k. \tag{5.63}$$

5.2.7

Walsh's theorem on symmetric linear forms *Let $L(z_1, \ldots, z_n)$ be a symmetric linear form such that $L(z, \ldots, z)$ has exact degree n and let R be a circular region. If $z_1, \ldots, z_n$ are points in R, then $\exists z \in R$ such that*

$$L(z, \ldots, z) = L(z_1, \ldots, z_n). \tag{5.64}$$

In other words, the values attained by the form as the variables range over R are exactly those attained by the form when the variables are all equal. Note that, if the theorem is to be true for all circular regions R, then the condition that $L(z, \ldots, z)$ have exact degree n is necessary. For example, if $L(z_1, \ldots, z_n) = z_1 + \cdots + z_n$ and R is the exterior of the unit circle, then 0 is clearly a value of $L(z_1, \ldots, z_n)$ but not a value of $L(z, \ldots, z)$.

Proof of theorem Since for any fixed w, $L - w$ is a symmetric linear form, it is enough to show that if for suitable $z_1, \ldots, z_n$ in R we have $L(z_1, \ldots, z_n) = 0$, then the polynomial $L(z, \ldots, z)$ has a zero in R. With this assumption let

$$Q(z) = \prod_{k=1}^{n}(z - z_k) = \sum_{k=0}^{n} L_{n-k}(z_1, \ldots, z_n)(-1)^{n-k} z^k. \tag{5.65}$$

Then it is immediately verified that the polynomials $Q(z)$ and $L(z, \ldots, z)$ are apolar. Since Q has all its zeros in R, Grace's apolarity theorem implies that $L(z, \ldots, z)$ has at least one zero in R, which is the required result.

5.2.8 *Weak apolarity*

Let P and Q be polynomials of degree $\leq n$ with coefficients A_k and B_k respectively ($0 \leq k \leq n$). We will say that P and Q are n-**weakly apolar** if the relation (5.59) is satisfied. From theorem 5.2.2 we obtain by repeating the argument of 5.2.5

5.2.8.1 Theorem *Let P and Q be n-weakly apolar and let P have all its zeros in a circular region R. Then Q has a zero in R if any one of the following conditions holds.*

(i) *R is a disc (open or closed) and P has degree n;*
(ii) *R is the exterior of a disc (open or closed) and Q has degree n;*
(iii) *R is a half-plane (open or closed) and either P or Q has degree n.*

The result is clear from the earlier results if (i) or (ii) holds; also if (iii) holds with R an open half-plane and P of degree n or a closed half-plane and Q of degree n. Suppose P has degree n and R is a closed half-plane, bounded by a line l say; then, if λ is any line not in R and parallel to l, λ bounds an open half-plane ρ containing the zeros of P and so Q has a zero in ρ. Since Q has at most n zeros, it follows that there is a zero in R as required. The remaining case follows from this.

5.2.9

We deduce from the above the following.

Theorem *Let $L(z_1, \ldots, z_n)$ be a symmetric linear form and let R be a disc (open or closed) or a half-plane (open or closed). Then, if $z_1, \ldots, z_n$ are points in R, $\exists z \in R$ such that*

$$L(z_1, \ldots, z_n) = L(z, \ldots, z). \tag{5.66}$$

5.2.10

We can also add to the information in theorem 5.2.2 as follows.

Theorem *Let Λ be a linear functional defined on the polynomials of degree n; let R be a half-plane (open or closed); let P be a polynomial of degree $\leq n$. Suppose that*

$$P(w) \neq 0\ (w \in R) \text{ and } \Lambda P = 0. \tag{5.67}$$

Then either $\Lambda 1 = 0$ or $\exists \zeta \notin R$ such that $\Lambda(w - \zeta)^n = 0$.

Proof Let $P(w) = \sum_{k=0}^{n} a_k w^k$, $c_k = \Lambda w^k$ $(0 \leq k \leq n)$. Assume that $c_0 = \Lambda 1 \neq 0$. We have

$$\sum_{k=0}^{n} a_k c_k = 0, \tag{5.68}$$

and therefore we see that the polynomials $P(w)$ and

$$Q(w) = \sum_{k=0}^{n} \binom{n}{k} (-1)^k c_{n-k} w^k \tag{5.69}$$

are weakly apolar; also $Q(w)$ has degree n. Hence $\exists \zeta \notin R$ such that $Q(\zeta) = 0$. This gives $\Lambda(w - \zeta)^n = 0$, as required.

Note that, if P has exact degree n, then we still obtain $\Lambda(w - \zeta)^n = 0$ for a suitable $\zeta \notin R$, even if $\Lambda 1 = 0$. Note also the interesting fact that, if P has degree $m < n$, then the choices for Λw^k $(k > m)$ are free. This strengthening of theorem 5.2.2 is false if R is the exterior of a disc.

5.2.11

Theorem *Let*

$$P(z) = \sum_{k=0}^{n} A_k z^k, \qquad Q(z) = \sum_{k=0}^{n} B_k z^k \tag{5.70}$$

be polynomials of degree n and set

$$T(z) = \sum_{k=0}^{n} \frac{A_k B_k}{\binom{n}{k}} z^k. \tag{5.71}$$

If P has all its zeros in a circular region R, then, if $\infty \notin R$, every zero ζ of T has the form

$$\zeta = -\alpha\beta, \tag{5.72}$$

where $\alpha \in R$ and β is a zero of Q. The result remains valid, if $\infty \in R$ and $Q(0) \neq 0$.

Proof Define a linear functional Λ by the relation

$$\Lambda\left(\sum_{k=0}^{n} C_k z^k\right) = \sum_{k=0}^{n} \frac{B_k C_k \zeta^k}{\binom{n}{k}} \tag{5.73}$$

so $\Lambda P = 0$. Hence from theorem 5.2.2, either $\infty \in R$ and $\Lambda 1 = 0$, or $\exists \alpha \in R$ such that $\Lambda(z-\alpha)^n = 0$. In the former case we obtain $Q(0) = 0$, contradicting the hypothesis. In the latter case we obtain

$$\sum_{k=0}^{n} B_k \zeta^k (-\alpha)^{n-k} = 0. \tag{5.74}$$

If $\alpha = 0$, this gives $B_n \zeta^n = 0$, so $\zeta = 0$. Otherwise we have $Q(-\frac{\zeta}{\alpha}) = 0$, and the result follows.

Note that the condition $Q(0) \neq 0$ can be dropped, if R is a half-plane.

5.3 The polar derivative

5.3.1

Every polynomial $P(z)$ of degree $\leq n$ can be written in the form

$$P(z) = L(z, \ldots, z), \tag{5.75}$$

where L is the symmetric linear form

$$L(z_1, \dots, z_n) = \sum_{k=0}^{n} A_k L_k(z_1, \dots, z_n) \tag{5.76}$$

and $\binom{n}{k} A_k$ $(0 \le k \le n)$ are the coefficients of P. Suppose that P has all its zeros in a circular region R. Then $L(z, \dots, z) \neq 0$ in the complement of R. If P has degree n, this implies that

$$L(z_1, \dots, z_n) \neq 0 \quad (z_1 \notin R, \dots, z_n \notin R). \tag{5.77}$$

If P has degree $< n$, this relation also holds provided that R is a half-plane or the exterior of a disc. Now consider the polynomial

$$P_1(z; z_1) = P_1(z) = L(z_1, z, \dots, z). \tag{5.78}$$

This is a polynomial of degree $\le n$; if under the above conditions $z_1 \notin R$, $z \notin R$, then $P_1(z) \neq 0$. Hence for $z_1 \notin R$, $P_1(z)$ has all its zeros in R. We call $P_1(z)$ the n**-polar derivative of** P **with respect to the point** z_1. A few calculations show that

$$P_1(z) = P(z) - \frac{1}{n}(z - z_1)P'(z). \tag{5.79}$$

If P has degree n, then P_1 has degree $n - 1$ and is called the **polar derivative of** P. Note that

$$\frac{nP_1(z)}{z_1 - z} \to P'(z) \text{ as } z_1 \to \infty, \tag{5.80}$$

the convergence being locally uniform. Since z_1 can approach ∞ in a half-plane or the exterior of a disc, it follows from Hurwitz's theorem that, if R is a closed half-plane or a closed disc containing all the zeros of P, then R contains all the zeros of P', i.e. all the critical points of P. From this we easily deduce a second proof of the Gauss–Lucas theorem: *the critical points of P lie in the convex hull of the set of zeros of P.*

5.3.2

We can use the properties of the polar derivative to give alternative proofs of some other of our earlier results. Let P be a polynomial of degree n which omits the value w in the unit disc $\mathbb{U}$. Then P_1 omits w when $z_1 \in \mathbb{U}$. Hence, if $|P(z)| \le M$ $(z \in \mathbb{U})$, then

$$\left| P(z) - \frac{1}{n}(z - z_1)P'(z) \right| \le M \quad (z_1, z \in \mathbb{U}). \tag{5.81}$$

It follows that

$$|nP(z) - zP'(z)| + |P'(z)| \leq nM \quad (z \in \mathbb{U}), \tag{5.82}$$

which is slightly stronger than inequality (4.166) of chapter 4, and as before implies Bernstein's inequality

$$\|P'\|_\infty \leq n\|P\|_\infty. \tag{5.83}$$

We can also use the polar derivative to establish Lax's theorem 4.4.4 of chapter 4. If $P(z)$ *has degree n and no zeros in* $\mathbb{U}$, *then*

$$\|P'\|_\infty \leq \frac{n}{2}\|P\|_\infty. \tag{5.84}$$

Indeed, since for $z_1 \in \mathbb{U}$, $P_1 \neq 0$ in $\mathbb{U}$, we have $nP(z) - zP'(z) \neq z_1P'(z)$ if $|z_1| < 1$. Hence

$$|nP(z) - zP'(z)| \geq |P'(z)| \quad (z \in \mathbb{U}). \tag{5.85}$$

Now (5.84) follows from (5.82) and (5.85). Note that we can further improve Bernstein's inequality, if we know that $P(z) \neq 0$ in $\{|z| < R\}$, where $R > 1$. For then we have $nP(z) - zP'(z) \neq z_1P'(z)$ if $|z_1| < R$, and so

$$|nP(z) - zP'(z)| \geq R|P'(z)| \quad (z \in \mathbb{U}). \tag{5.86}$$

We deduce from (5.82)

$$\|P'\|_\infty \leq \frac{n}{1+R}\|P\|_\infty. \tag{5.87}$$

5.4 Locating critical points

5.4.1

There are a number of applications of Grace's theorem which give very useful results in the problem of locating the critical points of a polynomial or a rational function. So far we know that the critical points of a polynomial P lie in the convex hull of the set of zeros. We can, of course, infer from this that the critical points of P lie in the convex hull of the set of roots of the equation $P(z) = a$, where a is an arbitrary complex number. Thus the critical points lie in the intersection of the family of all such convex hulls, which is in turn a convex set. However, it may be larger than the convex hull of the set of critical points. Perhaps the most useful aspect of this information is that the roots of the equation $P(z) = a$ cannot be separated by a line from the roots of $P(z) = b$.

5.4.2

A polynomial certainly has a critical point at a multiple zero and its multiplicity is one less than the multiplicity of the zero. The problem is to find the remaining critical points. If we disturb the polynomial slightly by scattering a multiple zero to simple zeros close to the original multiple zero, then the corresponding multiple critical point will also scatter locally into several critical points. The effect on other critical points is likely to be marginal (unless such critical points are themselves multiple) and any such critical point may still 'believe' it is a critical point of a polynomial with a multiple zero. This faith is justified by the next result.

5.4.3

Theorem *Let $C_1, C_2, \ldots, C_N$ be circular regions and let P be a polynomial whose zeros lie in these regions, say n_k zeros in C_k $(1 \le k \le N)$. Let ζ be a critical point of P which does not lie in any of the regions. Then $\exists \alpha_k \in C_k$ $(1 \le k \le N)$ such that*

$$\sum_{k=1}^{N} \frac{n_k}{\zeta - \alpha_k} = 0. \tag{5.88}$$

The point α_k is a finite point if C_k is a half-plane or disc, but may be ∞ if C_k is the exterior of a disc; in this case we take the relevant term in the sum to be zero.

5.4.4

To prove the above result we make use of the following lemma.

Lemma *Let S be a polynomial of degree n with all its zeros in a circular region C. Then, if $\zeta \notin C$, $\exists \alpha \in C$ such that*

$$\frac{S'(\zeta)}{S(\zeta)} = \frac{n}{\zeta - \alpha}. \tag{5.89}$$

Proof By theorem 5.2.3 on the ratio of two linear functionals, to find the possible values of the left-hand side we may replace $S(z)$ by one of the polynomials $(z-\alpha)^n$ for $\alpha \in C$, or by 1 if C is the exterior of a disc. We immediately obtain the result.

The proof of the theorem is now clear, for we can write $P(z) = \prod_{k=1}^{N} S_k(z)$, where each S_k has degree n_k with all its zeros in C_k. As the critical point ζ is not in any of the C_k, we have

$$0 = \frac{P'(\zeta)}{P(\zeta)} = \sum_{k=1}^{N} \frac{S_k'(\zeta)}{S_k(\zeta)} = \sum_{k=1}^{N} \frac{n_k}{\zeta - \alpha_k} \tag{5.90}$$

where $\alpha_k \in C_k$ $(1 \le k \le N)$.

The case $N = 1$ is effectively the Gauss–Lucas theorem. Of particular interest is the case $N = 2$. We first consider the case of two circles.

5.4.5

Walsh's two circles theorem *Let C_1 be a disc with centre c_1 and radius r_1 and C_2 a disc with centre c_2 and radius r_2. Let P be a polynomial of degree n with all its zeros in $C_1 \cup C_2$, say n_1 zeros in C_1 and n_2 zeros in C_2. Then P has all its critical points in $C_1 \cup C_2 \cup C_3$, where C_3 is the disc with centre c_3 and radius r_3 given by*

$$c_3 = \frac{n_1 c_2 + n_2 c_1}{n}, \qquad r_3 = \frac{n_1 r_2 + n_2 r_1}{n}. \tag{5.91}$$

Furthermore, if C_1, C_2 and C_3 are pairwise disjoint, then C_1 contains $n_1 - 1$ critical points, C_2 contains $n_2 - 1$ critical points and C_3 contains 1 critical point.

Proof If ζ is a critical point not in C_1 or C_2, then $\exists \alpha_1 \in C_1$ and $\alpha_2 \in C_2$ such that

$$\frac{n_1}{\zeta - \alpha_1} + \frac{n_2}{\zeta - \alpha_2} = 0. \tag{5.92}$$

This gives $\zeta = \frac{1}{n}(n_2\alpha_1 + n_1\alpha_2)$ and the first part of the result follows easily. For the second part, where C_1, C_2 and C_3 are mutually disjoint, we employ a continuity argument. We may move the zeros of P continuously in C_1 and C_2 obtaining critical points in C_1, C_2 and C_3 also moving continuously and which cannot leave the respective regions. Therefore the number of critical points in each region remains constant. To find the three numbers we need only consider the polynomial $P(z) = (z - c_1)^{n_1}(z - c_2)^{n_2}$. This has $n_1 - 1$ critical points at c_1, $n_2 - 1$ at c_2 and 1 at c_3.

Note that c_3 lies on the line segment joining c_1 to c_2. An interesting case occurs when C_2 reduces to a single point c_2, so $r_2 = 0$. Thus P has a single zero of multiplicity n_2 at c_2 and its remaining zeros in C_1. Then $r_3 = n_2 r_1/n$.

The three regions C_1, C_2 and C_3 are disjoint if

$$r_1 < \frac{n - n_2}{n + n_2}|c_1 - c_2| \tag{5.93}$$

and so in this case P has a single critical point within distance $n_2 r_1/n$ of c_3 and its remaining critical points in C_1.

5.4.6

More generally, we may also consider the case that the two regions C_1 and C_2 are circular but not both discs. Unless the regions are disjoint and one of them is a disc, we cannot improve on the Gauss–Lucas theorem. The result, whose proof is left to the reader, is as follows.

Theorem *Let C_1 be a disc with centre c_1, radius r_1 and let C_2 be a half-plane or the exterior of a disc disjoint from C_1. Suppose that P is a polynomial of degree n with n_1 zeros in C_1 and n_2 zeros in C_2. Then*

(i) *if C_2 is the exterior of the disc of centre c_2 and radius r_2, every critical point ζ of P not in C_1 or C_2 lies exterior to the disc C_3 with centre c_3 and radius r_3 given by*

$$c_3 = \frac{n_2 c_1 + n_1 c_2}{n}, \qquad r_3 = \frac{n_1 r_2 - n_2 r_1}{n} \tag{5.94}$$

(if $r_3 < 0$ we interpret this statement as giving no information); if $C_1 \subset C_3$, then P has exactly $n_1 - 1$ critical points in C_1;

(ii) *if C_2 is the half-plane $\{\Re((z - c_1)e^{i\phi}) > R\}$, then every critical point ζ not in C_1 or C_2 lies in the half-plane*

$$\left\{\Re((z - c_1)e^{i\phi}) > \frac{n_1 R - n_2 r_1}{n}\right\}; \tag{5.95}$$

if this latter half-plane is disjoint from C_1, then P has exactly $n_1 - 1$ critical points in C_1.

As an interesting example, we note that, if P is a polynomial of degree n with a zero of multiplicity n_1 at the origin and all its remaining zeros in $\{|z| \geq 1\}$, then the n_2 critical points not at the origin lie in $\{|z| \geq n_1/n\}$. If the remaining zeros lie in $\{\Re z \geq 1\}$, then the remaining critical points lie in $\{\Re z \geq n_1/n\}$. Note that both statements give better information than that contained in the Gauss–Lucas theorem.

5.5 Critical points of rational functions

5.5.1

Theorem *Let $R(z)$ be a rational function whose zeros are distributed over N circular regions C_k with n_k zeros in C_k and whose poles are distributed over M circular regions D_j with m_j poles in D_j. If ζ is a critical point of R not lying in any of the C_k or D_j, then $\exists \alpha_k \in C_k (1 \le k \le N)$ and $\beta_j \in D_j (1 \le j \le M)$ (which may be ∞ if C_k or D_j is the exterior of a disc) such that*

$$\sum_{k=1}^{N} \frac{n_k}{\zeta - \alpha_k} = \sum_{j=1}^{M} \frac{m_j}{\zeta - \beta_j}. \tag{5.96}$$

The proof is similar to that of theorem 5.4.3. Of particular interest is the case $M = N = 1$. Thus with a change of notation, if R is a rational function with its n zeros in C and its p poles in D (C and D circular regions), then a critical point $\zeta \notin C \cup D$ satisfies

$$\frac{n}{\zeta - \alpha} = \frac{p}{\zeta - \beta} \tag{5.97}$$

where $\alpha \in C$ and $\beta \in D$. This gives

$$\zeta = \frac{n\beta - p\alpha}{n - p}, \tag{5.98}$$

except in the case $n = p$, when we obtain $\alpha = \beta$, or in the case when both regions are the exterior of discs. Thus if C and D overlap and R has equally many zeros and poles, we obtain no information at all. On the other hand, if C and D are disjoint and R has equally many zeros and poles, then all the critical points lie in either C or D.

If $n \neq p$ and C and D are discs with centres c and d, radii ρ and σ respectively, then the critical points not in C or D lie in a disc G with centre g and radius τ given by

$$g = \frac{nd - pc}{n - p}, \qquad \tau = \frac{n\sigma + p\rho}{|n - p|}. \tag{5.99}$$

Note that the three centres are collinear and, if $n > p$, the order is c, d, g and, if $n < p$, then the order is g, c, d. If the three discs are disjoint, we determine the number of critical points in each region as follows. Firstly, let $R = P/Q$, where P and Q are polynomials of degrees n and p respectively. The critical points of R are the zeros of $T = P'Q - PQ'$ which are not zeros of Q. The number of such zeros is at most $n + p - 1$. If Q has a zero of multiplicity ν, then T has a zero of multiplicity $\nu - 1$ and R' has a pole of multiplicity $\nu + 1$. Thus R' loses $\nu - 1$ zeros of the $n + p - 1$ possible zeros at each pole of multiplicity ν.

These losses therefore occur in D, on the assumption that D is open, since if we scatter the pole locally, T will have $\nu - 1$ zeros in the local region, which are now critical points of R. The rational function $R(z) = (z-c)^n/(z-d)^p$ has $n-1$ critical points at c, 1 critical point at g and no critical points in D. Thus in general, if the discs C, D, G are disjoint, and D is open, then R has $n-1$ critical points in C, 1 critical point in G and at most $p-1$ critical points in D. If the poles are simple, then R has exactly $p-1$ critical points in D.

Let us now consider the case when C and D are disjoint half-planes, so the two regions are separated by a line or a parallel strip. Then from (5.98) we see that a critical point ζ not in C or D lies on a line passing through a point in C and a point in D, but does not lie on the line segment joining these points. Therefore it lies either in C or in D. In other words all critical points lie either in C or in D.

If C and D are disjoint and C is a disc and D is a half-plane or the exterior of a disc, then from (5.97) we have for a critical point $\zeta \notin C \cup D$,

$$\alpha = \frac{p-n}{p}\zeta + \frac{n}{p}\beta, \qquad \beta = \frac{p}{n}\alpha + \frac{n-p}{n}\zeta \tag{5.100}$$

where $\alpha \in C$ and $\beta \in D$. If $n > p$, β is a point on the line segment joining ζ to α; since ζ and α lie in the complement of D, which is convex, we obtain $\beta \notin D$, a contradiction. Hence for $n > p$, the critical points lie in $C \cup D$. Similarly, if C and D are disjoint and C is a half-plane or the exterior of a disc and D is a disc, then the critical points lie in $C \cup D$, when $n < p$.

Now in these cases, where we have $n \neq p$, $C \cap D = \emptyset$ and the critical points lie in $C \cup D$, the total degree of $T = P'Q - PQ'$ is $n + p - 1$, so that no critical points can be lost at ∞ (though, as we have seen, critical points can be lost in the poles of R according to the multiplicity of the pole). Therefore the continuity method shows that in all these cases there are exactly $n-1$ critical points in C and at most p critical points in D, the exact number depending on the multiplicities of the poles (at a pole of multiplicity ν we lose $\nu - 1$ critical points). Thus there is at least one critical point in D.

Lastly, we need to consider again the case $n = p$. If $C \cap D = \emptyset$, the critical points lie in $C \cup D$ and the degree of T is at most $2n-2$. A few calculations show that the degree of T is $< 2n-2$ exactly when the sum of the zeros is equal to the sum of the poles. This cannot occur if C and D are disjoint and convex. Therefore in this case we obtain exactly $n-1$ critical points in C and at most $n-1$ critical points in D, again the exact number depending on the multiplicities of the poles. If C is a disc and D is the exterior of a disc, we have exactly $n-1$ critical points in C and at most $n-1$ critical points in D, the exact number depending on the degree of T and the multiplicities of the poles. Finally, if C is the exterior of a disc and D is a disc, there are exactly $\deg T - n + 1$ critical points in C and at most $n - 1$ critical points in D depending on the multiplicities of the poles.

Note that, when we have a total of p poles of which m are distinct each of multiplicity ν_i $(1 \le i \le m)$, we lose $p - m$ critical points.

5.5.2

Bôcher–Walsh theorem *Let R be a rational function with its n zeros in a circular region C and its p poles in a circular region D, and assume that $C \cap D = \emptyset$. Suppose further that there are exactly m distinct poles each of multiplicity ν_i $(1 \le i \le m)$.*

(i) *The case $n \neq p$.*
  (a) *If C and D are discs with centres c and d, radii ρ and σ respectively, then the critical points of R not in C or D lie in a disc G with centre g and radius τ given by (5.99). (This is the case even when $C \cap D \neq \emptyset$.) If the discs C, D and G are mutually disjoint, then R has $n - 1$ critical points in C,1 in G and $m - 1$ in D.*
  (b) *If C and D are half-planes, then all the critical points of R lie in $C \cup D$. There are exactly $n - 1$ critical points in C and m critical points in D.*
  (c) *If C is a disc and D a half-plane or exterior of a disc and if $n > p$, then all the critical points of R lie in $C \cup D$. There are exactly $n - 1$ critical points in C and m critical points in D.*
  (d) *If C is a half-plane or exterior of a disc and D is a disc and if $n < p$, then all the critical points lie in $C \cup D$. There are exactly $n - 1$ critical points in C and m critical points in D.*

(ii) *The case $n = p$. In this case all the critical points lie in $C \cup D$.*
  (a) *If C is a half-plane or disc and D is a half-plane or disc, then there are exactly $n - 1$ critical points in C and $m - 1$ critical points in D.*
  (b) *If C is a disc and D is the exterior of a disc, then there are exactly $n - 1$ critical points in C and $\mu + m - 2n + 1$ critical points in D, where μ is the degree of $T = P'Q - PQ'$ $(R = P/Q)$.*
  (c) *If C is the exterior of a disc and D is a disc, then there are exactly $\mu - n + 1$ critical points in C and $m - 1$ critical points in D.*

A neat summary of this result can be given as follows. *Let R be a rational function whose zeros and poles are separated by a line L; then there are no zeros of R' on L. Similarly, if the zeros and poles are separated by a circle*

Γ (where ∞ is counted as a zero or pole when $R(z) \to 0$ or ∞ as $z \to \infty$), then there are no zeros of R' on Γ.

5.6 The Borwein–Erdélyi inequality

5.6.1

Borwein and Erdélyi [9] have established the following generalisation to rational functions of Bernstein's inequality for polynomials.

Theorem *Let $f(z) = p(z)/q(z)$ be a rational function with no poles on the unit circle with $q(z) = \prod_{k=1}^{n}(z - a_k)$ and $p(z)$ a polynomial of degree $\leq n$. If $|f(z)| \leq 1$ on the unit circle, then*

$$|f'(z)| \leq \max\left\{ \sum_{|a_k|>1} \frac{|a_k|^2 - 1}{|a_k - z|^2}, \sum_{|a_k|<1} \frac{1 - |a_k|^2}{|a_k - z|^2} \right\} \tag{5.101}$$

for $|z| = 1$.

5.6.2

We shall give a direct proof here of the following weaker result (it is equivalent in the case that q has its zeros either all in the unit disc or all outside the unit disc).

Theorem *Let $f(z) = p(z)/q(z)$ be a rational function with no poles on the unit circle with $q(z) = \prod_{k=1}^{n}(z - a_k)$ and $p(z)$ a polynomial of degree $\leq n$. If $|f(z)| \leq 1$ on the unit circle, then*

$$|f'(z)| \leq \sum_{|a_k|>1} \frac{|a_k|^2 - 1}{|a_k - z|^2} + \sum_{|a_k|<1} \frac{1 - |a_k|^2}{|a_k - z|^2} \tag{5.102}$$

for $|z| = 1$.

Proof Assume first that $|a_k| < 1$ for $1 \leq k \leq n$ and let

$$B(z) = \prod_{k=1}^{n} \frac{z - a_k}{1 - \overline{a_k} z} = \frac{q(z)}{q^*(z)} \tag{5.103}$$

denote the Blaschke product formed with the zeros of q, where $q^*(z) = z^n \overline{q(1/\overline{z})}$. Then we are required to prove that $|f'(z)| \leq |B'(z)|$ on $|z| = 1$. Now on the circle $|p(z)| \leq |q(z)| = |q^*(z)|$, and hence, since q^* has no zeros in the unit disc, it follows from the maximum principle that $|p(z)| \leq |q^*(z)|$ for

$|z| \leq 1$. Hence, if $|\zeta| < 1$, the polynomial $P(z) = \zeta p(z) - q^*(z)$ has no zeros in $|z| \leq 1$. Since $q(z)$ has all its zeros in $|z| < 1$, we obtain

$$\Re \frac{P'(1)}{P(1)} < \frac{n}{2} < \Re \frac{q'(1)}{q(1)}, \tag{5.104}$$

and hence $F'(1) \neq 0$, where $F(z) = P(z)/q(z) = \zeta f(z) - 1/B(z)$. We obtain $\zeta f'(1) + B'(1)/B^2(1) \neq 0$, from which we deduce that $|f'(1)| \leq |B'(1)|$, and the desired result follows.

Secondly, we assume that $|a_k| > 1$ for $1 \leq k \leq n$. Then the function $z \to \overline{f(1/\overline{z})}$ is a rational function of degree n with its poles inside the disc. Hence applying the previous case, we obtain $|f'(1)| \leq |B'(1)|$, where B is the Blaschke product formed with the a_k (the value of $|B'(1)|$ does not change if we replace each a_k by $1/\overline{a_k}$).

Finally, we consider the general case and assume that $|a_k| < 1$ for $1 \leq k \leq r$ and $|a_k| > 1$ for $r + 1 \leq k \leq n$. Let

$$C(z) = \prod_{k=1}^{r} \frac{z - a_k}{1 - \overline{a_k} z}, \qquad D(z) = 1 \Big/ \prod_{k=r+1}^{n} \frac{z - a_k}{1 - \overline{a_k} z}. \tag{5.105}$$

Also let

$$u = \frac{C'(1)}{C(1)} = |C'(1)|, \qquad v = \frac{D'(1)}{D(1)} = |D'(1)|. \tag{5.106}$$

The function $f(z)C(z)$ has all its poles outside the unit disc, and hence by the second part

$$|f'(1)C(1) + f(1)C'(1)| \leq u + v, \tag{5.107}$$

or equivalently

$$|f'(1) + f(1)u| \leq u + v. \tag{5.108}$$

The function $f(z)/D(z)$ has all its poles inside the unit disc, and hence by the first part

$$|f'(1)/D(1) - f(1)D'(1)/D(1)^2| \leq u + v, \tag{5.109}$$

or equivalently

$$|f'(1) - f(1)v| \leq u + v. \tag{5.110}$$

We thus have

$$f'(1) = -f(1)u + \rho(u + v) = f(1)v + \sigma(u + v), \tag{5.111}$$

where $|\rho| \leq 1$ and $|\sigma| \leq 1$. This gives

$$f'(1) = \rho v + \sigma u, \tag{5.112}$$

and therefore $|f'(1)| \leq u + v$, which is the desired result.

Since $f(1)$ may be $= 0$, we cannot use this argument to improve on $u + v$ without further information. However if we assume that $|f(1)| = 1$, then we can obtain the Borwein–Erdélyi estimate as follows. Since then $z = 1$ is a point of maximum value on the circle for $|f|$, we have that $f'(1)/f(1)$ is real. If $f'(1)/f(1) > 0$, we obtain from the above

$$|f'(1)| + u \leq u + v, \tag{5.113}$$

giving $|f'(1)| \leq v$. On the other hand, if $f'(1)/f(1) < 0$, then we obtain

$$|f'(1)| + v \leq u + v, \tag{5.114}$$

giving $|f'(1)| \leq u$. Thus in either case we obtain

$$|f'(1)| \leq \max(u, v), \tag{5.115}$$

which is the Borwein–Erdélyi result. To make this argument work requires a functional argument to the effect that the extremal case, maximising $|f'(1)|$ over all f satisfying the requirements, is attained for a function f satisfying $|f(1)| = 1$.

Note that the result gives the inequality

$$\|f'\|_\infty \leq n\frac{1+R}{1-R}\|f\|_\infty \tag{5.116}$$

for $f = p/q$, where p has degree $\leq n$ and q has degree n with zeros lying outside the annulus $\{R < |z| < \frac{1}{R}\}$. The norm here denotes the supremum value on the unit circle.

5.6.3

The method of proof adopted here can be adapted to give inequalities involving the zeros of p as well as of q. For example, suppose that q has zeros a_k satisfying $|a_k| < 1$ $(1 \leq k \leq n)$, and assume again that $f = p/q$, where p has degree $\leq n$, and $|f(z)| \leq 1$ for $|z| = 1$. As before we observe that, for each ζ satisfying $|\zeta| < 1$, the polynomial $P(z) = \zeta p(z) - q^*(z)$ has no zeros in $|z| \leq 1$. It follows that the function

$$H(z) = \frac{\overline{\zeta}\, p^*(z) - q(z)}{\zeta p(z) - q^*(z)} \tag{5.117}$$

is a Blaschke product with all its zeros in the unit disc. Hence

$$\frac{zH'(z)}{H(z)} > 0 \quad (|z| = 1). \tag{5.118}$$

When $z = 1$ we obtain

$$\frac{\overline{\zeta} p^{*\prime}(1) - q'(1)}{\overline{\zeta} p^*(1) - q(1)} - \frac{\zeta p'(1) - q^{*\prime}(1)}{\zeta p(1) - q^*(1)} > 0. \tag{5.119}$$

This gives

$$(\overline{\zeta} p^{*\prime}(1) - q'(1))(\zeta p(1) - q^*(1)) - (\zeta p'(1) - q^{*\prime}(1))(\overline{\zeta} p^*(1) - q(1)) > 0, \tag{5.120}$$

from which we obtain

$$\begin{aligned}(p^{*\prime}(1)p(1) - p'(1)p^*(1))|\zeta|^2 + 2\Re(\zeta(p'(1)q(1) - p(1)q'(1)))\\ > q^{*\prime}(1)q(1) - q'(1)q^*(1).\end{aligned} \tag{5.121}$$

As this is true for every ζ $(|\zeta| < 1)$, we obtain letting $|\zeta| \to 1$

$$\begin{aligned}|p'(1)q(1) - p(1)q'(1)| \leq \frac{1}{2}(q'(1)q^*(1) - q^{*\prime}(1)q(1))\\ + (p^{*\prime}(1)p(1) - p'(1)p^*(1)).\end{aligned} \tag{5.122}$$

Dividing by $|q(1)|^2$ this becomes

$$|f'(1)| \leq \frac{1}{2}(|B'(1)| - |f(1)|^2 C'(1)/C(1)) \leq \frac{1}{2}(|B'(1)| + |C'(1)|), \tag{5.123}$$

where $B(z)$ is the Blaschke product formed with the a_k and $C(z)$ is the Blaschke product formed with the zeros of p.

Furthermore, we observe that, if $p(z)$ has all its zeros in the unit disc, then $C'(1)/C(1) > 0$, and therefore

$$|f'(1)| \leq \frac{1}{2}|B'(1)|. \tag{5.124}$$

5.7 Univalence properties of polynomials

5.7.1

If $P' \neq 0$ in a region, then P is locally 1–1 in the region. We consider the problem of specifying regions in which P is 1–1. This leads to the following question. If a and b are points at which $P(a) = P(b)$, does this impose any restriction on the location of the critical points? In this section we will use Grace's theorem to elucidate this question. We begin with the result in the next subsection.

5.7.2

Theorem *Let P be a polynomial of degree n and suppose that P has no critical points in $\mathbb{U} = \{|z| < 1\}$. Suppose further that a and b are distinct points in $\mathbb{U}$ such that $P(a) = P(b)$. Then*

$$|a| + |b| \geq 2 \sin \frac{\pi}{n}. \tag{5.125}$$

Proof The linear functional

$$\Lambda P' = \int_a^b P'(t)dt = 0. \tag{5.126}$$

It follows from Grace's theorem that $\exists x$ ($|x| \leq 1$) such that $\Lambda(1 + xz)^{n-1} = 0$. If $x = 0$ we obtain $a = b$. Hence $x \neq 0$ and we obtain

$$(1 + bx)^n = (1 + ax)^n \tag{5.127}$$

which gives

$$\frac{1 + bx}{1 + ax} = e^{2\pi ik/n} \tag{5.128}$$

where k is an integer satisfying $1 \leq k \leq n - 1$. Since $|x| \leq 1$, we obtain

$$|1 - e^{2\pi ik/n}| \leq |ae^{2\pi ik/n} - b|. \tag{5.129}$$

It follows that $2|\sin \frac{\pi k}{n}| \leq |a| + |b|$ for some integer k ($1 \leq k \leq n - 1$). As the smallest value of $\sin(\pi k/n)$ occurs when $k = 1$, the result follows.

5.7.3

Corollary *Let P be a polynomial of degree n such that $P'(z) \neq 0$ for $z \in \mathbb{U}$. Then P is univalent in the disc $\{|z| < \sin(\pi/n)\}$. Furthermore, if $P(0) = 0$, then*

$$P(z) \neq 0 \textit{ for } 0 < |z| < 2 \sin \frac{\pi}{n}. \tag{5.130}$$

Both results are sharp for the polynomial $P(z) = (1 + z)^n - 1$.

Note that the statement of the theorem does not contain as much information as is implicitly contained in its proof. In the relation (5.128) as x varies over the disc $\{|x| \leq 1\}$ the left-hand expression describes the disc

$$\left| w - \frac{1 - \bar{a}b}{1 - |a|^2} \right| \leq \frac{|a - b|}{1 - |a|^2}. \tag{5.131}$$

Note that this disc contains the point $w=1$ and does not contain the point $w=0$. Under the hypotheses of the theorem this disc contains the point $w = e^{2\pi ik/n}$ for some integer k not divisible by n. Therefore it certainly contains either the point $w = e^{2\pi i/n}$ or its conjugate. Thus, if we wish to find the region of univalence which covers all polynomials P with no critical points in $\mathbb{U}$, this is given by the domain D with the following property: if $a, b \in D$ then, referring to (5.129)

$$2\sin\frac{\pi}{n} > |ae^{2\pi i/n} - b| \text{ and } |\overline{a}e^{2\pi i/n} - \overline{b}|; \tag{5.132}$$

for, if both these inequalities hold, then (5.129) cannot hold for any k, so $P(a) \neq P(b)$.

An important case occurs here when $b = -a$. The above inequality then becomes

$$|a| < \tan\frac{\pi}{n}, \tag{5.133}$$

and so for these values $P(a) \neq P(-a)$. This result is expressed in the celebrated theorem that we give next.

5.7.4

Grace–Heawood theorem *Let $P(z)$ be a polynomial of degree n and suppose that*

$$P(1) = P(-1). \tag{5.134}$$

Then P has a critical point in the disc $\{|z| \leq \cot\frac{\pi}{n}\}$. This radius is sharp.

For otherwise, writing $R = \tan\frac{\pi}{n}$, the polynomial $Q(z) = P(Rz)$ has no critical points in $\{|z| \leq 1\}$ and so $Q(R) \neq Q(-R)$, i.e. $P(1) \neq P(-1)$.

In particular, if P is a polynomial of degree n with two known zeros a, b then a critical point exists in the disc centred on the mean $\frac{1}{2}(a+b)$ of the two zeros with radius $\frac{1}{2}|a-b|\cot\frac{\pi}{n}$. Since the radius becomes large with n, this result does not give us a degree independent region other than the whole plane. To rectify this situation, we next consider half-planes free of critical points.

5.7.5

Lemma *Let $P(z)$ be a polynomial of degree n and suppose that P has no critical points in the half-plane $\{\Re z > 0\}$. Suppose further that a and b are* *distinct points such that* $P(a) = P(b)$. *Then* $\exists w$ *satisfying* $\Re w \leq 0$ *such that*

$$\frac{w-a}{w-b} = e^{2\pi ik/n}, \tag{5.135}$$

where $k = \pm 1$.

Proof As before we deduce from Grace's theorem the existence of w satisfying $\Re w \leq 0$ such that

$$\int_a^b (t-w)^{n-1} dt = 0. \tag{5.136}$$

This gives (5.135) with k an integer satisfying $1 \leq k \leq n-1$. Now for $\Re w \leq 0$ the left-hand expression in (5.135) describes a circular region containing the point 1 on its boundary ($w = \infty$). Therefore if it contains one of the points $e^{2\pi ik/n}$ $(1 \leq k \leq n-1)$, then it contains one of these points with $k = \pm 1$.

We can interpret this result geometrically. First of all it tells us that the line bisecting the line segment $[a, b]$ must meet the left half-plane. In fact this line is defined by

$$\frac{w-a}{w-b} = e^{it} \quad (-\pi \leq t \leq \pi), \tag{5.137}$$

and is angled so that one of the two values $e^{\pm 2\pi i/n}$ gives w in the left half-plane. As the value 1 gives $w = \infty$, this means that w lies in the left half-plane for either (a) $0 < t \leq 2\pi/n$ or (b) $-2\pi/n \leq t < 0$. Solving for w we have

$$w = \frac{a - be^{it}}{1 - e^{it}} = -\frac{ae^{-it/2} - be^{it/2}}{2i \sin \frac{1}{2}t} \tag{5.138}$$

from which we obtain

$$\Re w = \frac{\Im(\lambda e^{it/2})}{2 \sin \frac{1}{2}t}, \tag{5.139}$$

where $\lambda = \overline{a} + b$. We see easily that cases (a) and (b) together imply that

$$|\arg \lambda| \geq \frac{\pi}{n}. \tag{5.140}$$

We therefore have the next result.

5.7.6

Theorem *Let $P(z)$ be a polynomial of degree n such that*

$$P'(z) \neq 0 \text{ for } \Re z > 0. \tag{5.141}$$

If a and b are distinct points such that

$$|\arg(\overline{a} + b)| < \frac{\pi}{n}, \tag{5.142}$$

then $P(a) \neq P(b)$. In particular P is univalent in the sector $\{|\arg z| < \frac{\pi}{n}\}$.

For, if a and b are points in this sector, then $\overline{a} + b$ is also a point in the sector. In fact the theorem is easily expressed geometrically. If P has no critical points in the right half-plane, then $P(z) \neq P(a)$ in the sector $\{|\arg(z + \overline{a})| < \frac{\pi}{n}\}$. This is the sector with vertex at the point $-\overline{a}$, i.e. at the reflection of a through the imaginary axis, of angle $2\pi/n$ symmetrically placed about the line parallel to the real axis through a and oriented to contain numbers with large real part. Thus P is univalent in every such sector with a vertex on the imaginary axis. In particular P is 1–1 on the positive real axis and on every half-line parallel to it with endpoint on the imaginary axis.

5.7.7

It is clear that, if we strengthen the condition of the above theorem to P having no critical points in the closed half-plane $\{\Re z \geq 0\}$, then $P(a) \neq P(b)$ in the corresponding closed sector. In particular $P(-1) \neq P(1)$. This gives us the following very useful result.

The bisector theorem *Let P be a polynomial and let a and b be distinct points such that $P(a) = P(b)$. Then $\exists$ a critical point ζ of P such that*

$$|\zeta - a| \leq |\zeta - b|. \tag{5.143}$$

In other words, if we draw the line bisecting the segment $[a, b]$, then P has a critical point either on the line or on the same side of the line as a. Of course P also has a critical point on the other side of the line, or on it, i.e. a critical point for which the inequality is reversed.

This result implies an interesting inequality relating the zeros and critical points of a polynomial. Suppose that P is a polynomial of degree n with zeros z_k $(1 \leq k \leq n)$ and critical points ζ_j $(1 \leq j \leq n-1)$. Then

$$\max_{1\leq k\leq n} \min_{1\leq j\leq n-1} |\zeta_j - z_k| \leq \min_{1\leq k\leq n} \max_{1\leq j\leq n-1} |\zeta_j - z_k|. \tag{5.144}$$

This follows from the fact that, given two zeros z_r and z_s, $\exists\zeta_j$ such that

$$|\zeta_j - z_r| \leq |\zeta_j - z_s|. \tag{5.145}$$

A similar inequality follows from the Gauss–Lucas theorem. From this latter theorem it is clear that, given two critical points ζ_i and ζ_j, $\exists z_k$ such that

$$|\zeta_i - z_k| \leq |\zeta_j - z_k| \tag{5.146}$$

from which we deduce the inequality

$$\max_{1\leq j\leq n-1} \min_{1\leq k\leq n} |\zeta_j - z_k| \leq \min_{1\leq j\leq n-1} \max_{1\leq k\leq n} |\zeta_j - z_k|. \tag{5.147}$$

Each of the two inequalities is giving us general information about the $(n-1)\times n$ matrix $M_{j,k} = \{|\zeta_j - z_k|\}$. In terms of this matrix the bisector theorem asserts that *given any two columns (i.e. given k_1 and k_2) $\exists$ a row (i.e. $\exists j$) such that $M_{j,k_1} \leq M_{j,k_2}$*. Of course, by the same token, there is another row giving the reverse inequality. The inequality (5.144) is a consequence of this (but not equivalent to it). In a similar manner the Gauss–Lucas theorem tells us that the corresponding results hold for the transpose $n\times(n-1)$ matrix $M_{k,j}$. We will consider this matrix again in chapter 6.

5.7.8

In studying the relation between critical points and zeros, our point of view so far has been to determine regions containing critical points given information concerning the zeros. The bisector theorem is well suited for the converse problem, that is to determine the location of the zeros given the location of the critical points. It is clear that the zeros are determined if we know all the critical points plus the location of one zero. The following result illustrates this general observation.

Theorem *Let $P(z)$ be a polynomial of degree $n\geq 2$ with all its critical points in $\{|z|\leq 1\}$ and suppose that $P(a)=0$. Then every zero z of P satisfies the equation*

$$z = 2w - \bar{a}w^2, \tag{5.148}$$

where $|w|\leq 1$. In particular all the zeros of P lie in $\{|z|\leq 2+|a|\}$.

Proof According to the bisector theorem, since a and z are zeros of P, $\exists$ a critical point of P on each side of the line bisecting the segment $[a, z]$, and therefore this bisector must meet the unit circle $\mathbb{T}$. It follows that $\exists \alpha \in \mathbb{T}$ such that

$$|z - \alpha| = |a - \alpha| = |1 - \bar{a}\alpha|. \tag{5.149}$$

We deduce that $\exists \beta \in \mathbb{T}$ such that $z - \alpha = \beta(1 - \bar{a}\alpha)$, which gives the relation

$$z = \alpha + \beta - \bar{a}\alpha\beta. \tag{5.150}$$

This expresses z as a value of a symmetric linear form in the variables α and β taking their values on $\mathbb{T}$, and therefore in the closed unit disc. It follows from Walsh's theorem 5.2.7 that z is a value of the polynomial obtained by putting $\alpha = \beta = w$ with $|w| \le 1$. We obtain the desired result.

5.7.9

Corollary *Let $P(z)$ be a polynomial of degree $n \ge 2$ with all its critical points in $\{|z| \le 1\}$. Then all the zeros of P lie in an annulus $\{R \le |z| \le R + 2\}$ for some $R \ge 0$.*

Proof By the theorem, if z and ζ are zeros of P, then $|z| - |\zeta| \le 2$. In particular, if ζ is the zero of smallest modulus, then $|\zeta| \le |z| \le |\zeta| + 2$.

We note from the theorem that in the case $|a| \le 1$ the possible region of zeros of P is a Jordan region. This follows from the fact that the function $\phi(w) = 2w - \bar{a}w^2$ is univalent for $|w| \le 1$ and maps $\{|w| = 1\}$ onto a Jordan curve. For example, in the case $a = 1$ the curve is symmetric about the real axis meeting it at $z = 1$ and $z = -3$ and possessing a cusp at $z = 1$. The curve encloses the disc of radius 2, centre -1. We note further that the locus region for the zeros can be described also by the inequality

$$||z|^2 - |a|^2| \le 2|z - a|. \tag{5.151}$$

The method of proof of the above theorem has an obvious generalisation. If we know that the critical points of a polynomial P lie in a compact connected region K and that $P(a) = 0$, then every zero z of P satisfies the relation

$$|z - \zeta| = |\zeta - a| \tag{5.152}$$

for some $\zeta \in \partial K$, for the perpendicular bisector of the segment $[a, z]$ must meet K and therefore ∂K.

5.7.10 Problems

1. Let P be a polynomial with all its critical points in $\{|z| \leq 1\}$ and suppose that $|P(z)| > m$ for $|z| = 1$. Show that

either (a) $|P(z)| > m$ for $|z| \leq 1$
or (b) $|P(z)| > m$ for $|z| \geq 3$.

2. By considering the polynomials $P(z) = (z - \zeta)^n - (z - a)^n$, show that the region described by theorem 5.7.8 is the best possible degree independent connected region.

5.8 Linear operators

5.8.1 Polynomials in the unit disc

A linear operator Λ on the class of all polynomials is defined by a sequence

$$\lambda_k(z) = \Lambda z^k \quad (k = 0, 1, 2, \ldots) \tag{5.153}$$

of functions analytic in $\mathbb{U}$, for if $P(z) = \sum_{k=0}^{n} a_k z^k$, then $\Lambda(P)(z) = \sum_{k=0}^{n} a_k \lambda_k(z)$. We consider the following problem.

Which operators Λ have the property of mapping the class $\mathcal{V}_n$ into the class $\mathcal{V}$ of functions non-vanishing in $\mathbb{U}$ (see section 5.1.2)?

It is not difficult to see that a necessary and sufficient condition for this is that

$$\Lambda(1 + xz)^n \neq 0 \quad (|x| \leq 1, |z| < 1). \tag{5.154}$$

Indeed, this is clearly necessary. On the other hand, fixing $z \in \mathbb{U}$ the condition gives

$$\sum_{k=0}^{n} \binom{n}{k} \lambda_k(z) x^k \neq 0 \quad (|x| \leq 1) \tag{5.155}$$

and therefore, if P has degree $\leq n$ and no zeros in $\mathbb{U}$, $\Lambda P(z) \neq 0$ by the Grace–Szegö theorem.

It is interesting to note that the condition (5.154) can be weakened. For example, suppose that the condition holds for $|x| < 1$ and $|z| < 1$. Then it is certainly true by our earlier results that $\Lambda P(z) \neq 0$ for any polynomial P of degree $\leq n$ with no zeros in $\overline{\mathbb{U}}$. Furthermore, if $\{x_i\}$ is a sequence in $\mathbb{U}$ tending to a point $x \in \mathbb{T}$, then by Hurwitz's theorem, either (5.154) holds for this x and all $z \in \mathbb{U}$, or the expression vanishes identically for this x and all $z \in \mathbb{U}$. Thus a sufficient condition is that (5.154) holds for $|x| < 1$ and $|z| < 1$ and also for $|x| = 1$ and $z = 0$.

5.8.2

Interestingly, the above pair of conditions can be turned around. If (5.154) holds for $|x| = 1$ and $|z| < 1$ and also for $|x| < 1$ and $z = 0$, then it holds in general. This is a consequence of the following result on the analytic continuation of a function analytic in two variables.

Lemma *Let $f(x, z)$ be a function analytic for $|x| \leq 1$ and $|z| < 1$ and suppose that*

$$f(x, z) \neq 0 \tag{5.156}$$

for $|x| = 1$ and $|z| < 1$ and also for $|x| < 1$ and $z = 0$. Then the condition holds for $|x| \leq 1$ and $|z| < 1$.

Proof Choose $0 < r < 1$ and note that $\exists \delta(r) > 0$ such that $f(x, z)$ is analytic and $\neq 0$ for $1 - \delta(r) \leq |x| \leq 1 + \delta(r)$ and $|z| \leq r$ and therefore in this range the reciprocal $g(x, z) = 1/f(x, z)$ is analytic and has a Laurent expansion in the form

$$g(x, z) = \sum_{k=-\infty}^{\infty} A_k(z) x^k \tag{5.157}$$

where each $A_k(z)$ is analytic for $|z| \leq r$. On the other hand, $f(x, 0)$ is non-zero for $|x| \leq 1$ and therefore $\exists \rho > 0$ such that $f(x, z) \neq 0$ for $|x| \leq 1$ and $|z| \leq \rho$. Therefore in this range we can write g in the form

$$g(x, z) = \sum_{k=0}^{\infty} B_k(z) x^k \tag{5.158}$$

where each $B_k(z)$ is analytic for $|z| \leq \rho$. Now both these representations are valid for $1 - \delta(r) \leq |x| \leq 1$ and $|z| \leq \rho$ and therefore, as the Laurent expansion is unique, we obtain $B_k(z) = A_k(z)$ for $k \geq 0$ and $A_k(z) = 0$ for $k < 0$. This last will hold, by analytic continuation, for $|z| \leq r$ and so the Laurent expansion of g is a Taylor expansion converging for $|x| \leq 1$ and $|z| \leq r$. By analytic continuation $1/f$ remains analytic in this region and so $f \neq 0$ in the region. Hence result.

5.8.3

Summarising, we have the following result.

Theorem *Let* $\Lambda\colon \mathcal{A}_n \to \mathcal{A}$ *be a linear operator and suppose that*

$$\Lambda(1+xz)^n \neq 0 \tag{5.159}$$

either for (a) $|x| < 1, |z| < 1$ *and* $|x| = 1, z = 0$
or for (b) $|x| = 1, |z| < 1$ *and* $|x| < 1, z = 0$.

Then, if $P(z)$ *is a polynomial of degree* $\leq n$ *satisfying* $P(z) \neq 0$ $(z \in \mathbb{U})$,

$$\Lambda P(z) \neq 0 \quad (z \in \mathbb{U}). \tag{5.160}$$

6
The Ilieff–Sendov conjecture

6.1 Introduction

6.1.1

There are a number of problems on polynomials in chapter 4 of Hayman's celebrated collection *Research Problems in Function Theory* [20], and many of these have since been solved. We devote this chapter to the study of Problem 4.5 from which a substantial literature has evolved and yet which remains unsolved to this day. The problem is as follows.

Let $P(z)$ be a polynomial whose zeros $z_1, z_2, \ldots, z_n$ lie in $|z| \leq 1$. Is it true that $P'(z)$ always has a zero in $|z - z_1| \leq 1$?

This problem was communicated to Hayman, subsequent to the conference in Classical Function Theory held at Imperial College, London in 1964, by the Bulgarian mathematician L. Ilieff and was attributed to Ilieff in Hayman's book. As a result the problem was known for several years as 'Ilieff's conjecture' and much of the literature on the problem bears this title. However, word eventually circulated that the conjecture was made originally by Ilieff's colleague and compatriot Bl. Sendov. In the circumstances, and bearing in mind that Ilieff deserves a share of the credit for publicising the problem, we have compromised by naming the problem the 'Ilieff–Sendov conjecture'.

In this chapter we shall give a selection of the many partial results obtained in the attempt at settling the conjecture. Surveys of the problem have been given by M. Marden [31] and Bl. Sendov [51] and we refer the reader to these for further information and bibliographies. Most of the results obtained point strongly in favour of the conjecture being correct, although there are occasional hints in the opposite direction. My own view is that odds of 4–1 in favour of the conjecture are cautious odds in the circumstances.

If $P(z)$ has all its zeros in $\{|z| \leq 1\}$, then from the Gauss–Lucas theorem so does P', and hence all the critical points lie within distance $1 + |z_1|$ of z_1. Therefore the conjecture certainly holds in the case $z_1 = 0$. The conjecture arises by considering the polynomials $P(z) = z^n - 1\,(n = 2, 3, \ldots)$. These have their zeros on the unit circle and a single multiple critical point at $z=0$. Therefore, if the conjecture is correct, these are extremal polynomials for the problem. It is interesting to note that there are a number of unsolved problems in the literature, for which these polynomials are conjectured to be extremal.

6.2 Proof of the conjecture for those zeros on the unit circle

6.2.1

We shall consider a polynomial $P(z)$ of degree $n \geq 2$ with all its zeros in $\overline{\mathbb{U}}$ and with a given zero at $z = a$. The first major verification of the conjecture was given by Z. Rubinstein [43] who showed that the conjecture holds when $|a| = 1$. In particular, the conjecture is true for polynomials all of whose zeros lie on the unit circle. An improved version of this result was later given by Goodman, Rahman and Ratti [17] as in the next subsection.

6.2.2

Theorem *Suppose that $P(z)$ has all its zeros in $\overline{\mathbb{U}}$ and that $P(a) = 0$, where $|a| = 1$. Then P' has a zero in the disc $\{|z - \frac{1}{2}a| \leq \frac{1}{2}\}$.*

Proof We can write $P(z) = (z - a)S(z)$, where $S(z)$ is a polynomial of degree $n - 1$ with zeros $z_1, \ldots, z_{n-1}$ in $\overline{\mathbb{U}}$. We may assume that these zeros are distinct from a, as otherwise the result is trivial. Let $\zeta_1, \ldots, \zeta_{n-1}$ be the critical points of P and note the relation

$$\frac{P''(a)}{P'(a)} = 2\frac{S'(a)}{S(a)}. \tag{6.1}$$

This gives

$$\sum_{k=1}^{n-1} \frac{a}{a - \zeta_k} = 2\sum_{k=1}^{n-1} \frac{a}{a - z_k} \tag{6.2}$$

and therefore

$$\Re\left(\sum_{k=1}^{n-1} \frac{a}{a - \zeta_k}\right) = 2\Re\left(\sum_{k=1}^{n-1} \frac{1}{1 - z_k\bar{a}}\right) \geq n - 1. \tag{6.3}$$

It follows that, for at least one k, we have

$$\Re\left(\frac{a}{a-\zeta_k}\right) \geq 1. \tag{6.4}$$

This is equivalent to $|\zeta_k - \frac{1}{2}a| \leq \frac{1}{2}$, as required.

Note further that (6.3) gives the inequality

$$\sum_{k=1}^{n-1} \frac{1}{|a-\zeta_k|} \geq n-1, \tag{6.5}$$

so that the mean of the reciprocals of the distances of the critical points from a is at least 1. This is a stronger statement than the conjecture and it would be interesting to know whether the inequality continues to hold when $|a| < 1$. Clearly it holds when $a = 0$. What is clear is that this method of proof will not work for $0 < |a| < 1$, as it is possible to have $S'(a) = 0$. (Nevertheless, the conjecture holds for a when $S'(a) = 0$, for then $P''(a) = 0$ and so by the Gauss–Lucas theorem the point a lies in the convex hull of the set of critical points. This implies that at least one critical point ζ satisfies $\Re(\zeta/a) \geq 1$, and since $|\zeta| \leq 1$, we obtain $|\zeta - a| \leq \sqrt{1-|a|^2}$.)

6.2.3 Problem

1. Let $P(z) = (z-a)S(z)$ be a polynomial of degree $n \geq 2$ with all its critical points in $\{|z| \leq 1\}$, where $0 < a < 1$. Show that P has a critical point in $\{|z-a| \leq 1\}$ if

   $$\Re\frac{S'(a)}{S(a)} \leq \frac{(n-1)a}{4}.$$

   [Hint: Show that, if P' has no zeros in $\{|z-a| \leq 1\}$, then $P''(a)/P'(a) = (n-1)/(a-\zeta)$, where ζ satisfies $|\zeta + \frac{1-a^2}{a}| < \frac{1}{a}$; note that this defines a disc which excludes a, but contains all the zeros of P'.]

6.3 The direct application of Grace's theorem

6.3.1

We may state the Ilieff–Sendov conjecture in the following form. Let $S(z)$ be a polynomial of degree $n \geq 1$ with all its zeros in $\overline{\mathbb{U}}$. To show that, for each a ($|a| \leq 1$) the polynomial $z \mapsto S(z+a) + zS'(z+a)$ has at least one zero in $\overline{\mathbb{U}}$. If we assume that for a particular a and a particular n there is a polynomial S for which this is not true, then there is certainly such a polynomial with all its zeros in the open disc $\mathbb{U}$. Therefore we have two polynomials which are linearly related, one having all its zeros in $\mathbb{U}$ and the other having no zeros in $\overline{\mathbb{U}}$. We may therefore attempt to contradict this assertion by the application of Grace's theorem. In what follows we give an organised account of this method of attack on the problem.

6.3.2

Let Φ and Ψ be two linear functionals defined on the polynomials of degree n. Writing $T(z) = S(z + a) + zS'(z + a)$ and $Q(z) = S(z + a)$, we can find numbers c and d not both zero such that

$$c\Phi Q + d\Psi T = 0. \tag{6.6}$$

Since T is obtained from Q by a linear relation, this equation can be regarded as the statement that a linear functional acting on Q vanishes. From Grace's theorem we deduce that $\exists\alpha$ ($|\alpha + a| < 1$) such that

$$c\Phi(z-\alpha)^n + d\Psi((z-\alpha)^n + zn(z-\alpha)^{n-1}) = 0. \tag{6.7}$$

On the other hand the vanishing linear functional can also be regarded as a linear functional acting on T, and therefore a second application of Grace's theorem gives that $\exists\beta$ ($|\beta| < 1$) such that

$$c\Phi\left[\frac{1}{(n+1)\beta z}((1+\beta z)^{n+1} - 1)\right] + d\Psi(1+\beta z)^n = 0. \tag{6.8}$$

Eliminating c and d from these equations we obtain

$$\begin{aligned}\Phi(z-\alpha)^n\Psi(1+\beta z)^n = \Phi&\left[\frac{1}{(n+1)\beta z}((1+\beta z)^{n+1} - 1)\right]\\ &\times \Psi[(z-\alpha)^{n-1}((n+1)z - \alpha)].\end{aligned} \tag{6.9}$$

Writing $\gamma = -\alpha$ and expanding in powers of z, this gives

$$\begin{aligned}&\sum_{k=0}^{n}\binom{n}{k}A_k\gamma^{n-k}\sum_{k=0}^{n}\binom{n}{k}B_k\beta^k\\ &\quad = \sum_{k=0}^{n}(k+1)\binom{n}{k}B_k\gamma^{n-k}\sum_{k=0}^{n}\frac{1}{k+1}\binom{n}{k}A_k\beta^k,\end{aligned} \tag{6.10}$$

where $A_k = \Phi(z^k)$, $B_k = \Psi(z^k)$ ($0 \le k \le n$). Thus we have shown that for arbitrary complex numbers A_k and B_k we can find γ satisfying $|\gamma - a| < 1$ and β satisfying $|\beta| < 1$ such that (6.10) is satisfied. Now if $p(z)$, $q(z)$ are arbitrary polynomials of degree $\le n$, then we can write these polynomials in the form

$$p(z) = \sum_{k=0}^{n} \frac{1}{k+1} \binom{n}{k} A_k z^k, \qquad q(z) = \sum_{k=0}^{n} \binom{n}{k} B_k z^k, \tag{6.11}$$

and the condition (6.10) becomes

$$p^*(\gamma) q(\beta) = p(\beta) q^*(\gamma), \tag{6.12}$$

where, if $s(z)$ is a polynomial of degree $\le n$,

$$s^*(z) = z^n s(1/z) + z^{n-1} s'(1/z). \tag{6.13}$$

We have thus established the following result. *Suppose that there exists a polynomial P of degree $n+1$ with the following properties: (a) P has all its zeros in the closed unit disc; (b) $P(a) = 0$; (c) P has no critical points in the disc $\{|z - a| \le 1\}$. Then for each pair of polynomials p and q, each of degree $\le n$, we can find a point β satisfying $|\beta| < 1$ and a point γ satisfying $|\gamma - a| < 1$ such that $p^*(\gamma) q(\beta) = p(\beta) q^*(\gamma)$.*

6.3.3

Let us consider, for example, the case where $q(z) \equiv 1$. The equation becomes

$$p^*(\gamma) = \gamma^n p(\beta). \tag{6.14}$$

If we apply this to the polynomial $p(z) = z$, we obtain $2\gamma^{n-1} = \gamma^n \beta$. If $|a| = 1$, then $\gamma = 0$ is not a solution and so the equation becomes $\beta\gamma = 2$. This has no solutions with $|\beta| < 1$ and $|\gamma - a| < 1$. Therefore, the conjecture is true for zeros a lying on the unit circle; a fact which we have already proved. Notice however that this argument fails when $|a| < 1$ and $n \ge 2$, for then $\gamma = 0$ gives a solution to the equation. Indeed the equation (6.14) is satisfied when $\gamma = 0$, whenever p is a polynomial of degree $< n$.

Thus our strategy is the following. For a particular value a ($|a| < 1$) and n we should like to show that the assertion of the conjecture holds. We will have achieved this, if we can find a polynomial p of degree n such that the equation (6.14) has no solutions with $|\beta| < 1$ and $|\gamma - a| < 1$.

For example, consider the case $p(z) = z^n$. We obtain the equation

$$(\beta\gamma)^n = n + 1, \tag{6.15}$$

which gives

$$|\gamma| > (n+1)^{1/n}. \tag{6.16}$$

If $|\gamma - a| < 1$, this implies that $|a| > (n+1)^{1/n} - 1$. Therefore the conjecture is true for polynomials P of degree $n+1$, when the zero a satisfies $|a| \leq (n+1)^{1/n} - 1$.

6.3.4

We shall apply this method to prove the conjecture for polynomials P of degree 2, 3, 4 or 5; i.e. in the above notation for the cases $n = 1, 2, 3, 4$. We assume $|a| < 1$ and choose

$$p(z) = \frac{1}{(n+1)z}\left(\left(z + \frac{\bar{a}}{1-|a|^2}\right)^{n+1} - \left(\frac{\bar{a}}{1-|a|^2}\right)^{n+1}\right) \tag{6.17}$$

which gives

$$p(z) + zp'(z) = \left(z + \frac{\bar{a}}{1-|a|^2}\right)^{n}. \tag{6.18}$$

For $|\beta| < 1$ we obtain

$$|p(\beta)| < \frac{1}{n+1}\left(\left(1 + \frac{|a|}{1-|a|^2}\right)^{n+1} - \left(\frac{|a|}{1-|a|^2}\right)^{n+1}\right). \tag{6.19}$$

On the other hand, writing $\sigma = 1/\gamma$ it is easily seen that, if $|\gamma - a| < 1$, then

$$\left|\sigma + \frac{\bar{a}}{1-|a|^2}\right| > \frac{1}{1-|a|^2}. \tag{6.20}$$

It follows that for $|\gamma - a| < 1$ we have

$$|p(\sigma) + \sigma p'(\sigma)| > \left(\frac{1}{1-|a|^2}\right)^{n}. \tag{6.21}$$

Now our criterion (6.14) gives for suitable β and γ in the above ranges

$$p(\beta) = p(\sigma) + \sigma p'(\sigma). \tag{6.22}$$

We deduce therefore the inequality

$$\left(\frac{1}{1-|a|^2}\right)^{n} < \frac{1}{n+1}\left(\left(1 + \frac{|a|}{1-|a|^2}\right)^{n+1} - \left(\frac{|a|}{1-|a|^2}\right)^{n+1}\right). \tag{6.23}$$

To prove the conjecture for any particular values of a and n, it is sufficient to show that this inequality is false for those values. We show that it is false for $0 \leq |a| < 1$ and $n = 1, 2, 3, 4$. Let $t = \frac{|a|}{1-|a|^2}$, so $t \geq 0$ and the inequality reduces to

$$\left(\frac{1+s}{2}\right)^{n} < \frac{(1+t)^{n+1} - t^{n+1}}{n+1}, \tag{6.24}$$

where $s = \sqrt{1+4t^2}$. For the case $n = 1$ this reduces to $s < 2t$, which is clearly false. For the case $n = 2$ we obtain

$$3(1 + 1 + 4t^2 + 2s) < 4(1 + 3t + 3t^2), \tag{6.25}$$

i.e. $6s < 12t - 2$, so again $s < 2t$. For the case $n = 3$ we obtain

$$4(1 + 3s + 3(1 + 4t^2) + s(1 + 4t^2)) < 8(1 + 4t + 6t^2 + 4t^3), \tag{6.26}$$

i.e. $16s(1 + t^2) < 32t(1 + t^2) - 8$, which again implies $s < 2t$. For the case $n = 4$ we obtain

$$\begin{aligned} 5(1 + 4s + 6(1 + 4t^2) + 4s(1 + 4t^2) + (1 + 4t^2)^2) \\ < 16(1 + 5t + 10t^2 + 10t^3 + 5t^4), \end{aligned} \tag{6.27}$$

i.e. $40s(1 + 2t^2) < 80t(1 + 2t^2) - 24$, again implying $s < 2t$.

6.3.5

Theorem *The Ilieff–Sendov conjecture is correct for polynomials of degree* 2, 3, 4 *or* 5.

The above argument does not work for all values of $a \in \mathbb{U}$ when $n \geq 5$. Indeed for large $t > 0$ we have

$$\left(\frac{1+s}{2}\right)^n = t^n + \frac{n}{2}t^{n-1} + \frac{n^2}{8}t^{n-2} + \cdots \tag{6.28}$$

and

$$\frac{(1+t)^{n+1} - t^{n+1}}{n+1} = t^n + \frac{n}{2}t^{n-1} + \frac{n(n-1)}{6}t^{n-2} + \cdots. \tag{6.29}$$

The difference between these is $\frac{n(4-n)}{24}t^{n-2} + \cdots$ which is negative for large t when $n > 4$. Thus for polynomials of degree ≥ 6 the argument fails when $|a|$ is close to 1. It is also easily seen that the method fails for fixed a ($0 < |a| < 1$) when n is large.

However, we still obtain a proof for values of a near to 0. Indeed, for polynomials of degree 6 we obtain a proof of the conjecture for zeros a satisfying $|a| \leq 0.7477$.

6.3.6

Another approach using this method is to choose p^* and q^* so that q^*/p^* is a Blaschke product relative to the disc $\{|z - a| < 1\}$ and then attempt to show that $|p/q|$ is <1 in the unit disc. For example, let

$$p^*(z) = (1 + b(z - a))^n, \qquad q^*(z) = (z - a + b)^n, \tag{6.30}$$

where $0 < a < 1$ and where $0 < b < 1$ is a parameter to be chosen. Then

$$p(z) = \frac{1}{(n+1)z(1-ab)}[((1-ab)z + b)^{n+1} - b^{n+1}] \tag{6.31}$$

and

$$q(z) = \frac{1}{(n+1)z(b-a)}[(1 + (b-a)z)^{n+1} - 1]. \tag{6.32}$$

Clearly

$$\left|\frac{p^*(z)}{q^*(z)}\right| > 1 \quad (|z - a| < 1). \tag{6.33}$$

Thus to prove the conjecture it is sufficient to show that $q(z) \neq 0\,\{|z| < 1\}$ and that $|p(z)/q(z)| < 1$ for $|z| < 1$. At $z = 0$, $p(0) = b$ and $q(0) = 1$. Otherwise, if $q(z) = 0$, then $1 + (b-a)z = \omega$, where ω is an $(n+1)$th root of unity. To avoid this we choose b so that $0 \leq b - a < 2\sin\frac{\pi}{n+1}$. We thus require

$$\frac{b-a}{1-ab}\left|\sum_{k=1}^{n+1}(1-ab)^k z^k \binom{n+1}{k} b^{n+1-k}\right| \leq \left|\sum_{k=1}^{n+1}\binom{n+1}{k}(b-a)^k z^k\right|, \tag{6.34}$$

for $|z| < 1$. This reduces to

$$\left|\sum_{k=0}^{n}\binom{n+1}{k+1}(1-ab)^k b^{n-k} z^k\right| \leq \left|\sum_{k=0}^{n}\binom{n+1}{k+1}(b-a)^k z^k\right|. \tag{6.35}$$

6.4 A global upper bound

6.4.1

Failing a general proof (or disproof) of the conjecture we may proceed in one of two ways. Either we may seek additional conditions on P under which the conjecture will hold or we may ask for an upper bound on the distance from a zero a of the nearest critical point to a. The best general bound to date has been given by Bojanov, Rahman and Szynal [8] and in this section we present a slight modification of their method.

Without loss of generality we assume that $P(a) = 0$, where $0 < a < 1$, and that P has degree $n \geq 2$ with all its remaining zeros in the closed unit disc. Let d denote the distance from a of the nearest critical point to a. Assuming P is monic we have

$$P'(z) = n\prod_{k=1}^{n-1}(z - \zeta_k), \tag{6.36}$$

where ζ_k are the critical points of P, and therefore $|\zeta_k - a| \geq d$ for $1 \leq k \leq n-1$. We therefore obtain

$$|P'(z)| \geq n(d-r)^{n-1} \quad (|z-a| \leq r \leq d). \tag{6.37}$$

The basic idea is to integrate this inequality along a suitably chosen path. To do this we consider that branch of the inverse function $Q(w) = P^{-1}(w)$ corresponding to the zero of P at a. P is 1–1 near a and so the branch of the inverse function exists near 0 and satisfies $Q(0) = a$. This branch may be analytically continued along any path from 0 which does not meet a branch point ϕ. Each such branch point ϕ has the form $\phi = P(\zeta)$, where ζ is a critical point of P. We continue Q along rays from the origin stopping only at such branch points. In this way we find that Q can be continued into a domain D which consists of the whole plane cut along at most $n-1$ radial slits. As D is simply-connected, the monodromy theorem ensures that this process defines a single-valued analytic branch Q of the inverse function in D. P is then univalent in the domain $Q(D)$ and furthermore there is at least one critical point ζ on the boundary of $Q(D)$. The point $P(\zeta) = \phi$ is the endpoint of one of the radial slits. Let $z \in Q(D)$ and let Γ denote the line segment $[0, P(z)]$ and $\gamma = Q(\Gamma)$, so γ is a path in $Q(D)$ joining a to z. Then

$$|P(z)| = \int_\Gamma |dw| = \int_\gamma |P'(z)||dz| \geq \int_{\gamma'} n(d - |z-a|)^{n-1}|dz|, \tag{6.38}$$

where γ' is that part of γ lying in the disc $\{|z-a| \leq d\}$. We deduce that

$$|P(z)| \geq \int_0^r n(d-\rho)^{n-1}d\rho = d^n - (d-r)^n, \tag{6.39}$$

where $r = |z-a|$. In particular, letting $z \to \zeta$, we deduce the existence of a critical point ζ of P such that

$$|P(\zeta)| \geq d^n. \tag{6.40}$$

Now, since P is monic and has all its zeros in the unit disc, $|P(0)|$ is the product of the absolute values of these zeros and so satisfies

$$|P(0)| \leq a. \tag{6.41}$$

On the other hand, either $d^n \leq a$ or $d^n > a$. In the latter case the domain D contains the point $P(0)$ and so there exists $p \in Q(D)$ such that $P(p) = P(0)$.

If $p = 0$, we deduce from (6.39)

$$d^n - (d-a)^n \leq |P(0)| \leq a, \tag{6.42}$$

which gives

$$\sum_{k=0}^{n-1} d^{n-1-k}(d-a)^k \leq 1 \tag{6.43}$$

and therefore $d \leq 1$. Thus the conjecture holds in this case. If $d > 1$, then $p \neq 0$ and so p and 0 are distinct points at which P takes the same value. It follows from the bisector theorem (chapter 5, theorem 5.7.7) that $\exists$ a critical point ω of P such that

$$|\omega - p| \leq |\omega|. \tag{6.44}$$

Hence

$$d \leq |\omega - a| \leq |\omega| + |p - a| \leq 1 + \rho, \tag{6.45}$$

where $\rho = |p - a|$. Also, as $p \in Q(D)$, we have from this and (6.39)

$$|P(0)| = |P(p)| \geq d^n - (d-\rho)^n \geq d^n - 1. \tag{6.46}$$

We obtain

$$d \leq (1 + |P(0)|)^{\frac{1}{n}} \leq (1 + a)^{\frac{1}{n}} < 2^{\frac{1}{n}}. \tag{6.47}$$

Since the conjecture is proved for polynomials up to degree 5, we obtain the global upper bound

$$d \leq 2^{\frac{1}{6}} < 1.123. \tag{6.48}$$

With the help of a more refined version of this argument Bojanov, Rahman and Szynal [8] have improved this bound to $d \leq 1.08\ldots$.

6.4.2

Theorem *Let $P(z)$ be a polynomial of degree $n \geq 2$ with all its zeros in $\{|z| \leq 1\}$. Then for each zero of P there is a critical point of P within distance $2^{1/n}$ of the zero.*

6.4.3

In addition to giving a global bound the method of this section proves the conjecture in certain cases. The most notable is the case $P(0) = 0$, which follows directly from (6.47). Thus *the Ilieff–Sendov conjecture is true for all polynomials P which have a zero at the origin* (in other words the distance from any zero of P to the set of critical points is at most 1). This result is due to G. Schmeisser. More generally, the bisector theorem shows that, if $P(a) = 0$ where $0 < a < 1$, and if P has a second zero in $[0, 1]$, then P has a critical point within distance 1 of a; for $\exists$ a critical point ζ such that $\Re\zeta \geq \frac{1}{2}a$ and since also $|\zeta| \leq 1$, we obtain $|\zeta - a| \leq 1$.

The conjecture also holds for real zeros a if the remaining zeros of P are symmetric relative to the real axis, i.e. when P is a real polynomial. For if P is real and $P(a) = 0$ where $0 < a < 1$, then with d defined as before we have

$$|P'(x)| \geq n(d - x)^{n-1} > 0 \quad (0 \leq x \leq a) \tag{6.49}$$

(assuming $d > a$) and so, since $P'(x)$ is real, it has constant sign. It follows that

$$a \geq |P(0)| = \int_0^a |P'(x)|dx \geq d^n - (d - a)^n \geq ad^{n-1} \tag{6.50}$$

and so $d \leq 1$.

6.5 Inequalities relating the nearest critical point to the nearest second zero

6.5.1

Let P be a polynomial of degree $n \geq 2$ with a zero at $z = a$. Let $d = d(a, P)$ denote the distance from a of the nearest critical point of P, and let $r = r(a, P)$ denote the distance from a of the nearest second zero of P. It follows from Grace's theorem (see chapter 5, corollary 5.7.3) that

$$d \leq \tfrac{1}{2}r \operatorname{cosec} \frac{\pi}{n} \tag{6.51}$$

and therefore

$$n \geq \frac{4d}{r}. \tag{6.52}$$

In this section we establish some inequalities in the reverse direction. These will show that for large values of n, if d is too large, r must be small and *vice versa*. As an application we will show that, if a is fixed $(0 < a < 1)$ and n is large, then the Ilieff–Sendov conjecture is true for a if there are no other zeros of P close to a. We prove the two results given next.

6.5.2

Theorem *Let P be a polynomial of degree $n \geq 2$ with all its critical points in $\{|z| \leq 1\}$ and satisfying $P(a) = 0$; let $d = d(a, P)$, $r = r(a, P)$.*

(a) *If $|a| > 1$, then*

$$n \leq 2 + 3(1 + |a|)^2 \left(\frac{1}{d^2} + \frac{4}{r^2} \right). \tag{6.53}$$

(b) *If $|a| \leq 1$, then* **either** *$d \leq \sqrt{1 - |a|^2}$,* **or**

$$n \leq 2 + \frac{12|a|^2 d^2 (r^2 + 4d^2)}{r^2 (|a|^2 + d^2 - 1)^2}. \tag{6.54}$$

6.5.3

Theorem *Let P be a polynomial of degree $n \geq 2$ with all its zeros in $\{|z| \leq 1\}$ and satisfying $P(a) = 0$; let $d = d(a, P)$, $r = r(a, P)$.*

(a) *If $|a| > 1$, then*

$$n \leq 1 + (1 + |a|)^2 \left(\frac{1}{d^2} + \frac{3}{r^2} \right). \tag{6.55}$$

(b) *If $|a| \leq 1$, then* **either** *$r \leq \sqrt{1 - |a|^2}$,* **or**

$$n \leq 1 + \frac{4|a|^2 r^2 (r^2 + 3d^2)}{d^2 (r^2 + |a|^2 - 1)^2}. \tag{6.56}$$

6.5.4

The proofs of the two theorems above are based on the following lemma, which is the case $a = 1$ of the theorems.

Lemma *Let P have degree n with $P(1) = 0$ and all its critical points in $\{|z| \leq 1\}$. Then*

$$n \leq 2 + \frac{12}{d^2} + \frac{48}{r^2}. \tag{6.57}$$

If, further, P has all its zeros in $\{|z| \leq 1\}$, then

$$n \leq 1 + \frac{4}{d^2} + \frac{12}{r^2}. \tag{6.58}$$

Proof Write $P(z) = (z-1)S(z)$ and let z_k denote the zeros of S and ζ_k the zeros of P' $(1 \le k \le n-1)$. Then

$$P^{(k+1)}(1) = (k+1)S^{(k)}(1) \quad (0 \le k \le n). \tag{6.59}$$

We may assume no $\zeta_k = 1$, as otherwise $d = r = 0$. Then

$$\Re\frac{P''(1)}{P'(1)} = \sum_{k=1}^{n-1} \Re\frac{1}{1-\zeta_k} \ge \frac{n-1}{2}. \tag{6.60}$$

Now

$$\frac{S'(z)}{S(z)} = \sum_{k=1}^{n-1} \frac{1}{z-z_k} \tag{6.61}$$

and so differentiating

$$\frac{S''(z)}{S(z)} - \left(\frac{S'(z)}{S(z)}\right)^2 = -\sum_{k=1}^{n-1} \frac{1}{(z-z_k)^2}. \tag{6.62}$$

Hence from (6.62)

$$\frac{1}{3}\frac{P'''(1)}{P'(1)} - \frac{1}{4}\left(\frac{P''(1)}{P'(1)}\right)^2 = -\sum_{k=1}^{n-1} \frac{1}{(1-z_k)^2}. \tag{6.63}$$

Applying the identity (6.62) to P' we obtain

$$\frac{P'''(1)}{P'(1)} - \left(\frac{P''(1)}{P'(1)}\right)^2 = -\sum_{k=1}^{n-1} \frac{1}{(1-\zeta_k)^2}. \tag{6.64}$$

From (6.63/64) we obtain

$$\frac{1}{3}\frac{P'''(1)}{P'(1)} = \sum_{k=1}^{n-1} \frac{1}{(1-\zeta_k)^2} - 4\sum_{k=1}^{n-1} \frac{1}{(1-z_k)^2}. \tag{6.65}$$

Now, since P' has all its zeros in $\overline{\mathbb{U}}$, so does P''. Also $P''(1) \ne 0$, as otherwise $P'(1) = 0$. Hence

$$\Re\frac{P'''(1)}{P''(1)} \ge \frac{n-2}{2}. \tag{6.66}$$

From (6.60/65/66) we deduce that

$$\begin{aligned}\frac{(n-1)(n-2)}{12} &\le \frac{1}{3}\left|\frac{P'''(1)}{P''(1)}\right|\left|\frac{P''(1)}{P'(1)}\right| \\ &\le \sum_{k=1}^{n-1} \frac{1}{|1-\zeta_k|^2} + 4\sum_{k=1}^{n-1} \frac{1}{|1-z_k|^2} \le \frac{n-1}{d^2} + \frac{4(n-1)}{r^2}.\end{aligned} \tag{6.67}$$

This gives (6.57). To prove (6.58) we observe that, if S has all its zeros in $\overline{\mathbb{U}}$, then

$$\Re \frac{P''(1)}{P'(1)} = 2\Re \frac{S'(1)}{S(1)} \geq n - 1. \tag{6.68}$$

Also from (6.63/64)

$$\frac{1}{4}\left(\frac{P''(1)}{P'(1)}\right)^2 = \sum_{k=1}^{n-1} \frac{1}{(1-\zeta_k)^2} - 3\sum_{k=1}^{n-1} \frac{1}{(1-z_k)^2}. \tag{6.69}$$

Hence

$$\frac{(n-1)^2}{4} \leq \frac{1}{4}\left|\frac{P''(1)}{P'(1)}\right|^2 \leq \frac{n-1}{d^2} + \frac{3(n-1)}{r^2} \tag{6.70}$$

and the result follows.

6.5.5 Proof of theorems 6.5.2 and 6.5.3

Note first that, if P' has all its zeros in $\{|z-c| \leq \rho\}$ and $P(c+\rho) = 0$, we may apply the lemma to $Q(z) = P(\rho z + c)$. We obtain

$$n - 2 \leq 12\rho^2\left(\frac{1}{d^2} + \frac{4}{r^2}\right). \tag{6.71}$$

In case (a) of theorem 5.2 P' has all its zeros inside the smallest disc passing through a and containing the unit disc. This gives $\rho = \frac{1}{2}(1+|a|)$ and (6.53) follows. In case (b) P' has all its zeros satisfying $|z| \leq 1$ and $|z-a| \geq d$. Assuming $0 \leq a < 1$, either $d \leq \sqrt{1-a^2}$ or P' has all its zeros in

$$\left|z + \frac{a(1-a^2)}{d^2+a^2-1}\right| \leq \frac{ad^2}{d^2+a^2-1} \tag{6.72}$$

and

$$P\left(\frac{a(a^2-1)}{d^2+a^2-1} + \frac{ad^2}{d^2+a^2-1}\right) = P(a) = 0. \tag{6.73}$$

Hence putting

$$\rho = \frac{ad^2}{d^2+a^2-1}, \tag{6.74}$$

we deduce (6.54). The proof of theorem 6.5.3 is similar.

6.5.6

Note in particular that, if P has all its critical points in the unit disc and $P(a) = 0$, where $0 < a < 1$, then if $d \geq 1$ we obtain from 6.5.2

$$n \leq 2 + \frac{12(r^2 + 4)}{r^2 a^2} \tag{6.75}$$

and therefore

$$r \leq \frac{K(a)}{\sqrt{n}}. \tag{6.76}$$

Thus for large n, P has a second zero close to a. This result has some significance for the nature of the extremal polynomials for the Ilieff–Sendov problem. Fixing a ($0 < a < 1$) let $\delta_n(a)$ denote the maximal distance of the nearest critical point to a for polynomials of degree n with zeros in $\overline{\mathbb{U}}$. This value will be attained for some extremal polynomial $\Pi(z)$ of degree n (i.e. $\delta_n(a) = d(a, \Pi)$). There are two possibilities. Either $\limsup_{n\to\infty} \delta_n(a) \leq \sqrt{1 - a^2}$, or for some large n, Π_n has a zero within distance $c(a)/\sqrt{n}$ of a, where $c(a)$ depends only on a. In the latter case the extremal polynomials Π_n do not have their remaining zeros on the unit circle, a fact which is perhaps something of a surprise. We think that this is probably the situation, since it seems likely that $\delta_n(a) \to 1$ as $n \to \infty$.

6.5.7

Theorem *Suppose that P is a polynomial of degree n with $P(a) = 0$, where $0 < a < 1$, and suppose further that P has its remaining zeros on the unit circle. Then*

$$d(a, P) \leq 1 \textit{ for } n > 2 + \frac{60 - 12a^2}{a^2(1 - a^2)}. \tag{6.77}$$

Proof We may assume $d \geq 1$. Then we have the inequality (6.75). We obtain the desired result from the inequality $r^2 \geq 1 - a^2$, which can be seen as follows. Let $w = u + iv$ denote the nearest second zero of P to a, so $|w|^2 = 1$. By theorem 6.2.2 $\exists$ a critical point z satisfying $|z - \frac{1}{2}w| \leq \frac{1}{2}$. We then have

$$1 \leq |z - a| \leq \left|z - \tfrac{1}{2}w\right| + \left|\tfrac{1}{2}w - a\right| \leq \tfrac{1}{2} + \left|\tfrac{1}{2}w - a\right| \tag{6.78}$$

and therefore $|w - 2a| \geq 1$; i.e. $|w|^2 - 4au + 4a^2 \geq 1$. This gives $u \leq a$ and so $r^2 = 1 - 2au + a^2 \geq 1 - a^2$.

6.6 The extremal distance

6.6.1

Theorem *For* $0 \leq a \leq 1$ *let*

$$\delta(a) = \sup_P d(a, P), \tag{6.79}$$

where P ranges over all polynomials of degree $n \geq 2$ *with their zeros in* $\{|z| \leq 1\}$ *and satisfying* $P(a) = 0$. *Then* $\delta(a)$ *is a continuous function of bounded variation in* $[0, 1]$ *satisfying* $\delta(0) = \delta(1) = 1$.

Proof Rubinstein's result verifying the conjecture when $a = 1$ together with the polynomials $1 - z^n$ shows that $\delta(1) = 1$. It is clear that $\delta(0) \leq 1$; to show that $\delta(0) = 1$, consider the polynomials $z(z^n - 1)$; the critical points lie on the circle $\{|z| = (n+1)^{-1/n}\}$, whose radius $\to 1$ as $n \to \infty$.

For the main part of the proof we shall use a rather trivial method of variation and prove the following two facts:

(a) *the function* $\delta(a)/(1 + a)$ *is decreasing in* $[0, 1]$;
(b) *the function* $\delta(a)/(1 - a)$ *is increasing in* $(0, 1)$.

(a) The Gauss–Lucas theorem gives $\delta(a) \leq 1 + a$ and the polynomial $(z + 1)(z - a)$ gives $\delta(a) \geq \frac{1}{2}(1 + a)$; these verify the decreasing property at the endpoints of the interval. Now choose $0 < a < 1$ and $\eta > 0$. There exists a polynomial with all its zeros in the disc, with $P(a) = 0$ and such that $d(a, P) > \delta(a) - \eta$. Choose ϵ $(0 < \epsilon < a)$. Then P has all its zeros in $\{|z - \epsilon| \leq 1 + \epsilon\}$ and so

$$Q(\zeta) = P(\epsilon + (1 + \epsilon)\zeta) \tag{6.80}$$

has all its zeros in $\{|\zeta| \leq 1\}$ and $Q(\frac{a-\epsilon}{1+\epsilon}) = 0$. Hence

$$\delta\left(\frac{a - \epsilon}{1 + \epsilon}\right) \geq d\left(\frac{a - \epsilon}{1 + \epsilon}, Q\right) \geq \frac{\delta(a) - \eta}{1 + \epsilon} \tag{6.81}$$

and letting $\eta \to 0$ we deduce that

$$\delta\left(\frac{a - \epsilon}{1 + \epsilon}\right) \geq \frac{\delta(a)}{1 + \epsilon}. \tag{6.82}$$

This is equivalent to

$$\frac{\delta(b)}{1 + b} \geq \frac{\delta(a)}{1 + a} \text{ if } 0 < b < a \tag{6.83}$$

and therefore (a) follows.

(b) With a, P and η as before we choose ϵ $(0 < \epsilon < 1 - a)$ and apply a similar argument to $R(\zeta) = P(-\epsilon + (1 + \epsilon)\zeta)$.

It is immediate from (a) that $\delta(a)$ is a function of bounded variation on $[0, 1]$, and indeed the total variation is at most 2. It follows that the left and right limits $\delta(a-)$ and $\delta(a+)$ exist in $(0, 1)$ and $\delta(0+)$ and $\delta(1-)$ exist. Furthermore we have from (a)

$$\delta(a+) \leq \delta(a) \leq \delta(a-) \quad (0 < a < 1) \tag{6.84}$$

and from (b)

$$\delta(a-) \leq \delta(a) \leq \delta(a+) \quad (0 < a < 1) \tag{6.85}$$

and so $\delta(a)$ is continuous on $(0, 1)$. It remains to prove the continuity at the endpoints. To prove continuity at $a = 0$ we consider the polynomials $P_n(z) = (z - a)(z^n - 1)$; the critical points are the solutions of the equation $(n + 1)z^n - naz^{n-1} = 1$, whose roots satisfy

$$z^n = \frac{1}{n + 1 - naz} \tag{6.86}$$

and $|z| \leq 1$. We obtain

$$|z|^n \geq \frac{1}{2n + 1} \tag{6.87}$$

so $|z| \geq (2n + 1)^{-1/n} \to 1$ as $n \to \infty$. It follows that $\delta(a) \geq 1 - a$, and since $\delta(a) \leq 1 + a$, we deduce that $\delta(a) \to 1 = \delta(0)$ as $a \to 0$.

Finally, we show that $\delta(a) \to 1$ as $a \to 1$. Since $\delta(a) \geq \frac{1}{2}(1+a)$, we may assume that for some a, $\delta(a) > 1$ and consider a polynomial P satisfying the conditions with $d = d(a, P) > 1$. Let $\epsilon = \sqrt{1 - a}$. Either $d \leq a + \sqrt{1 - a} \to 1$ as $a \to 1$ or $a + \epsilon < d$. Assume the latter. Then P has a zero b satisfying $|b + \epsilon| > a + \epsilon$; for otherwise P has all its zeros in $\{|z + \epsilon| \leq a + \epsilon\}$ and, as a lies on the boundary of this disc, it follows from Rubinstein's result that P has a critical point within distance $a + \epsilon < d$ of a. We may assume that b is the zero of P furthest from the point $-\epsilon$; then P has all its zeros in the disc $\{|z+\epsilon| \leq |b+\epsilon|\}$ and, as b is a zero on the boundary of this disc, it follows from theorem 6.2.2 that P has a critical point ζ satisfying $|\zeta - \frac{1}{2}(b-\epsilon)| \leq \frac{1}{2}|b+\epsilon|$. We deduce that

$$\begin{aligned} d \leq |\zeta - a| &\leq \tfrac{1}{2}|b + \epsilon| + \tfrac{1}{2}|b - \epsilon - 2a| \leq \tfrac{1}{2}(1 + \epsilon) + \tfrac{1}{2}|b - a| + \tfrac{1}{2}(a + \epsilon) \\ &\leq \tfrac{1}{2}(1 + \epsilon) + \tfrac{1}{2}|b - a| + \tfrac{1}{2}d \end{aligned} \tag{6.88}$$

and so

$$d \leq 1 + \epsilon + |b - a|. \tag{6.89}$$

Now, since $|b+\epsilon| > a+\epsilon$ and $|b| \leq 1$, we deduce that $1+2\epsilon\Re b > a^2+2\epsilon a$, which gives

$$\Re b > a - \frac{1-a^2}{2\epsilon} \tag{6.90}$$

and so

$$|b-a|^2 \leq 1 - 2a\Re b + a^2 \leq 1 - a^2 + a\frac{1-a^2}{\epsilon} \leq 1 - a^2 + 2\sqrt{1-a}. \tag{6.91}$$

We thus see that in all cases $d \leq 1 + o(1)$ as $a \to 1$ and so the result follows.

6.7 Further remarks on the conjecture

6.7.1

It follows from the result of the previous section that, if the Ilieff–Sendov conjecture is false, then the function $\delta(a)$ attains its maximum value $\rho > 1$ say at a point $\alpha \in (0, 1)$; $\rho = \delta(\alpha)$. Now, if P is a polynomial of degree n with all zeros in $\overline{\mathbb{U}}$ and satisfying $P(\alpha) = 0$, then by (6.47) $d(\alpha, P) \leq (1+\alpha)^{1/n}$. Choosing P so that $d(\alpha, P) > \rho - \epsilon > 1$, where $\epsilon > 0$ is small, we obtain $n \leq \log(1+\alpha)/\log(\rho-\epsilon)$. Thus $\exists N$ such that $\rho = \sup d(\alpha, P)$ taken over all monic polynomials P of degree $\leq N$ with all zeros in $\overline{\mathbb{U}}$ and satisfying $P(\alpha) = 0$. Since the class of such polynomials is compact and $d(\alpha, P)$ varies continuously with P, it follows that $\exists \Pi$ in the class such that $\rho = d(\alpha, \Pi)$; in other words we have at least one extremal polynomial for the Ilieff–Sendov problem. Indeed, we have at least two extremal polynomials, for $\overline{\Pi(\bar{z})}$ is also extremal and is distinct from Π (otherwise Π is real and by 6.4.3 $d(\alpha, \Pi) \leq 1$).

6.7.2

Let $P(z) = (z-a)S(z)$, where $S(z)$ has n distinct zeros z_k lying in the closed unit disc, but not equal to a. Then the zeros of P' are the solutions of the equation

$$R(z) = z + \frac{S(z)}{S'(z)} = a,$$

whose solutions are given by the global analytic function $z = R^{-1}(a)$, which is an algebraic function of a. The conjecture amounts to the assertion that, for each a satisfying $|a| \leq 1$, at least one branch of the function R^{-1} satisfies

$$|R^{-1}(a) - a| \leq 1.$$

This is certainly true when $|a| = 1$, when indeed we have

$$\left|R^{-1}(a) - \tfrac{1}{2}a\right| \le \tfrac{1}{2}$$

for a suitable branch. However, a counter-example has been constructed by M.J. Miller [32] to show that this inequality does not extend to $|a| < 1$, and indeed it is not even true that

$$\left|R^{-1}(a) - \tfrac{1}{2}a\right| \le 1 - \tfrac{1}{2}|a|$$

when $|a| < 1$. Thus because of the existence of branch points we cannot automatically extend boundary inequalities into the interior region. Note that branch points correspond to critical points of $R(z)$ in the z-plane, i.e. zeros of

$$R'(z) = 2 - \frac{S''(z)S(z)}{S'^2(z)}$$

which are the zeros of the polynomial

$$2S'^2(z) - S''(z)S(z).$$

There are $2n-2$ such points in the plane. Writing $P_a(z) = P(z) = (z-a)S(z)$, the existence of such a critical point means that we can find a such that $\exists z$ satisfying $P'(z) = P''(z) = 0$, i.e. P has a multiple critical point.

6.7.3

A direct approach to the problem is to consider the equation

$$\frac{(z-a)S'(z)}{S(z)} = -1, \tag{6.92}$$

i.e.

$$\sum_{k=1}^{n} \frac{a-z}{z_k - z} = -1. \tag{6.93}$$

We observe that, if $|z| > 1$, then the left-hand expression cannot take on any negative value, since then $\Re(1 - b/z) > 0$ for the values $b = a$ and $b = z_k$, and therefore the expression is the product of two values having positive real part. Since the value of the expression is 0 at $z = a$, we may consider the curve emanating from $z = a$ on which the expression is negative. The curve will remain in the unit disc and the problem is to show that the value -1 is attained on the curve before it meets the circle $|z - a| = 1$.

6.7.4

Rather similar reasoning is as follows. Let b be a second zero of P and assume further that P is an extremal polynomial for the problem, so that the nearest critical point of P to a is at distance d from a and every polynomial of degree n with its zeros in the disc and a zero at a has a critical point within distance d from a. Writing $P(z) = (z-b)T(z)$, we see that T has degree $n-1$, $T(a) = 0$ and T has its remaining zeros in the unit disc. If $P'(c) = 0$, then either T, and so P, has a multiple zero at c or $R(c) = b$, where

$$R(z) = z + \frac{T(z)}{T'(z)}. \tag{6.94}$$

Assume that P has no multiple zeros in $|z-a| \le d$ and that c satisfies $|c-a| = d$. Now consider the image of $|z-a| \le d$ under the mapping R. If this image does not contain the point p, then there is no critical point of the polynomial $(z-p)T(z)$ in $|z-a| \le d$. It follows that $|p| > 1$, for otherwise the extremal property of d is not satisfied. Thus $R(\{|z-a| \le d\}) \supset \{|w| \le 1\}$. On the other hand $b \notin R(\{|z-a| < d\})$, as otherwise P is not an extremal polynomial. Now if the extremal polynomial P has a zero, other than a, in the open unit disc, then we may choose b so that $|b| < 1$. We then find that $R(\{|z-a| < d\})$ omits the value b in the open unit disc but its closure contains every value in the closed unit disc. We believe this implies either $R'(c) = 0$, which gives $P''(c) = 0$, or that R attains the value b at least twice on the circle $|z-a| = d$. Thus in either case P has at least two critical points on the extremal circle. An interesting speculation is that the extremal polynomial has all its critical points on the extremal circle. This question has been studied by M.J. Miller [33], who has obtained estimates for the number of critical points on the extremal circle for an extremal polynomial.

6.7.5

We may seek a form of the conjecture, which eliminates considering a particular point a. The conjecture may be expressed in the following form: let $P(z)$ be a polynomial of degree n satisfying $P(0) = 0$ and $P'(z) \ne 0$ for $|z| \le 1$; then it is required to prove that for every complex a, $P(z)$ has a zero satisfying $|z-a| > 1$; in other words, there is no disc of radius 1 containing all the zeros of P. Now if this is the case, then some of the $n-1$ remaining zeros of P must account for all the values a in the unit disc. On grounds of symmetry it seems likely that one zero would account for the values a satisfying $-1 \le a \le 0$. If so, then the following region D omits at least one zero of P. D is the union of the discs $\{|z+a| \le 1\}$ $(0 \le a \le 1)$, i.e. the convex hull of the union of the two discs $\{|z| \leq 1\}$ and $\{|z+1| \leq 1\}$. It is therefore bounded by the convex curve consisting of

$$\left\{z = e^{i\theta} \colon -\frac{\pi}{2} \leq \theta \leq \frac{\pi}{2}\right\} \cup \{(x, 1) \colon -1 \leq x \leq 0\}$$
$$\cup \left\{z = -1 + e^{i\theta} \colon \frac{\pi}{2} \leq \theta \leq \frac{3\pi}{2}\right\} \cup \{(x, -1) \colon -1 \leq x \leq 0\}. \quad (6.95)$$

Thus the following assertion implies the conjecture: let $P(z)$ be a polynomial with all its zeros in D and satisfying $P(0) = 0$; then $P'(z)$ has a zero in $\{|z| \leq 1\}$.

6.7.6

Assume now that P' has all its zeros in the disc $\{|z| \leq \epsilon\}$, where $0 < \epsilon < 1$. Then clearly $d \leq a + \epsilon$, and this is ≤ 1 if $a \leq 1 - \epsilon$. We can improve on this using the method of Bojanov, Rahman and Szynal (section 4). With the notations of that section we have from (6.45), $d \leq \epsilon + \rho$ and therefore as for (6.46)

$$d^n \leq a + \epsilon^n. \quad (6.96)$$

Thus we obtain $d \leq 1$ if $a \leq 1 - \epsilon^n$.

6.7.7

Returning to the matrix $M_{i,j}$ discussed in chapter 5, section 5.7.7, the Ilieff–Sendov conjecture can be expressed as the proposed inequality

$$\max_{1 \leq k \leq n} \min_{1 \leq j \leq n-1} |\zeta_j - z_k| \leq \min_{a \in \mathbb{C}} \max_{1 \leq k \leq n} |z_k - a|. \quad (6.97)$$

Compare this with (5.144).

6.7.8

It has been shown by M.J. Miller [34] and V. Vâjâitu and A. Zaharescu [65] that, if P has degree n and all its zeros in $\{|z| \leq 1\}$ and if a is a zero of P where $|a|$ is close to 1 (closeness depending only on n), then P has a critical point satisfying $|z - a| \leq 1$.

The following weaker statement is easily proved: *let $S(z)$ be a polynomial of degree $n - 1$ with all zeros satisfying $|z| \leq 1$; then, if $|a| < 1$ is sufficiently close to 1 (closeness depending on S), the polynomial $P_a(z) = (z - a)S(z)$ has a critical point satisfying $|z - a| \leq 1$.*

We need only consider $0 < a < 1$ and we may assume that S is monic (i.e. the coefficient of z^{n-1} is 1). The proof of the Goodman–Rahman–Ratti theorem of section 6.2 shows that the critical points ζ_k of the polynomial $(z-1)S(z)$ satisfy either (i) $\zeta_k = 0$ for $1 \le k \le n-1$ or (ii) $\exists k$ such that $|\zeta_k - 1| < 1$. In case (ii), continuity of zeros implies that, if $a < 1$ is sufficiently close to 1, then P_a has a critical point ζ satisfying $|\zeta - 1| < 1$; but then (since $|\zeta| < 1$) $|\zeta - a| < 1$, as required. In case (i) we obtain $S(z) = 1 + z + z^2 + \cdots + z^{n-1}$, which implies $|S(a)| \le n$. Then $|P_a'(a)| \le n$ and so, if P_a' has zeros ξ_k, we deduce that $\prod_k |\xi_k - a| \le 1$, so for at least one k, $|\xi_k - a| \le 1$.

7
Self-inversive polynomials

7.1 Introduction

7.1.1

Self-inversive polynomials were introduced in chapter 4 (see subsection 4.4.2). We recall that a polynomial P of degree n is self-inversive if

$$P^*(z) = z^n \overline{P\left(\frac{1}{\bar{z}}\right)} = P(z) \tag{7.1}$$

for all $z \neq 0$. Equivalently, writing $P(z) = \sum_{k=0}^{n} a_k z^k$ we have for a self-inversive polynomial

$$\sum_{k=0}^{n} \overline{a_{n-k}}\, z^k = \sum_{k=0}^{n} a_k z^k \tag{7.2}$$

and therefore $\overline{a_{n-k}} = a_k$ for $0 \leq k \leq n$. Conversely, this coefficient relation implies that P is self-inversive. In particular, since P has degree n, $a_0 = \overline{a_n} \neq 0$, and so P does not have a zero at the origin.

This definition of self-inversive is tied to the degree of P, so is a degree-relative concept. The relation $P = P^*$ as given by (7.1) may well hold for a polynomial P of degree smaller than n. Indeed, suppose P has degree $n - m$ and yet (7.1) is satisfied. Then P^* has a zero of order m at the origin, and so P does; i.e. $P = z^m Q$, where Q is a polynomial of degree $n - 2m$. We then see that Q is a self-inversive polynomial relative to its degree $n - 2m$. Conversely, for such a self-inversive polynomial Q, the polynomial $P = z^m Q$ satisfies (7.1). Nevertheless, our convention is that we do *not* regard P as being self-inversive relative to degree n.

The zeros of a self-inversive polynomial either lie on the unit circle $\mathbb{T}$ or occur in pairs conjugate to $\mathbb{T}$; thus ζ is a zero of multiplicity $m \Rightarrow 1/\overline{\zeta}$ is a zero of multiplicity m. Conversely, this property of the zeros of P implies that $P = cQ$, where c is a constant of absolute value 1 and where Q is self-inversive. In particular, up to a multiplicative constant, polynomials with all their zeros on $\mathbb{T}$ are self-inversive.

We have also proved for self-inversive polynomials of degree n the remarkable relation

$$\max_{z\in\mathbb{T}} |P'(z)| = \frac{n}{2} \max_{z\in\mathbb{T}} |P(z)| \tag{7.3}$$

(see chapter 4, theorem 4.4.3). Furthermore, every point of maximum modulus of P on $\mathbb{T}$ is also a point of maximum modulus of P'. In particular, this relation holds for all polynomials of degree n whose zeros lie on $\mathbb{T}$.

7.1.2 Relations on the unit circle

If P is a self-inversive polynomial of degree n, then

$$e^{-in\theta/2} P(e^{i\theta}) \text{ is real} \quad (0 \le \theta \le 2\pi) \tag{7.4}$$

and differentiating we deduce that

$$\Re\left(\frac{e^{i\theta} P'(e^{i\theta})}{P(e^{i\theta})}\right) = \frac{n}{2} \tag{7.5}$$

for all real θ such that $P(e^{i\theta}) \neq 0$. This implies that

$$\frac{zP'(z)}{P(z)} = \frac{n\omega(z)}{1+\omega(z)} \tag{7.6}$$

where $\omega(z)$ is a rational function satisfying $\omega(0) = 0$ and

$$|\omega(z)| = 1 \quad (z \in \mathbb{T}). \tag{7.7}$$

In fact

$$\omega(z) = \frac{zP'(z)}{nP(z) - zP'(z)}. \tag{7.8}$$

Because P is self-inversive, $nP(z) - zP'(z) = (zP'(z))^*$, and we deduce that

$$\omega(z) = cz \prod_{k=1}^{m-1} \frac{z - \zeta_k}{1 - \overline{\zeta_k} z} \tag{7.9}$$

where ζ_k are the zeros of P'/P, i.e. those critical points of P which are not zeros of P, and where c is a constant satisfying $|c| = 1$. Thus $m \le n$ with equality when the zeros of P are simple. We prove next a relationship between the number of critical points of P in the unit disc and the number of zeros of P on the unit circle.

7.1.3

Theorem *Let P be a self-inversive polynomial of degree n. Suppose that P has exactly τ zeros on the unit circle (counted according to multiplicity) and exactly ν critical points in the closed unit disc (counted according to multiplicity). Then*

$$\tau = 2(\nu + 1) - n. \tag{7.10}$$

Proof Since the zeros of P off the circle occur in conjugate pairs, we have

$$\tau + 2\sigma = n \tag{7.11}$$

where σ denotes the number of zeros of P in $\mathbb{U}$. We now show that

$$\tau + \sigma = \nu + 1 \tag{7.12}$$

which clearly implies the result. Now $\tau + \sigma$ is the number of zeros of P in the closed unit disc, and $\nu + 1$ is the number of zeros of $zP'(z)$ in the same region. We thus have to show that these numbers are equal. This will follow from the argument principle and the inequality

$$\Re\left(\frac{zP'(z)}{P(z)}\right) > 0 \quad (1 < |z| = r < 1 + \delta) \tag{7.13}$$

which holds for a suitable $\delta > 0$. To prove the inequality let $z_1, z_2, \ldots, z_n$ be the zeros of P so that we have

$$\frac{zP'(z)}{P(z)} = \sum_{k=1}^{n} \frac{z}{z - z_k}. \tag{7.14}$$

If $|z_k| = 1$, then

$$\Re\frac{z}{z - z_k} = \Re\frac{1}{1 - z_k/z} > \frac{1}{2} \tag{7.15}$$

for $|z| > 1$. If $|z_k| \neq 1$, then for some j, $z_j = 1/\overline{z_k}$ and we have on $|z| = 1$

$$\Re\left(\frac{z}{z - z_k} + \frac{z}{z - z_j}\right) = 1. \tag{7.16}$$

It follows by continuity that for a suitable $\delta_k > 0$ the left-hand expression is positive for $1 - \delta_k < |z| < 1 + \delta_k$. Thus choosing δ as the smallest of the δ_k the result follows.

Notice in particular that P has all its zeros on the unit circle iff P has all its critical points in the closed unit disc. Furthermore the critical points of P on the unit circle are necessarily multiple zeros of P (this follows from (7.5) which shows that P'/P has no zeros on $\mathbb{T}$).

We make a further observation concerning the function $\omega(z)$ of (7.8). The zeros of ω are exactly the zeros of zP'/P. This function has equally many zeros and poles in $\overline{\mathbb{U}}$; as the poles are simple poles occurring at the distinct zeros of P, *the number of zeros of $\omega(z)$ in $\mathbb{U}$ is exactly equal to the number of distinct zeros of P in $\overline{\mathbb{U}}$*. In particular, if ω has all its zeros in $\mathbb{U}$, then ω has no poles in $\mathbb{U}$, so is analytic in $\mathbb{U}$. It follows from Schwarz's lemma that $|\omega(z)| \leq |z|$ in $\mathbb{U}$ and so $\omega(z) \neq -1$ in $\mathbb{U}$. Therefore $zP'(z)/P(z)$ is analytic in $\mathbb{U}$ and this implies that $P(z)$ has all its zeros on $\mathbb{T}$.

7.1.4

Conversely, if P has all its zeros on $\mathbb{T}$, then ω has all its zeros in $\mathbb{U}$. In this case ω takes the form of a Blaschke product of degree m. We have $|\omega(z)| \leq |z|$ for $|z| < 1$ and $|\omega(z)| \leq |z|$ for $|z| > 1$. In particular, for each $\zeta\,(|\zeta| = 1)$ the equation $\omega(z) = \zeta$ has all its roots on $\mathbb{T}$. There are exactly m such roots. Consider the function

$$F(z) = F(\zeta; z) = \frac{1}{1 - \overline{\zeta}\omega(z)}. \tag{7.17}$$

F is a rational function whose poles are the m roots on $\mathbb{T}$ of the equation $\omega(z) = \zeta$. Observe that

$$\begin{aligned} \Re F(z) &> \frac{1}{2} \quad (|z| < 1), \\ \Re F(z) &< \frac{1}{2} \quad (|z| > 1), \end{aligned} \tag{7.18}$$

and $\Re F(z) = \frac{1}{2}$ at all points on $\mathbb{T}$ other than the zeros of $\omega(z) - \zeta$. We show that we can write

$$F(\zeta; z) = \sum_{k=1}^{m} \frac{t_k}{1 - \overline{\tau_k} z} \tag{7.19}$$

where τ_k are the roots of $\omega(z) = \zeta$ and where t_k are positive numbers satisfying $\sum_1^m t_k = 1$. To see this we make use of the representation (7.9) for ω. Differentiating this equation logarithmically we obtain

$$\frac{z\omega'(z)}{\omega(z)} = 1 + \sum_{k=1}^{m-1} \left(\frac{z}{z - \zeta_k} + \frac{\overline{\zeta_k} z}{1 - \overline{\zeta_k} z} \right) = 1 + \sum_{k=1}^{m-1} \frac{z(1 - |\zeta_k|^2)}{(z - \zeta_k)(1 - \overline{\zeta_k} z)}. \tag{7.20}$$

In particular, when $|z| = 1$ we obtain

$$\frac{z\omega'(z)}{\omega(z)} = 1 + \sum_{k=1}^{m-1} \frac{1 - |\zeta_k|^2}{|1 - \overline{\zeta_k} z|^2} > 1. \tag{7.21}$$

Next we consider

$$\frac{1-\overline{\tau_k}z}{1-\overline{\zeta}\omega(z)}=\frac{\zeta}{\tau_k}\frac{z-\tau_k}{\omega(z)-\zeta}\to\frac{\omega(\tau_k)}{\tau_k\omega'(\tau_k)}\text{ as } z\to\tau_k. \tag{7.22}$$

It follows that the function F has simple poles at the points τ_k and, if we set

$$t_k=\frac{\omega(\tau_k)}{\tau_k\omega'(\tau_k)}\quad (1\le k\le m), \tag{7.23}$$

then the numbers $t_k\in(0,1)$ and the function

$$F(\zeta;z)-\sum_{k=1}^{m}\frac{t_k}{1-\overline{\tau_k}z} \tag{7.24}$$

is a rational function with no poles and therefore either identically zero or a polynomial. Since $\omega(z)\to\infty$ as $z\to\infty$, we see that this function $\to 0$ as $z\to\infty$. It is therefore identically zero, as required. Finally, putting $z=0$ we obtain $\sum_1^m t_k=1$.

Note that the number m is the number of *distinct* zeros of P.

7.1.5 Problems

1. Let $P(z)$ be a polynomial of degree n with all its zeros in $\overline{\mathbb{U}}$. Show that, for each $\zeta\in\mathbb{T}$, the polynomial $P(z)+\zeta P^*(z)$ has all its zeros on $\mathbb{T}$. Show also that, if $Q(z)$ is a polynomial of degree n with no zeros in $\mathbb{U}$, then $Q(z)+\zeta Q^*(z)$ has all its zeros on $\mathbb{T}$.
2. Let $P(z)$ be a polynomial of degree n and suppose that for each $\zeta\in\mathbb{T}$ the polynomial $P(z)+\zeta P^*(z)$ has all its zeros on $\mathbb{T}$. Show that either (a) $P(z)$ has all its zeros in $\overline{\mathbb{U}}$, or (b) $P(z)$ has no zeros in $\mathbb{U}$. [Hint: If α and β are points in $\mathbb{U}$ with $P(\alpha)=0$ and $P^*(\beta)=0$, then at some point on the line segment $[\alpha,\beta]$ the rational function $P(z)/P^*(z)$ has absolute value $=1$.]

7.2 Polynomials with interspersed zeros on the unit circle

7.2.1

Lemma *Let $\alpha_k,\beta_k\,(1\le k\le n)$ be real numbers such that*

$$\alpha_1<\beta_1<\alpha_2<\beta_2<\cdots<\alpha_n<\beta_n<\alpha_1+2\pi. \tag{7.25}$$

Then

$$\Im\left(\prod_{k=1}^{n}\frac{e^{i\alpha_k/2}-ze^{-i\alpha_k/2}}{e^{i\beta_k/2}-ze^{-i\beta_k/2}}\right)<0\quad(|z|<1). \tag{7.26}$$

Proof Let

$$P(z) = \prod_{k=1}^{n} (e^{i\alpha_k/2} - ze^{-i\alpha_k/2}), \qquad Q(z) = \prod_{k=1}^{n} (e^{i\beta_k/2} - ze^{-i\beta_k/2}) \quad (7.27)$$

and $F(z) = P(z)/Q(z)$. Then, if $|z| = 1$,

$$F(z) = \prod_{k=1}^{n} \frac{(e^{i\alpha_k/2} - ze^{-i\alpha_k/2})(e^{-i\beta_k/2} - \overline{z}e^{i\beta_k/2})}{|e^{i\beta_k/2} - ze^{-i\beta_k/2}|^2} \quad (7.28)$$

is easily seen to be a product of real numbers, and therefore real, provided that $z \neq e^{i\beta_k}$ $(1 \leq k \leq n)$. For $\theta \in (\beta_k, \beta_{k+1})$ we have $|F(e^{i\theta})| \to +\infty$ as $\theta \to \beta_k$ or β_{k+1}. $F(e^{i\theta})$ is real in the interval and has a single simple zero at $\theta = \alpha_{k+1}$. It follows by continuity that $F(e^{i\theta})$ takes all real values in the interval and therefore $F(z)$ takes each real value at least n times on $\mathbb{T}$. On the other hand F can take no value more than n times. In particular F takes no real values in $\mathbb{U}$, so has constant imaginary part in $\mathbb{U}$. It only remains to show that $\Im F(0) < 0$. But

$$\Im F(0) = \Im \prod_{k=1}^{n} e^{i(\alpha_k - \beta_k)/2} = -\sin \sum_{k=1}^{n} \frac{1}{2}(\beta_k - \alpha_k) < 0, \quad (7.29)$$

since the hypothesis implies that

$$0 < \sum_{k=1}^{n} \frac{1}{2}(\beta_k - \alpha_k) < \pi. \quad (7.30)$$

7.2.2

From the lemma we see that

$$\frac{F(z)}{F(0)} = \frac{1 - e^{i\gamma}\omega(z)}{1 - \omega(z)} \quad (7.31)$$

where

$$\gamma = \sum_{k=1}^{n} (\beta_k - \alpha_k), \quad (7.32)$$

and where $\omega(z)$ is a rational function of degree n satisfying

$$\omega(0) = 0, \ |\omega(z)| = 1 \text{ on } \mathbb{T} \text{ and } |\omega(z)| < 1 \text{ on } \mathbb{U}. \quad (7.33)$$

The only functions with these properties are Blaschke products, which have the form

$$\omega(z) = cz^m \prod_{k=1}^{n-m} \frac{z - z_k}{1 - \overline{z_k}z} \quad (7.34)$$

where $|c| = 1$ and $|z_k| < 1\,(1 \leq k \leq n - m)$. To see this we argue as follows. Let ζ be a zero in $\mathbb{U}$ of ω; then the function $z \mapsto \omega(\frac{z+\zeta}{1+\bar{\zeta}z})$ satisfies the conditions of Schwarz's lemma, and so we obtain

$$|\omega(z)| \leq \left|\frac{z - \zeta}{1 - \bar{\zeta}z}\right| \quad (z \in \mathbb{U}). \tag{7.35}$$

By induction we divide out the Blaschke function $B(z)$ whose factors correspond to the zeros of ω in $\mathbb{U}$, including the zeros at the origin. Then the function $\Omega(z) = \omega(z)/B(z)$ satisfies $|\Omega(z)| \leq 1$ in $\mathbb{U}$ and $|\Omega(z)| = 1$ on $\mathbb{T}$. Also Ω has no zeros in $\mathbb{U}$ so the function $1/\Omega$ is analytic in $\mathbb{U}$ and has absolute value 1 on $\mathbb{T}$. By the maximum principle, $|\Omega(z)| \geq 1$ in $\mathbb{U}$. It follows that Ω is a constant of absolute value 1 in $\mathbb{U}$ and therefore throughout $\mathbb{C}$.

The argument given in subsection 7.1.4 then shows that we can write

$$\frac{F(z)}{F(0)} = \sum_{k=1}^{n} t_k \frac{1 - ze^{i\gamma}e^{-i\beta_k}}{1 - ze^{-i\beta_k}} \tag{7.36}$$

where the t_k are positive numbers satisfying $\sum_1^n t_k = 1$.

7.2.3

Conversely, if $0 < \gamma < 2\pi$, a rational function $R(z)$ of the form

$$R(z) = \sum_{k=1}^{n} t_k \frac{1 - ze^{i\gamma}e^{-i\beta_k}}{1 - ze^{-i\beta_k}}, \tag{7.37}$$

where the t_k are positive numbers satisfying $\sum_1^n t_k = 1$, has all its zeros on $\mathbb{T}$ and of the form $e^{i\alpha_k}$, where the α_k satisfy (7.25) and where (7.32) holds; in other words the zeros and poles of R are **interspersed**. This can be seen by observing that R is a convex combination of mappings which send $\mathbb{U} \mapsto H_\gamma$, $\mathbb{T} \mapsto L_\gamma \cup \{\infty\}$ and $\mathbb{C} - \overline{\mathbb{U}} \mapsto \mathbb{C} - \overline{H_\gamma}$, where H_γ is the half-plane $\{w\colon \Im(e^{-i\gamma/2}w) < 0\}$ and L_γ is the line bounding H_γ. As all three image regions are convex, R retains the three mapping properties and in particular has all its zeros on $\mathbb{T}$. Furthermore, it follows from the subordination principle that we can write

$$R(z) = \frac{1 - e^{i\gamma}\omega(z)}{1 - \omega(z)} \tag{7.38}$$

where ω is a rational function of degree n satisfying the three properties (7.33); $\omega(z)$ is therefore a Blaschke product of the form (7.34). Now on $\mathbb{T}$, ω has the form

$$\omega(e^{i\theta}) = e^{i\phi(\theta)}, \tag{7.39}$$

where $\phi(\theta)$ is a differentiable and strictly increasing function satisfying $\phi(2\pi) - \phi(0) = 2\pi n$. This follows from the fact that on $\mathbb{T}$, each of the n Blaschke factors of ω gives a 1–1 positively oriented mapping of $\mathbb{T}$ onto itself. We thus see that, as θ goes between 0 and 2π, ω rotates positively around $\mathbb{T}$ exactly n times. In particular, between two successive roots of $\omega = 1$ there is exactly one root of $\omega = e^{-i\gamma}$. Thus as stated the zeros are interspersed.

7.2.4

Interspersion theorem *Let $P(z)$ and $Q(z)$ be polynomials of degree n satisfying $P(0) = Q(0) = 1$ and having all their zeros $a_k = e^{i\alpha_k}$ and $b_k = e^{i\beta_k}$ respectively on $\mathbb{T}$. The following conditions are equivalent.*

(a) *P and Q have their zeros interspersed, i.e.*

$$\alpha_1 \leq \beta_1 \leq \alpha_2 \leq \beta_2 \leq \cdots \leq \alpha_n \leq \beta_n \leq \alpha_1 + 2\pi; \tag{7.40}$$

(b) *writing $R(z) = P(z)/Q(z)$ we have*

$$\Im(e^{-i\gamma/2}R(z)) < 0 \quad (z \in \mathbb{U}) \tag{7.41}$$

for suitable γ real; in this case we have $\gamma = \sum_{k=1}^{n}(\beta_k - \alpha_k)$ and

$$R(z) = \frac{1 - e^{i\gamma}\omega(z)}{1 - \omega(z)} = \sum_{k=1}^{n} t_k \frac{1 - ze^{i\gamma}e^{-i\beta_k}}{1 - ze^{-i\beta_k}} \tag{7.42}$$

where $\omega(z)$ is a Blaschke product of degree $\leq n$ satisfying $\omega(0) = 0$ and where t_k are non-negative numbers satisfying $\sum_1^n t_k = 1$. Conversely, any rational function R of either of these two forms (with the properties stated) is necessarily the ratio of two polynomials with interspersed zeros.

Proof It only remains to show that, if R maps $\mathbb{U}$ into a half-plane excluding the origin, then the zeros of P and Q are interspersed. We may assume that P and Q have no common zeros. The hypothesis on R can be stated as follows. Each branch of $\arg R(z)$ is harmonic in $\mathbb{U}$ and satisfies, for a suitable real λ

$$|\arg R(z) - \lambda| < \frac{\pi}{2} \quad (z \in \mathbb{U}). \tag{7.43}$$

Now

$$\frac{R'(z)}{R(z)} = \sum_{k=1}^{n}\left(\frac{1}{z-a_k} - \frac{1}{z-b_k}\right) = \sum_{k=1}^{2n}\frac{\epsilon_k}{z-c_k} \tag{7.44}$$

where the c_k appear in cyclic order so run through the zeros and poles as they occur, and where $\epsilon_k = 1$ when c_k is a zero and $\epsilon_k = -1$ when c_k is a pole. Let $\rho_1 = \rho_{2n+1} = 0$, and set

$$\rho_k = -2\pi i \sum_{j=1}^{k-1} \epsilon_j \quad (2 \le k \le 2n+1). \tag{7.45}$$

Then we see that

$$\rho_k - \rho_{k+1} = 2\pi i \epsilon_k \quad (1 \le k \le 2n) \tag{7.46}$$

and so, taking $c_0 = c_{2n}$,

$$\frac{R'(z)}{R(z)} = \frac{1}{2\pi i}\sum_{k=1}^{2n}\frac{\rho_k - \rho_{k+1}}{z-c_k} = \frac{1}{2\pi i}\sum_{k=1}^{2n}\rho_k\left(\frac{1}{z-c_k} - \frac{1}{z-c_{k-1}}\right)$$

$$= \frac{1}{2\pi}\sum_{k=1}^{2n}\rho_k \int_{t_{k-1}}^{t_k}\frac{e^{-it}}{(1-ze^{-it})^2}dt, \tag{7.47}$$

where $c_k = e^{it_k}$ for all k. This gives

$$\log R(z) = \frac{1}{2\pi}\sum_{k=1}^{2n}\rho_k \int_{t_{k-1}}^{t_k}\frac{ze^{-it}}{1-ze^{-it}}dt. \tag{7.48}$$

Since the ρ_k are purely imaginary, we obtain

$$\arg R(z) = \frac{1}{2\pi}\int_0^{2\pi}\Re\left(\frac{1+ze^{-it}}{1-ze^{-it}}\right)\phi(e^{it})dt - \sigma, \tag{7.49}$$

where σ is a real constant and where $\phi(e^{it})$ is a step function on $\mathbb{T}$ given by

$$\phi(e^{it}) = \frac{\rho_k}{2i} \quad (t_{k-1} < t < t_k\ (1 \le k \le 2n)). \tag{7.50}$$

We have thus expressed $\arg R(z)$ as the harmonic extension into $\mathbb{U}$ of a step function ϕ on $\mathbb{T}$ with jumps occurring at the zeros and poles of R. The condition (7.43) implies that

$$|\phi(e^{it}) - \lambda| \le \frac{\pi}{2} \tag{7.51}$$

at all points other than the jump points, and therefore for any pair of such points

$$|\phi(e^{it}) - \phi(e^{is})| \le \pi. \tag{7.52}$$

This implies that

$$\left|\sum_{k=\mu}^{\nu} \epsilon_k\right| \le 1 \tag{7.53}$$

for any μ and ν. It is immediately clear that the ϵ_k alternate in sign, which means that zeros and poles alternate, and are simple. The result follows.

Since $\epsilon_k = \pm 1$, the same conclusion could be drawn from the inequality

$$\left|\sum_{k=\mu}^{\nu} \epsilon_k\right| < 2, \tag{7.54}$$

and therefore it would still be true that the zeros and poles of R were interspersed if we had

$$|\arg R(z) - \lambda| \le \pi - \delta \quad (z \in \mathbb{U}) \tag{7.55}$$

for a small $\delta > 0$ and real λ.

7.2.5 Problems

1. Let $P(z) = \sum_0^n a_k z^k$ and $Q(z) = \sum_0^n b_k z^k$ be polynomials of degree n with their zeros interspersed on $\mathbb{T}$ and with $a_0 = b_0 = 1$. Show that for a suitable real γ we have

$$\left|\sum_{k=1}^{n} (a_k - b_k) z^{k-1}\right| \le \left|\sum_{k=0}^{n-1} (a_k - e^{i\gamma} b_k) z^k\right| \quad (z \in \mathbb{U}). \tag{7.56}$$

Hence show that

$$|a_N - b_N|^2 \le 4 \sum_{k=0}^{N-1} |a_k||b_k| \quad (N = 1, 2, \ldots, n). \tag{7.57}$$

2. Let $P(z) = \sum_0^n a_k z^k$ be a polynomial of degree n with all its zeros on $\mathbb{T}$ and satisfying $a_0 = a_n = 1$. Let

$$S(z) = \sum_{k=0}^{n} \frac{a_k}{\binom{n}{k}} z^k. \tag{7.58}$$

Show that S has all its zeros on $\mathbb{T}$ and show further that the rational function

$$\frac{S(z) - 1}{S(z) - z^n} \tag{7.59}$$

is a Blaschke product (so has all its zeros in $\mathbb{U}$).

7.3 Relations with the maximum modulus

7.3.1

Let $P(z)$ be a polynomial of degree n with its zeros $a_k = e^{i\alpha_k}$ on $\mathbb{T}$ and assume the α_k are in ascending order, so the a_k are cyclically ordered on $\mathbb{T}$. At a point $e^{i\theta}$ on $\mathbb{T}$ where $|P(e^{i\theta})|$ has a local maximum,

$$\frac{d}{d\theta}|P(e^{i\theta})| = 0. \tag{7.60}$$

Writing $z = e^{i\theta}$, this implies that $zP'(z)/P(z)$ is real. Therefore applying (7.5) we deduce that

$$\frac{zP'(z)}{P(z)} = \frac{n}{2}. \tag{7.61}$$

Thus the local maxima on $\mathbb{T}$ occur at the zeros of $Q(z) = P(z) - \frac{2}{n}zP'(z)$. This polynomial has all its zeros on $\mathbb{T}$ and

$$\Re\left(\frac{Q(z)}{P(z)}\right) > 0 \quad (z \in \mathbb{U}). \tag{7.62}$$

This is a typical case of a pair of polynomials with interspersed zeros. If the zeros $b_k = e^{i\beta_k}$ of Q are labelled so that (7.40) is satisfied, then we have

$$\sum_{k=1}^{n}(\beta_k - \alpha_k) = \pi, \tag{7.63}$$

so *on average* the local maxima occur half-way between the zeros. We have

$$\frac{Q(z)}{P(z)} = \frac{1 - \omega(z)}{1 + \omega(z)} \tag{7.64}$$

where $\omega(z)$ is a Blaschke product satisfying $\omega(0) = 0$ and whose remaining zeros are the critical points of P not on $\mathbb{T}$.

We observe that this gives

$$|P(e^{i\theta})| = \frac{1}{n}|P'(e^{i\theta})||1 + \omega(e^{i\theta})| \quad (\theta \text{ real}). \tag{7.65}$$

Now suppose that $|P(z)| \le M$ in $\mathbb{U}$. Then from (7.3) we have $|P'(z)| \le \frac{n}{2}M$

in $\mathbb{U}$; we therefore obtain

$$|P(e^{i\theta})| \leq \frac{M}{2}|1+\omega(e^{i\theta})| \quad (\theta \text{ real}). \tag{7.66}$$

We now apply the important Littlewood subordination inequality; $1+\omega(z)$ is subordinate to $1+z$. It follows that for every $\lambda > 0$

$$\begin{aligned}\frac{1}{2\pi}\int_0^{2\pi} |P(e^{i\theta})|^\lambda d\theta &\leq \left(\frac{M}{2}\right)^\lambda \frac{1}{2\pi}\int_0^{2\pi} |1+e^{i\theta}|^\lambda d\theta = \frac{2M^\lambda}{\pi}\int_0^{\pi/2} (\cos\theta)^\lambda d\theta \\ &= \frac{2M^\lambda}{\lambda\sqrt{\pi}}\frac{\Gamma\left(\frac{1}{2}\lambda+\frac{1}{2}\right)}{\Gamma\left(\frac{1}{2}\lambda\right)}\end{aligned} \tag{7.67}$$

where Γ is Euler's gamma function. (See Whittaker and Watson [70] page 256 for the evaluation of the integral.) We deduce the next result, which is proved in Saff and Sheil-Small [50].

7.3.2

Theorem *Let $P(z)$ be a polynomial with all its zeros on $\mathbb{T}$ and suppose that $|P(z)| \leq M$ on $\mathbb{T}$. Then for $\lambda > 0$ we have*

$$\frac{1}{2\pi}\int_0^{2\pi} |P(e^{i\theta})|^\lambda d\theta \leq \frac{2M^\lambda}{\lambda\sqrt{\pi}}\frac{\Gamma\left(\frac{1}{2}\lambda+\frac{1}{2}\right)}{\Gamma\left(\frac{1}{2}\lambda\right)}. \tag{7.68}$$

In particular

$$\frac{1}{2\pi}\int_0^{2\pi} |P(e^{i\theta})| d\theta \leq \frac{2M}{\pi} \tag{7.69}$$

and

$$\frac{1}{2\pi}\int_0^{2\pi} |P(e^{i\theta})|^2 d\theta \leq \frac{M^2}{2}. \tag{7.70}$$

7.3.3

Corollary *Let $P(z) = \sum_{k=0}^n a_k z^k$ be a polynomial of degree n with all its zeros on $\mathbb{T}$ and suppose that $|P(z)| \leq M$ on $\mathbb{T}$. Then*

$$\sum_{k=0}^n |a_k|^2 \leq \frac{M^2}{2} \tag{7.71}$$

and

$$|a_k| \leq \frac{M}{2} \quad \left(k \neq \frac{n}{2}\right). \tag{7.72}$$

The relation (7.71) follows from (7.70) and Parseval's formula; the relation (7.72) follows from (7.71) and the fact that $|a_k| = |a_{n-k}|$. Thus $M/2$ is an upper bound for the coefficients when P has odd degree, and an upper bound for all coefficients except possibly the middle term $a_{n/2}$ when P has even degree. In fact it is also true that

$$|a_{n/2}| \leq \frac{M}{2} \quad (n \text{ even}). \tag{7.73}$$

This was proved by G.K. Kristiansen [24] using a variational argument. Kristiansen also obtains several other interesting inequalities for polynomials with their zeros on $\mathbb{T}$.

7.3.4 Problems

1. Let $P(z) = \sum_0^n a_k z^k$ be a polynomial of even degree $n = 2m$ with all its zeros on $\mathbb{T}$. Show that

$$|a_m|^2 \leq 4 \sum_{k=0}^{m-1} \frac{m-k}{m} |a_k|^2 \tag{7.74}$$

and deduce that

$$|a_m| < \frac{M}{\sqrt{3}} \tag{7.75}$$

where $M = \max_{|z|=1} |P(z)|$.

2. Let $P(z) = \sum_{k=0}^n a_k z^k$ be a self-inversive polynomial of degree n with $M = \max_{|z|=1} |P(z)|$. Show that

$$|a_k| \leq \frac{M}{2} \text{ provided that } k \notin \left[\frac{n}{3}, \frac{2n}{3}\right]. \tag{7.76}$$

[Hint: use Problem 4.2.9.1 of chapter 4.]

3. Let $P(z) = \sum_{k=0}^n a_k z^k$ be a self-inversive polynomial of even degree $n = 2m$ with $M = \max_{|z|=1} |P(z)|$. Show that

$$|a_k| + |a_m| \leq M \quad (k \neq m). \tag{7.77}$$

Hence show that either (i) $|a_m| \leq \frac{1}{2}M$ or (ii) $|a_k| \leq \frac{1}{2}M$ for every $k \neq m$. [Hint: if $f(z) = \frac{1}{2}z^r + z^j + \frac{1}{2}z^{2j-r}$ where $0 \leq r < j$, then $f*$ is norm preserving on H^∞.]

4. Let $P(z) = \sum_{k=0}^n a_k z^k$ be a polynomial of degree n with all its zeros in $\overline{\mathbb{U}}$. Show that

$$|a_k| \leq \frac{M}{2} \quad (0 \leq k \leq \tfrac{1}{2}n) \tag{7.78}$$

where $M = \max_{|z|=1} |P(z)|$. [Hint: use Kristiansen's result; the polynomials $z^N P + \epsilon P^*$ have all their zeros on $\mathbb{T}$ for $N = 0, 1, 2, \ldots$ and $|\epsilon| = 1$.]

7.4 Univalent polynomials

7.4.1

Let $P(z)$ be a polynomial of degree n whose critical points all lie on $\mathbb{T}$. Then $P'(z) \neq 0$ in $\mathbb{U}$, so P is locally 1–1 in $\mathbb{U}$. We may ask whether there are conditions on the location of the critical points which imply that P is globally 1–1, i.e. univalent in $\mathbb{U}$. It is easily seen that this is not possible, if P' has a multiple zero. This suggests that we look for a condition that implies that the zeros of P' are suitably separated on $\mathbb{T}$. We begin with a simple example.

The polynomial $Q(z) = z - \frac{1}{n}z^n$ is univalent in $\mathbb{U}$ and maps $\mathbb{U}$ onto a domain D which is starlike with respect to the origin. The condition for this is

$$\Re\left(\frac{zQ'(z)}{Q(z)}\right) > 0 \quad (z \in \mathbb{U}), \tag{7.79}$$

i.e.

$$\Re\left(\frac{1 - z^{n-1}}{1 - \frac{1}{n}z^{n-1}}\right) > 0 \quad (z \in \mathbb{U}). \tag{7.80}$$

Writing $\zeta = z^{n-1}$ and $\alpha = \frac{1}{n}$, we have for $|\zeta| < 1$

$$\Re\left(\frac{1-\zeta}{1-\alpha\zeta}\right) = 1 - \frac{1-\alpha}{\alpha}\Re\left(\frac{\alpha\zeta}{1-\alpha\zeta}\right) > 1 - \frac{1-\alpha}{\alpha}\frac{\alpha}{1-\alpha} = 0. \tag{7.81}$$

Notice that the critical points of Q are the $(n-1)$th roots of unity, so are separated on $\mathbb{T}$ by arcs of exact length $2\pi/(n-1)$. This is an example of a general phenomenon. If P is a polynomial of degree n with its critical points on $\mathbb{T}$, and if these critical points are separated by arcs whose minimum length is at least $2\pi/(n+1)$, then P is univalent in $\mathbb{U}$. This is one of T.J. Suffridge's many remarkable results on polynomials with zeros on $\mathbb{T}$ separated by a minimum angle. Suffridge [62] shows that such polynomials can be used to approximate some important classes of functions analytic in $\mathbb{U}$. In addition every univalent function in $\mathbb{U}$ has a derivative which is the local uniform limit in $\mathbb{U}$ of a sequence of polynomials whose zeros lie on $\mathbb{T}$.

The remainder of this chapter will be devoted to an exposition of Suffridge's theory.

7.4.2 Close-to-convex functions

In 1952 W. Kaplan [22] introduced an elegant analytic condition on a function f analytic in $\mathbb{U}$ which was sufficient to imply the univalence of f in $\mathbb{U}$; the condition is that there exist a univalent mapping $\phi(z)$ of $\mathbb{U}$ onto a convex domain such that

$$\Re\left(\frac{f'(z)}{\phi'(z)}\right) > 0 \quad (z \in \mathbb{U}). \tag{7.82}$$

Kaplan called such functions 'close-to-convex', presumably on the basis that the derivative of f differs multiplicatively from that of ϕ by at most a nice function, namely a function of positive real part. Certainly analytic information about f flows readily from this condition. That the condition implies univalence can be seen as follows. Consider the mapping $w = \phi(z)$ of $\mathbb{U}$ onto a convex domain D and let $z = \psi(w)$ denote the inverse mapping. Define $g(w) = f(\psi(w))$. Then the condition becomes

$$\Re g'(w) > 0 \quad (w \in D). \tag{7.83}$$

It will be sufficient to show that this implies the univalence of g in D. Let a and b be distinct points of D. As D is convex, the line segment $[a, b] \subset D$ and so

$$g(b) - g(a) = \int_a^b g'(w)dw = (b - a)\int_0^1 g'((1-t)a + tb)dt. \tag{7.84}$$

We deduce

$$\Re\left(\frac{g(b) - g(a)}{b - a}\right) = \int_0^1 \Re g'((1-t)a + tb)dt > 0 \tag{7.85}$$

and so $g(a) \neq g(b)$. Thus g is univalent.

Kaplan went on to show that the domain onto which $\mathbb{U}$ is mapped by a close-to-convex function f (called a close-to-convex domain) can be described geometrically as follows:

$$d(zf', \gamma) > -\tfrac{1}{2} \tag{7.86}$$

for every positively described arc γ of the circle $\{|z| = r\}$ $(0 < r < 1)$ (see chapter 2 for a description of the degree notation). In other words, if Γ_r denotes the image under f of the circle $\{|z| = r\}$, then the tangent to the curve Γ_r cannot swing backwards from a previous direction by more than an angle π. A more direct geometric criterion due to Lewandowski [26 and 27] is that a domain W is close-to-convex iff the complement of W can be written as a union of non-crossing half-lines. Equivalently (Sheil-Small [52]), from any pair of boundary points infinity can be seen along straight paths not passing through W and not intersecting (unless they eventually coincide).

7.4.3

Suffridge [62] proved the following result.

Theorem *Let $P(z)$ be a polynomial of degree n with all its critical points on $\mathbb{T}$ and suppose that each pair of critical points is separated by an angle of at least $2\pi/(n+1)$. Then P is close-to-convex, and hence univalent, in $\mathbb{U}$. Conversely, if P is close-to-convex in $\mathbb{U}$, then the critical points have an angle separation of at least $2\pi/(n+1)$.*

Before giving the proof of this specific result we will find it convenient to discuss in detail the behaviour of the argument on $\mathbb{T}$ of a polynomial with its zeros on $\mathbb{T}$. We begin with the observation that

$$\arg(1-e^{i\theta}) = \frac{\theta-\pi}{2} \quad (0<\theta<2\pi), \tag{7.87}$$

if, by the left-hand side, we refer to the continuous extension to $\mathbb{T}-\{1\}$ of the principal branch in $\mathbb{U}$ of $\arg(1-z)$. It follows that, if $0\leq\alpha<2\pi$, then

$$\arg\big(1-e^{i(\theta-\alpha)}\big) = \begin{cases} \frac{1}{2}(\theta-\alpha+\pi) & (0<\theta<\alpha), \\ \frac{1}{2}(\theta-\alpha-\pi) & (\alpha<\theta<2\pi). \end{cases} \tag{7.88}$$

Consider now a polynomial Q of degree m with simple zeros $e^{i\alpha_k}$ on $\mathbb{T}$ satisfying

$$0\leq\alpha_1<\alpha_2<\cdots<\alpha_m<2\pi \tag{7.89}$$

and assume $Q(0)=1$. Then, taking $\alpha_{m+1}=\alpha_1+2\pi$,

$$\begin{aligned}\arg Q(e^{i\theta}) &= \sum_{k=1}^{m}\arg\big(1-e^{i(\theta-\alpha_k)}\big) \\ &= \frac{1}{2}\sum_{i=1}^{m}(\theta-\alpha_i+\pi)-k\pi \quad (\alpha_k<\theta<\alpha_{k+1});\end{aligned} \tag{7.90}$$

in other words we obtain a piece-wise linear function of the angle with jumps of $-\pi$ at the zeros. Consider now an arc γ of $\mathbb{T}$, whose endpoints are not zeros and having its parametrisation in the range (α_1,α_{m+1}). Then

$$d(Q,\gamma) = \frac{m}{2}d(z,\gamma)-\frac{\nu}{2}, \tag{7.91}$$

where ν is the number of zeros of Q on γ. (Recall that $d(f,\gamma)$ is the change in $\arg f$ on γ divided by 2π.) Now in Suffridge's theorem we have $Q=P'$ and $m=n-1$. Thus we have

$$d(zP',\gamma) = \frac{n+1}{2}d(z,\gamma)-\frac{\nu}{2} > -\frac{1}{2} \tag{7.92}$$

iff $d(z,\gamma) > \frac{\nu-1}{n+1}$. Now $d(z,\gamma) > \sum_i d(z,\gamma_i)$, where γ_i are successive arcs containing the ν zeros. Since $2\pi d(z,\gamma_i)$ is the length of the arc γ_i, it is clear that the desired inequality holds on $\mathbb{T}$ iff every arc separating two successive zeros has length $\geq 2\pi/(n+1)$. The proof that it is sufficient to perform our calculations on $\mathbb{T}$ will be left as an exercise.

7.4.4 *Kaplan classes*

It is clear that this argument will give us a result for any choice of minimum length of separating arcs. Indeed, suppose that our polynomial Q is such that its m zeros on $\mathbb{T}$ are simple and each pair is separated by an angle of at least λ. For this to be possible we must have $m\lambda \leq 2\pi$, i.e $\lambda \leq 2\pi/m$. Applying (7.91) we see that, if α and β are non-negative numbers such that $\alpha - \beta < m$, then

$$d(Q,\gamma) > \frac{1}{2}(\alpha-\beta)d(z,\gamma) - \frac{\alpha}{2} \tag{7.93}$$

provided that

$$(m-\alpha+\beta)d(z,\gamma) - \nu + \alpha > 0. \tag{7.94}$$

Now, if $\nu = 0$ this is trivial. If $\nu \geq 1$ we have

$$d(z,\gamma) > \frac{(\nu-1)\lambda}{2\pi}, \tag{7.95}$$

so we require

$$(m-\alpha+\beta)(\nu-1)\lambda - 2\pi(\nu-\alpha) \geq 0. \tag{7.96}$$

When $\nu = 1$, we see that this is possible only if $\alpha \geq 1$. If $\alpha = 1$, the condition reduces to

$$\lambda \geq \frac{2\pi}{m-1+\beta} \tag{7.97}$$

and this is a sensible condition provided that $\beta \geq 1$.

7.4.5

Our motivation for performing these calculations lies in the following definition.

Definition Let $f(z)$ be a function analytic and non-zero in $\mathbb{U}$. For non-negative numbers α and β we say that f belongs to the **Kaplan class** $K(\alpha,\beta)$ if

$$d(f,\gamma) \geq \tfrac{1}{2}(\alpha-\beta)d(z,\gamma) - \tfrac{1}{2}\alpha \tag{7.98}$$

for every arc γ of $\{|z| = r\}$ $(0 < r < 1)$.

The meaning of this definition and the choice of notation will become clear in what follows. We note first that the inequality implies an inequality in the reverse direction. For let γ^c denote the complementary arc to γ, so that we have

$$d(z,\gamma)+d(z,\gamma^c)=1, \qquad d(f,\gamma)+d(f,\gamma^c)=0, \tag{7.99}$$

the second equality following from the argument principle ($f(z)\neq 0$ in $\mathbb{U}$). Substituting we obtain

$$d(f,\gamma^c)\leq \tfrac{1}{2}(\alpha-\beta)d(z,\gamma^c)+\tfrac{1}{2}\beta \tag{7.100}$$

and therefore since the original definition applies equally to γ^c, we deduce that

$$\tfrac{1}{2}(\alpha-\beta)d(z,\gamma)-\tfrac{1}{2}\alpha\leq d(f,\gamma)\leq \tfrac{1}{2}(\alpha-\beta)d(z,\gamma)+\tfrac{1}{2}\beta. \tag{7.101}$$

Thus either of the two inequalities here will serve for the definition. From the left-hand inequality we see that, if we increase β to β' say, then the definition holds for α and β'. On the other hand from the right-hand inequality, we can increase α to α' say, and the inequality holds for α' and β. Thus in general we have

$$K(\alpha,\beta)\subset K(\alpha',\beta') \quad (\alpha'\geq\alpha,\ \beta'\geq\beta); \tag{7.102}$$

i.e. the classes are increasing with the variables.

Next we note that we have immediately from the definition: (i) if $f\in K(\alpha,\beta)$ and $g\in K(\lambda,\mu)$, then $fg\in K(\alpha+\lambda,\beta+\mu)$; (ii) if $f\in K(\alpha,\beta)$, then $1/f\in K(\beta,\alpha)$ and for each $p\geq 0$, $f^p\in K(p\alpha,p\beta)$.

There is an important converse to (i) giving a factorisation theorem for functions in $K(\alpha,\beta)$. Before giving this we consider some particular cases. From (7.98) we see that $f\in K(\alpha,\alpha)\Leftrightarrow |d(f,\gamma)|\leq\frac{1}{2}\alpha$, and it is clear that this holds iff for a suitable real μ

$$|\arg(e^{i\mu}f(z))|\leq\frac{\alpha\pi}{2} \quad (z\in\mathbb{U}). \tag{7.103}$$

Thus the Kaplan classes contain all functions of bounded argument in $\mathbb{U}$. In particular a function f is in $K(1,1)$ iff it is at most a rotation of a function with positive real part.

Next we consider the class $K(\alpha,0)$. Clearly $K(0,0)$ consists entirely of the non-zero constants. If $\alpha>0$, then from (7.98) we see that $f\in K(\alpha,0)$ iff $d(f,\gamma)\leq\frac{1}{2}\alpha d(z,\gamma)$. Writing $\phi(\theta)=\arg f(re^{i\theta})$ this implies that $\phi'(\theta)\leq\frac{1}{2}\alpha$. Conversely the latter inequality implies the inequality between the degrees. It follows that $f \in K(\alpha, 0)$ iff

$$\Re \frac{zf'(z)}{f(z)} \le \frac{\alpha}{2} \quad (z \in \mathbb{U}). \tag{7.104}$$

Again, $f \in K(0, \beta) \Leftrightarrow 1/f \in K(\beta, 0)) \Leftrightarrow$

$$\Re \frac{zf'(z)}{f(z)} \ge -\frac{\beta}{2} \quad (z \in \mathbb{U}). \tag{7.105}$$

It follows easily from (7.101) that a function f of the form

$$f(z) = \prod_{k=1}^{n} (1 + x_k z)^{\alpha_k}, \tag{7.106}$$

where the α_k are positive numbers satisfying $\sum_1^n \alpha_k = \alpha$ and where the x_k are complex numbers satisfying $|x_k| \le 1$, lies in the class $K(\alpha, 0)$. Furthermore $K(\alpha, 0)$ (with a normalisation so that $f(0) = 1$) is the closure of the set of such products in the topology of local uniform convergence in $\mathbb{U}$. A proof of this in a sharp form will be given in chapter 8 (subsection 8.3.2). In particular, we note that the class of polynomials of degree n with no zeros in $\mathbb{U}$ lies in the class $K(n, 0)$. It is convenient to introduce a notation which reflects this similarity to finite products. We define for λ real

$$\Pi_\lambda = \begin{cases} K(\lambda, 0) & (\lambda \ge 0), \\ K(0, -\lambda) & (\lambda < 0). \end{cases} \tag{7.107}$$

7.4.6

We can now state our factorisation theorem.

Theorem *A function $f \in K(\alpha, \beta)$ iff f can be factorised in the form*

$$f = gH \tag{7.108}$$

where $g \in \Pi_{\alpha-\beta}$ and $H \in K(\mu, \mu)$ $(\mu = \min(\alpha, \beta))$.

Proof It is clear from (i) in 7.4.5 that, if g and H have the given form, then $gH \in K(\alpha, \beta)$. Thus the main problem is to prove the possibility of such a factorisation for an arbitrary $f \in K(\alpha, \beta)$. If $\alpha = \beta$, the result is clear. Assume that $\alpha > \beta$ and let $\phi(\theta) = \arg f(re^{i\theta})$, where $0 < r < 1$; then by hypothesis we have from (7.101)

$$\phi(\theta_2) - \phi(\theta_1) - \tfrac{1}{2}(\alpha - \beta)(\theta_2 - \theta_1) \le \beta\pi \quad (\theta_1 < \theta_2 < \theta_1 + 2\pi). \tag{7.109}$$

Now set $\psi(\theta) = \theta - 2\phi(\theta)/(\alpha - \beta)$; then the previous inequality becomes

$$\psi(\theta_2) - \psi(\theta_1) \geq -\frac{2\beta\pi}{\alpha - \beta}. \tag{7.110}$$

Furthermore, $\psi(\theta + 2\pi) = \psi(\theta) + 2\pi$, and therefore the previous inequality holds without restriction, whenever $\theta_2 > \theta_1$. We now define

$$\omega(\theta) = \inf\{\psi(\theta') : \theta' \geq \theta\} + \frac{\beta\pi}{\alpha - \beta} \tag{7.111}$$

and observe that $\omega(\theta)$ is an increasing function satisfying $\omega(\theta + 2\pi) = \omega(\theta) + 2\pi$ and

$$|\omega(\theta) - \psi(\theta)| \geq \frac{\beta\pi}{\alpha - \beta}. \tag{7.112}$$

Consider next the analytic function given by

$$-iF(z) = \frac{1}{2\pi}\int_0^{2\pi} \frac{1 + ze^{-it}}{1 - ze^{-it}}(\omega(t) - t)dt; \tag{7.113}$$

we have

$$\Im F(z) = \frac{1}{2\pi}\int_0^{2\pi} \Re\left(\frac{1 + ze^{-it}}{1 - ze^{-it}}\right)(\omega(t) - t)dt, \tag{7.114}$$

and this is the harmonic extension into $\mathbb{U}$ of the periodic function $\omega(\theta) - \theta$ interpreted as a function of angle on $\mathbb{T}$. Now differentiating we have

$$zF'(z) = \frac{1}{2\pi}\int_0^{2\pi} \frac{2ize^{-it}}{(1 - ze^{-it})^2}(\omega(t) - t)dt \tag{7.115}$$

and therefore, integrating by parts we obtain

$$zF'(z) = \frac{1}{2\pi}\int_0^{2\pi} \frac{1 + ze^{-it}}{1 - ze^{-it}}d\omega(t) - 1 \tag{7.116}$$

and writing $g(z) = e^{F(z)}$ we obtain for $z \in \mathbb{U}$

$$\Re\frac{zg'(z)}{g(z)} = \frac{1}{2\pi}\int_0^{2\pi} \Re\left(\frac{1 + ze^{-it}}{1 - ze^{-it}}\right)d\omega(t) - 1 > -1 \tag{7.117}$$

and so $g \in \Pi_{-2}$. Furthermore $\arg g(z) = \Im F(z)$ is the harmonic extension of $\omega(\theta) - \theta$ to $\mathbb{U}$. Now let $h(z) = f(rz)^{-2/(\alpha-\beta)}$ and observe that for $z = e^{i\theta}$ we have

$$\arg h(e^{i\theta}) = \psi(\theta) - \theta, \tag{7.118}$$

so that $\arg h(z)$ is the harmonic extension of $\psi(\theta) - \theta$ to $\mathbb{U}$. Therefore (7.112) and the maximum principle for harmonic functions give

$$\left| \arg \frac{h(z)}{g(z)} \right| \leq \frac{\beta\pi}{\alpha - \beta} \quad (z \in \mathbb{U}). \tag{7.119}$$

From this we obtain

$$\left| \arg \frac{f(rz)}{G_r(z)} \right| \leq \frac{\beta\pi}{2} \quad (z \in \mathbb{U}) \tag{7.120}$$

where $G_r \in \Pi_{\alpha-\beta}$. Since $\Pi_{\alpha-\beta}$ is a compact class we obtain letting $r \to 1$ the representation $f(z) = p(z)L(z)$, where $p \in \Pi_{\alpha-\beta}$ and $L \in K(\beta, \beta)$, as required. Finally, if $\beta > \alpha$ we just apply this result to $1/f$.

7.4.7

Returning to Suffridge's method we have the following result.

Theorem *Let $P(z)$ be a polynomial of degree n with all its zeros on $\mathbb{T}$. Then each pair of zeros of P on $\mathbb{T}$ is separated by an angle of at least λ $(0 < \lambda \leq \frac{2\pi}{n})$ iff $P \in K(1, \frac{2\pi}{\lambda} - n + 1)$.*

For the proof refer back to 7.4.3, in particular (7.91). In fact the calculations there show that, if the zeros of P are separated on $\mathbb{T}$ by an angle λ, then

$$P \in K\left(\alpha, \frac{n-\alpha}{n-1}\left(\frac{2\pi}{\lambda} - n + 1\right)\right) \quad (1 \leq \alpha \leq n). \tag{7.121}$$

On the other hand, if $P \in K(\alpha, \beta)$, then if we choose an arc γ containing just two zeros (so $\nu = 2$) and slightly exceeding in length the arc joining the two zeros, we easily deduce that

$$\lambda \geq \frac{2\pi(2-\alpha)}{n - \alpha + \beta}. \tag{7.122}$$

In the case $\alpha = 1$ this gives us the required converse to the theorem. We do not get the converse, when $1 < \alpha < 2$, but in this case we are still able to deduce the existence of a separating angle. For $\alpha \geq 2$ we cannot deduce the existence of such an angle. The example $P(z) = (1 - z^n)^2 \in K(2, 2)$ but has only double zeros. The case $\alpha < 1$ cannot occur; i.e. there are no polynomials with their zeros on $\mathbb{T}$ in any class $K(\alpha, \beta)$ if $\alpha < 1$. (Consider a small arc γ containing just one zero.)

7.5 A second necessary and sufficient condition for angular separation of zeros

7.5.1

Suffridge [62] proves the following interesting result.

Theorem *Let $P(z)$ be a self-inversive polynomial of degree n. P has all its zeros on $\mathbb{T}$ and each pair of zeros is separated by an angle of at least λ $(0 < \lambda < \frac{2\pi}{n})$ iff the polynomial*

$$P(ze^{i\lambda/2}) - P(ze^{-i\lambda/2}) \tag{7.123}$$

has all its zeros in $\overline{\mathbb{U}}$.

It is easily verified that in the case $\lambda = 2\pi/n$ the necessary and sufficient condition is that the function (7.123) be identically zero. We also recall from theorem 7.1.3 that in the limiting case $\lambda = 0$ the necessary and sufficient condition is that P' have all its zeros in $\overline{\mathbb{U}}$.

To prove the theorem suppose firstly that P has all its zeros on $\mathbb{T}$ and that these have a separating angle λ. Then certainly $P'(z)$ has all its zeros in $\overline{\mathbb{U}}$; furthermore P' has a zero on $\mathbb{T}$ only at a point where P has a multiple zero. Since the zeros are separated on $\mathbb{T}$, P' has all its zeros in $\mathbb{U}$.

Consider now, for $0 < t < 2\pi$, the polynomials

$$Q(z,t) = \frac{P(ze^{it}) - P(ze^{-it})}{2iz \sin t}. \tag{7.124}$$

As $t \to 0$, $Q(z,t) \to P'(z)$. Suppose that for a particular t $(0 < t < 2\pi)$ $Q(z,t)$ has a zero on $\mathbb{T}$, say at $z = e^{i\theta}$. Since P is self-inversive, the function $\theta \mapsto e^{-in\theta/2}P(e^{i\theta})$ is real and $= T(\theta)$ say. Therefore

$$e^{in(\theta+t)/2}T(\theta+t) = e^{in(\theta-t)/2}T(\theta-t) \tag{7.125}$$

and so either $T(\theta+t) = T(\theta-t) = 0$, or e^{int} is real. In the latter case we have $e^{2int} = 1$ and so nt is a multiple of π. It follows that, if $0 < t < \frac{\pi}{n}$, then $Q(z,t)$ has a zero on $\mathbb{T}$ iff P has two zeros on $\mathbb{T}$ separated by an angle of exactly $2t$. In particular our hypothesis implies that $Q(z,t)$ has no zeros on $\mathbb{T}$ if $0 < t < \frac{\lambda}{2}$. As we have shown, this also holds when $t = 0$. Now fix t in this range and consider the curves $\gamma_s = Q(e^{i\theta}, s)$ $(0 \le \theta \le 2\pi, 0 \le s \le t)$ in the domain $D = \mathbb{C} - \{0\}$. These form a homotopy in D between γ_0 and γ_t and the function z is $\ne 0$ in D. By the homotopic degree principle (chapter 2, theorem 2.4.2) $d(z, \gamma_0) = d(z, \gamma_t)$; in other words

$$d(Q(z,0), \mathbb{T}) = d(Q(z,t), \mathbb{T}) \tag{7.126}$$

and so, by the argument principle, $Q(z, 0) = P'(z)$ and $Q(z, t)$ have equally many zeros in $\mathbb{U}$, and this is exactly $n - 1$. Thus $Q(z, t)$ has all its zeros in $\mathbb{U}$. By considering a slightly bigger circle than $\mathbb{T}$ and arguing similarly, we deduce that $Q(z, \frac{\lambda}{2})$ has all its zeros in $\overline{\mathbb{U}}$; furthermore a zero on $\mathbb{T}$ will occur precisely when there is an exact angle λ separating two zeros of P on $\mathbb{T}$ and will bisect this angle.

For the converse, let P be a self-inversive polynomial of degree n and suppose that for a given $t \in (0, \frac{\pi}{n})$, the polynomial $P(ze^{it}) - P(ze^{-it})$ has all its zeros in $\overline{\mathbb{U}}$. Then from Problem 7.1.5.1, for each $\zeta\,(|\zeta| = 1)$ the polynomial

$$\begin{aligned} &P(ze^{it}) - P(ze^{-it}) + \zeta(P(ze^{it}) - P(ze^{-it}))^* \\ &\quad = (1 + \zeta e^{-int})P(ze^{it}) - (1 + \zeta e^{int})P(ze^{-it}) \end{aligned} \tag{7.127}$$

has all its zeros on $\mathbb{T}$. In particular, when $\zeta = -e^{int}$ we deduce that P has all its zeros on $\mathbb{T}$. Furthermore, we find that the rational function

$$R(z, t) = \frac{P(ze^{it})}{P(ze^{-it})} \tag{7.128}$$

takes the values

$$w = \frac{1 + \zeta e^{int}}{1 + \zeta e^{-int}} \quad (\zeta \in \mathbb{T}) \tag{7.129}$$

only on $\mathbb{T}$. This set of values is the line $w = \rho e^{int}$ (ρ real). Thus $e^{-int}R(z, t)$ is analytic in $\mathbb{U}$ and takes no real values; it therefore maps $\mathbb{U}$ into either the upper half-plane or the lower half-plane. At the origin it takes the value e^{-int}, and since $0 < nt < \pi$, $\sin nt > 0$. We deduce

$$\Im(e^{-int}R(z, t)) < 0 \quad (z \in \mathbb{U}). \tag{7.130}$$

Now from our interspersion theorem 7.2.4 this means that the zeros of the polynomials $P(ze^{it})$ and $P(ze^{-it})$ are interspersed on $\mathbb{T}$. Equivalently, $P(z)$ and $P(ze^{-2it})$ have interspersed zeros. Furthermore, we have from theorem 7.2.4 that, if $P(z)$ has zeros $e^{i\alpha_k}$ and $P(ze^{-2it})$ has zeros $e^{i\beta_k}$, labelled so that $\alpha_1 \leq \beta_1 \leq \alpha_2 \leq \cdots \leq \alpha_n \leq \beta_n \leq \alpha_1 + 2\pi$, then

$$\sum_{k=1}^{n}(\beta_k - \alpha_k) = 2nt. \tag{7.131}$$

On the other hand, the zeros of $P(ze^{-2it})$ are those of $P(z)$ rotated through the positive angle $2t$, and therefore each $\beta_k = \alpha_{j+k} + 2t$ for some fixed j, where $\alpha_m = \alpha_{m-n} + 2\pi$ when $n + 1 \leq m \leq 2n$. This gives

$$\sum_{k=1}^{n}(\beta_k - \alpha_k) = 2nt + 2j\pi \tag{7.132}$$

and so $j = 0$. Thus we have

$$\alpha_1 \leq \alpha_1 + 2t \leq \alpha_2 \leq \alpha_2 + 2t \leq \cdots \leq \alpha_n \leq \alpha_n + 2t \leq \alpha_1 + 2\pi \qquad (7.133)$$

and therefore each pair of zeros of P is separated by an angle of at least $2t$, as required.

On the other hand it is clear that, if the zeros are separated by an angle $2t$, then the inequalities of (7.133) hold and therefore the zeros of $P(ze^{it})$ and $P(ze^{-it})$ are interspersed. Thus the interspersion theorem gives us a third necessary and sufficient condition for angular separation of zeros, which we give next.

7.5.2

Theorem *Let $P(z)$ be a self-inversive polynomial of degree n. P has all its zeros on $\mathbb{T}$ and each pair of zeros is separated by an angle of at least λ $(0 < \lambda < \frac{2\pi}{n})$ iff*

$$\Im\left(e^{-in\lambda/2}\frac{P(ze^{i\lambda/2})}{P(ze^{-i\lambda/2})}\right) < 0 \quad (z \in \mathbb{U}). \qquad (7.134)$$

In this case we have

$$R\left(z, \frac{\lambda}{2}\right) = \frac{P(ze^{i\lambda/2})}{P(ze^{-i\lambda/2})} = \frac{1 - e^{in\lambda}\omega(z)}{1 - \omega(z)} = \sum_{k=1}^{n} t_k \frac{1 - ze^{i(n-\frac{1}{2})\lambda}e^{-i\alpha_k}}{1 - ze^{-i\lambda/2}e^{-i\alpha_k}} \qquad (7.135)$$

where $e^{i\alpha_k}$ are the zeros of $P(z)$, $\omega(z)$ is a Blaschke product of degree $\leq n$ satisfying $\omega(0) = 0$ and t_k are non-negative numbers satisfying $\sum_1^n t_k = 1$.

7.6 Suffridge's extremal polynomials

7.6.1

The natural extremal polynomials are polynomials $Q_n(\lambda; z)$ which have zeros on $\mathbb{T}$ separated by exact angles λ, except for one gap of $2\pi - (n-1)\lambda$. We assume that the zeros are rotated so that Q_n has zeros at $e^{i\alpha_k}$, where

$$\alpha_k = \pi + \left(k - \tfrac{1}{2}(n+1)\right)\lambda \quad (1 \leq k \leq n). \qquad (7.136)$$

Then the polynomial $Q_n(\lambda; ze^{-i\lambda})$ has zeros at $e^{i\alpha_k}$ $(2 \leq k \leq n)$ and a zero at $e^{i(\alpha_n+\lambda)}$. Thus writing $t = \frac{1}{2}\lambda$ we have

$$\frac{Q_n(\lambda; ze^{it})}{Q_n(\lambda; ze^{-it})} = \frac{1 + ze^{int}}{1 + ze^{-int}} \qquad (7.137)$$

and therefore

$$(1+ze^{-int})Q_n(\lambda;ze^{it})=(1+ze^{int})Q_n(\lambda;ze^{-it}). \tag{7.138}$$

Writing

$$Q_n(\lambda;z)=\sum_{k=0}^{n}C_k^{(n)}(\lambda)z^k \quad \left(C_0^{(n)}(\lambda)=1\right) \tag{7.139}$$

and equating coefficients, we obtain the recursion relationship

$$C_k^{(n)}(\lambda)\sin kt = C_{k-1}^{(n)}(\lambda)\sin(n-k+1)t \quad (1\le k\le n). \tag{7.140}$$

Solving this gives

$$C_k^{(n)}(\lambda)=\prod_{j=1}^{k}\frac{\sin(n-j+1)t}{\sin jt}=\frac{F_n(t)}{F_k(t)F_{n-k}(t)} \quad (1\le k\le n), \tag{7.141}$$

where $F_m(t)=\prod_{j=1}^{m}\sin jt$ and $2t=\lambda$. We also have the product representation

$$Q_n(2t;z)=\prod_{k=1}^{n}\left(1+ze^{i(n+1)t}e^{-2ikt}\right). \tag{7.142}$$

We also leave it to the reader to verify the identity

$$\frac{e^{int}Q_n(2t;ze^{-it})-e^{-int}Q_n(2t;ze^{it})}{2i\sin nt}=Q_{n-1}(2t;z). \tag{7.143}$$

This relation has the following significance. Suffridge's theorem 7.5.1 states that a self-inversive polynomial P satisfies our gap condition with $2t$ iff the polynomial (7.123) has all its zeros in $\overline{\mathbb{U}}$. Equivalently the conjugate of this polynomial has no zeros in $\mathbb{U}$. However, the conjugate of (7.123) is, after a normalisation, the polynomial

$$\frac{e^{int}P(ze^{-it})-e^{-int}P(ze^{it})}{2i\sin nt}. \tag{7.144}$$

Suppose now we write P in the convolution form

$$P(z)=p(z)*Q_n(2t;z). \tag{7.145}$$

Then it is immediately verified that, because of the identity, Suffridge's condition takes the form

$$p(z)*Q_{n-1}(2t;z)\neq 0 \quad (z\in\mathbb{U}). \tag{7.146}$$

Thus we can classify the class of all polynomials $p(z)$, which correspond to $P(z)$ satisfying the $2t$ gap condition, as those self-inversive polynomials of degree n which lie in the dual of the particular polynomial $Q_{n-1}(2t;z)$. Suffridge obtains the fundamental result that this class of polynomials p is closed under convolution.

It is convenient at this point to set up some notations. For $0 < \lambda < \frac{2\pi}{n}$ we define $\mathcal{P}_n(\lambda)$ as the class of self-inversive polynomials $P(z) = 1 + a_1 z + \cdots + z^n$ whose zeros lie on $\mathbb{T}$ and are separated by angles of at least λ. We define the **Suffridge class** $\mathcal{S}_n(\lambda)$ as those polynomials $p(z)$ such that $p(z) * Q_n(\lambda; z) \in \mathcal{P}_n(\lambda)$. Theorems 7.5.1 and 7.5.2 can then be translated into convolution form as in the next subsection.

7.6.2

Theorem *Let $p(z)$ be a self-inversive polynomial of degree n satisfying $p(0) = 1$. Then $p \in$ the Suffridge class $\mathcal{S}_n(\lambda)$ iff*

$$p(z) * Q_{n-1}(\lambda; z) \neq 0 \quad (z \in \mathbb{U}). \tag{7.147}$$

Furthermore, this condition holds iff

$$p(z) * (1 + xz) Q_{n-1}(\lambda; z) \neq 0 \quad (|x| \leq 1,\ z \in \mathbb{U}). \tag{7.148}$$

To prove this last statement we make use of the algebraic relation

$$p(z) * zq(z) = (p * q)^*(z) \tag{7.149}$$

which holds for any self-inversive polynomial p of degree n and any self-inversive polynomial q of degree $n - 1$. Therefore (7.148) is equivalent to

$$(p * Q_{n-1}) + x(p * Q_{n-1})^* \neq 0 \quad (|x| \leq 1, z \in \mathbb{U}), \tag{7.150}$$

which certainly holds if $p * Q_{n-1} \neq 0$ in $\mathbb{U}$. An immediate consequence of this result is the following interesting fact. If $p \in \mathcal{G}_n$ (the Grace class, see chapter 5, subsection 5.1.2), then $p \in \mathcal{S}_n(\lambda)$ for every λ $(0 < \lambda < \frac{2\pi}{n})$; for $Q_{n-1}(\lambda; z) \neq 0$ in $\mathbb{U}$, so by Grace's theorem $p * Q_{n-1}(\lambda; z) \neq 0$ in $\mathbb{U}$. Thus every self-inversive member of $\mathcal{G}_n$ is in $\mathcal{S}_n(\lambda)$. On the other hand, $Q_{n-1}(\lambda; z) \to (1 + z)^{n-1}$ as $\lambda \to 0$ and therefore by Hurwitz's theorem

$$\widetilde{\mathcal{G}}_n = \bigcap_{0 < \lambda < \frac{2\pi}{n}} \mathcal{S}_n(\lambda), \tag{7.151}$$

where $\widetilde{\mathcal{G}}_n$ denotes the set of self-inversive members of $\mathcal{G}_n$. Under these circumstances it is reasonable to define $\mathcal{S}_n(0) = \widetilde{\mathcal{G}}_n$. We can also formulate a definition of $\mathcal{S}_n(\frac{2\pi}{n})$ as follows. A calculation shows that

$$Q_{n-1}\left(\frac{2\pi}{n}; z\right) = e_{n-1}(z) = 1 + z + z^2 + \cdots + z^{n-1} \tag{7.152}$$

and so we say $p \in \mathcal{S}_n(\frac{2\pi}{n})$ if p is self-inversive of degree n with $p(0) = 1$ and

$$p(z) * e_{n-1}(z) = p(z) - z^n \neq 0 \quad (z \in \mathbb{U}). \tag{7.153}$$

7.6.3

Suffridge's [62] main theorem now takes the following form.

Suffridge's theorem (a) $\mathcal{S}_n(\lambda_1) \subset \mathcal{S}_n(\lambda_2)$ *for* $0 \leq \lambda_1 < \lambda_2 \leq \frac{2\pi}{n}$.

(b) *If* p_1 *and* $p_2 \in \mathcal{S}_n(\lambda)$, *where* $0 \leq \lambda \leq \frac{2\pi}{n}$, *then* $p_1 * p_2 \in \mathcal{S}_n(\lambda)$.

(c) $p \in \mathcal{S}_n(\frac{2\pi}{n})$ *iff* $\exists a_k \geq 0$ *with* $\sum_1^n a_k = 1$ *such that*

$$p(z) = \sum_{k=1}^{n} a_k e_n(\omega_k z), \tag{7.154}$$

where $\omega_k = e^{2\pi i k/n}$ *are the nth roots of unity.*

7.6.4

Lemma *Let* $S(z)$ *be a self-inversive polynomial of degree n with* $S(0)$ *real. Then* $\exists$ *real numbers* a_k *such that*

$$S(z) = \sum_{k=1}^{n} a_k e_n(\omega_k z). \tag{7.155}$$

Furthermore, we have

$$a_k = \frac{1}{n}(S(\overline{\omega_k}) - S(0)) \quad (1 \leq k \leq n). \tag{7.156}$$

Proof Let $S(z) = \sum_0^n c_j z^j$, so $c_0 = c_n$ is real and $c_{n-j} = \overline{c_j}$ $(0 \leq j \leq n)$. Consider the polynomial

$$P(z) = \frac{1}{n} \sum_{j=0}^{n-1} c_j \overline{\omega_j} z e_{n-1}(\overline{\omega_j} z). \tag{7.157}$$

Note that $P(0) = 0$ and $P(\omega_j) = c_j$ $(0 \leq j \leq n)$. Also

$$\overline{P(\overline{z})} = \frac{1}{n} \sum_{j=0}^{n-1} \overline{c_j} \omega_j z e_{n-1}(\omega_j z) = \frac{1}{n} \sum_{j=1}^{n} c_j \omega_{n-j} z e_{n-1}(\omega_{n-j} z) = P(z) \tag{7.158}$$

and so P is a real polynomial. Writing $P(z) = \sum_{k=1}^{n} a_k z^k$, the a_k are real and

$$S(z) = \sum_{j=0}^{n} P(\omega_j) z^j = \sum_{j=0}^{n} \sum_{k=1}^{n} a_k \omega_j^k z^j = \sum_{k=1}^{n} a_k \sum_{j=0}^{n} \omega_k^j z^j = \sum_{k=1}^{n} a_k e_n(\omega_k z), \tag{7.159}$$

as required. To find a_k we observe that

$$a_k z^k = z^k * P(z) = \frac{1}{n} \sum_{j=0}^{n-1} c_j (z^k * e_{n-1}(\overline{\omega_j} z)) = \frac{1}{n} \sum_{j=0}^{n-1} c_j \overline{\omega_j^k} z^k \tag{7.160}$$

which gives (7.156).

An alternative proof that the a_k are real is as follows. Since $S(0)$ is real, we need to show that the values $S(\overline{\omega_k})$ are real; in other words we need to show that a self-inversive polynomial S is real at the nth roots of unity. However, if S is self-inversive, then $e^{-in\theta/2} S(e^{i\theta})$ is real for θ real. When $\theta = 2\pi k/n$, $e^{-in\theta/2} = (-1)^k$ and hence the result.

7.6.5

We now prove part (c) of Suffridge's theorem. Firstly consider a self-inversive polynomial p of degree n such that $p(0) = 1$. Then we have from the previous result

$$p(z) = \sum_{k=1}^{n} a_k e_n(\omega_k z), \tag{7.161}$$

where the a_k are real and $\sum_1^n a_k = 1$. Now from the identity $e_n(z) = (1 - z^{n+1})/(1 - z)$ we easily obtain the relation

$$e_n(\omega_k z) = \frac{1 - z^n}{1 - \omega_k z} + z^n \tag{7.162}$$

and therefore we obtain

$$\frac{p(z) - z^n}{1 - z^n} = \sum_{k=1}^{n} \frac{a_k}{1 - \omega_k z}. \tag{7.163}$$

Suppose now that $a_k \geq 0\,(1 \leq k \leq n)$. Then we deduce that

$$\Re\left(\frac{p(z) - z^n}{1 - z^n}\right) > \frac{1}{2} \quad (z \in \mathbb{U}) \tag{7.164}$$

and, in particular, $p(z) - z^n \neq 0$ in $\mathbb{U}$. Thus $p \in \mathcal{S}_n(\frac{2\pi}{n})$. Conversely, if $p(z) - z^n \neq 0$ in $\mathbb{U}$, then because p is self-inversive we deduce that

$$\frac{p(z) - 1}{p(z) - z^n} = B(z) \tag{7.165}$$

is a Blaschke product satisfying $B(0) = 0$. We then obtain

$$\frac{p(z) - z^n}{1 - z^n} = \frac{1}{1 - B(z)} \tag{7.166}$$

which again implies (7.164). From (7.163) we obtain for $z \in \mathbb{U}$

$$\Re\left(\sum_{k=1}^{n} a_k \frac{1 + \omega_k z}{1 - \omega_k z}\right) = \Re\left(\frac{1 + B(z)}{1 - B(z)}\right) > 0 \tag{7.167}$$

and so

$$\sum_{k=1}^{n} \frac{a_k}{|1 - \omega_k z|^2} > 0 \quad (z \in \mathbb{U}). \tag{7.168}$$

We deduce that

$$a_k = \lim_{z \to \overline{\omega}_k} \sum_{j=1}^{n} |1 - \omega_k z|^2 \frac{a_j}{|1 - \omega_j z|^2} \geq 0. \tag{7.169}$$

This completes the proof.

Notice that a consequence of the above representation is that for a polynomial $p \in \mathcal{S}_n(\frac{2\pi}{n})$, the polynomials $p(z) * (1+xz)e_{n-1}(z)$ $(|x| = 1)$ have all their zeros on $\mathbb{T}$ interspersed between the nth roots of unity. Note also that p has all its zeros on $\mathbb{T}$.

Another consequence of this theorem is part (b) of Suffridge's theorem in the case $\lambda = \frac{2\pi}{n}$. To prove this let p and $q \in \mathcal{S}_n(\frac{2\pi}{n})$. Then we can write

$$p(z) = \sum_{k=1}^{n} a_k e_n(\omega_k z), \qquad q(z) = \sum_{k=1}^{n} b_k e_n(\omega_k z), \tag{7.170}$$

where $a_k \geq 0$, $b_k \geq 0$. We obtain

$$\begin{aligned}(p * q)(z) &= \sum_{j=1}^{n} \sum_{k=1}^{n} a_j b_k e_n(\omega_j z) * e_n(\omega_k z) = \sum_{j=1}^{n} \sum_{k=1}^{n} a_j b_k e_n(\omega_{j+k} z) \\ &= \sum_{r=1}^{n} \left(\sum_{j+k \equiv r \,(\mathrm{mod}\ n)} a_j b_k \right) e_n(\omega_r z)\end{aligned} \tag{7.171}$$

which has the required form. Hence $p * q \in \mathcal{S}_n(\frac{2\pi}{n})$.

7.6.6

As the proofs of parts (a) and (b) of Suffridge's theorem are long and technically difficult, we shall not give his arguments here, but refer the reader to his original paper Suffridge [62]. However, we point out the following argument, which could provide a much shortened proof of part (a). Suppose that $p \in \mathcal{S}_n(\lambda)$, so that $p * Q_{n-1}(\lambda) \neq 0$ in $\mathbb{U}$. Then we can write

$$p * Q_{n-1}(\lambda) = q * (1+z)^{n-1}, \tag{7.172}$$

where $q \in \mathcal{G}_{n-1}$, the Grace class. Then

$$p * Q_{n-1}(\mu) = q * Q_{n-1}^{(i)}(\lambda) * Q_{n-1}(\mu) * (1+z)^{n-1}, \tag{7.173}$$

where $Q_{n-1}^{(i)}(\lambda)$ denotes the convolution inverse of $Q_{n-1}(\lambda)$. By Grace's theorem the expression is $\neq 0$ in $\mathbb{U}$ provided that

$$Q_{n-1}^{(i)}(\lambda) * Q_{n-1}(\mu) * (1+z)^{n-1} \neq 0. \tag{7.174}$$

Note that this is a self-inversive polynomial of degree $n-1$. To prove part (a) it is sufficient to prove the above inequality for $0 \leq \lambda < \mu \leq \frac{2\pi}{n}$.

7.6.7

An important sufficient condition for the self-inversive polynomial P to have its zeros on $\mathbb{T}$ separated by an angle λ is obtained as follows. If P is a self-inversive polynomial of degree n, then we can write P in the form

$$P(z) = S(z) + z^n \overline{S\left(\frac{1}{\bar{z}}\right)} \tag{7.175}$$

where S is a polynomial of degree $\frac{1}{2}n$ when n is even and degree $\frac{1}{2}n - \frac{1}{2}$ when n is odd. Suppose now that $S \in \Pi_{-\gamma}$ so that

$$\Re\left(\frac{zS'(z)}{S(z)}\right) > -\frac{\gamma}{2} \quad (z \in \mathbb{U}). \tag{7.176}$$

Then $S \neq 0$ in $\mathbb{U}$ and so

$$B(z) = \frac{z^n \overline{S\left(\frac{1}{\bar{z}}\right)}}{S(z)} \tag{7.177}$$

satisfies $|B(z)| < 1$ in $\mathbb{U}$. We have

$$\Re\left(\frac{P(z)}{S(z)}\right) = \Re(1 + B(z)) > 0 \tag{7.178}$$

and so $P = S(1+B)$ factorises into the product of a function in $K(0,\gamma)$ and a function in $K(1,1)$, so is a function in $K(1,\gamma+1)$. It follows from theorem 7.4.7 that P has its zeros on $\mathbb{T}$ separated by an angle of at least $\lambda = \frac{2\pi}{n+\gamma}$.

7.6.8

The above argument leads to an important approximation result.

Theorem *Let $f \in \Pi_{-\gamma}$, where $\gamma > 0$. There exists a sequence of polynomials $P_n \in \mathcal{P}_n(2\pi/(n+\gamma))$, which converges locally uniformly in $\mathbb{U}$ to f.*

Proof Since $\Re((zf'(z))/f(z)) > -\gamma/2$ for $z \in \mathbb{U}$, we have the inequality

$$\Re\frac{zf'(z)}{f(z)} + \frac{\gamma}{2} \geq \frac{\gamma}{2}\frac{1-|z|}{1+|z|}, \tag{7.179}$$

and therefore, for a given R ($0 < R < 1$), the Nth partial sum $f_N(Rz)$ of $f(Rz)$ is a polynomial of degree N in $\Pi_{-\gamma}$ for sufficiently large N. Picking such an N we form the polynomial $P_{2N}(z) = f_N(Rz) + z^{2N}\overline{f_N(R/\overline{z})}$, which, by the above, is a polynomial in $\mathcal{P}_{2N}(2\pi/(2N+\gamma))$. We approximate f by such a polynomial by choosing R close to 1 and N sufficiently large.

As an important case we show that

$$Q_n\left(\frac{2\pi}{n+\gamma}; z\right) \to \frac{1}{(1-z)^\gamma} \text{ as } n \to \infty, \tag{7.180}$$

the convergence being locally uniform in $\mathbb{U}$. From the formula (7.141) for the coefficients of Q_n we have

$$C_k^{(n)}\left(\frac{2\pi}{n+\gamma}\right) = \prod_{j=1}^{k} \frac{\sin\left(\dfrac{n-j+1}{n+\gamma}\right)\pi}{\sin\left(\dfrac{j}{n+\gamma}\right)\pi} = \prod_{j=1}^{k} \frac{\sin\left(\dfrac{\gamma+j-1}{n+\gamma}\right)\pi}{\sin\left(\dfrac{j}{n+\gamma}\right)\pi}.$$

$$\to \prod_{j=1}^{k} \frac{\gamma+j-1}{j} \text{ as } n \to \infty, \tag{7.181}$$

and thus the kth coefficient of $Q_n(\frac{2\pi}{n+\gamma}; z)$ converges to the kth coefficient of $(1-z)^{-\gamma}$ for each $k = 0, 1, \ldots$. The desired result will follow, if we show that the sequence $\{Q_n(\frac{2\pi}{n+\gamma}; z)\}$ is a normal family in $\mathbb{U}$, i.e. is locally bounded in $\mathbb{U}$. This follows easily from the above formula for $C_k^{(n)}(\frac{2\pi}{n+\gamma})$; for if we make use of the fact that the function $\sin x/x$ is decreasing on $[0, \pi]$, we find that for $1 \leq k \leq n$

$$0 < C_k^{(n)}\left(\frac{2\pi}{n+\gamma}\right) \leq \begin{cases} \displaystyle\prod_{j=1}^{k} \frac{\gamma+j-1}{j} & \text{if } \gamma \geq 1, \\ 1 \quad \text{if } 0 \leq \gamma < 1. \end{cases} \tag{7.182}$$

Thus for $|z| \leq r < 1$ we obtain

$$\left| Q_n\left(\frac{2\pi}{n+\gamma}; z\right)\right| \leq \frac{1}{(1-r)^{\gamma'}} \quad (n = 1, 2, \ldots), \tag{7.183}$$

where $\gamma' = \max(1, \gamma)$.

7.6.9

The convolution characterisation of $\mathcal{S}_n(\lambda)$ leads to the following result.

Theorem *Let p be a self-inversive polynomial of degree n satisfying* $p(0) = 1$. *Then if* $p \in \mathcal{S}_n(\lambda)$, *where* $0 \leq \lambda \leq \frac{2\pi}{n}$, *we have*

$$\Re\left(\frac{p(z) * Q_{n-1}(\lambda; z)}{p(z) * Q_n(\lambda; ze^{i\lambda/2})}\right) > \frac{1}{2} \quad (z \in \mathbb{U}). \tag{7.184}$$

Proof We note first that

$$Q_{n-1}(\lambda; z) = \frac{Q_n(\lambda; ze^{i\lambda/2})}{1 + ze^{in\lambda/2}}. \tag{7.185}$$

Let $u = e^{i\lambda/2}$, $y = e^{in\lambda/2}$. Then by theorem 7.6.2 $p \in \mathcal{S}_n(\lambda)$ iff

$$p(z) * \frac{1+xz}{1+yz} Q_n(uz) \neq 0 \quad (z \in \mathbb{U}, |x| \leq 1). \tag{7.186}$$

This can be rewritten in the form

$$p(z) * \left(\frac{1 - x\overline{y}}{1+yz} Q_n(uz) + x\overline{y} Q_n(uz)\right) \neq 0 \quad (z \in \mathbb{U}) \tag{7.187}$$

and so

$$\frac{p(z) * Q_{n-1}(z)}{p(z) * Q_n(uz)} \neq -\frac{x\overline{y}}{1 - x\overline{y}}. \tag{7.188}$$

As x varies over $\overline{\mathbb{U}}$ we deduce the required inequality.

7.6.10

In fact, if we refer to the relation (7.135) we can infer this result and more; for we have, writing $p(z) * Q_n(z) = P(z)$, that

$$p(z) * Q_{n-1}(z) = \frac{e^{int} P(ze^{-it}) - e^{-int} P(ze^{it})}{2i \sin nt} = \sum_{k=1}^{n} t_k \frac{P(ze^{-it})}{1 - ze^{-it} e^{-i\alpha_k}} \tag{7.189}$$

where t_k are non-negative numbers satisfying $\sum_{k=1}^{n} t_k = 1$ and $e^{i\alpha_k}$ are the zeros of $P(z)$. Now each

$$P_k(z) = \frac{P(ze^{-it})}{1 - ze^{-it}e^{-i\alpha_k}} \tag{7.190}$$

is a polynomial of degree $n-1$ whose zeros on $\mathbb{T}$ are separated by angles of at least λ. It follows that $P_k(z)$ has the form $(p_k * Q_{n-1})(ze^{i\lambda_k})$, where $p_k \in \mathcal{S}_{n-1}(\lambda)$. We therefore obtain the representation

$$p(z) = \sum_{k=1}^{n} t_k p_k(ze^{i\lambda_k}) + z^n \tag{7.191}$$

where the numbers λ_k satisfy the relation

$$e^{i(n-1)\lambda_k} = -e^{-i\alpha_k} e^{i(n-1)t}. \tag{7.192}$$

It follows by induction that the coefficients of p are bounded by 1. This is trivial when $n = 1$ and the induction hypothesis means that we may assume it true up to $n-1$ for $n \geq 2$. The above representation then gives the result for n.

7.6.11

Theorem *Let* $p(z) = 1 + \sum_{k=1}^{n} c_k z^k \in \mathcal{S}_n(\lambda)$, *where* $0 \leq \lambda \leq \frac{2\pi}{n}$. *Then*

$$|c_k| \leq 1 \quad (1 \leq k \leq n). \tag{7.193}$$

If $P(z) = 1 + \sum_{k=1}^{n} a_k z^k \in \mathcal{P}_n(\lambda)$, *then writing* $\gamma = \frac{2\pi}{\lambda} - n$, *we have*

$$|a_k| \leq C_k^{(n)}(\lambda) \leq \begin{cases} \prod_{j=1}^{k} \frac{\gamma + j - 1}{j} & \text{if } \gamma \geq 1, \\ 1 \quad \text{if } 0 \leq \gamma < 1. \end{cases} \tag{7.194}$$

In particular, for $|z| \leq r < 1$ *we have*

$$|P(z)| \leq \frac{1}{(1-r)^{\gamma'}}, \tag{7.195}$$

where $\gamma' = \max(1, \gamma)$.

7.6.12 Limiting versions of Suffridge's convolution theorem

Suffridge showed that the convolution theorem 7.6.3 gives a new proof of the conjecture of Pólya and Schoenberg [40] that the class of 1–1 conformal mappings of $\mathbb{U}$ onto convex domains is a class closed under convolution. The study of this conjecture formed a substantial part of their paper on de la Vallée

Poussin means; the proof of the main theorem of their paper was given in chapter 4 above. The proof of the conjecture was given by Ruscheweyh and Sheil-Small [49] and an account of this theory and its generalisations will be given in chapter 8 below. However, Suffridge's unique polynomial approach to the problem produced the first major generalisation of the original result.

To explain this we need firstly to state the conjecture in an analytic form, which ties in with our present notations. Consider a smooth convex curve given by a parametric representation $w = w(t)$; as the curve is positively traversed, the tangent vector is increasing; thus $\arg dw = \arg w'(t)$ is an increasing function of t; i.e.

$$\Im \frac{w''(t)}{w'(t)} \geq 0. \tag{7.196}$$

If $w = f(e^{it})$ represents the boundary values of a conformal mapping of $\mathbb{U}$ onto the interior of the curve, we obtain the condition

$$\Re\left(1 + \frac{zf''(z)}{f'(z)}\right) \geq 0 \quad (z \in \mathbb{T}). \tag{7.197}$$

The maximum principle implies that this inequality will hold throughout $\mathbb{U}$ and leads to *the necessary and sufficient condition for an analytic function $f(z)$ to be a 1–1 conformal mapping of $\mathbb{U}$ onto a convex domain* as

$$\Re\left(1 + \frac{zf''(z)}{f'(z)}\right) > 0 \quad (z \in \mathbb{U}). \tag{7.198}$$

In our notations one can immediately check that this condition is equivalent to the condition that $f' \in \Pi_{-2}$. In addition, as we show in chapter 8, subsection 8.2.8, the condition can be expressed in the duality form

$$\frac{f(z) - f(0)}{z} * \frac{1 + xz}{(1 - z)^3} \neq 0 \tag{7.199}$$

for $|x| \leq 1$ and $z \in \mathbb{U}$. Thus the conjecture of Pólya and Schoenberg can be expressed in the form: the dual of the class of functions $\{\frac{1+xz}{(1-z)^3}\colon |x| \leq 1\}$ is closed under convolution.

Suffridge proves that *for each $\gamma > 0$, the dual of the class of functions $\{\frac{1+xz}{(1-z)^{\gamma+1}}\colon |x| \leq 1\}$ is closed under convolution; furthermore, as γ increases these classes decrease.*

For suppose that ϕ and ψ are two such functions; then each of the two functions $f = \phi * (1-z)^{-\gamma}$ and $g = \psi * (1-z)^{-\gamma}$ lies in the class $\Pi_{-\gamma}$. By the approximation result we can write

$$f = \lim_{n\to\infty} P_n \quad \text{and} \quad g = \lim_{n\to\infty} R_n, \tag{7.200}$$

where P_n and R_n are polynomials in $\mathcal{P}_n\left(\frac{2\pi}{n+\gamma}\right)$. We can write $P_n = p_n * Q_n$ and $R_n = q_n * Q_n$ where p_n and q_n lie in $\mathcal{S}_n\left(\frac{2\pi}{n+\gamma}\right)$ and Q_n is the Suffridge extremal polynomial for $\lambda = 2\pi/(n+\gamma)$. We consider the sequence $\{r_n = p_n * q_n\}$; each $r_n \in \mathcal{S}_n(\frac{2\pi}{n+\gamma})$. Furthermore $r_n \to \phi * \psi$ as $n \to \infty$; for $P_n \to f$ and $Q_n \to (1-z)^{-\gamma}$, so $p_n \to \phi$; similarly $q_n \to \psi$. Finally $r_n * (1 + xz)Q_{n-1} \neq 0$ in $\mathbb{U}$ and $Q_{n-1} \to (1-z)^{-\gamma-1}$. Thus by Hurwitz's theorem $\phi * \psi * (1 + xz)(1-z)^{-\gamma-1} \neq 0$ in $\mathbb{U}$, and the dual class is therefore closed under convolution. The containment result is proved similarly.

8

Duality and an extension of Grace's theorem to rational functions

8.1 Linear operators and rational functions

8.1.1

We saw in chapter 5 that Grace's theorem is equivalent to the assertion that the polynomials of degree at most n, which have no zeros in the open unit disc $\mathbb{U}$, lie in the second dual of the singleton function $(1+z)^n$. In other words, if $\phi(z)$ is a function satisfying $\phi(z) * (1+z)^n \neq 0$ in $\mathbb{U}$, then $\phi(z) * p(z) \neq 0$ in $\mathbb{U}$ for every polynomial p of degree at most n satisfying $p(z) \neq 0$ in $\mathbb{U}$. As we saw, this result has a number of interesting interpretations as well as many applications in the theory. In this chapter we will develop a similar result for rational functions, whose zeros and poles lie outside $\mathbb{U}$.

The result can be very simply stated as in the next subsection.

8.1.2

Theorem *Let $R(z) = p(z)/q(z)$ be a rational function with no zeros or poles in $\mathbb{U}$, and suppose that p is a polynomial of degree at most m and q is a polynomial of degree at most n, where $m \geq 1$ and $n \geq 0$. Then $R(z)$ belongs to the second dual of the class of functions*

$$\left\{ \frac{(1+xz)^m}{(1-z)^n} : |x| = 1 \right\}. \tag{8.1}$$

In other words, if $\phi(z)$ is a function analytic in $\mathbb{U}$ and satisfying

$$\phi(z) * \frac{(1+xz)^m}{(1-z)^n} \neq 0 \quad (|x| = 1, |z| < 1), \tag{8.2}$$

then $\phi(z) * R(z) \neq 0$ for $|z| < 1$. The case $n = 0$ is the Grace–Szegö theorem. Except for the trivial case $m = n = 0$, the theorem is false in the case $m = 0$. For example, we leave it as a problem to prove that, *if* $n \geq 2$, *then* $\phi(z) * (1-z)^{-n}$ $\neq 0$ *in* $\mathbb{U}$ *does not imply* $\phi(z) \neq 0$ *in* $\mathbb{U}$. *In other words, for a suitably chosen* ϕ *satisfying the condition,* $\phi(z) * (1-z)^{-1}$ *has a zero in* $\mathbb{U}$.

8.1.3

There are some immediate consequences of the theorem. For example we have

Corollary *Let* Λ *denote a linear operator on* $\mathcal{A}$ *satisfying the condition*

$$\Lambda\left(\frac{(1+xz)^m}{(1-yz)^n}\right) \neq 0 \tag{8.3}$$

for $|x| \leq 1$, $|y| \leq 1$ *and* $|z| < 1$, *where* $m \geq 1$ *and* $n \geq 0$ *are integers. Let* $R(z) = p(z)/q(z)$, *where p is a polynomial non-vanishing in* $\mathbb{U}$ *and of degree at most m and q is a polynomial non-vanishing in* $\mathbb{U}$ *and of degree at most n. Then*

$$(\Lambda R)(z) \neq 0 \quad (z \in \mathbb{U}). \tag{8.4}$$

To deduce the corollary from the theorem, we recall that, for any function f analytic in $\mathbb{U}$, $(\Lambda f)(z) = (H(z,\zeta) *_\zeta f(\zeta))_{\zeta=1}$ where $H(z,\zeta)$ is the kernel of the operator defined by $H(z,\zeta) = \Lambda(\frac{1}{1-z\zeta})$. Thus the hypothesis of the corollary states that for each $z \in \mathbb{U}$

$$H(z,\zeta) * \frac{(1+x\zeta)^m}{(1-y\zeta)^n} \neq 0 \tag{8.5}$$

when $\zeta = 1$ and when $|x| \leq 1$ and $|y| \leq 1$. This condition holds, not only when $\zeta = 1$, but for all ζ satisfying $|\zeta| \leq 1$. This follows from the fact that the left term of the expression is just $\Lambda(\frac{(1+x\zeta z)^m}{(1-y\zeta z)^n})$. Now $\zeta \mapsto H(z,\zeta)$ is analytic for $|\zeta| < 1+\delta$, where $\delta > 0$, and therefore the left expression remains analytic for $|\zeta| < 1+\delta$, $|x| \leq 1$, $|y| \leq 1$; the expression therefore has no zeros for $|\zeta| < \rho$, where $\rho > 1$. Hence

$$H(z,\rho\zeta) * \frac{(1+x\zeta)^m}{(1-y\zeta)^n} \neq 0 \tag{8.6}$$

for $\zeta \in \mathbb{U}$. From the theorem we obtain $H(z,\rho\zeta) * R(\zeta) \neq 0$ for $\zeta \in \mathbb{U}$, and putting $\zeta = 1/\rho$ we obtain $\Lambda R \neq 0$, as required.

8.1.4

From the above result we deduce

Corollary *Let* Λ *denote a continuous linear functional on* $\mathcal{A}$ *and let* $R(z) = p(z)/q(z)$ *be a rational with no zeros or poles in* $\mathbb{U}$, *where p has degree at most*

$m \geq 1$ and q has degree at most n. If $R(0) = 1$, then $\exists x, y$ satisfying $|x| \leq 1$, $|y| \leq 1$ such that

$$\Lambda R = \Lambda\left(\frac{(1+xz)^m}{(1-yz)^n}\right). \tag{8.7}$$

For, if $\Lambda R = w$, then the functional $\Lambda_1 R = 0$, where Λ_1 is defined by $\Lambda_1 f = \Lambda f - wf(0)$. By the previous corollary applied to Λ_1, this gives $\Lambda_1((1+xz)^m(1-yz)^{-n}) = 0$ for suitable x, y in the given ranges. In other words $\Lambda((1+xz)^m(1-yz)^{-n}) = w$, as required.

8.1.5

Another interesting conclusion to be deduced from the theorem is the following.

Theorem *Let $R(z) = p(z)/q(z)$ be a rational function with no poles in $\mathbb{U}$ and let*

$$m = \deg p, \qquad n = \deg q, \qquad N = \max(m, n). \tag{8.8}$$

Suppose that $\phi(z)$ is a function analytic in $\mathbb{U}$ satisfying $\phi(0) = 1$ and

$$\phi(z) * \frac{(1+xz)^N}{(1-z)^n} \neq 0 \tag{8.9}$$

for $z \in \mathbb{U}$ and $|x| = 1$. Then

$$(\phi * R)(\mathbb{U}) \subset R(\mathbb{U}). \tag{8.10}$$

Proof If $R(z) \neq w$ in $\mathbb{U}$, then $S = (p - wq)/q$ is a rational function with no zeros or poles in $\mathbb{U}$, whose numerator has degree at most N and whose denominator has degree at most n. Either $N = n = 0$ or $N \geq 1$. Therefore we can apply the previous theorem to S to deduce that $(\phi * S)(z) \neq 0$ for $z \in \mathbb{U}$. Thus $(\phi * R)(z) \neq w$ in $\mathbb{U}$, and the set containment result follows.

8.1.6

The proof of our extension of Grace's theorem to rational functions will, as for the proof of Grace's theorem itself, make use of the finite product lemma 5.1.3 of chapter 5. Indeed, this lemma will play a key role throughout this chapter. However, the lemma itself is vacuous, unless we can start the ball rolling with an initial duality result. Recall that Grace's theorem was proved by first showing that the function $1 + z$ lies in the second dual of each of the functions

$(1 + z)^n$, and this result requires an application of the fundamental theorem of algebra.

8.1.7

The starting point which we require here is the following result.

Lemma *Let $\beta \geq 1$ be a real number and suppose that $\phi(z)$ is analytic in $\mathbb{U}$ and satisfies in $\mathbb{U}$*

$$\phi(z) * \frac{1 + xz}{(1 - z)^\beta} \neq 0 \quad (z \in \mathbb{U},\ |x| = 1). \tag{8.11}$$

Then

$$\phi(z) * \frac{1 + xz}{(1 - z)^{\beta - 1}} \neq 0 \quad (z \in \mathbb{U},\ |x| = 1). \tag{8.12}$$

In other words, the class of functions $\{\frac{1+xz}{(1-z)^{\beta-1}} : |x| = 1\}$ is contained within the second dual of the corresponding class of functions obtained by replacing $\beta - 1$ by β. It will be observed that this lemma is a special case of the theorem, when β is a natural number, and is also of a type required in order to apply the finite product lemma. The proof will employ an interesting application of the maximum modulus principle, which was first achieved by J. Clunie and has since had many other applications. The method has become known as the Clunie–Jack lemma, because of its initial appearance in a paper of I.S. Jack [21].

8.1.8

Clunie–Jack lemma *Let $\omega(z)$ be analytic for $|z| \leq r$ and suppose that ω has a zero of order m at the origin (if $\omega(0) \neq 0$,* we take $m = 0$). *Then $\exists \zeta$ satisfying $|\zeta| = r$ such that*

$$|\omega(z)| \leq |\omega(\zeta)| \text{ for } |z| \leq r \text{ and } \zeta\omega'(\zeta) = t\omega(\zeta), \tag{8.13}$$

where t is a real number satisfying $t \geq m$.

In polar coordinates we have $\omega(r, \theta) = R(r, \theta)e^{i\phi(r,\theta)}$, and from the Cauchy–Riemann equations we obtain

$$\Re\frac{z\omega'(z)}{\omega(z)} = \frac{\partial\phi}{\partial\theta} = \frac{r\partial R}{R\partial r}, \qquad \Im\frac{z\omega'(z)}{\omega(z)} = r\frac{\partial\phi}{\partial r} = -\frac{1}{R}\frac{\partial R}{\partial\theta}. \tag{8.14}$$

Therefore, choosing ζ as a point on $|z| = r$ at which $|\omega|$ attains a maximum value on the circle, the first inequality of the Clunie–Jack lemma follows from the maximum principle. Furthermore, at ζ we have $\partial R/\partial\theta = 0$ and so $\zeta\omega'(\zeta)/\omega(\zeta)$ is real. It is therefore equal to $(r/R)\partial R/\partial r$. However, it is geometrically clear that at a point of maximum modulus on the circle, $|\omega| = R$ is increasing with the radius r. Thus $\zeta\omega'(\zeta)/\omega(\zeta) \geq 0$. This is true without any assumptions about ω at the origin, and proves the Clunie–Jack lemma in the case $m = 0$. However, if ω has a zero of order $m \geq 1$ at 0, then $\sigma(z) = \omega(z)/z^m$ is analytic for $|z| \leq r$ and furthermore ζ is a point of maximum modulus for σ. Therefore we may apply the above result to σ, and this gives us the desired conclusion.

8.1.9 Proof of lemma 8.1.7

According to lemma 5.8.2 of chapter 5 the condition (8.11) will hold for $|x| \leq 1$, and in particular when $x = 0$. Hence $\phi(z) * (1 - z)^{-\beta} \neq 0$ for $z \in \mathbb{U}$. Now in general, a convolution condition of the form

$$\phi(z) * (1 + xz)F(z) \neq 0 \quad (|x| \leq 1, z \in \mathbb{U}) \tag{8.15}$$

is equivalent to

$$|\phi(z) * zF(z)| < |\phi(z) * F(z)| \quad (z \in \mathbb{U}). \tag{8.16}$$

Thus, if we set

$$\omega(z) = \frac{\phi(z) * z(1 - z)^{-\beta}}{\phi(z) * (1 - z)^{-\beta}}, \tag{8.17}$$

then $\omega(z)$ is analytic in $\mathbb{U}$, $\omega(0) = 0$ and $|\omega(z)| < 1$ in $\mathbb{U}$. This then is our hypothesis. The conclusion, which we wish to draw, is that the function

$$\tau(z) = \frac{\phi(z) * z(1 - z)^{1-\beta}}{\phi(z) * (1 - z)^{1-\beta}} \tag{8.18}$$

also satisfies the same conditions. Clearly $\tau(0) = 0$ and τ is meromorphic in $\mathbb{U}$, since the denominator is analytic and not identically zero (it is $=1$ at $z = 0$). Thus we are required to show that $|\tau(z)| < 1$ for $z \in \mathbb{U}$. If this is not the case, then, since $\tau(0) = 0$, there is a first circle when $|\tau|$ takes the value 1. Thus by the Clunie–Jack lemma $\exists\zeta \in \mathbb{U}$ such that $|\tau(\zeta)| = 1$ and $\zeta\tau'(\zeta) = t\tau(\zeta)$, where $t \geq 1$.

To simplify the calculations we define $F_\gamma = (1 - z)^{-\gamma}$ and note the following easily proved algebraic relations.

$$zF_\beta = F_\beta - F_{\beta-1}; \qquad z^2F_\beta = F_\beta - (1 + z)F_{\beta-1}; \qquad F'_{\beta-1} = (\beta - 1)F_\beta. \tag{8.19}$$

Furthermore we have

$$\omega = \frac{\phi * zF_\beta}{\phi * F_\beta} = 1 - \frac{\phi * F_{\beta-1}}{\phi * F_\beta}, \qquad \tau = \frac{\phi * zF_{\beta-1}}{\phi * F_{\beta-1}}. \tag{8.20}$$

We therefore obtain

$$\begin{aligned}
\frac{z\tau'(z)}{\tau(z)} &= \frac{\phi * (zF_{\beta-1} + z^2F'_{\beta-1})}{\phi * zF_{\beta-1}} - \frac{\phi * zF'_{\beta-1}}{\phi * F_{\beta-1}} \\
&= 1 + (\beta - 1)\left(\frac{\phi * (F_\beta - (1+z)F_{\beta-1})}{\phi * zF_{\beta-1}} - \frac{\phi * (F_\beta - F_{\beta-1})}{\phi * F_{\beta-1}}\right) \\
&= 1 + (\beta - 1)\left(\frac{\phi * (F_\beta - (1+z)F_{\beta-1})}{(\phi * F_{\beta-1})\tau} - \frac{\phi * (F_\beta - F_{\beta-1})}{\phi * F_{\beta-1}}\right) \\
&= 1 + (\beta - 1)\left(\frac{1}{\tau(1-\omega)} - \frac{1}{\tau} - \frac{1}{1-\omega}\right) \\
&= 1 + (\beta - 1)\frac{\omega - \tau}{\tau(1-\omega)}.
\end{aligned} \tag{8.21}$$

At $z = \zeta$ we have $|\tau(\zeta)| = 1$ and therefore

$$t = \frac{\zeta\tau'(\zeta)}{\tau(\zeta)} = 1 - (\beta - 1)\frac{1 - \overline{\tau}\omega}{1 - \omega} \geq 1.$$

In the case that $\beta > 1$, this implies that $(1 - \overline{\tau}\omega)/(1 - \omega)$ has a non-positive real value. However, as $|\omega| < 1$, both $1 - \overline{\tau}\omega$ and $1 - \omega$ have positive real part, and therefore their ratio can be neither zero nor negative. This contradiction proves the lemma for $\beta > 1$. If $\beta = 1$, then the hypothesis is equivalent to

$$\phi(z) * \frac{1 + xz}{1 - yz} \neq 0 \quad (z \in \mathbb{U}, |x| = |y| = 1). \tag{8.22}$$

Fixing x and applying lemma 5.8.2 of chapter 5 to the function of y and z, the convolution condition will hold for $|y| \leq 1$, and in particular for $y = 0$; i.e. $\phi(z) * 1 + xz \neq 0$, which is the required result.

8.1.10

We have given this last argument because it is the natural one in the context. However, the result itself is equivalent to the statement that, if $\phi(0) = 1$ and $\Re\phi(z) > \frac{1}{2}$ in $\mathbb{U}$, then $|\phi'(0)| \leq 1$, which is a well-known inequality. A general result of a type which we will require frequently is the following.

Lemma *Let $F(z_1, z_2, \ldots, z_n)$ be a function analytic in the polydisc $|z_k| < 1$ $(1 \le k \le n)$ and suppose that $\phi(z)$ is analytic in $\mathbb{U}$ and satisfies*

$$\phi(z) * F(x_1 z, x_2 z, \ldots, x_n z) \neq 0 \quad (|x_k| = 1,\ 1 \le k \le n,\ z \in \mathbb{U}). \tag{8.23}$$

Then this inequality continues to hold for all x_k with $|x_k| \le 1$ $(1 \le k \le n)$ and $z \in \mathbb{U}$.

For $n = 2$ this is a consequence of lemma 5.8.2 of chapter 5. The general case follows easily by induction.

8.1.11

We turn now to the proof of theorem 8.1.2. To simplify the presentation and for later use, we make the following definition.

Definition For non-negative α and β we define $T(\alpha, \beta)$ as the dual of the class of functions

$$\left\{ \frac{(1 + xz)^\alpha}{(1 - z)^\beta} : |x| = 1 \right\} \tag{8.24}$$

normalised by the condition $\phi(0) = 1$. In other words, $\phi \in T(\alpha, \beta)$ iff $\phi(0) = 1$ and

$$\phi(z) * \frac{(1 + xz)^\alpha}{(1 - z)^\beta} \neq 0 \quad (|x| = 1, z \in \mathbb{U}). \tag{8.25}$$

With this terminology lemma 8.1.7 states that *the class $T(1, \beta) \subset T(1, \beta - 1)$ for $\beta \ge 1$.*

One of the consequences of the methods developed in this chapter is a general containment property for the $T(\alpha, \beta)$ classes.

8.1.12 Proof of theorem 8.1.2

Let $R = p/q$ satisfy the hypotheses of the theorem. We show first that, if $\phi \in T(1, n)$, then

$$\phi(z) * \frac{1 + xz}{q(z)} \neq 0 \quad (|x| = 1, z \in \mathbb{U}). \tag{8.26}$$

This is trivial in the case $n = 1$, and therefore we may assume that $n \ge 2$ and that for $\phi \in T(1, n - 1)$

$$\phi(z) * \frac{1 + xz}{q_0(z)} \neq 0 \quad (|x| = 1, z \in \mathbb{U}) \tag{8.27}$$

for any polynomial q_0 of degree at most $n-1$ satisfying $q_0(z) \neq 0$ in $\mathbb{U}$. Applying the above containment lemma, this inequality will therefore hold for $\phi \in T(1,n)$. Thus we have

$$\phi(z) * \frac{1+xz}{(1-yz)^n} \neq 0 \quad \Rightarrow \quad \phi(z) * \frac{1+xz}{\prod_1^{n-1}(1-y_k z)} \neq 0, \tag{8.28}$$

for values of the x, y, y_k parameters in the closed unit disc. This implication is of precisely the type required for the application of the finite product lemma. As we intend to make considerable use of this lemma, we remark that the method states that, given such an implication, we get a new implication obtained by adding in one extra term whose effect is to restore the original **order** of the product; the order of such a product is the net sum of the powers of the terms. The above implication has a product of order $1-n$ on the left and a product of order $2-n$ on the right. We can therefore add in one extra term of order -1. This gives

$$\phi(z) * \frac{1+xz}{\prod_1^{n}(1-y_k z)} \neq 0, \tag{8.29}$$

and since by the fundamental theorem of algebra $q(z) = c\prod_1^n(1-y_k z)$ for a non-zero constant c and suitable y_k, we obtain (8.26) by induction.

Suppose now that $\phi \in T(m,n)$, where $m \geq 1$, and consider the linear operator Λ defined by $\Lambda f = \phi(z) * (f(z)/(1-z)^n)$. Then we have $\Lambda(1+xz)^m \neq 0$ for $|x| \leq 1$ and $z \in \mathbb{U}$. Applying lemma 5.8.2 of chapter 5, we obtain $\Lambda P(z) \neq 0$ in $\mathbb{U}$ for any polynomial P of degree at most m possessing no zeros in $\mathbb{U}$. In particular $\Lambda(1+xz) \neq 0$; in other words $\phi \in T(1,n)$. From the first part we obtain $\phi(z) * (1+xz)/q(z) \neq 0$ for any q with no zeros in $\mathbb{U}$ and of degree at most n. Therefore by the finite product lemma we obtain $\phi(z) * (1+xz)(1+yz)^{m-1}/q(z) \neq 0$, and when $x = y$ we get $\Lambda_1(1+xz)^m \neq 0$, where $\Lambda_1 f = \phi(z) * (f(z)/q(z))$. Again by theorem 5.7.2 of chapter 5 we obtain $\Lambda_1 p(z) \neq 0$. This is $\phi(z) * R(z) \neq 0$, the desired result.

8.2 Interpretations of the convolution conditions

8.2.1

The above result and the arguments used are the product of a long process of generalisation and refinement of some classical results on conformal mappings and, as we shall see in later sections of this chapter, the methods can be taken a good deal further. For example, we will show that each of the dual classes $T(m, \beta)$, for $m = 1, 2, 3, \ldots$ and $\beta \geq 1$, is closed under convolution; i.e. the convolution of two functions in $T(m, \beta)$ is itself in $T(m, \beta)$. However, the significance of this and other results can only be appreciated when we have explained the nature of these classes in more familiar terms.

8.2.2 The class $T(1,1)$

Lemma *If $\phi(z)$ is analytic in $\mathbb{U}$ and satisfies $\phi(0) = 1$, then $\phi \in T(1, 1)$ if, and only if,*

$$\Re\phi(z) > \frac{1}{2} \quad (z \in \mathbb{U}). \tag{8.30}$$

Proof Since $\phi(0) = 1$, we have

$$\phi(z) * \frac{1 + xz}{1 - z} = \phi(z) + x(\phi(z) - 1), \tag{8.31}$$

and therefore $\phi \in T(1, 1)$ iff $|\phi(z) - 1| < |\phi(z)|$ for $z \in \mathbb{U}$, which is equivalent to (8.30).

8.2.3

Lemma *If $\phi \in T(1, 1)$ and f is analytic in $\mathbb{U}$ and satisfies $\Re f(z) > 0$ in $\mathbb{U}$, then*

$$\Re(\phi(z) * f(z)) > 0 \quad (z \in \mathbb{U}). \tag{8.32}$$

Proof As the convolution of two positive harmonic functions is positive, we have $(2\Re\phi(z) - 1) * \Re f(z) > 0$ and therefore

$$(\phi + \overline{\phi} - 1) * (f + \overline{f}) > 0. \tag{8.33}$$

However, $\phi * \overline{f} = \overline{f(0)}$ and $\overline{\phi} * f = f(0)$, and so this inequality reduces to

$$(\phi * f) + (\overline{\phi} * \overline{f}) > 0. \tag{8.34}$$

Since $\overline{\phi} * \overline{f} = \overline{\phi * f}$, we obtain the result.

8.2.4

Theorem *The dual of $T(1, 1)$ is the Kaplan class $K(1, 1)$.*

The Kaplan classes, whose definition and characterising properties were given in chapter 7, subsections (7.4.4) and (7.4.5), will play a highly important role in the convolution theory which follows. (See chapter 7, (7.103) for the characterisation of $K(1, 1)$).

Proof of theorem An $f(z)$ analytic in $\mathbb{U}$ belongs to $K(1, 1)$ iff, for a suitable real μ, $\Re(e^{i\mu} f(z)) > 0$ in $\mathbb{U}$. Hence from the above, if $\phi \in T(1, 1)$, then $\phi * f \neq 0$, and so f belongs to the dual of $T(1, 1)$. Conversely, suppose that f belongs to the dual of $T(1, 1)$. Let p and q be two values attained by $f(z)$ as z varies over $\mathbb{U}$; say $p = f(a)$ and $q = f(b)$. Choose r so that $|a| < r < 1$ and $|b| < r$ and put $\alpha = a/r, \beta = b/r$. Then $p = f(\alpha r)$ and $q = f(\beta r)$; therefore, for $0 \le t \le 1$ we have

$$(1-t)p + tq = \left(f(z) * \left((1-t)\frac{1}{1-\alpha z} + t\frac{1}{1-\beta z} \right) \right)_{z=r}. \tag{8.35}$$

However, the function $\phi(z) = (1-t)\frac{1}{1-\alpha z} + t\frac{1}{1-\beta z}$ satisfies $\phi(0) = 1$ and $\Re\phi(z) > \frac{1}{2}$ for $z \in \mathbb{U}$; i.e. $\phi \in T(1, 1)$. Thus by hypothesis $f(z) * \phi(z) \neq 0$ in $\mathbb{U}$ and so in particular $(1-t)p + tq \neq 0$. In other words, we have shown that 0 lies outside the convex hull of the set $f(\mathbb{U})$, and therefore this set lies in a half-plane which excludes the origin. This is equivalent to $f \in K(1, 1)$.

8.2.5

Theorem *If $\phi \in T(1, 1)$ and $f \in K(1, 1)$, then $\phi * f \in K(1, 1)$. If ϕ and $\psi \in T(1, 1)$, then $\phi * \psi \in T(1, 1)$.*

Both results clearly follow from 8.2.3. They indicate the structure preserving properties for convolution of the class $T(1, 1)$. Indeed for $\phi \in T(1, 1)$, the operator $\phi *$ is *convexity preserving* in the sense described in chapter 4, subsection 4.2.8.

8.2.6

A result for general operators, of which we shall make considerable use, is the following.

Lemma *Let Λ denote a linear operator on $\mathcal{A}$ satisfying the condition*

$$\Lambda\left(\frac{1+xz}{1-yz}\right) \neq 0 \tag{8.36}$$

for $|x| = 1$, $|y| = 1$, $z \in \mathbb{U}$. *Then, if* $F(z)$ *is analytic in* $\mathbb{U}$ *and satisfies* $\Re F(z) > 0$ *in* $\mathbb{U}$, *we have*

$$\Re \frac{(\Lambda F)(z)}{(\Lambda 1)(z)} > 0 \quad (z \in \mathbb{U}). \tag{8.37}$$

Proof The kernel of the operator is given by $H(z, y) = \Lambda(1/(1 - yz))$ and therefore, when $x = -y$ we obtain $H(z, 0) = \Lambda 1 \neq 0$. Also

$$H(z, y) + x\overline{y}(H(z, y) - H(z, 0)) \neq 0 \quad (z \in \mathbb{U}). \tag{8.38}$$

This implies

$$\Re \frac{H(z, y)}{H(z, 0)} \neq \frac{1}{2} \quad (|y| = 1, z \in \mathbb{U}). \tag{8.39}$$

Fixing $z \in \mathbb{U}$, the function $y \to H(z, y)/H(z, 0)$ is analytic for $|y| < 1 + \delta$, where $\delta > 0$, and takes the value 1 at $y = 0$. Therefore by continuity and the maximum principle, we have for small $\delta > 0$

$$\Re \frac{H(z, y)}{H(z, 0)} > \frac{1}{2} \quad (|y| < 1 + \delta). \tag{8.40}$$

As in 8.2.3 above we deduce that

$$\Re \frac{H(z, y) * F(y)}{H(z, 0)} > 0 \quad (|y| \leq 1), \tag{8.41}$$

which when $y = 1$ gives the desired result.

8.2.7 The classes $T(1, \beta)$

We are going to show in the next section that the classes $T(1, \beta)$ decrease as $\beta \geq 1$ increases. In particular they are all subclasses of $T(1, 1)$. Furthermore, if $\phi \in T(1, \beta)$, then the operator $\phi*$ preserves the structure of $K(1, \beta)$, i.e. $\phi * f \in K(1, \beta)$ for every $f \in K(1, \beta)\,(\beta \geq 1)$. Since the classes $K(1, \beta)$ increase with β, we see that the structural preservation properties under convolution of $T(1, \beta)$ increase with β.

Although these results are hinted at in our extension of Grace's theorem to rational functions, they are far from obvious and in fact lie much deeper. Furthermore such structure preserving properties make it highly desirable to express these classes in terms more traditional than that of a dual class.

8.2.8

Choose $\beta > 1$ and let $\phi \in T(1,\beta)$. We set

$$\psi(z) = \phi(z) * \frac{1}{(1-z)^{\beta-1}} \tag{8.42}$$

and observe that

$$z\psi'(z) = (\beta-1)\left(\phi(z) * \frac{z}{(1-z)^{\beta}}\right) = (\beta-1)\left(\phi(z) * \frac{1}{(1-z)^{\beta}} - \psi(z)\right). \tag{8.43}$$

Now $\phi \in T(1,\beta)$ iff

$$\left|\phi(z) * \frac{z}{(1-z)^{\beta}}\right| < \left|\phi(z) * \frac{1}{(1-z)^{\beta}}\right|, \tag{8.44}$$

for $z \in \mathbb{U}$, and this translates to

$$|z\psi'(z)| < |z\psi'(z) + (\beta-1)\psi(z)|. \tag{8.45}$$

This is equivalent to

$$\Re \frac{z\psi'(z)}{\psi(z)} > -\frac{\beta-1}{2} \quad (z \in \mathbb{U}), \tag{8.46}$$

which is the defining condition for the class $\Pi_{1-\beta} = K(0, \beta-1)$ (see chapter 7, (7.107). Thus for $\beta > 1, \phi \in T(1,\beta)$ iff $\psi \in \Pi_{1-\beta}$, where ψ is given by (8.42).

This gives a useful way of constructing members of $T(1,\beta)$: any product of the form

$$\psi(z) = \prod_{k=1}^{n} \frac{1}{(1-x_k z)^{\lambda_k}} \tag{8.47}$$

where $|x_k| \le 1$ and $\lambda_k > 0$ satisfy $\sum_{k=1}^{n} \lambda_k = \beta - 1$, is a function in $\Pi_{1-\beta}$; indeed such products are dense in $\Pi_{1-\beta}$; that is any function in $\Pi_{1-\beta}$ is a local uniform limit in $\mathbb{U}$ of a sequence of these products. The notation is designed to reflect this fact. We obtain a member of $T(1,\beta)$ by forming the convolution of ψ with the convolution inverse of $(1-z)^{1-\beta}$; that is, for $\beta > 1$,

$$\psi(z) * \left(1 + \sum_{n=1}^{\infty} \frac{n!}{(\beta-1)(\beta)\cdots(\beta+n-2)} z^n\right) \in T(1,\beta) \tag{8.48}$$

for $\psi \in \Pi_{1-\beta}$.

8.2.9 The cases $\beta = 2$ *and* $\beta = 3$

These cases are of particular interest. When $\beta = 2$, we obtain $T(1, 2) = \Pi_{-1}$; thus a function $\phi(z) \in T(1, 2)$ iff $\phi(0) = 1$ and

$$\Re \frac{z\phi'(z)}{\phi(z)} > -\frac{1}{2} \quad (z \in \mathbb{U}). \tag{8.49}$$

This is equivalent to asserting that $\phi(z) = g(z)/z$, where $g(z) = z + \cdots$ is **starlike of order** $\frac{1}{2}$ (i.e. satisfies $\Re(zg'(z)/g(z)) > \frac{1}{2}$).

For $\beta = 3$ we obtain the correspondence $\phi \in T(1, 3)$ iff $\phi(z) + z\phi'(z) \in \Pi_{-2}$; writing $g(z) = z\phi(z)$, this is equivalent to $g'(z) \in \Pi_{-2}$, which in turn is equivalent to the assertion that the function $zg'(z)$ is **starlike**; i.e. is a 1–1 mapping of $\mathbb{U}$ onto a domain starlike with respect to the origin. This last condition is equivalent to the assertion that $g(z)$ is **convex**; i.e. is a 1–1 mapping of $\mathbb{U}$ onto a convex domain. Thus $\phi \in T(1, 3)$ iff $z\phi(z)$ is a normalised convex mapping of $\mathbb{U}$. It is in this case that the origins of the theory of this chapter lie. In their celebrated paper on de la Vallée Poussin means, which we studied in chapter 4, Pólya and Schoenberg [40] conjectured (with considerable evidence) that *the convolution of two convex mappings remains convex*. This is equivalent to our assertion, still to be proved, that $T(1, 3)$ is closed under convolution. The first proof of this was given by Ruscheweyh and Sheil-Small [49] and the theory developed in several papers (Ruscheweyh [45, 44, 47], Sheil-Small [54]) over the succeeding years.

8.3 The duality theorem for $T(1, \beta)$

8.3.1

Theorem *If* $\phi \in T(1, \beta)$ *and* $f \in K(1, \beta)$, *where* $\beta \geq 1$, *then* $(\phi * f)(z) \neq 0$ *for* $z \in \mathbb{U}$.

To prove this we begin with the observation that by lemma 8.1.7, if $\phi \in T(1, \beta)$ then $\phi \in T(1, \beta - 1)$. Therefore, just as for the proof of theorem 8.1.2, we obtain using an induction argument together with the finite product lemma

$$\phi(z) * \frac{1 + xz}{(1 - yz)^{\gamma} \prod_{k=1}^{n} (1 + x_k z)} \neq 0 \quad (z \in \mathbb{U}) \tag{8.50}$$

for $|x| \leq 1, |x_k| \leq 1, |y| \leq 1$, where $n = [\beta] =$ largest integer $\leq \beta$ and $\gamma = \{\beta\} =$ fractional part of β; so $0 \leq \gamma < 1$ and $n + \gamma = \beta$.

We observe next that this can be written as

$$\phi(z) * \frac{1+xz}{1+x_1 z} G(z) \neq 0, \tag{8.51}$$

where in this case $G(z) = 1/(1-yz)^\gamma \prod_{k=2}^{n}(1+x_k z)$. The left side of this expression has the form $\Lambda(\frac{1+xz}{1+x_1 z})$, where Λ is a linear operator on $\mathcal{A}$; therefore applying lemma 8.2.6 we deduce that $\Lambda F \neq 0$ for any $F \in \mathcal{A}$ satisfying $\Re F(z) > 0$ in $\mathbb{U}$. Therefore we have

$$\phi(z) * \frac{F(z)}{(1-yz)^\gamma \prod_{k=2}^{n}(1+x_k z)} \neq 0 \quad (z \in \mathbb{U}), \tag{8.52}$$

for $|x_k| \leq 1$, $|y| \leq 1$ and any $F \in \mathcal{A}$ satisfying $\Re F(z) > 0$ in $\mathbb{U}$; indeed any $F \in K(1,1)$.

This result takes us one step further and the use of lemma 8.2.6 in this way will recur. Now we wish to prove theorem 8.3.1, which requires a general $f \in K(1,\beta)$ to replace the expression to the right of $\phi(z)*$. Notice that the given expression is in fact in $K(1,\beta)$. The general form of a function $f \in K(1,\beta)$ is given in theorem 7.4.6 of chapter 7; f has the factorisation $f = Fg$, where $F \in K(1,1)$ and $g \in \Pi_{1-\beta}$. Therefore by applying the technique of the previous paragraph, we can prove the theorem provided that we can first prove that

$$\phi(z) * \frac{1+xz}{1-yz} g(z) \neq 0 \quad (z \in \mathbb{U}), \tag{8.53}$$

for $|x| = |y| = 1$ and any $g \in \Pi_{1-\beta}$. On the face of it this looks rather a long way from what we have managed to prove in (8.52) above. Nevertheless, it turns out that the statement proved and the statement required are essentially equivalent; historically it was this discovery which constituted the breakthrough in proving the conjecture of Pólya and Schoenberg. The result, which is the last of our interlocking lemmas, is given next.

8.3.2

The starlike approximation lemma *Let $h(z)$ be a starlike mapping of $\mathbb{U}$ and let n be a natural number. Then for each $\zeta_1 \in \mathbb{T}$ (the unit circle), we can find $\zeta_2, \ldots, \zeta_n$ on $\mathbb{T}$ and μ real such that*

$$\left| \arg\left(e^{i\mu} \frac{h(z)}{P(z)} \right) \right| < \frac{\pi}{n} \quad (z \in \mathbb{U}), \tag{8.54}$$

where

$$P(z) = z \prod_{k=1}^{n} (1 - \zeta_k z)^{-2/n}. \qquad (8.55)$$

Proof By hypothesis h is a 1–1 mapping of $\mathbb{U}$ onto a domain starlike with respect to the origin, which satisfies $h(0) = 0$. There is no loss of generality in assuming that h extends smoothly to $\mathbb{T}$ and maps $\mathbb{T}$ onto a Jordan curve, which is starlike with respect to the origin; so that $\arg h(e^{i\theta})$ is strictly increasing with θ. (A general starlike h can be approximated by such a smooth mapping; furthermore, for fixed n and ζ_1, the class of possible P is compact; thus it is clear that the desired inequality will hold in general if it holds in the smooth case.) Let $D = h(\mathbb{U})$ and let $w_1 = h(\overline{\zeta_1})$. Then $w_1 \in \partial D$ and the ray $L_1 = [0, w_1]$ lies in $\overline{D}$. We construct the rays $L_2, \ldots, L_n$ from the origin to make equal angles of $2\pi/n$, with L_1 as the starting ray. Then each ray L_k meets ∂D at one point w_k and hence $w_k = h(\overline{\zeta_k})$ for some point $\overline{\zeta_k} \in \mathbb{T}$ $(2 \le k \le n)$. The points $\overline{\zeta_1}, \overline{\zeta_2}, \ldots, \overline{\zeta_n}, \overline{\zeta_{n+1}} = \overline{\zeta_1}$ are in order on $\mathbb{T}$, positively described. With ζ_k thus defined, we define $P(z)$ by (8.55). Then P is a 1–1 mapping of $\mathbb{U}$ onto the plane cut along n radial slits making equal angles of $2\pi/n$ at the origin; well-known properties of such mappings imply that $\arg P$ extends to $\mathbb{T}$ as a step function with jumps at the points $\overline{\zeta_k}$ of $2\pi/n$; on the arc $(\overline{\zeta_k}, \overline{\zeta_{k+1}})P$ travels along the relevant slit and the value of $\arg P$ is the angle of the slit, so is constant. Over the same arc $\arg h$ increases smoothly through an angle $2\pi/n$. Consider now an arbitrary arc γ of $\mathbb{T}$. We wish to compare the change over γ in $\arg P$ and $\arg h$ (i.e. the quantities $2\pi d(P, \gamma)$ and $2\pi d(h, \gamma)$). The change in $\arg P$ is exactly $2r\pi/n$, where r is the number of points $\overline{\zeta_k}$ lying in the interior of γ, assuming for the moment that neither endpoint of γ is a $\overline{\zeta_k}$. The corresponding change in $\arg h$ is $2(r-1)\pi/n + a + b$, where a and b are the changes in $\arg h$ over the two end arcs of γ containing no $\overline{\zeta_k}$. The values of a and b lie in $[0, 2\pi/n]$ and so $0 \le a+b \le 4\pi/n$. It follows that $|\Delta \arg h - \Delta \arg P| \le 2\pi/n$, where $\Delta \arg$ denotes the change in arg over γ. If $\overline{\zeta_k}$ is an endpoint of γ, then $\arg P$ is not strictly defined at the point; however, the limiting values from inside the disc all lie in the range of the jump, and a similar argument shows that whatever such value we take, the inequality remains valid. Such an inequality holding for all arcs γ implies the existence of a μ such that $|\arg P - \arg h - \mu| \le \pi/n$ at all points of $\mathbb{T}$. Applying the maximum principle to the harmonic function $\arg(e^{i\mu}h(z)/P(z))$, we deduce the desired inequality.

For the applications of this result we need to interpret it in the notations of the Kaplan classes. A function h is starlike iff $h(z)/z \in \Pi_{-2}$. Therefore we may rewrite the result as follows. *If $G \in \Pi_{-2}$ and $\zeta_1 \in \mathbb{T}$, then $\exists \zeta_2, \ldots, \zeta_n \in \mathbb{T}$ such that*

$$G(z) = k(z) \prod_{k=1}^{n} \left(1 - \zeta_k z\right)^{-2/n}, \tag{8.56}$$

where $k(z) \in K(\frac{2}{n}, \frac{2}{n})$.

Raising both sides of this equation to the power $n/2$, we obtain the result: *if $H \in \Pi_{-n}$ and $\zeta_1 \in \mathbb{T}$, then $\exists \zeta_2, \ldots, \zeta_n \in \mathbb{T}$ such that*

$$H(z) = K(z) \prod_{k=1}^{n} (1 - \zeta_k z)^{-1}, \tag{8.57}$$

where $K(z) \in K(1, 1)$.

8.3.3

We can now complete the proof of the duality theorem for $T(1, \beta)$. Let $\phi \in T(1, \beta)$; we are required to prove that $\phi(z) * f(z) \neq 0$ in $\mathbb{U}$ for any $f \in K(1, \beta)$. We have shown that it is sufficient to prove that

$$\phi(z) * \frac{1 + xz}{1 - yz} g(z) \neq 0 \quad (z \in \mathbb{U}), \tag{8.58}$$

for $|x| = |y| = 1$ and any $g \in \Pi_{1-\beta}$. This result will follow from (8.52), if we can show that, given $g \in \Pi_{1-\beta}$ and $|x| = |y| = 1$, we can find u and x_k on $\mathbb{T}$ such that

$$\frac{1 + xz}{1 - yz} g(z) = \frac{F(z)}{(1 - uz)^{\gamma} \prod_{k=2}^{n} (1 + x_k z)}, \tag{8.59}$$

where $F \in K(1, 1)$, $\gamma = \{\beta\}$, $n = [\beta]$. Choose $u = y$; then we need to show that

$$\frac{g(z)}{(1 - yz)^{1-\gamma}} = \frac{F(z)}{(1 + xz) \prod_{k=2}^{n} (1 + x_k z)}. \tag{8.60}$$

Now $g(z)(1 - yz)^{\gamma - 1}$ is a function in Π_{-n} and therefore the result follows from (8.57).

8.3.4 The convolution inequality

Theorem *If $\phi(z) \in T(1, \beta)$ and $g(z) \in \Pi_{1-\beta}$ ($\beta \geq 1$), then for every function $F \in \mathcal{A}$ satisfying $\Re F(z) > 0$ in $\mathbb{U}$, we have*

$$\Re \frac{\phi(z) * g(z)F(z)}{\phi(z) * g(z)} > 0 \quad (z \in \mathbb{U}). \tag{8.61}$$

Proof Denote by Λf the linear operator $f \to \phi * gf$. Then from the duality theorem we have that $\Lambda(\frac{1+xz}{1-yz}) \neq 0$ in $\mathbb{U}$ for $|x| = |y| = 1$. Therefore we can apply lemma 8.2.6 to deduce (8.37). This is precisely the required inequality (8.61).

We remark that this result asserts that the linear operator P defined by

$$\mathrm{P}f = \frac{\phi * gf}{\phi * g} \tag{8.62}$$

is a *convexity preserving* linear operator on $\mathcal{A}$, as described in chapter 4. Furthermore, it is clearly a *generalised convolution operator*, and therefore all the results established for such operators, including coefficient inequalities and integral mean inequalities, can be applied to P.

8.3.5

The convolution theorem *Let $\phi \in T(1, \beta)$, where $\beta \geq 1$. Then*

(i) *if $f \in K(1, \beta)$, then $\phi * f \in K(1, \beta)$,*
(ii) *if $\psi \in T(1, \beta)$, then $\phi * \psi \in T(1, \beta)$,*
(iii) *if $g \in \Pi_{1-\beta}$, then $\phi * g \in \Pi_{1-\beta}$.*

Proof If $f \in K(1, \beta)$, then we can factorise f in the form $f = gF$, where $g \in \Pi_{1-\beta}$ and $F \in K(1, 1)$. From the convolution inequality we deduce that $\phi * f = (\phi * g)K$, where $K \in K(1, 1)$. Therefore part (i) will follow, when we have proved that $\phi * g \in \Pi_{1-\beta}$, which is part (iii). Now $g \in \Pi_{1-\beta}$ iff $zg'(z) = g(z)h(z)$, where $\Re h(z) > -\beta/2$ in $\mathbb{U}$. From the convolution inequality we have that, in this case, $\Re(\mathrm{P}h)(z) > -\beta/2$; thus we obtain

$$\Re \frac{z(\phi * g)'(z)}{\phi(z) * g(z)} = \Re \frac{\phi(z) * zg'(z)}{\phi(z) * g(z)} > -\frac{\beta}{2}, \tag{8.63}$$

which is the desired result that $\phi * g \in \Pi_{1-\beta}$. Finally part (ii) requires us to prove that $\phi * \psi * (1+xz)(1-z)^{-\beta} \neq 0$ in $\mathbb{U}$. Since $(1+xz)(1-z)^{-\beta} \in K(1, \beta)$, part (i) implies that $\psi * (1+xz)(1-z)^{-\beta} \in K(1,\beta)$, and therefore the desired result follows from the duality theorem.

8.3.6

Corollary *If $1 \leq \beta < \gamma$, then $T(1,\beta) \supset T(1,\gamma)$.*

Proof If $\phi \in T(1,\gamma)$, then, since $(1+xz)(1-z)^{-\beta} \in K(1,\beta) \subset K(1,\gamma)$, it follows from the duality theorem that $\phi(z) * (1+xz)(1-z)^{-\beta} \neq 0$ in $\mathbb{U}$. Hence $\phi \in T(1,\beta)$.

8.3.7

Perhaps the most important aspect of this theory lies in the manner in which it unifies much of the pre-existing theory of these classes of analytic functions. Many known results, previously obtained by often quite difficult *ad hoc* arguments, turn out to be simple corollaries of the general convolution theorem. Furthermore this theorem establishes quite new and unsuspected results. For example, if $f(z) = \sum_0^\infty a_n z^n$ is a univalent mapping of $\mathbb{U}$ onto a convex domain, then the same is true of $g(z) = \sum_0^\infty |a_n|^2 z^n$. For $g(z) = f(z) * \overline{f(\overline{z})}$ and as both $f(z)$ and $\overline{f(\overline{z})}$ are convex, the conclusion follows from the convolution theorem part (ii) in the case $\beta = 3$.

As a second example, let us prove the result of Kaplan that a function in $K(1,3)$ is the derivative of a function univalent in $\mathbb{U}$, which by his definition are called **close-to-convex**. We shall give a proof based on the important fact that the class of univalent functions in $\mathbb{U}$ is a dual class. To be precise, $f(z) = z + a_2 z^2 + \cdots$ is univalent in $\mathbb{U}$ iff

$$\frac{f(z)}{z} * \frac{1}{(1-xz)(1-yz)} \neq 0 \quad (z \in \mathbb{U}) \tag{8.64}$$

for $|x| = |y| = 1$. Although this is a completely trivial transformation of the condition for univalence, it immediately raises many interesting questions. This is not a topic which we can pursue here, but in the author's view is well worth future study. To prove our assertion, we need to show that, if $h \in K(1,3)$, then

$$h(z) * \frac{1}{z} \log \frac{1}{1-z} * \frac{1}{(1-xz)(1-yz)} \neq 0, \tag{8.65}$$

which by the convolution theorem reduces to showing that

$$\frac{1}{z} \log \frac{1}{1-z} * \frac{1}{(1-xz)(1-yz)} \in T(1,3). \tag{8.66}$$

However, this follows immediately from the fact that $1/(1-xz)(1-yz) \in \Pi_{-2}$.

8.3.8 A generalisation of the Eneström–Kakeya theorem

Theorem *Let $a_0, a_1, \ldots, a_n$ be real numbers satisfying*

$$1 = a_0 \geq a_1 \geq \cdots \geq a_n \geq 0. \tag{8.67}$$

Then the polynomial $P(z) = \sum_{k=0}^{n} a_k z^k$ has no zeros in $\mathbb{U}$.

This useful and elegant theorem is very easy to prove. Writing $\alpha_k = a_k - a_{k+1}$, we note that $\alpha_k \geq 0$ $(0 \leq k \leq n)$, where $a_{n+1} = 0$, and $\sum_0^n \alpha_k = 1$. Furthermore

$$P(z) = \sum_{k=0}^{n} \alpha_k P_k(z), \tag{8.68}$$

where $P_k(z) = 1 + z + \cdots + z^k = \frac{1-z^{k+1}}{1-z}$. Thus $P(z) = \sum_0^n \alpha_k (1 - z^{k+1})/(1-z)$, which, for $z \in \mathbb{U}$, is a convex combination of points in the disc of radius 1, centre 1, all divided by $(1 - z)$. Therefore P has the form

$$P(z) = \frac{1 - \omega(z)}{1 - z}, \tag{8.69}$$

where $|\omega(z)| < 1$ in $\mathbb{U}$.

8.3.9

Ruscheweyh [46] has generalised the above argument to prove the following.

Theorem *Let $\phi(z) = 1 + \sum_{k=1}^{\infty} c_k z^k$ be a function in the class Π_{-1} and let $a_0, a_1, \ldots, a_n$ be real numbers satisfying (8.67). Then the polynomial*

$$P(z) = 1 + \sum_{k=1}^{n} a_k c_k z^k \tag{8.70}$$

can be written in the form

$$P(z) = (1 - \omega(z))\phi(z), \tag{8.71}$$

where $|\omega(z)| < 1$ for $z \in \mathbb{U}$. In particular P has no zeros in $\mathbb{U}$.

Proof As above, since P is a convex combination of partial sums of ϕ, it will be enough to prove this representation for each of the partial sums $\phi_j(z) = 1 + \sum_{k=1}^{j} c_k z^k$. This is a simple consequence of the convolution inequality in the case $\alpha = 1$, $\beta = 2$. Since $\phi \in T(1, 2)$, for any $g \in \Pi_{-1}$ the operator $F \mapsto (\phi * gF)/(\phi * g)$ is convexity preserving. In particular, taking $g = 1/(1 - z)$ and $F = 1 - z^{j+1}$, we deduce that $\phi_j(z)/\phi(z)$ maps $\mathbb{U}$ into the convex hull of $F(\mathbb{U})$; i.e. $\phi_j/\phi = 1 - \omega_j$, where $|\omega_j| < 1$. This proves the result.

8.4 The duality theorem for $T(m, \beta)$

8.4.1

Theorem *If $\phi \in T(m, \beta)$ and $f \in K(m, \beta)$, where $\beta \geq 1$ and m is an integer ≥ 1, then $(\phi * f)(z) \neq 0$ for $z \in \mathbb{U}$.*

We shall find that with this result we can extend the entire theory of the previous section to the cases $m = 2, 3, \ldots$ with one significant difference. Although $T(m, \beta)$ is clearly defined by (8.25), there is no longer an alternative form of the condition such as we found in the case $m = 1$, where we related it to starlikeness and the class $\Pi_{1-\beta}$ (see subsection 8.2.8). This has two consequences. There is virtually no classical theory to be generalised and the theorems constitute something quite new on previously unstudied classes. However, the class $T(m, \beta)$ is a very difficult one to work with in any direct way and there are few known members. Consequently it is desirable to restructure theorems so that there is no mention of an underlying duality class. This will be achieved by couching the results in terms which give sufficient conditions for linear operators to have certain structure preserving properties.

8.4.2

For the proof of the duality theorem, we will first show that, if $\phi \in T(m, \beta)$, then for $g \in \Pi_{1-\beta}$, $h \in \Pi_{m-1}$ and $F \in K(1, 1)$ we have

$$\phi(z) * g(z)h(z)F(z) \neq 0 \quad (z \in \mathbb{U}). \tag{8.72}$$

For the first step, and several others, we require the following remark. If $\phi(z) * (1+xz)^m G(z) \neq 0$ in $\mathbb{U}$ for $|x| = 1$ and a given function G satisfying $G(0) \neq 0$, then $\phi(z) * P(z)G(z) \neq 0$ in $\mathbb{U}$ for any polynomial P of degree at most m with no zeros in $\mathbb{U}$. This follows from theorem 5.8.3 part (b) of chapter 5 applied to the linear operator Λ defined by $\Lambda f = \phi * fG$ (a version of Grace's theorem). In particular we obtain $\phi(z) * (1 + xz)G(z) \neq 0$ in $\mathbb{U}$ for $|x| \leq 1$. We thus see, putting $G(z) = (1 - z)^{-\beta}$, that $\phi \in T(m, \beta) \Rightarrow \phi \in T(1, \beta)$. From the duality theorem for $T(1, \beta)$ we deduce that for $f \in K(1, \beta)$, $\phi * f \neq 0$. In particular we have

$$\phi(z) * \frac{1 + xz}{1 - yz} \prod_{k=1}^{N} (1 - y_k z)^{-\beta_k} \neq 0 \quad (z \in \mathbb{U}), \tag{8.73}$$

for values of the parameters x, y, y_k in the closed disc, where β_k are positive numbers whose sum is $\beta - 1$. It follows from the finite product lemma that

$$\phi(z) * \frac{(1 + xz)^m}{1 - yz} \prod_{k=1}^{N} (1 - y_k z)^{-\beta_k} \neq 0 \quad (z \in \mathbb{U}), \tag{8.74}$$

and therefore applying the above remark we obtain

$$\phi(z) * \frac{1+xz}{1-yz} \frac{\prod_{j=1}^{m-1}(1+x_j z)}{\prod_{k=1}^{N}(1-y_k z)^{\beta_k}} \neq 0 \quad (z \in \mathbb{U}) \tag{8.75}$$

for values of the parameters x, y, x_j and y_k in the closed disc. From lemma 8.2.6 we can replace the term $(1+xz)/(1-yz)$ in this last expression by any function $F \in K(1, 1)$. Furthermore it is easily shown (e.g. using the starlike approximation lemma) that any function $g \in \Pi_{1-\beta}$ can be obtained as a local uniform limit in $\mathbb{U}$ of products of the form $\prod_{k=1}^{N}(1-y_k z)^{-\beta_k}$. Therefore applying Hurwitz's theorem we deduce that

$$\phi(z) * g(z)F(z) \prod_{j=1}^{m-1}(1+x_j z) \neq 0 \quad (z \in \mathbb{U}) \tag{8.76}$$

for $g \in \Pi_{1-\beta}$ and $F \in K(1, 1)$. Now to prove (8.72) it will be sufficient to show that

$$\phi(z) * \frac{1+xz}{1-yz} g(z)h(z) \neq 0 \quad (z \in \mathbb{U}) \tag{8.77}$$

for $g \in \Pi_{1-\beta}$, $h \in \Pi_{m-1}$ and $|x| = |y| = 1$; for we can then apply lemma 8.2.6. This last result will follow from (8.76) if we can show that given x, y on $\mathbb{T}$ and $h \in \Pi_{m-1}$ we can find $x_j \in \mathbb{T}$ and $F \in K(1, 1)$ such that

$$\frac{1+xz}{1-yz} h(z) = F(z) \prod_{j=1}^{m-1}(1+x_j z). \tag{8.78}$$

Choosing $x_1 = x$, we require

$$h(z) = (1-yz) \prod_{j=2}^{m-1}(1+x_j z)F(z), \tag{8.79}$$

which follows from the starlike approximation theorem in the form (8.57)

8.4.3

To complete the proof of the theorem consider two cases: (i) $\beta \geq m$, (ii) $\beta < m$. In case (i) we have from (8.72) that for $g \in \Pi_{m-\beta}$

$$\phi(z) * g(z) \prod_{k=1}^{m} \frac{1+x_k z}{1-y_k z} \neq 0 \quad (z \in \mathbb{U}), \tag{8.80}$$

since $g(z)/\prod_2^m(1-y_kz) \in \Pi_{1-\beta}$ and $\prod_2^m(1+x_k) \in \Pi_{m-1}$. Now if $f \in K(m,\beta)$, then we can write $f = gH$, where $g \in \Pi_{m-\beta}$ and $H \in K(m,m)$ (see chapter 7, theorem 7.4.6). On the other hand, $H \in K(m,m)$ iff $H = F^m$ where $F \in K(1,1)$. Therefore we see that m applications of lemma 8.2.6 to (8.80) gives $\phi * gH \neq 0$, which is the desired result.

In case (ii) we set $n = [\beta]$ and $\gamma = \{\beta\}$, so $0 \leq \gamma < 1$ and $n + \gamma = \beta$. Applying (8.72) we have for $u \in \Pi_{-\gamma}$ and $v \in \Pi_{m-n}$

$$\phi(z) * u(z)v(z)\prod_{k=1}^{n}\frac{1+x_kz}{1-y_kz} \neq 0 \quad (z \in \mathbb{U}). \tag{8.81}$$

Applying lemma 8.2.6 n times we obtain for $L \in K(n,n)$

$$\phi(z) * u(z)v(z)L(z) \neq 0 \quad (z \in \mathbb{U}). \tag{8.82}$$

Now $f \in K(m,\beta)$ iff $f = Hg$, where $H \in K(\beta,\beta)$ and $g \in \Pi_{m-\beta}$. Consider such g and $G \in K(\beta-1,\beta-1)$. Then

$$\frac{1+xz}{1-yz}Gg = \left(\frac{1+xz}{1-yz}\right)^{1-\gamma} G(1-yz)^{-\gamma}(1+xz)^{\gamma}g = Luv, \tag{8.83}$$

where $L = (\frac{1+xz}{1-yz})^{1-\gamma}G \in K(n,n)$, $u = (1-yz)^{-\gamma} \in \Pi_{-\gamma}$ and $v = (1+xz)^{\gamma}g \in \Pi_{m-n}$. Thus by (8.82) we have

$$\phi(z) * \frac{1+xz}{1-yz}Gg \neq 0. \tag{8.84}$$

Finally applying lemma 8.2.6 we obtain $\phi * Hg \neq 0$, which completes the proof of the theorem.

8.4.4 The convolution inequality

Theorem *Let $\phi \in T(m,\beta)$, where $m \geq 1$ is an integer and $\beta \geq 1$ is real. Choose λ and δ satisfying $1 \leq \lambda \leq \min(m,\beta)$ and $0 \leq \delta \leq \lambda$. Then, if $h \in K(m-\lambda,\beta-\lambda)$ and $L \in \mathcal{A}$ satisfies $|\arg L(z)| \leq \delta\pi/2$ in $\mathbb{U}$, we have*

$$\left|\arg\left(\frac{\phi(z) * h(z)L(z)}{\phi(z) * h(z)}\right)\right| \leq \frac{\delta\pi}{2} \quad (z \in \mathbb{U}). \tag{8.85}$$

Proof Let $\rho = \min(m,\beta)$. If $|x| = |y| = 1$, $g \in \Pi_{m-\beta}$ and $|\arg G| \leq (\rho-1)\times \pi/2$, then

$$\frac{1+xz}{1-yz}gG \in K(m,\beta) \tag{8.86}$$

and therefore by the duality theorem and lemma 8.2.6 we have

$$\Re\left(\frac{\phi(z) * g(z)G(z)F(z)}{\phi(z) * g(z)G(z)}\right) > 0 \quad (z \in \mathbb{U}) \tag{8.87}$$

for every function $F \in \mathcal{A}$ satisfying $\Re F(z) > 0$ in $\mathbb{U}$. Choose h and L satisfying the hypotheses of the theorem. Then by the factorisation theorem 7.4.6 of chapter 7, $h = gM$ where $g \in \Pi_{m-\beta}$ and $|\arg M| \le (\rho - \lambda)\pi/2$ in $\mathbb{U}$. Let $k = [\delta]$ and $\sigma = \{\delta\}$. We first consider the case when $k = 0$, giving $0 \le \delta < 1$. Since $|\arg M| \le (\rho - \lambda)\pi/2 \le (\rho - 1)\pi/2$, we can apply (8.87), replacing G by M, to deduce that the operator $f \to (\phi * gMf)/(\phi * gM)$ is convexity preserving on $\mathcal{A}$. Since the set $\{w\colon |\arg w| \le \delta\pi/2\}$ is convex, it follows that

$$\left|\arg\left(\frac{\phi * gML}{\phi * gM}\right)\right| \le \frac{\delta\pi}{2}, \tag{8.88}$$

which is the desired result (8.85). Assume now that $k \ge 1$; then we can write $L = L_1 \ldots L_k R$, where $\Re L_i > 0$ in $\mathbb{U}$ $(1 \le i \le k)$ and $|\arg R| \le \sigma\pi/2$ in $\mathbb{U}$. Now the condition (8.87) can be written $\phi * gGF = \widetilde{F}(\phi * gG)$, where $\Re\widetilde{F} > 0$; thus $\phi * gML = \widetilde{L_1}(\phi * gML_2 \ldots L_k R) = \widetilde{L_1}\widetilde{L_2}(\phi * gML_3 \ldots L_k R) = \cdots = \widetilde{L_1}\widetilde{L_2} \ldots \widetilde{L_k}(\phi * gMR)$, where $\Re\widetilde{L_i} > 0 (1 \le i \le k)$. Also from the case $k = 0$ we have $\phi * gMR = \widetilde{R}(\phi * gM)$, where $|\arg\widetilde{R}| \le \sigma\pi/2$. Thus we obtain $|\arg(\phi * gML)/(\phi * gM)| \le (k + \sigma)\pi/2 = \delta\pi/2$, and this is the desired result.

8.4.5

The convolution theorem *Let $\phi \in T(m, \beta)$, where $m \ge 1$ is an integer and $\beta \ge 1$ is real. Then*

(i) *if $0 \le \epsilon \le \min(m, \beta)$ and $f \in K(m - \epsilon, \beta - \epsilon)$, then $\phi * f \in K(m - \epsilon, \beta - \epsilon)$,*
(ii) *if $\psi \in T(m, \beta)$, then $\phi * \psi \in T(m, \beta)$.*

Proof If $f \in K(m - \epsilon, \beta - \epsilon)$, then we can factorise f in the form $f = gF$, where $g \in \Pi_{m-\beta}$ and $F \in K(\rho - \epsilon, \rho - \epsilon)$, where $\rho = \min(m, \beta)$. From the convolution inequality, applied in the case $\lambda = \rho$ and $\delta = \rho - \epsilon$, we deduce that $\phi * f = (\phi * g)K$, where $K \in K(\rho - \epsilon, \rho - \epsilon)$. Therefore part (i) will follow, when we have proved that $\phi * g \in \Pi_{m-\beta}$, which is the case $\epsilon = \rho$. We apply the convolution inequality in the case $\lambda = \rho$ and $\delta = 1$ to obtain for $F \in \mathcal{A}$ satisfying $\Re F > 0$ in $\mathbb{U}$

$$\Re\left(\frac{\phi(z) * g(z)F(z)}{\phi(z) * g(z)}\right) > 0 \quad (z \in \mathbf{U}). \tag{8.89}$$

Equivalently, the operator $f \to (\phi * gf)/(\phi * g)$ is convexity preserving on $\mathcal{A}$. In particular putting $f = zg'/g$, we deduce that $(\phi * zg')/(\phi * g)$ takes its values for $z \in \mathbb{U}$ in the same half-plane that zg'/g maps to. In other words $\phi * g \in \Pi_{m-\beta}$ as required.

Part (ii) requires us to prove that $\phi * \psi * (1 + xz)^m(1 - z)^{-\beta} \neq 0$ in $\mathbb{U}$. Since $(1 + xz)^m(1 - z)^{-\beta} \in K(m, \beta)$, part (i), in the case $\epsilon = 0$, implies that $\psi * (1 + xz)^m(1 - z)^{-\beta} \in K(m, \beta)$, and therefore the desired result follows from the duality theorem.

8.4.6

Corollary *If m and n are integers such that* $1 \leq m \leq n$ *and* β *and* γ *are real such that* $1 \leq \beta \leq \gamma$, *then* $T(m, \beta) \supset T(n, \gamma)$.

Proof If $\phi \in T(n, \gamma)$, then, since $(1 + xz)^m(1 - z)^{-\beta} \in K(m, \beta) \subset K(n, \gamma)$, it follows from the duality theorem that $\phi(z) * (1 + xz)^m(1 - z)^{-\beta} \neq 0$ in $\mathbb{U}$. Hence $\phi \in T(m, \beta)$.

8.5 The duality principle

8.5.1

It is clear from these and other results which we have established that significant information flows, when we are able to show that a non-trivial class lies in the second dual of a relatively simple class. It is therefore worthwhile to consider in general terms what can be said concerning the relationship between a class and its second dual. A result which we have observed on several occasions in particular cases is the following general duality principle.

Theorem *Let* $\mathcal{M}$ *denote a compact class of functions g in* $\mathcal{A}$ *satisfying* $g(0) = 1$, *which is circled (i.e. if* $g(z) \in \mathcal{M}$, *then* $g(xz) \in \mathcal{M}$ *for each x satisfying* $|x| \leq 1$). *Let* Λ *be a linear functional on* $\mathcal{A}$. *Then for each* $f \in \mathcal{M}^{**}$ *there exists* $g \in \mathcal{M}$ *such that* $\Lambda f = \Lambda g$.

Proof Let $\Lambda f = w$ and assume that for all $g \in \mathcal{M}$ we have $\Lambda g \neq w$. Thus, if we define the functional Λ_1 by $\Lambda_1 F = \Lambda F - wF(0)$, we have $\Lambda_1 f = 0$ but $\Lambda_1 g \neq 0$ (recall that by definition the functions f in the second dual satisfy $f(0) = 1$: see subsection 5.1.1 of chapter 5). Now the kernel of Λ_1 is a function ϕ analytic in $|z| \leq 1$ and we have for $F \in \mathcal{A}$, $\Lambda_1 F = (\phi * F)(1)$. Thus $(\phi * g)(1) \neq 0 \, \forall g \in \mathcal{M}$. Since $\mathcal{M}$ is circled, we can apply this to $g(xz)$ to deduce that

$(\phi * g)(x) \neq 0 \, \forall x$ such that $|x| \leq 1$. Because $\mathcal{M}$ is compact, we can deduce from this that $\exists R > 1$ such that $(\phi * g)(x) \neq 0$ for $|x| < R$ and all $g \in \mathcal{M}$. Equivalently, $\phi(Rz) * g(z) \neq 0$ for $z \in \mathbb{U}$; since this implies that the function $\phi(rz) \in \mathcal{M}^*$, it follows that $\phi(Rz) * f(z) \neq 0$ in $\mathbb{U}$. In particular, when $z = 1/R$ we obtain $\Lambda_1 f \neq 0$, a contradiction. This completes the proof.

An important conclusion can be drawn from this result, which can be seen when we apply the general convexity theory of linear topological spaces. The space $\mathcal{A}$ of functions analytic in $\mathbb{U}$ is a locally convex linear topological space. If a function f does not belong to a closed convex subset $\mathcal{K} \subset \mathcal{A}$, then there exists a continuous linear functional Λ on $\mathcal{A}$ such that $|\Lambda f| < \inf\{|\Lambda g| : g \in \mathcal{K}\}$ (see e.g. Kelley, Namioka *et al.* [23], page 119). If we now take $\mathcal{K}$ to be the closed convex hull of $\mathcal{M}$ and $f \notin \mathcal{K}$, then we obtain for a suitable functional Λ, $\Lambda f \neq \Lambda g$ for all $g \in \mathcal{K}$, and therefore for all $g \in \mathcal{M}$. From the duality principle it follows that $f \notin \mathcal{M}^{**}$. We deduce that $\mathcal{M}^{**} \subset \mathcal{K}$. We have proved the next theorem.

8.5.2

Theorem *Let $\mathcal{M}$ denote a compact and circled class of functions g in $\mathcal{A}$ satisfying $g(0) = 1$. Then each function $f \in \mathcal{M}^{**}$ lies in the closed convex hull of $\mathcal{M}$.*

According to the Krein–Milman theorem (Kelley, Namioka *et al.* [23], pages 131–2), the set of extreme points of the closed convex hull of $\mathcal{M}$ lies in $\mathcal{M}$. Furthermore, if we wish to maximise a real linear functional, or the absolute value of a complex linear functional, over the closed convex hull of $\mathcal{M}$, it is sufficient to do so over the set of extreme points and therefore over $\mathcal{M}$ itself. Thus by the theorem, the extremal value for a linear functional on $\mathcal{M}^{**}$ is attained for a function in $\mathcal{M}$.

8.5.3

Our final convexity observation arises from a classic theorem of Choquet which enables one to represent explicitly an element of the closed convex hull by an integral against a probability measure over the set of extreme points. For our purposes the following theorem will suffice.

Theorem *Let $F(z_1, z_2, \ldots, z_n)$ be a complex-valued function analytic in the polydisc $|z_i| < 1$ $(1 \leq i \leq n)$ and satisfying $F(0, 0, \ldots, 0) = 1$. Let $\mathcal{M}$ denote* *the class of functions* $\{F(x_1z, x_2z, \ldots, x_nz)\colon |x_i| \le 1 \text{ for } 1 \le i \le n\}$. *Then every extreme point of the closed convex hull of* $\mathcal{M}$ *lies in the class* $\{F(x_1z, x_2z, \ldots, x_nz)\colon |x_i| = 1 \text{ for } 1 \le i \le n\}$. *Furthermore, if* f *belongs to the closed convex hull of* $\mathcal{M}$, *then we can write*

$$f(z) = \int_{\mathbb{T}^n} F(x_1z, x_2z, \ldots, x_nz)d\mu, \tag{8.90}$$

where μ *is a probability measure on the torus* $\mathbb{T}^n = \mathbb{T} \times \mathbb{T} \times \cdots \times \mathbb{T}$ *(n times). In particular, every function* $f \in \mathcal{M}^{**}$ *has such a representation.*

The reason that the parameters x_i can be confined to the circle $\mathbb{T}$ lies in the fact that for any function $g(z_1, \ldots, z_n)$ analytic in the polydisc and parameters y_k satisfying $|y_k| \le 1\,(1 \le k \le n)$, there exists a probability measure μ on $\mathbb{T}^n$ such that

$$g(y_1z, \ldots, y_nz) = \int_{\mathbb{T}^n} g(x_1z, x_2z, \ldots, x_nz)d\mu. \tag{8.91}$$

This is easily proved by developing the convolution theory for several variables in the polydisc in a completely analogous way to the one variable case. It can then be shown (e.g. Sheil-Small [53]) that the function $\mathbb{U}^n \to \mathbb{C}$ defined by

$$(z_1, \ldots, z_n) \mapsto \prod_{k=1}^{n} \frac{1}{1 - y_kz_k} \tag{8.92}$$

has a representation in the form

$$\prod_{k=1}^{n} \frac{1}{1 - y_kz_k} = \int_{\mathbb{T}^n} \prod_{k=1}^{n} \frac{1}{1 - x_kz_k} d\mu. \tag{8.93}$$

Then using several variable convolution we have

$$g(y_1z, \ldots, y_nz) = g(z_1, \ldots, z_n) * \prod_{k=1}^{n} \frac{1}{1 - y_kz_k} = \int_{\mathbb{T}^n} g(x_1z, x_2z, \ldots, x_nz)d\mu. \tag{8.94}$$

These convexity observations are important, yet are by no means the full story. The essential point about the duality principle is that the actual set of values which a linear functional takes over $\mathcal{M}$ is identical with the set of values which the same functional takes over the potentially much larger class $\mathcal{M}^{**}$. It was this fact in the context of Grace's theorem of chapter 5, which we were able to exploit in a variety of ways in studying the zeros and critical points of polynomials.

8.6 Duality and the class $T(\alpha, \beta)$

8.6.1

The theory we have developed will remain incomplete without some attempt to generalise the above convolution theorems to the case when α is a general real parameter ≥ 1. The natural generalisation would be to have the same results holding for $T(\alpha, \beta,)$ as for $T(m, \beta)$. However, what is actually known about the class $T(\alpha, \beta)$ is very little; even whether the class forms a normal family of functions in the unit disc is an unsolved problem. The central difficulty is that we have been unable to make the first step in breaking down the convolution condition; as we have seen the theory flows from this first step. In the case $\alpha = m$ an integer, we were able to employ Grace's theorem to start the breakdown process. However, for general real α, the condition $f(z) * (1+z)^\alpha \neq 0$ has far fewer implications for f than when α is a natural number.

Nevertheless, progress in the theory can be made if, in essence, we make the first step part of our assumption. To achieve this we modify the definition as follows. We say $\phi \in T_0(\alpha, \beta)$ if ϕ is analytic in $\mathbb{U}$ with $\phi(0) = 1$ and satisfies

$$\phi(z) * \frac{(1+xz)^m(1+uz)^\gamma}{(1-z)^\beta} \neq 0 \quad (z \in \mathbb{U}), \tag{8.95}$$

for $|x| = |u| = 1$, where $m = [\alpha]$ and $\gamma = \{\alpha\}$. We then have the next result.

8.6.2

Theorem *If $\alpha \geq 1$ and $\beta \geq 1$ and $\phi \in T_0(\alpha, \beta)$, then for $f \in K(\alpha, \beta)$ we have*

$$\phi(z) * f(z) \neq 0 \quad (z \in \mathbb{U}). \tag{8.96}$$

Proof The defining condition for $T_0(\alpha, \beta)$ will hold for $|x| \leq 1$ and $|u| \leq 1$. Therefore, if $\phi \in T_0(\alpha, \beta)$, then

$$\frac{(1+xw)^m}{(1-yw)^\beta} *_w \left\{\phi(z) *_z \frac{(1+uz)^\gamma}{1-wz}\right\} = \phi(z) *_z \frac{(1+uz)^\gamma(1+xwz)^m}{(1-ywz)^\beta} \neq 0, \tag{8.97}$$

for $|w| < 1$, $|x| \leq 1$, $|y| \leq 1$, $|u| \leq 1$ and $|z| < 1$. It follows that the function

$$w \to \phi(z) *_z \frac{(1+uz)^\gamma}{1-wz} \in T(m, \beta) \tag{8.98}$$

for each fixed $z \in \mathbb{U}$. Therefore, if $f \in K(m, \beta)$, we obtain by convolving this with $f(w)$

$$\phi(z) * (1+uz)^\gamma f(wz) \neq 0 \tag{8.99}$$

for $z \in \mathbb{U}$, $w \in \mathbb{U}$ and $|u| \leq 1$. Applying Hurwitz's theorem this result will extend to values w satisfying $|w| \leq 1$.

Next, let $g \in \Pi_\gamma$ and consider

$$\phi(z) * \frac{(1+xz)^m}{(1-yz)^\beta} g(z). \tag{8.100}$$

This will be non-zero in $\mathbb{U}$ for $|x| \leq 1$ and $|y| \leq 1$ provided that we can find u with $|u| \leq 1$ and $f \in K(m, \beta)$ such that

$$\frac{(1+xz)^m}{(1-yz)^\beta} g(z) = (1+uz)^\gamma f(z). \tag{8.101}$$

Choose $u = x$. Then the relation follows, since $(1+xz)^{m-\gamma}(1-yz)^{-\beta} g(z) \in K(m, \beta)$. Thus (8.100) is non-zero and in the same way that we proved the implication (8.97)⇒(8.99), we obtain

$$\phi(z) * g(z) f(z) \neq 0 \quad (z \in \mathbb{U}) \tag{8.102}$$

for $g \in \Pi_\gamma$ and $f \in K(m, \beta)$.

To complete the duality proof we need to show that $\phi * F \neq 0$ for arbitrary $F \in K(\alpha, \beta)$. This is already done when F has a factorisation $F = fg$, as above. In general a factorisation in this form may not hold. What we do know is that F can be factorised in the form $F = HP$, where $H \in K(\rho, \rho)$ and $P \in \Pi_{\alpha-\beta}$ ($\rho = \min(\alpha, \beta)$). Furthermore, since $\rho \geq 1$, $H \in K(\rho, \rho)$ can be factorised in the form $H = KG$, where $K \in K(1, 1)$ and $G \in K(\rho - 1, \rho - 1)$. Therefore, by lemma 8.2.6, it is sufficient to show that, if $G \in K(\rho - 1, \rho - 1)$ and $P \in \Pi_{\alpha-\beta}$, we can write for each pair $x, y \in \mathbb{T}$

$$\frac{1+xz}{1-yz} G(z) P(z) = h(z) L(z) g(z), \tag{8.103}$$

where $h \in \Pi_{m-\beta}$, $L \in K(\sigma, \sigma)$ and $g \in \Pi_\gamma$ ($\sigma = \min(m, \beta)$); for then $hL \in K(m, \beta)$ and the result follows from (8.102). We consider three cases (assuming $\gamma > 0$):

(i) $\beta \leq m$. Choose

$$g = P^{\gamma/(\alpha-\beta)}, \qquad h = P^{(m-\beta)/(\alpha-\beta)}, \qquad L = \frac{1+xz}{1-yz} G. \tag{8.104}$$

(ii) $m < \beta < \alpha$. Choose

$$g = (1 + xz)^{\beta - m} P, \qquad h = (1 - yz)^{m - \beta}, \qquad L = \left(\frac{1 + xz}{1 - yz}\right)^{m - \beta + 1} G. \tag{8.105}$$

(iii) $\alpha \le \beta$. Choose

$$g = (1 + xz)^{\gamma}, \qquad h = (1 - yz)^{-\gamma} P, \qquad L = \left(\frac{1 + xz}{1 - yz}\right)^{1 - \gamma} G. \tag{8.106}$$

The proof is complete.

8.6.3 The convolution inequality

Theorem *Let $\phi \in T_0(\alpha, \beta)$, where $\alpha \ge 1$ and $\beta \ge 1$. Choose λ and δ satisfying $1 \le \lambda \le \min(\alpha, \beta)$ and $0 \le \delta \le \lambda$. Then, if $h \in K(\alpha - \lambda, \beta - \lambda)$ and $L \in \mathcal{A}$ satisfies $|\arg L(z)| \le \delta\pi/2$ in $\mathbb{U}$, we have*

$$\left|\arg\left(\frac{\phi(z) * h(z)L(z)}{\phi(z) * h(z)}\right)\right| \le \frac{\delta\pi}{2} \quad (z \in \mathbb{U}). \tag{8.107}$$

Proof Let $\rho = \min(\alpha, \beta)$. If $|x| = |y| = 1$, $g \in \Pi_{\alpha-\beta}$ and $|\arg G| \le (\rho - 1) \times \pi/2$, then

$$\frac{1 + xz}{1 - yz} gG \in K(\alpha, \beta) \tag{8.108}$$

and therefore by the duality theorem and lemma 8.2.6 we have

$$\Re\left(\frac{\phi(z) * g(z)G(z)F(z)}{\phi(z) * g(z)G(z)}\right) > 0 \quad (z \in \mathbb{U}) \tag{8.109}$$

for every function $F \in \mathcal{A}$ satisfying $\Re F(z) > 0$ in $\mathbb{U}$. Choose h and L satisfying the hypotheses of the theorem. Then by the factorisation theorem 7.4.6 of chapter 7, $h = gM$ where $g \in \Pi_{\alpha-\beta}$ and $|\arg M| \le (\rho - \lambda)\pi/2$ in $\mathbb{U}$. Let $k = [\delta]$ and $\sigma = \{\delta\}$. We first consider the case when $k = 0$, giving $0 \le \delta < 1$. Since $|\arg M| \le (\rho - \lambda)\pi/2 \le (\rho - 1)\pi/2$, we can apply (8.109), replacing G by M, to deduce that the operator $f \mapsto (\phi * gMf)/(\phi * gM)$ is convexity preserving on $\mathcal{A}$. Since the set $\{w\colon |\arg w| \le \delta\pi/2\}$ is convex, it follows that

$$\left|\arg\left(\frac{\phi * gML}{\phi * gM}\right)\right| \le \frac{\delta\pi}{2}, \tag{8.110}$$

which is the desired result (8.107). Assume now that $k \ge 1$; then we can write $L = L_1 \ldots L_k R$, where $\Re L_i > 0$ in $\mathbb{U}\,(1 \le i \le k)$ and $|\arg R| \le \sigma\pi/2$ in $\mathbb{U}$.

Now the condition (8.109) can be written $\phi * gGF = \widetilde{F}(\phi * gG)$, where $\Re\widetilde{F} > 0$; thus $\phi * gML = \widetilde{L_1}(\phi * gML_2 \ldots L_kR) = \widetilde{L_1}\widetilde{L_2}(\phi * gML_3 \ldots L_kR) = \cdots = \widetilde{L_1}\widetilde{L_2} \ldots \widetilde{L_k}(\phi * gMR)$, where $\Re\widetilde{L_i} > 0$ $(1 \le i \le k)$. Also from the case $k = 0$ we have $\phi * gMR = \widetilde{R}(\phi * gM)$, where $|\arg\widetilde{R}| \le \sigma\pi/2$. Thus we obtain $|\arg(\phi * gML)/(\phi * gM)| \le (k+\sigma)\pi/2 = \delta\pi/2$, and this is the desired result.

8.6.4

The convolution theorem *Let $\phi \in T_0(\alpha, \beta)$, where $\alpha \ge 1$ and $\beta \ge 1$. Then*

(i) *if $0 \le \epsilon \le \min(\alpha, \beta)$ and $f \in K(\alpha-\epsilon, \beta-\epsilon)$, then $\phi * f \in K(\alpha-\epsilon, \beta-\epsilon)$;*
(ii) *if $\psi \in T_0(\alpha, \beta)$, then $\phi * \psi \in T_0(\alpha, \beta)$.*

Proof If $f \in K(\alpha-\epsilon, \beta-\epsilon)$, then we can factorise f in the form $f = gF$, where $g \in \Pi_{\alpha-\beta}$ and $F \in K(\rho-\epsilon, \rho-\epsilon)$, where $\rho = \min(\alpha, \beta)$. From the convolution inequality, applied in the case $\lambda = \rho$ and $\delta = \rho - \epsilon$, we deduce that $\phi * f = (\phi * g)K$, where $K \in K(\rho-\epsilon, \rho-\epsilon)$. Therefore part (i) will follow, when we have proved that $\phi * g \in \Pi_{\alpha-\beta}$, which is the case $\epsilon = \rho$. We apply the convolution inequality in the case $\lambda = \rho$ and $\delta = 1$ to obtain for $F \in \mathcal{A}$ satisfying $\Re F > 0$ in $\mathbb{U}$

$$\Re\left(\frac{\phi(z) * g(z)F(z)}{\phi(z) * g(z)}\right) > 0 \quad (z \in \mathbb{U}). \tag{8.111}$$

Equivalently, the operator $f \mapsto (\phi * gf)/(\phi * g)$ is convexity preserving on $\mathcal{A}$. In particular putting $f = zg'/g$, we deduce that $(\phi * zg')/(\phi * g)$ takes its values for $z \in \mathbb{U}$ in the same half-plane that zg'/g maps to. In other words $\phi * g \in \Pi_{\alpha-\beta}$ as required.

Part (ii) requires us to prove that $\phi * \psi * (1+xz)^m(1+uz)^\gamma(1-z)^{-\beta} \ne 0$ in $\mathbb{U}$. Since $(1+xz)^m(1+uz)^\gamma(1-z)^{-\beta} \in K(\alpha, \beta)$, part (i), in the case $\epsilon = 0$, implies that $\psi * (1+xz)^m(1+uz)^\gamma(1-z)^{-\beta} \in K(\alpha, \beta)$, and therefore the desired result follows from the duality theorem.

8.6.5

Corollary *If $1 \le \alpha \le \delta$ and $1 \le \beta \le \gamma$, then $T_0(\alpha, \beta) \supset T_0(\delta, \gamma)$.*

Proof If $\phi \in T_0(\delta, \gamma)$, then, since $(1+xz)^m(1+uz)^\gamma(1-z)^{-\beta} \in K(\alpha, \beta) \subset K(\delta, \gamma)$, it follows from the duality theorem that $\phi(z) * (1+xz)^m(1+uz)^\gamma \times (1-z)^{-\beta} \ne 0$ in $\mathbb{U}$. Hence $\phi \in T_0(\alpha, \beta)$.

8.7 Properties of the Kaplan classes

8.7.1

Theorem *Let $f(z) \in K(\alpha, \beta)$, where $\alpha \geq 1$ and $\beta \geq 1$ and assume that $f(0) = 1$. If Λ is a linear functional on $\mathcal{A}$, then for suitable x, u and $y \in \overline{\mathbb{U}}$,*

$$\Lambda f = \Lambda\left(\frac{(1+xz)^m(1+uz)^\gamma}{(1-yz)^\beta}\right), \tag{8.112}$$

where $m = [\alpha]$ and $\gamma = \{\alpha\}$.

This is an immediate consequence of the duality theorem: $K(\alpha, \beta)$ lies in the second dual of the class of functions in parentheses. Observe that this formulation makes no mention of the class $T_0(\alpha, \beta)$; it simply asserts that in studying a linear functional on $K(\alpha, \beta)$, it is sufficient to consider the same functional on a small subclass of $K(\alpha, \beta)$.

8.7.2

Theorem *The extreme points of the closed convex hull of the class $\{f \in K(\alpha, \beta)\colon f(0) = 1\}$, where $\alpha \geq 1$ and $\beta \geq 1$, lie among the functions $\{(1+xz)^\alpha \times (1-yz)^{-\beta}\colon |x| = |y| = 1\}$. In particular, every such function can be represented in the form*

$$f(z) = \int_{\mathbb{T}^2} \frac{(1+xz)^\alpha}{(1-yz)^\beta}\, d\mu, \tag{8.113}$$

where μ is a probability measure on the torus.

It is clear from the previous section that the extreme points lie among the functions $(1+xz)^m(1+uz)^\gamma(1-yz)^{-\beta}$ with the parameters lying on $\mathbb{T}$. Thus to prove the theorem we need to show that the only actual extreme points are when $x = u$. Now we can write

$$\begin{aligned} h(z) &= \frac{(1+xz)^m(1+uz)^\gamma}{(1-yz)^\beta} = \frac{(1+xz)^{\alpha-1}}{(1-yz)^{\beta-1}} \frac{(1+xz)^{1-\gamma}(1+uz)^\gamma}{1-yz} \\ &= \int_{\mathbb{T}} \frac{(1+xz)^{\alpha-1}}{(1-yz)^{\beta-1}} \frac{1+\zeta vz}{1-vz}\, d\mu(v), \end{aligned} \tag{8.114}$$

where μ is a probability measure on $\mathbb{T}$ and $|\zeta| = 1$. This follows from the fact that $(1+xz)^{1-\gamma}(1+uz)^\gamma(1-yz)^{-1} \in K(1, 1)$. It follows that we have represented h as a convex combination of elements in $K(\alpha, \beta)$, and so if h is an extreme point, then the measure μ must be a point mass. Thus for suitable

ζ and $v \in \mathbb{T}$ we have

$$h(z) = \frac{(1+xz)^{\alpha-1}}{(1-yz)^{\beta-1}} \frac{1+\zeta vz}{1-vz}. \tag{8.115}$$

Putting the two formulae for h together we obtain

$$(1+xz)^{1-\gamma}(1+uz)^{\gamma}(1-vz) = (1-yz)(1+\zeta vz). \tag{8.116}$$

As the left side $\to 0$ as $z \to \overline{v}$, so must the right side, and hence either $v = y$ or $\zeta = -1$. In either case we deduce that $(1+xz)^{1-\gamma}(1+uz)^{\gamma} = 1 - wz$, where $w \in \mathbb{T}$. The right side has one zero, and therefore so does the left side. Hence, as $0 \le \gamma < 1$, either $\gamma = 0$ or $x = u$.

8.7.3 Coefficient bounds for $K(\alpha, \beta)$

Theorem *Let $f(z) = 1 + \sum_{n=1}^{\infty} a_n z^n \in K(\alpha, \beta)$, where $\alpha + \beta \ge 2$. If either (i) $0 \le \alpha \le 1$ or (ii) $\alpha \ge 1$ and $\beta \ge 1$, then*

$$|a_n| \le A_n(\alpha, \beta) \quad (n = 1, 2, \dots), \tag{8.117}$$

where

$$\frac{(1+z)^{\alpha}}{(1-z)^{\beta}} = 1 + \sum_{n=1}^{\infty} A_n(\alpha, \beta) z^n. \tag{8.118}$$

Before embarking on the proof we make a number of preliminary observations. Firstly, to simplify the notation and the methods of argument, we make use of the standard notation $f \ll F$ to signify that F has non-negative coefficients and that the absolute value of each coefficient of f is bounded by the corresponding coefficient of F. Thus we wish to prove that $f \ll (1+z)^{\alpha}(1-z)^{-\beta}$. Secondly, we shall use the simple but useful fact that, if $f \ll F$ and $g \ll G$, then $fg \ll FG$.

Thirdly, we will use two coefficient results: (a) the theorem is true when $\alpha = \beta = 1$; indeed the extreme points of $K(1, 1)$ lie among the functions $(1+xz)(1-yz)^{-1}$; since $1+xz \ll 1+z$ and $(1-yz)^{-1} \ll (1-z)^{-1}$, the result follows from the second observation; (b) the theorem is true when $\alpha = 0$ for every $\beta \ge 0$; in other words functions in $\Pi_{-\beta}$ have coefficients dominated by the function $(1-z)^{-\beta}$; this can be proved by modifying a classical starlikeness argument; it also follows from the observation already made that each function in $\Pi_{-\beta}$ is a limit of products of the form $\prod_{k=1}^{n}(1 - x_k z)^{-\beta_k}$, where β_k are positive numbers whose sum is β and $|x_k| \le 1$. Since for each term we have $(1 - x_k z)^{-\beta_k} \ll (1-z)^{-\beta_k}$, the result follows from the second observation.

We can now establish the theorem in case (i). Firstly we can write $f = gF$, where $g \in \Pi_{\alpha-\beta}$ and $F \in K(\alpha, \alpha)$. Rewriting this we have

$$f(z) = \left(F(z)g(z)^{(\beta-1)/(\beta-\alpha)}g(-z)^{(\alpha-1)/(\beta-\alpha)}\right)(g(z)g(-z))^{(1-\alpha)/(\beta-\alpha)}. \quad (8.119)$$

Now $g(z)^{(\beta-1)/(\beta-\alpha)} \in K(0, \beta - 1)$ and $g(-z)^{(\alpha-1)/(\beta-\alpha)} \in K(1 - \alpha, 0)$. Hence $F(z)g(z)^{(\beta-1)/(\beta-\alpha)}g(-z)^{(\alpha-1)/(\beta-\alpha)} \in K(1, \alpha + \beta - 1)$, and so can be written in the form Hp, where $H \in K(1, 1)$ and $p \in \Pi_{2-\alpha-\beta}$. Applying our above results we get

$$H(z)p(z) \ll \frac{1+z}{1-z}\frac{1}{(1-z)^{\alpha+\beta-2}} = \frac{1+z}{(1-z)^{\alpha+\beta-1}}. \quad (8.120)$$

Secondly, it can be seen, either directly from the definition or using the above approximation method for Π functions, that

$$(g(z)g(-z))^{(1-\alpha)/(\beta-\alpha)} = k(z^2), \quad (8.121)$$

where $k(z) \in \Pi_{\alpha-1}$. This gives

$$k(z^2) \ll \frac{1}{(1-z^2)^{1-\alpha}}. \quad (8.122)$$

From (8.120) and (8.122) we obtain

$$f(z) \ll \frac{1+z}{(1-z)^{\alpha+\beta-1}}\frac{1}{(1-z^2)^{1-\alpha}} = \frac{(1+z)^\alpha}{(1-z)^\beta} \quad (8.123)$$

which is the required result. The ingenious use of the superfluous terms in $g(-z)$ is based on an idea of D.A. Brannan [10]. It reduces to a completely elementary argument what had once been a very difficult problem. For example, this result includes the solution to the coefficient problem for functions of bounded boundary rotation, first solved by Brannan, Clunie and Kirwan [11]; the derivatives of such functions are contained in the classes $K(\alpha, \alpha + 2)$ (see the next section).

8.7.4

To prove part (ii) of our theorem we use the duality result. From the extreme point theorem 8.7.2 it is sufficient to prove that

$$\frac{(1+xz)^\alpha}{(1-z)^\beta} \ll \frac{(1+z)^\alpha}{(1-z)^\beta} \quad (|x| = 1), \quad (8.124)$$

when $\alpha \geq 1$ and $\beta \geq 1$.

To prove this we note that, since $(1+xz)^m \ll (1+z)^m$ for any non-negative integer m, we may assume that $1 < \alpha < 2$ and $\beta = 1$. Thus we consider

$$g(z) = \frac{(1+xz)^{1+\gamma}}{1-z}, \tag{8.125}$$

where $|x| = 1$ and $0 < \gamma < 1$. Differentiating gives

$$g'(z) = \frac{(1+xz)^\gamma}{(1-z)^{2-\gamma}} \frac{1+(\gamma+1)x-\gamma xz}{(1-z)^\gamma} \tag{8.126}$$

and it will be sufficient to show that the coefficients of this function are maximised in absolute value when $x = 1$. By part (i) we have

$$\frac{(1+xz)^\gamma}{(1-z)^{2-\gamma}} \ll \frac{(1+z)^\gamma}{(1-z)^{2-\gamma}}. \tag{8.127}$$

Therefore it remains to prove that

$$\frac{1+(\gamma+1)x-\gamma xz}{(1-z)^\gamma} \ll \frac{2+\gamma-\gamma z}{(1-z)^\gamma} \tag{8.128}$$

with the right-hand function having non-negative coefficients. Now it is clear that

$$\frac{1+(\gamma+1)x-\gamma xz}{(1-z)^\gamma} \ll \frac{1}{(1-z)^\gamma} + \frac{\gamma+1-\gamma z}{(1-z)^\gamma}, \tag{8.129}$$

which is the desired result, provided that the function $(\gamma+1-\gamma z)(1-z)^{-\gamma}$ has non-negative coefficients. This is certainly true of $\gamma(1-z)^{-\gamma}$, and therefore the proof is completed by observing that

$$\frac{d}{dz}\left(\frac{1-\gamma z}{(1-z)^\gamma}\right) = \frac{\gamma(1-\gamma)z}{(1-z)^{\gamma+1}}, \tag{8.130}$$

which has non-negative coefficients for $0 < \gamma < 1$.

8.8 The class $S(\alpha, \beta)$

For $\alpha \geq 0$ and $\beta \geq 0$ we say $f \in S(\alpha, \beta)$ if $f = g/h$ where $g \in \Pi_\alpha$ and $h \in \Pi_\beta$. Although $S(\alpha, \beta) \subset K(\alpha, \beta)$, it is not true in general that $K(\alpha, \beta) \subset S(\alpha, \beta)$. For example, functions in $K(1, 1)$ need not be ratios of Π_1 functions. Nevertheless, rational functions of the form $R = P/Q$, where P is a polynomial of degree m with no zeros in $\mathbb{U}$ and Q is a polynomial of degree n with no zeros in $\mathbb{U}$, lie in the class $S(m, n)$.

8.8.1

The necessary and sufficient condition that $f \in S(\alpha, \beta)$ is that $f(0) \neq 0$ and we can write

$$\frac{zf'(z)}{f(z)} = G(z) - H(z), \tag{8.131}$$

where $G(0) = H(0) = 0$, $\Re G(z) < \frac{\alpha}{2}$ in $\mathbb{U}$ and $\Re H(z) < \frac{\beta}{2}$ in $\mathbb{U}$. This is clearly necessary by putting $G = zg'/g$ and $H = zh'/h$. Conversely, given this representation, we define

$$g(z) = \exp\left(\int_o^z \frac{G(w)}{w} dw\right); \qquad h(z) = \exp\left(\int_o^z \frac{H(w)}{w} dw\right), \tag{8.132}$$

obtaining $f'/f = g'/g - h'/h$. Writing $k = fh/g$, we obtain $k'/k = 0$, and so k is a non-zero constant. With the normalisation $f(0) = 1$, we obtain $f = g/h$, where $g \in \Pi_\alpha$ and $h \in \Pi_\beta$.

8.8.2

Theorem (Ruscheweyh [47]) *If $f \in S(\alpha, \beta)$, where $\alpha \geq 0$ and $\beta \geq 0$, and $\phi \in T_0(\alpha + 1, \beta + 1)$, then $\phi * f \in S(\alpha, \beta)$.*

Proof We apply the convolution inequality (8.117) for $T_0(\alpha + 1, \beta + 1)$ in the case $\lambda = \delta = 1$, to deduce that for $\phi \in T_0(\alpha + 1, \beta + 1)$, $f \in K(\alpha, \beta)$ and $\Re L > 0$, we have

$$\Re \frac{\phi * fL}{\phi * f} > 0 \text{ in } \mathbb{U}. \tag{8.133}$$

Thus the linear operator $F \mapsto \frac{\phi * fF}{\phi * f}$ is convexity preserving on $\mathcal{A}$. We wish to prove that $\phi * f \in S(\alpha, \beta)$ when $f \in S(\alpha, \beta)$. Writing $f = g/h$, as in the definition, we have

$$\begin{aligned}\frac{z(\phi * f)'}{\phi * f} &= \frac{\phi * zf'}{\phi * f} = \frac{\phi * f(zf'/f)}{\phi * f} = \frac{\phi * f(zg'/g)}{\phi * f} - \frac{\phi * f(zh'/h)}{\phi * f} \\ &= \frac{\phi * fG}{\phi * f} - \frac{\phi * fH}{\phi * f},\end{aligned} \tag{8.134}$$

where $\Re G < \frac{\alpha}{2}$ and $\Re H < \frac{\beta}{2}$. By the convexity preserving property, the last expression has the form $P - Q$, where $P(0) = Q(0) = 0$, $\Re P < \frac{\alpha}{2}$ and $\Re Q < \frac{\beta}{2}$. Thus, as shown above, $\phi * f \in S(\alpha, \beta)$.

8.8.3

Of particular geometric interest are the classes $S(\alpha, \alpha+2)$. A function $v(z) = z + v_2 z^2 + \cdots$ is said to have **bounded boundary rotation at most** $k\pi$ if $v'(z) \in S(\frac{1}{2}k - 1, \frac{1}{2}k + 1)$. Here we must take $k \geq 2$. The geometric meaning is as follows. For $0 < r < 1$ consider the curve $\{w(\theta) = v(re^{i\theta})\colon 0 \leq \theta \leq 2\pi\}$; the direction of the tangent vector is given by $\arg dw = \arg w'(\theta) = \arg ire^{i\theta} v'(re^{i\theta})$. *Then v has bounded boundary rotation at most $k\pi$ iff* $\arg dw$ *has total variation not exceeding $k\pi$ for* $0 < r < 1$. The net variation around the curve is exactly 2π, since $v(0) = 0$ and $v'(0) = 1$; this follows from the argument principle, since part of the definition is that $v'(z) \neq 0$ in $\mathbb{U}$. Thus the definition puts a bound on the amount of tangential swing as the curve is traversed.

To prove our assertion, assume first that $v'(z) \in S(\frac{1}{2}k - 1, \frac{1}{2}k + 1)$. Then $v' = p/q$, where $p \in \Pi_{\frac{1}{2}k-1}$ and $q \in \Pi_{\frac{1}{2}k+1}$. With $z = re^{i\theta}$ we have

$$\frac{\partial}{\partial\theta} \arg re^{i\theta} v'(re^{i\theta}) = \Re\left(1 + \frac{zv''(z)}{v'(z)}\right) = 1 + \Re\frac{zp'(z)}{p(z)} - \Re\frac{zq'(z)}{q(z)} = 1 + P - Q, \tag{8.135}$$

where $P < \frac{1}{4}k - \frac{1}{2}$ and $Q < \frac{1}{4}k + \frac{1}{2}$. We set

$$B(\theta) = \int_0^\theta \left(\frac{1}{4}k + \frac{1}{2} - Q(t)\right) dt, \qquad A(\theta) = \int_0^\theta \left(\frac{1}{4}k - \frac{1}{2} - P(t)\right) dt, \tag{8.136}$$

which gives $\arg dw = B - A +$ constant, where A and B are increasing functions. Thus $\arg dw$ is a function of bounded variation, whose total variation does not exceed

$$A(2\pi) + B(2\pi) = \int_0^{2\pi} \left(\frac{1}{2}k - P(t) - Q(t)\right) dt = k\pi. \tag{8.137}$$

For the converse result, let us assume that on $\{|z| = r\}$ the function $\arg dw$ has total variation $k\pi$; then, since the net variation of $\arg dw$ is 2π, we can write

$$\arg dw = B(\theta) - A(\theta), \tag{8.138}$$

where $B(\theta)$ and $A(\theta)$ are increasing functions satisfying

$$\begin{aligned} B(2\pi) - B(0) + A(2\pi) - A(0) &= k\pi, \\ B(2\pi) - B(0) - A(2\pi) + A(0) &= 2\pi. \end{aligned} \tag{8.139}$$

(The decomposition we are using here for a function of bounded variation f is $f = \frac{1}{2}(V + f) - \frac{1}{2}(V - f)$, where V is the total variation function.) From these relations we obtain

$$A(2\pi) - A(0) = \left(\tfrac{1}{2}k - 1\right)\pi, \qquad B(2\pi) - B(0) = \left(\tfrac{1}{2}k + 1\right)\pi. \tag{8.140}$$

We now consider the harmonic function

$$\rho(z) = \frac{-1}{2\pi}\int_0^{2\pi} \Re\frac{r + ze^{-it}}{r - ze^{-it}}\left(\left(\frac{1}{4}k + \frac{1}{2}\right)t - B(t)\right)dt \tag{8.141}$$

defined for $|z| < r$. Then

$$\rho(z) = \Im\left(\frac{-i}{2\pi}\int_0^{2\pi} \frac{r + ze^{-it}}{r - ze^{-it}}\left(\left(\frac{1}{4}k + \frac{1}{2}\right)t - B(t)\right)dt\right). \tag{8.142}$$

The function

$$q_r(z) = \exp\left(\frac{-i}{2\pi}\int_0^{2\pi} \frac{r + ze^{-it}}{r - ze^{-it}}\left(\left(\frac{1}{4}k + \frac{1}{2}\right)t - B(t)\right)dt\right) \tag{8.143}$$

is analytic and $\neq 0$ for $|z| < r$ with $\rho(z) = \arg q_r(z)$. We have

$$\begin{aligned}\frac{zq_r'(z)}{q_r(z)} &= \frac{-i}{2\pi}\int_0^{2\pi} \frac{2rze^{it}}{(re^{it} - z)^2}\left(\left(\frac{1}{4}k + \frac{1}{2}\right)t - B(t)\right)dt \\ &= \frac{1}{2\pi}\int_0^{2\pi} \frac{d}{dt}\frac{r + ze^{-it}}{r - ze^{-it}}\left(\left(\frac{1}{4}k + \frac{1}{2}\right)t - B(t)\right)dt \\ &= \frac{-1}{2\pi}\int_0^{2\pi} \frac{r + ze^{-it}}{r - ze^{-it}}d\left(\left(\frac{1}{4}k + \frac{1}{2}\right)t - B(t)\right),\end{aligned} \tag{8.144}$$

the last expression coming by integrating by parts; here we have used the fact that the function $(\frac{1}{4}k + \frac{1}{2})t - B(t)$ takes the same values at $t = 0$ and $t = 2\pi$. We thus obtain

$$\Re\frac{zq_r'(z)}{q_r(z)} = -\left(\frac{1}{4}k + \frac{1}{2}\right) + \frac{1}{2\pi}\int_0^{2\pi} \Re\frac{r + ze^{-it}}{r - ze^{-it}}\,dB(t) > -\frac{1}{4}k - \frac{1}{2}. \tag{8.145}$$

In a similar manner we define

$$p_r(z) = \exp\left(\frac{-i}{2\pi}\int_0^{2\pi} \frac{r + ze^{-it}}{r - ze^{-it}}\left(\left(\frac{1}{4}k - \frac{1}{2}\right)t - A(t)\right)dt\right) \tag{8.146}$$

obtaining

$$\Re\frac{zp_r'(z)}{p_r(z)} = -\left(\frac{1}{4}k - \frac{1}{2}\right) + \frac{1}{2\pi}\int_0^{2\pi} \Re\frac{r + ze^{-it}}{r - ze^{-it}}\,dA(t) > -\frac{1}{4}k + \frac{1}{2}. \tag{8.147}$$

On the other hand

$$\frac{q_r(z)}{p_r(z)} = \exp\left(\frac{-i}{2\pi}\int_0^{2\pi} \frac{r+ze^{-it}}{r-ze^{-it}}(t - B(t) + A(t))dt\right). \tag{8.148}$$

This gives

$$\arg\frac{q_r}{p_r} = \frac{1}{2\pi}\int_0^{2\pi} \Re\frac{r+ze^{-it}}{r-ze^{-it}}\,(B(t) - A(t) - t)dt, \tag{8.149}$$

whose boundary values on $|z| = r$ are $B(t) - A(t) - t = \arg v'(re^{it})$. Thus from Poisson's formula we deduce that up to a multiplicative constant

$$\frac{q_r(z)}{p_r(z)} = v'(z) \quad (|z| < r). \tag{8.150}$$

For $|z| < r, q_r \in \Pi_{-\frac{1}{2}k-1}$ and $p_r \in \Pi_{-\frac{1}{2}k+1}$. Hence $v' \in S(\frac{1}{2}k - 1, \frac{1}{2}k + 1)$ for $|z| < r$. The containment results for the Kaplan classes imply that this conclusion still holds if the total variation of $\arg dw$ is smaller than $k\pi$. Because of the compactness of the class, we obtain the result on letting $r \to 1$.

8.9 The classes $T_0(\alpha, \beta)$

8.9.1 Robinson's conjectures

The theory which we have developed shows that these classes of functions have extensive structure preserving properties as convolution operators. However, only in the cases when $\alpha = 1$ do we have a definitive method for constructing members of the classes. When $\alpha > 1$, very little is known. Finding even one non-trivial function in any of these classes is likely to yield significant results. An interesting problem considered by S. Robinson [42] is the construction of polynomial approximate identities for each of the classes $T_0(\alpha, \beta)$. By this we mean a sequence of polynomials $\{\phi_n(z)\}$, where each ϕ_n has degree n and belongs to $T_0(\alpha, \beta)$, such that $\phi_n(z) \to (1 - z)^{-1}$ as $n \to \infty$ locally uniformly in $\mathbb{U}$. We refer to chapter 4, subsection 4.3.2 and following for a detailed account of this concept.

On the basis of numerical and other investigations Robinson has proposed the following very interesting conjectures.

8.9.1.1 Conjecture For $\alpha \geq 1, \beta \geq 1$ the Cesàro polynomials $C_n^{(\alpha+\beta-1)}(z) \in T_0(\alpha, \beta)$ $(n = 1, 2, \ldots)$.

We recall that these polynomials are given by

$$C_n^{(\alpha)}(z) = 1 + \sum_{k=1}^{n} c_n^{(k)}(\alpha) z^k \tag{8.151}$$

where

$$c_k^{(n)}(\alpha) = \prod_{j=1}^{k} \left(1 + \frac{\alpha}{n-k+j}\right)^{-1} = \frac{A_{n-k}^{(\alpha)}}{A_n^{(\alpha)}} \tag{8.152}$$

and

$$(1-z)^{-\alpha-1} = \sum_{n=0}^{\infty} A_n^{(\alpha)} z^n. \tag{8.153}$$

The conjecture asserts that the sequence $\{C_n^{(\alpha+\beta-1)}\}$ is an approximate identity for $T_0(\alpha, \beta)$. Applying the convolution theorem this would imply

(1) if $\phi \in T_0(\alpha, \beta)$, then the sequence of Cesàro means $C_n^{(\alpha+\beta-1)} * \phi$ of ϕ is a sequence of polynomials in $T_0(\alpha, \beta)$ converging locally uniformly in $\mathbb{U}$ to ϕ,
(2) if $f \in K(\alpha, \beta)$, then the sequence of Cesàro means $C_n^{(\alpha+\beta-1)} * f$ of f is a sequence of polynomials in $K(\alpha, \beta)$ converging locally uniformly in $\mathbb{U}$ to f.

In the case $\alpha = 1$ the conjecture asserts that the polynomials $C_n^{(\beta)} \in T(1, \beta)$ for $\beta \geq 1$ and $n = 1, 2, \ldots$. This result has been proved by Ruscheweyh [48].

8.9.1.2 Conjecture Define

$$P_n^{\beta}(z) = 1 + \sum_{k=1}^{n} \left(\prod_{j=0}^{k-1} \frac{n-j}{\beta+n+j} \right) z^k. \tag{8.154}$$

Robinson conjectures that $P_n^{\beta} \in T(1, \beta)$ for $n = 1, 2, \ldots$ and $\beta \geq 1$. He shows that $P_n^{\beta} \in T(1, 1)$ and that the sequence $\{P_n^{\beta}\}$ is an approximate identity for each $\beta \geq 1$. The conjecture follows therefore for $\beta = 1$. It is also true for $\beta = 3$, since P_n^3 is a renormalisation of the de la Vallée Poussin polynomials (see chapter 4, section 4.6). As Pólya and Schoenberg showed, the variation diminishing property of the de la Vallée Poussin means implies that they preserve the convexity of a curve and therefore $P_n^3 \in T(1, 3)$. Robinson gives substantial numerical evidence to support his conjecture and shows that the polynomials would be extremal in the classes $T(1, \beta)$.

8.10 The class $T(2, 2)$

The cases $\alpha = 2, \beta \geq 1$ lead to quadratic conditions, which are explicitly solvable, if not elegantly. The case $\alpha = 2, \beta = 2$ is particularly interesting because of its geometric interpretation. If f is a function analytic in $\mathbb{U}$, which omits all values on a half-line, then $\phi * f$ omits these same values for any function $\phi \in T(2, 2)$. Indeed, if a is the endpoint of the half-line, then, for a suitable real μ, the function $g(z) = e^{i\mu}(f(z) - a)$ omits all values on the negative real axis including 0 and therefore satisfies $|\arg g(z)| < \pi$ for $z \in \mathbb{U}$. It follows from the convolution inequality that for $\phi \in T(2, 2)$ we have $|\arg(\phi(z) * g(z))| < \pi$, i.e. $\phi * g$ takes no non-positive real values, which is equivalent to our assertion. We deduce the next result.

8.10.1

Theorem *Let D be a linearly accessible domain and let $f(z)$ be a function analytic in $\mathbb{U}$, which takes all its values in D. Then, if $\phi(z) \in T(2, 2)$, the function $\phi(z) * f(z)$ takes all its values in D.*

A domain is **linearly accessible** if, for each $w \notin D$, $\exists$ a half-line with endpoint w not meeting D. Thus ∞ can be 'seen' from every point on the boundary of D. A starlike domain has this property, and more generally so does the image $f(\mathbb{U})$ of a univalent close-to-convex function. Thus the functions in $T(2, 2)$ have a much stronger geometric preservation property than the convexity preservation discussed in chapter 2 and true of any $T(1, 1)$ function.

Because of this, it becomes interesting to give explicit characterisations of the class and to find ways of constructing members. For the first task we consider the defining condition

$$\phi(z) * \frac{(1 + xz)^2}{(1 - z)^2} \neq 0 \quad (|x| \leq 1, z \in \mathbb{U}). \tag{8.155}$$

Expanding in powers of x we obtain

$$\phi(z) * \frac{1}{(1 - z)^2} + 2x\left(\phi(z) * \frac{z}{(1 - z)^2}\right) + x^2\left(\phi(z) * \frac{z^2}{(1 - z)^2}\right) \neq 0 \quad (|x| \leq 1), \tag{8.156}$$

which gives, since $\phi(0) = 1$,

$$(\phi(z) + z\phi'(z)) + 2xz\phi'(z) + x^2(z\phi'(z) - \phi(z) + 1) \neq 0 \quad (|x| \leq 1). \tag{8.157}$$

To eliminate x, we require a necessary and sufficient condition on the coefficients of a quadratic polynomial $p(x) = ax^2 + bx + c$ that $p(x) \neq 0$ when

$|x| \le 1$. This is given by

$$|a\overline{b} - b\overline{c}| < |c|^2 - |a|^2. \tag{8.158}$$

To prove this we apply Walsh's theorem 5.2.7 of chapter 5 to deduce that an equivalent condition is

$$axy + \tfrac{1}{2}b(x+y) + c \neq 0 \quad (|x| \le 1, |y| \le 1). \tag{8.159}$$

This gives $(ay + \frac{1}{2}b)x + (\frac{1}{2}by + c) \neq 0$, or equivalently,

$$\left|ay + \tfrac{1}{2}b\right| < \left|\tfrac{1}{2}by + c\right| \quad (|y| \le 1). \tag{8.160}$$

Squaring and multiplying out we obtain

$$\Re(y(a\overline{b} - b\overline{c})) < |c|^2 - |a|^2|y|^2 - \tfrac{1}{4}|b|^2(1 - |y|^2), \tag{8.161}$$

and when $|y| = 1$ we obtain (8.158). Conversely, (8.158) implies (8.160) when $|y| = 1$. The result will follow for $|y| \le 1$ by the maximum principle, provided that $\frac{1}{2}by + c \neq 0$ for $|y| \le 1$. This is equivalent to $|b| < 2|c|$, which is easily deduced from (8.158)

Here we have

$$a = z\phi' - \phi + 1, \qquad b = 2z\phi', \qquad c = \phi + z\phi'. \tag{8.162}$$

The condition (8.158) reduces to

$$|\Re(\overline{b}\alpha) + i\Im b| < \Re(\overline{b}\alpha) + \Re\alpha, \tag{8.163}$$

where $\alpha = 2\phi - 1$. This implies $\Re\alpha > 0$, i.e. $\phi \in T(1, 1)$, which of course we already know. Squaring and multiplying out this inequality, we obtain

$$(\Im b)^2 < \Re\alpha(\Re(\alpha(1 + 2\overline{b}))), \tag{8.164}$$

or equivalently

$$4(\Im z\phi')^2 < \Re(2\phi - 1)\,\Re\{(1 + 4z\phi')(2\overline{\phi} - 1)\}. \tag{8.165}$$

Note that this implies the inequality

$$\Re\left(\frac{1 + 4z\phi'(z)}{2\phi(z) - 1}\right) > 0 \quad (z \in \mathbb{U}) \tag{8.166}$$

for $\phi \in T(2, 2)$. However, S. Robinson [42] has shown that this condition is not a sufficient condition for $\phi \in T(2, 2)$. Nevertheless, using the Clunie–Jack lemma 8.1.8, one can show that this inequality does imply that $\Re(2\phi - 1) > 0$ in $\mathbb{U}$.

9
Real polynomials

9.1 Real polynomials

9.1.1

A real polynomial is a polynomial which is real on the real axis, i.e. $P(x)$ is real when x is real. This is true iff the coefficients of P are real numbers. A real polynomial can also be defined by the property that $\overline{P(\overline{z})} = P(z)\ \forall z \in \mathbb{C}$. From this we see that $P(z) = 0$ iff $P(\overline{z}) = 0$, and therefore the zeros of P either are real or occur in conjugate pairs. Indeed, if a is a zero of P of multiplicity m, then $\overline{a}$ is a zero of P of multiplicity m. This follows from the fact that, if a is non-real, then $P(z)/(z-a)(z-\overline{a})$ is still a real polynomial, which, if $m > 1$, has a zero a of multiplicity $m-1$, and therefore the asssertion follows easily by induction.

From this we see that the number of non-real zeros of a real polynomial P is an even number, whether counted as distinct zeros or according to multiplicity. Therefore an odd degree polynomial has at least one real zero.

The derivative P' of a real polynomial P is real and therefore the non-real critical points of P also occur in conjugate pairs. By Rolle's theorem, if $P(a) = P(b)$, where $a < b$, then P' has a zero in (a, b). It follows that, if P has ν real zeros, then P has at least $\nu - 1$ real critical points. In particular, if all the zeros of P are real, then all the critical points of P are real.

By the Gauss–Lucas theorem 2.1.6 of chapter 2, the critical points of P lie in the convex hull of the set of zeros of P. For real polynomials we can supplement this result as follows. Let a and $\overline{a}$ be a pair of conjugate zeros of the real polynomial P. The **Jensen circle** associated with this pair is the circle whose diameter is $[a, \overline{a}]$, i.e. writing $a = a_1 + ia_2$, the circle $\{|z - a_1| = |a_2|\}$. We then have the next result.

9.1.2

Theorem *Let ζ be a non-real critical point of a real polynomial P. Then ζ lies inside or on one of the Jensen circles associated with the pairs of non-real zeros of P.*

Proof Writing $a = a_1 + ia_2$, we have

$$\begin{aligned}
\Im\left(\frac{1}{z-a} + \frac{1}{z-\overline{a}}\right) &= -\Im\left(\frac{z-a}{|z-a|^2} + \frac{z-\overline{a}}{|z-\overline{a}|^2}\right) \\
&= -\Im\left(\frac{(z-a)(z-\overline{a})(\overline{z}-a) + (z-\overline{a})(z-a)(\overline{z}-\overline{a})}{|z-a|^2|z-\overline{a}|^2}\right) \\
&= -2\Im\left(\frac{(z-a)(z-\overline{a})(\overline{z}-a_1)}{|z-a|^2|z-\overline{a}|^2}\right) \\
&= -2\Im\left(\frac{(z-a_1-ia_2)(z-a_1+ia_2)(\overline{z}-a_1)}{|z-a|^2|z-\overline{a}|^2}\right) \\
&= -2y\frac{|z-a_1|^2 - a_2^2}{|z-a|^2|z-\overline{a}|^2},
\end{aligned} \tag{9.1}$$

where $y = \Im z$. We see that this expression is negative if $y > 0$ and z lies exterior to the Jensen circle associated with a. It follows that, *if z is a point in the upper half-plane lying exterior to all the Jensen circles associated with the zeros of P, then*

$$\Im\left(\frac{P'(z)}{P(z)}\right) < 0. \tag{9.2}$$

In particular, $P'(z) \neq 0$ at such a point. Thus either ζ or $\overline{\zeta}$ (and hence both) lies inside or on a Jensen circle, as asserted.

9.1.3

Walsh [68] has pointed out the following interesting corollary of (9.2).

Corollary *Let P be a real polynomial and let $[a, b]$ be an interval of the real axis lying exterior to all the Jensen circles associated with the non-real zeros of P. If P has no zeros in $[a, b]$, then P has at most one critical point in $[a, b]$.*

Proof On a small rectangle R containing $[a, b]$ in its interior P has no zeros and contains only those critical points lying in $[a, b]$. Let Γ denote the positively oriented perimeter of R. By the argument principle, the number of critical points of P in $[a, b]$ is $d(P'/P, \Gamma)$. Now $\Im P'/P < 0$ on that part Γ_1 of Γ

above the axis, and hence $d(P'/P, \Gamma_1) = 0, \frac{1}{2}$ or $-\frac{1}{2}$. On the part Γ_2 below the axis, $\Im P'/P > 0$ and therefore $d(P'/P, \Gamma_2) = 0, \frac{1}{2}$ or $-\frac{1}{2}$. It follows that $|d(P'/P, \Gamma)| \leq 1$ and the conclusion follows.

9.1.4

The assertion (9.2) raises a number of interesting questions. It implies that the level region $\{z \in H^+ : \Im(P'(z)/P(z)) \geq 0\}$, where H^+ denotes the upper half-plane, lies inside the union of the Jensen circles associated with the zeros of P. The structure of this level region is more closely tied to the behaviour of P than the rather crude Jensen circles, which merely provide a useful bound. The possible topological structures in H^+ created by the components of this level region are *a priori* quite complicated, and could divide the half-plane in a variety of ways. However, computer evidence suggests that the actual topological possibilities are limited. This has been made precise in an interesting conjecture of Craven, Csordas and Smith [15], which we give next.

9.1.5

Craven–Csordas–Smith conjecture Let P be a real polynomial and let R denote the rational function P'/P. The conjecture states that the number of real critical points of R does not exceed the number of non-real zeros of P.

If $z_1, \ldots, z_n$ are the zeros of P, then

$$\frac{P'(z)}{P(z)} = \sum_{k=1}^{n} \frac{1}{z - z_k}, \tag{9.3}$$

and therefore

$$R'(z) = -\sum_{k=1}^{n} \frac{1}{(z - z_k)^2}. \tag{9.4}$$

If all the zeros of P are real, then clearly $R'(x) < 0$ for x real, and therefore R has no real critical points (observe that the poles of R are simple). Thus the conjecture holds in this case.

In general, the real critical points of R are precisely those points on the real axis at which the level set $\{z \in H^+ : \Im R(z) = 0\}$ meets the axis, i.e. the branch points of the level curves $\Im R(z) = 0$. Each component of the level set will consist of either a level curve with two (possibly coincident) endpoints on the axis, which are critical points of R, or a loop in H^+. In the latter case the loop will contain equally many zeros and poles of R (i.e. non-real critical points and zeros of P) and at least one of each. Any component of the level region $\{z \in H^+ : \Im R(z) \neq 0\}$ is bounded by such level curves together with part of the real axis. The number of poles of R on the boundary of such a region determines the valence of R in the region: depending upon whether $\Im R(z) > 0$ or < 0 in the region, R attains all values in H^+ or H^- equally often in the region. For further information on the structure of level regions we refer to chapter 10.

An important observation is the following. Since R has a positive residue at each pole, $\Im R(z) < 0$ to the immediate north of the pole, when the pole is real or lies in H^+. In particular, the poles on the boundary of the level regions $\{z \in H^+ : \Im R(z) > 0\}$ are non-real.

9.1.6

Remark We can establish the existence of non-real zeros of $(P'/P)'$ in the following case.

If (i) the zeros of P are simple and (ii) P has at least one real zero, then P'/P has at least one pair of non-real critical points.

Since P has simple zeros, the critical points of P'/P are the zeros of the polynomial $f = PP'' - P'^2$. If all the zeros of f are real, then f'/f has no real critical points. Therefore the zeros of $ff'' - f'^2$ are either non-real or multiple real zeros of f. However, a calculation gives

$$ff'' - f'^2 = P[PP''P'''' - P''^3 - P'^2P'''' - PP'''^2 + 2P'P''P'''], \quad (9.5)$$

and so P is a factor of $ff'' - f'^2$. It follows that the zeros of P are either non-real or multiple real zeros of f. But a common zero of f and P is a zero of P', and therefore a multiple zero of P. Thus, if P has only simple zeros, it has no real zeros.

9.1.7

Remark An interesting observation arises from considering an invariant form of the problem. If P is a real polynomial and w is real, we may consider the problem for the polynomial $P(z) - w$. This leads to the equation

$$PP'' - P'^2 = wP'' \quad (9.6)$$

or equivalently

$$P - \frac{P'^2}{P''} = w. \tag{9.7}$$

In other words we are interested in the level regions $\Im f = 0$ for the rational function $f = P - \frac{P'^2}{P''}$. Note that near ∞ f has the expansion $f(z) = -\frac{a_n}{n-1}z^n + \cdots$, where $a_n z^n$ is the leading term of P. Thus we see that large real values of w are attained at most twice near ∞ on the real axis. More precisely, if n is odd, each of $P - w$ and $f - w$ has one real zero near ∞; if n is even, either $P - w$ has two real zeros near ∞ and $f - w$ has none, or $f - w$ has two real zeros near ∞ and $P - w$ has none. The remaining poles of f are the zeros of P''. As there are $n - 2$ such points, there are at most $n - 2$ other real solutions of the equation $f = w$ for w large. Thus for w large there are at most n real points which are solutions of either $f = w$ or $P = w$. Therefore the conjecture holds for the polynomial $P - w$, when w is sufficiently large.

Of course we are asked to show that this is true for every real w. This leads to looking at the level curves $\Im f = 0$: $n - 1$ distinct curves emanate from ∞ in the upper half-plane. One of these curves can meet the real axis only at a critical point of f. We now note that

$$f' = P'\left(\frac{P'P'''}{P''^2} - 1\right), \tag{9.8}$$

from which we see that the critical points of f are either zeros of P' or critical points of P''/P'. Now the original conjecture applied to P' is that there are at most $n - 1$ such points on the real axis: for the number of real critical points of P''/P' is supposed not to exceed the number of *non-real* zeros of P'. Thus we are led to a restatement of the conjecture: f has at most $n - 1$ real critical points.

9.1.8 The conjecture for polynomials with exactly two non-real zeros

For this case there is no loss of generality in assuming that P has the form $P(z) = (z^2 + 1)Q(z)$, where Q has $n - 2$ real zeros. We are asked to show that $(P'/P)'$ has at most two real zeros. The Jensen circle is the unit circle, and therefore the problem reduces to showing that the equation

$$\sum_{k=1}^{n-2} \frac{1}{(x - a_k)^2} = \frac{2(1 - x^2)}{(1 + x^2)^2} \tag{9.9}$$

has at most two real solutions for $-1 < x < 1$, where the a_k are arbitrary real numbers.

Suppose that α and β are two solutions of this equation with $-1 < \alpha < \beta < 1$, and assume that there is a zero $a = a_k$ of P satisfying $\alpha < a < \beta$. Then we have

$$\frac{1}{(\alpha - a)^2} \leq \frac{2(1-\alpha^2)}{(1+\alpha^2)^2}, \tag{9.10}$$

from which we obtain

$$a - \alpha \geq \frac{1+\alpha^2}{\sqrt{2(1-\alpha^2)}}. \tag{9.11}$$

Similarly we obtain

$$\beta - a \geq \frac{1+\beta^2}{\sqrt{2(1-\beta^2)}}. \tag{9.12}$$

From these two inequalities we deduce that

$$\alpha + \frac{1+\alpha^2}{\sqrt{2(1-\alpha^2)}} \leq \beta - \frac{1+\beta^2}{\sqrt{2(1-\beta^2)}}. \tag{9.13}$$

On the other hand, it is easily checked that for $-1 < x < 1$ we have

$$\frac{1+x^2}{\sqrt{2(1-x^2)}} > |x|, \tag{9.14}$$

and therefore the left side of (9.13) is positive, whereas the right side is negative. This contradiction shows that there are no zeros of P between two real critical points of P'/P. It follows that the $n-2$ real zeros of P lie on the boundary of the unbounded component of $\{z \in H^+ : \Im P'(z)/P(z) < 0\}$. Furthermore the non-real zero of P at i lies on the boundary of this component. Hence the valence of P'/P in this component is $n-1$, and therefore by the Riemann–Hurwitz formula (see chapter 10), the component contains at least $n-2$ critical points of P'/P. Thus there are at least $2n-4$ non-real critical points, and therefore at most 2 real critical points of P'/P. The conjecture follows in this case.

9.1.9

This argument can be sharpened up as follows. For x real, x_j denoting the real zeros and $a_k = b_k + ic_k$ denoting the non-real zeros of P in H^+, we have

$$-\left(\frac{P'(x)}{P(x)}\right)' = \sum_j \frac{1}{(x-x_j)^2} + \sum_k \frac{2\left((x-b_k)^2 - c_k^2\right)}{\left((x-b_k)^2 + c_k^2\right)^2}. \tag{9.15}$$

Now for every real $x \neq b_k$ it is easily checked that

$$\frac{2\left(c_k^2 - (x-b_k)^2\right)}{\left((x-b_k)^2 + c_k^2\right)^2} \leq \frac{1}{4(x-b_k)^2}. \tag{9.16}$$

Therefore, if P'/P has a critical point at x real, then

$$\sum_j \frac{1}{(x-x_j)^2} \leq \sum_k \frac{1}{4(x-b_k)^2}. \tag{9.17}$$

Indeed this inequality holds at all points x for which $(P'(x)/P(x))' \geq 0$. Suppose, at a particular such x, $\nu - 1$ of the zeros x_k are closer to x than the zero x_j; i.e. $|x - x_j| \geq |x - x_k|$. Then the left-hand sum is $\geq \nu/(x-x_j)^2$. On the other hand, if b_k is the nearest of the b_i to x, then the right-hand sum is $\leq d/4(x-b_k)^2$, where $2d$ is the number of non-real zeros of P. Thus in these circumstances we obtain

$$|x - x_j| \geq \sqrt{\frac{4\nu}{d}}|x - b_k|. \tag{9.18}$$

If this inequality is satisfied at $x = \alpha$ and $x = \beta$, where $\alpha < x_j < \beta$, then we deduce

$$x_j - \alpha \geq \sqrt{\frac{4\nu}{d}}|\alpha - b_k|, \qquad \beta - x_j \geq \sqrt{\frac{4\nu}{d}}|\beta - b_k|, \tag{9.19}$$

from which we obtain

$$\beta - \alpha \geq \sqrt{\frac{4\nu}{d}}(|\alpha - b_k| + |\beta - b_k|) \geq \sqrt{\frac{4\nu}{d}}(\beta - \alpha). \tag{9.20}$$

Hence $d \geq 4\nu$.

9.1.10 *Geometry of the level curves*

The curves $\{z \in H^+ \colon P'(z)/P(z) \text{ is real}\}$ are either loops or half-curves (humps) with endpoints on the real axis. These last are precisely the real critical points of P'/P. Every such curve will contain at least one point maximising $\Im z$, i.e. a highest point. We will show that these points are zeros of P or zeros of P' or points at which P''/P' is real.

We may set $z(t)$ as a parametrisation of such a curve with

$$R(z(t)) = \frac{P'(z(t))}{P(z(t))} = t. \tag{9.21}$$

Suppose that $t = \tau$ is a maximum point with $z(\tau)$ not a zero of P. Then $z'(\tau)R'(z(\tau)) = 1$ and $z'(\tau)$ is real (since $y'(\tau) = 0$). Thus $R'(z(\tau))$ is real, and so either $z(\tau)$ is a zero of P' or $(R'/R)(z(\tau))$ is real. But $R'/R = P''/P' - R$, and hence in the latter case P''/P' is real at $z(\tau)$. This proves our assertion, and indeed shows that any point on the level curve, which is a turning point of the height function y, is of one of the three types.

Conversely, let a be a zero of P in H^+ and consider the level curve $\Im(P'/P) = 0$ passing through a. We may choose a parametrisation $z(t)$ for which $P(z(t)) = tP'(z(t))$, where $z(0) = a$. Then $z'(0)P'(a) = P'(a)$, and so if the zero a is simple, we obtain $z'(0) = 1$. Thus $y'(0) = 0$, and we have a turning point of the height function. Furthermore, a second differentiation gives $z''(0) = P''(a)/P'(a)$ and so $y''(0) = \Im(P''(a)/P'(a))$. When this is negative, the turning point is a local maximum, and when it is positive, the turning point is a local minimum. Either case may occur.

In a similar manner, at a point in H^+ on the level curve at which P''/P' is real and finite, we obtain $y' = 0$, and so the height has a turning point. However, a zero of P' need not occur at a turning point.

As an application of these considerations, let us show that *the conjecture holds when P' has all its zeros real.* For in this case we have $\Im(P''(z)/P'(z)) < 0$ for all $z \in H^+$. Thus there are no points in H^+ at which either $P' = 0$ or P''/P' is real. It follows that every turning point on a level curve corresponds to a zero of P. Further, every such turning point is a maximum point of y. From this it is clear that there is an exact correspondence between the level curves in H^+ and the non-real zeros in H^+. In particular *the number of real critical points of P'/P is exactly the number of non-real zeros of P.*

9.1.11 The case for real polynomials with purely imaginary zeros

An argument similar to those above proves the conjecture in the case that *the level set $\{z \in H^+ : \Im(P'(z)/P(z)) < 0\}$ is connected.* For in this case every real zero of P lies on the boundary of this set, as does every non-real zero of P in H^+. It follows that the valence of P'/P in this set is $\nu + \mu$, where ν denotes the number of real zeros of P and μ denotes the number of non-real zeros of P in H^+; thus $\nu + 2\mu = n$. By the Riemann–Hurwitz formula, P'/P has at least $\nu + \mu - 1$ critical points in the level set, and therefore at least $2(\nu + \mu - 1)$ non-real critical points. It follows that the number of real critical points is at most $2n-2-2(\nu+\mu-1)=2\mu$, which is the number of non-real zeros of P, the desired result.

This fact, together with other considerations, suggests that we consider polynomials with no real zeros in testing for counter-examples. The simplest general case occurs when all the zeros are imaginary.

We consider the case where P has even degree $2n$ and imaginary zeros $\pm ic_k$, where $c_k > 0$ for $1 \le k \le n$. Then

$$R(z) = \frac{P'(z)}{P(z)} = \sum_{k=1}^{n} \frac{2z}{z^2 + c_k^2}. \tag{9.22}$$

This gives

$$R'(z) = -\sum_{k=1}^{n} \frac{2\left(z^2 - c_k^2\right)}{\left(z^2 + c_k^2\right)^2} \tag{9.23}$$

and therefore replacing z^2 by z and c_k^2 by a_k the problem reduces to estimating the number of positive zeros of

$$F(z) = \sum_{k=1}^{n} \frac{z - a_k}{(z + a_k)^2}, \tag{9.24}$$

where $a_k > 0$ for $1 \le k \le n$. According to the conjecture there should be at most n positive zeros (note that there are no negative zeros). The maximum number of zeros is $2n-1$. The result is trivial for $n = 1$. For $n \ge 2$ we will show that F has at most $2n-3$ positive zeros.

Replacing z by $1/z$ we may prove the same result for the function

$$H(z) = \sum_{k=1}^{n} \frac{1 - a_k z}{(1 + a_k z)^2}. \tag{9.25}$$

Let $G(z) = \prod_{k=1}^{n}(1 + a_k z)$. Then

$$H(z) = \sum_{k=1}^{n} \left\{ \frac{1}{1 + a_k z} - \frac{2a_k z}{(1 + a_k z)^2} \right\} = n - \frac{zG'(z)}{G(z)} - 2z\left(\frac{zG'(z)}{G(z)}\right)', \tag{9.26}$$

and the zeros of H occur at the zeros of the polynomial

$$Q(z) = G^2(z)H(z) = nG^2(z) - 3zG'(z)G(z) + 2z^2(G'^2(z) - G(z)G''(z)). \tag{9.27}$$

Writing $G(z) = 1 + \sum_1^n A_k z^k$ a calculation gives

$$Q(z) = n + (2n-3)A_1 z + \cdots - A_n A_{n-1} z^{2n-1} \tag{9.28}$$

from which it follows that Q has degree $2n-1$ with its first two coefficients positive and its last coefficient negative for $n > 1$. Thus the number of sign changes in the sequence of coefficients of Q cannot exceed $2n-3$, and so by Descartes' rule of signs (see section 9.2), Q has at most $2n-3$ positive zeros.

9.1.12

For the general case (with imaginary zeros) we need to show that Q has at most n positive zeros. Since the number of positive zeros is odd, we need to show that, (i) when n is even, the number of positive zeros is at most $n-1$ and (ii) when n is odd, the number of positive zeros is at most n. Using Descartes' rule, this would follow if in case (i) we showed that the coefficients for z^k $(0 \le k \le n-1)$ were positive and in case (ii) the coefficients for $0 \le k \le n-2$ were positive.

To study this possibility we expand Q in terms of the coefficients A_k of G, obtaining

$$Q(z) = \sum_{r=0}^{2n-1} \left\{ \sum_{j+k=r} (n-k-2k^2+2jk)A_j A_k \right\} z^r \tag{9.29}$$

and so we should like to show that

$$\sum_{j+k=r} (n-k-2k^2+2jk)A_j A_k > 0 \tag{9.30}$$

for $0 \le r \le n-1$ (n even) and for $0 \le r \le n-2$ (n odd). So far we have proved this for $n=2$ and $n=3$, which confirms the conjecture for polynomials of degrees 4 and 6 which have purely imaginary zeros.

To extend this consider the case $r=2$. The coefficient of z^2 is $2(n-5)A_2 + (n-1)A_1^2$ and, since the coefficients A_k are all positive, this is clearly positive for $n \ge 5$. More specifically, the particular form of G, which has only negative zeros, leads to the inequality

$$A_1^2 - 2A_2 = \sum_1^n a_k^2 > 0. \tag{9.31}$$

Thus $2(n-5)A_2 + (n-1)A_1^2 > 4(n-3)A_2 \ge 0$ for $n \ge 3$.

9.1.13

To proceed further we relate the expressions

$$\Sigma_k = \sum_{m=1}^{n} a_m^k \tag{9.32}$$

to the coefficients A_k. We have

$$\frac{zG'(z)}{G(z)} = \sum_{j=1}^{n} \frac{a_j z}{1 + a_j z} = \sum_{j=1}^{n} \sum_{p=1}^{\infty} a_j^p (-1)^{p-1} z^p = \sum_{p=1}^{\infty} \Sigma_p (-1)^{p-1} z^p, \tag{9.33}$$

and therefore

$$rA_r = \sum_{k=1}^{r} A_{r-k} \Sigma_k (-1)^{k-1} \quad (1 \leq r \leq n). \tag{9.34}$$

9.1.14

We next consider the case $r = 3$. We have

$$\sum_{j+k=3} (n - k - 2k^2 + 2jk) A_j A_k = (2n - 21) A_3 + (2n - 5) A_1 A_2. \tag{9.35}$$

From the above we have $3A_3 = A_2 \Sigma_1 - A_1 \Sigma_2 + \Sigma_3$; furthermore it is clear that $A_1 \Sigma_2 > \Sigma_3$. Hence $3A_3 < A_1 A_2$ and therefore

$$\sum_{j+k=3} (n - k - 2k^2 + 2jk) A_j A_k > (8n - 36) A_3 \tag{9.36}$$

and this is > 0 for $n \geq 5$. For the case $n = 4$ we have from the above

$$\begin{aligned} 3A_1 A_2 - 13A_3 = {} & 3a_1^2(a_2 + a_3 + a_4) + 3a_2^2(a_1 + a_3 + a_4) \\ & + 3a_3^2(a_1 + a_2 + a_4) + 3a_4^2(a_1 + a_2 + a_3) - a_1 a_2 a_3 \\ & - a_1 a_2 a_4 - a_1 a_3 a_4 - a_2 a_3 a_4. \end{aligned} \tag{9.37}$$

Now we may assume that $0 < a_1 \leq a_2 \leq a_3 \leq a_4$. We can then absorb the negative terms into the sum of non-negative terms

$$a_4 a_1 (a_4 - a_2) + a_4 a_2 (a_4 - a_3) + a_4 a_3 (a_4 - a_1) + a_3 a_2 (a_2 - a_1), \tag{9.38}$$

leaving a residue of positive terms. Thus $3A_1 A_2 - 13A_3 > 0$.

We have thus shown that *the Q coefficient of z^2 is positive for $n \geq 3$ and the coefficient of z^3 is positive for $n \geq 4$. In particular the conjecture holds for polynomials with purely imaginary zeros of degrees* 2, 4, 6, 8 *and* 10.

9.1.15 A general result

We can write $H(z)$ in the form

$$H(z) = n - \frac{3zG'(z)}{G(z)} - 2z^2\left(\frac{G'(z)}{G(z)}\right)' = n - 3\sum_{k=1}^{n}\frac{a_k z}{1+a_k z} + 2\sum_{k=1}^{n}\frac{a_k^2 z^2}{(1+a_k z)^2} \tag{9.39}$$

which gives

$$Q(z) = nG^2(z) - 3zG'(z)G(z) + R(z), \tag{9.40}$$

where $R(z) = 2\sum_{k=1}^{n}(a_k^2 z^2 G^2(z)/(1+a_k z)^2)$ is a polynomial with non-negative coefficients. Writing $G^2(z) = F(z) = \sum_{k=0}^{2n} B_k z^k$, we obtain

$$Q(z) = \sum_{k=0}^{2n}\left(n - \frac{3}{2}k\right)B_k z^k + R(z), \tag{9.41}$$

and, since each $B_k > 0$, the Q-coefficient of z^r is > 0 for $0 \le r \le \frac{2}{3}n$.

9.1.16 The case for real polynomials with purely imaginary zeros except for a zero at 0

We consider the case where P has odd degree $2n+1$ with a simple zero at 0 and imaginary zeros $\pm ic_k$, where $c_k > 0$ for $1 \le k \le n$. Then

$$R(z) = \frac{P'(z)}{P(z)} = \frac{1}{z} + \sum_{k=1}^{n}\frac{2z}{z^2 + c_k^2}. \tag{9.42}$$

This gives

$$R'(z) = -\frac{1}{z^2} - \sum_{k=1}^{n}\frac{2(z^2 - c_k^2)}{\left(z^2 + c_k^2\right)^2} \tag{9.43}$$

and therefore replacing z^2 by z and c_k^2 by a_k the problem reduces to estimating the number of positive zeros of

$$F(z) = \frac{1}{2z} + \sum_{k=1}^{n}\frac{z - a_k}{(z + a_k)^2}, \tag{9.44}$$

where $a_k > 0$ for $1 \le k \le n$. According to the conjecture there should be at most n positive zeros (note that there are no negative zeros). The maximum number of zeros is $2n$. We will show that F has at most $2n-2$ positive zeros.

Replacing z by $1/z$ we may prove the same result for the function

$$H(z) = \frac{1}{2} + \sum_{k=1}^{n}\frac{1 - a_k z}{(1 + a_k z)^2}. \tag{9.45}$$

Let $G(z) = \prod_{k=1}^{n}(1 + a_k z)$. Then

$$H(z) = \frac{1}{2} + \sum_{k=1}^{n}\left\{\frac{1}{1+a_k z} - \frac{2a_k z}{(1+a_k z)^2}\right\} = n + \frac{1}{2} - \frac{zG'(z)}{G(z)} - 2z\left(\frac{zG'(z)}{G(z)}\right)', \tag{9.46}$$

and the zeros of H occur at the zeros of the polynomial

$$Q(z) = G^2(z)H(z) = \left(n + \frac{1}{2}\right)G^2(z) - 3zG'(z)G(z) + 2z^2(G'^2(z) - G(z)G''(z)). \tag{9.47}$$

Writing $G(z) = 1 + \sum_1^n A_k z^k$ a calculation gives

$$Q(z) = n + \frac{1}{2} + (2n-2)A_1 z + \cdots + 0z^{2n-1} + \frac{1}{2}A_n^2 z^{2n} \tag{9.48}$$

from which it follows that Q has degree $2n$ with its first two coefficients positive and its last coefficient positive for $n > 1$. Thus the number of sign changes in the sequence of coefficients of Q cannot exceed $2n - 2$, and so by Descartes' rule of signs, Q has at most $2n - 2$ positive zeros.

9.1.17

For the general case (with imaginary zeros) we need to show that Q has at most n positive zeros. Since the number of positive zeros is even, we need to show that, (i) when n is even, the number of positive zeros is at most n and (ii) when n is odd, the number of positive zeros is at most $n - 1$. Using Descartes' rule, this would follow if in case (i) we showed that the coefficients for z^k $(0 \le k \le n-2)$ were positive and in case (ii) the coefficients for $0 \le k \le n - 1$ were positive.

To study this possibility we expand Q in terms of the coefficients A_k of G, obtaining

$$Q(z) = \sum_{r=0}^{2n}\left\{\sum_{j+k=r}\left(n + \frac{1}{2} - k - 2k^2 + 2jk\right)A_j A_k\right\}z^r \tag{9.49}$$

and so we should like to show that

$$\sum_{j+k=r}\left(n + \frac{1}{2} - k - 2k^2 + 2jk\right)A_j A_k > 0 \tag{9.50}$$

for $0 \le r \le n-2$ (n even) and for $0 \le r \le n-1$ (n odd). So far we have proved this for $n = 1$ and $n = 2$, which confirms the conjecture for polynomials of degrees 3 and 5 which have a simple zero at the origin and otherwise purely imaginary zeros.

To extend this consider the case $r = 2$. The coefficient of z^2 is $(2n-9)A_2 + (n-\frac{1}{2})A_1^2$ and, since the coefficients A_k are all positive, this is clearly positive for $n \geq 5$. More specifically, the particular form of G, which has only negative zeros, leads to the inequality

$$A_1^2 - 2A_2 = \sum_{1}^{n} a_k^2 > 0. \tag{9.51}$$

Thus $(2n-9)A_2 + (n-\frac{1}{2})A_1^2 > (4n-10)A_2 \geq 0$ for $n \geq 3$.

It follows that the conjecture holds for polynomials of degree 7 or 9 with a simple zero at the origin and otherwise purely imaginary zeros.

9.2 Descartes' rule of signs

9.2.1

Theorem: Descartes' rule *Let $P(z) = a_0 + a_1 z + \cdots + a_n z^n$ be a polynomial of degree n with real coefficients a_k. The number of positive real zeros of P does not exceed the number of sign changes in the sequence $a_0, a_1, \ldots, a_n$.*

In counting the sign changes, zero terms are ignored; if a_k and a_{k+j} are successive non-vanishing terms, then a sign change occurs iff $a_k a_{k+j} < 0$. For example, the polynomial $-3 + x^2 + 7x^5 - 2x^8$ has two sign changes, and therefore, according to Descartes' rule, at most two real zeros > 0. By applying the rule to both $P(z)$ and $P(-z)$, we can obtain an upper bound for the number of real zeros of P. For example, the above polynomial becomes $-3 + x^2 - 7x^5 - 2x^8$, which again has two sign changes. Therefore, the polynomial $-3 + x^2 + 7x^5 - 2x^8$ has at most four real zeros.

For the proof of Descartes' rule we proceed by induction. A polynomial of degree 0 has no zeros and no sign changes, as there is only one term. Let P have degree $n \geq 1$ and assume the result has been proved for polynomials up to degree $n-1$. Suppose that the coefficient sequence of P has m sign changes; clearly $m \leq n$. We consider the polynomial P', which has degree $n-1$. P' is obtained by deleting a_0 but, otherwise, not changing the sign of any other term. Therefore the coefficient sequence for P' has either (i) $m-1$ sign changes or (ii) m sign changes. In case (i) the induction hypothesis implies that P' has at most $m-1$ positive zeros. Applying lemma 9.2.2 below we deduce that P has at most m positive zeros. In case (ii) the induction hypothesis implies that P' has at most m positive zeros. Furthermore, as the coefficient sequences for both P and P' have the same number of sign changes, it follows that either (iii) $a_0 = 0$ or (iv) a_0 and a_j have the same sign, where a_j is the first non-vanishing coefficient for $j \geq 1$. In case (iii) $P(x)$ and $P(x)/x$ have the same positive zeros; since $P(x)/x$ has degree $n - 1$ and m sign changes, it has at most m positive zeros by the induction hypothesis. In case (iv) either $P(0) > 0$ and $P(x)$ is increasing for small $x > 0$, or $P(0) < 0$ and $P(x)$ is decreasing for small $x > 0$. In either case $P(x)$ has no zero in the interval $(0, a]$, where a is the first positive zero of P'. It follows from the lemma below that P has at most m positive zeros. Thus the result follows by induction.

9.2.2

Lemma *Let $f(x)$ be real and analytic in a closed interval $[a, b]$ and suppose that $f'(x)$ has m zeros in (a, b) counted according to multiplicity. Then $f(x)$ has at most $m + 1$ zeros in (a, b) counted according to multiplicity.*

Proof Rolle's theorem implies that f has at most finitely many zeros in (a, b). To be precise, let d denote the number of distinct zeros of f in the interval and ν the number of zeros of f counted according to multiplicity. At a zero of f of multiplicity α_j the derivative f' has $\alpha_j - 1$ zeros; furthermore, by Rolle's theorem f' has at least one zero between two successive zeros of f. It follows that

$$m \geq \sum_{j=1}^{d} (\alpha_j - 1) + d - 1 = \nu - d + d - 1 = \nu - 1, \tag{9.52}$$

and so $\nu \leq m + 1$, as required.

9.2.3

Descartes' rule of signs can be sharpened in the case that P has all its zeros real.

Theorem *If $P(z)$ is a polynomial with only real zeros, then the number of positive real zeros of P is equal to the number of sign changes in the coefficient sequence $a_0, a_1, \ldots, a_n$.*

For the proof we again proceed by induction, assuming that P has degree n and the result is proved for polynomials of degree $n - 1$. If $a_0 = 0$, the result is immediate by considering $P(x)/x$. Otherwise we may assume $a_0 > 0$ and let a_j be the next non-vanishing coefficient. Suppose the coefficient sequence for P has m sign changes. As before either (i) P' generates m sign changes or (ii) P' generates $m - 1$ sign changes. In case (i) $a_j > 0$ and $P(x)$ is initially increasing from a positive value $P(0) = a_0$. Thus P has no zeros in $(0, a]$, where a is the first positive zero of P'. Now as P has all its zeros real, the zeros of P' are interspersed between those of P; therefore between each successive pair of zeros of P' there is just one zero of P; moreover after the last zero of P, there is no zero of P'. Thus P has exactly m positive zeros. In case (ii) $a_j < 0$ and P is initially decreasing from a positive value. From the preceding interspersion considerations, it follows that P has one zero in $(0, a)$. Arguing as before it follows that P has exactly m positive zeros.

9.3 Strongly real rational functions

9.3.1

Let $R(z)$ be a real rational function whose zeros and poles are real (∞ will be regarded as a real point). The zeros and poles are said to be **interspersed** if between any two zeros there is a pole and between any two poles there is a zero. The word **between** should be interpreted in the following sense: if a and b are zeros, then there is a pole on the line segment joining a to b – and in addition a pole off this line segment (which may be ∞); effectively we regard the extended line as being circular. From this we see that R has equally many zeros and poles and the zeros and poles are simple.

9.3.2

Lemma *If $R(z)$ is a real rational function with interspersed zeros and poles, then $R(z)$ is real only on the real axis.*

Perhaps the most attractive proof of this is purely geometric. Let z be a point in the upper half-plane; if $a < b$ are real points, then $\arg\left(\frac{z-b}{z-a}\right)$ measures the angle subtended at z by the segment $[a, b]$; a sum of such angles for pairs of points succeeding each other down the axis clearly cannot exceed the full angle π in absolute value. But $|\arg R(z)|$ is exactly a sum of such angles by suitably pairing off the zeros with succeeding (or preceding) poles. Thus $|\arg R(z)| < \pi$ in the upper half-plane and so R is non-real there.

A rational function which is real only on the real axis will be called **strongly real**.

9.3.3

Lemma *Let $R(z)$ be a real rational function of degree n satisfying $\Im R(z) < 0$ for z in the upper half-plane. Then $R(z)$ can be written in the form*

$$R(z) = -az + b + \sum_{k=1}^{m} \frac{c_k}{z - x_k} \tag{9.53}$$

where x_k and b are real, $a \geq 0$ and $c_k > 0$. If $a = 0$, then $m = n$; otherwise $m = n - 1$.

Proof The finite poles of R are clearly real and every pole is simple, as otherwise, by considering the local behaviour near a pole, R would take real values in the upper half-plane. Thus R has the above form with a, b, x_k and c_k all real. When $z = x_k + iy$ $(y > 0)$, then the kth term in the sum is c_k/iy, which is imaginary and large in absolute value as $y \to 0$. It then dominates $\Im R$ and so $c_k > 0$, as $\Im R < 0$ in the upper half-plane. A similar argument shows that, if R has a pole at ∞, then $a > 0$.

9.3.4

Lemma *A rational function R of the form 9.53 has its zeros and poles interspersed on the real axis.*

For we have

$$R'(x) = -a - \sum_{k=1}^{m} \frac{c_k}{(x - x_k)^2} < 0, \tag{9.54}$$

for x real and $\neq x_k$ $(1 \leq k \leq m)$. The function R is therefore strictly decreasing in each of the intervals between successive poles. Furthermore R clearly decreases from $+\infty$ to $-\infty$ in each such interval and therefore takes every real value, including 0, exactly once in the interval. Thus the zeros and poles of R are interspersed.

9.3.5

Summarising, we obtain the following theorem.

Theorem *Let $R(z)$ be a real rational function of degree n. The following conditions are equivalent.*

(i) *The zeros and poles of R are interspersed on the real axis;*
(ii) *$R(z)$ is strongly real;*
(iii) *$R(z)$ can be written in the form*

$$R(z) = -az + b + \sum_{k=1}^{m} \frac{c_k}{z - x_k} \tag{9.55}$$

where all x_k and b are real, $a \geq 0$ and all $c_k > 0$ (or $a \leq 0$ and all $c_k < 0$). If $a = 0$, then $m = n$; otherwise $m = n - 1$.

9.3.6

Notice that we have also shown that a strongly real rational function R has no real critical points (including ∞). Conversely, if R is a real rational function with no real critical points (including ∞), then, if all the poles of R are real, R is strongly real. For the absence of real critical points implies that any level curve (other than the axis) on which R is real is a loop in either the upper or lower half-plane, and such a curve must contain a pole, contradicting the hypothesis. Therefore R is real only on the real axis.

9.3.7

Example Let $f(z)$ be an entire function of the form

$$f(z) = P(z)e^{-az^2+bz+c}, \tag{9.56}$$

where $a \geq 0$, b and c are real and $P(z)$ is a real polynomial all of whose zeros are real. Then $R(z) = f'(z)/f(z)$ is a strongly real rational function.

9.3.8

The following result gives a general representation for a real rational function in terms of two strongly real rational functions. A strongly real rational function is said to have **positive (negative) type** if it has positive (negative) imaginary part in the upper half-plane.

Theorem *Let $R(z)$ be a non-constant real rational function. Either R is strongly real or there exist two strongly real rational functions S and T of positive type such that*

$$R(w) = \frac{S(w)T(w)+1}{T(w)-S(w)}. \tag{9.57}$$

Proof Let $\omega(z) = i\frac{1+z}{1-z}$ and $\phi(w) = \frac{w-i}{w+i}$; ϕ is the inverse of ω, where ω is a 1–1 mapping of the unit disc onto the upper half-plane; the real axis corresponds to the unit circle with ∞ corresponding to $z = 1$. Since R is real on the real axis, the function $F = \phi \circ R \circ \omega$ is a rational function mapping the unit circle to the unit circle. It follows that $F = B/C$, where B and C are Blaschke products. If C is a constant of modulus 1, then R is strongly real of positive type; if B is a constant of modulus 1, then R is strongly real of negative type. Otherwise the converse argument shows that a Blaschke product B can be written in the form $B = \phi \circ S \circ \omega$, where S is strongly real of positive type; similarly $C = \phi \circ T \circ \omega$, where T is strongly real of positive type. We therefore have

$$R \circ \omega = \omega \circ F = \omega \circ \left(\frac{\phi \circ S \circ \omega}{\phi \circ T \circ \omega} \right) \tag{9.58}$$

which gives

$$R(w) = \omega \circ \left(\frac{\phi(S(w))}{\phi(T(w))} \right) = \frac{S(w)T(w) + 1}{T(w) - S(w)} \tag{9.59}$$

after some simple algebra.

In the above argument we may assume that B and C have no common zeros or poles; this means that S and T do not take the values $\pm i$ simultaneously. Writing $S = b/a$ and $T = d/c$ where a, b are relatively prime real polynomials with real zeros interspersed and similarly for d and c, we obtain

$$R = \frac{ac + bd}{ad - bc}. \tag{9.60}$$

Now if $ad - bc$ and $ac + bd$ have a common zero, then at that zero we obtain $a^2 + b^2 = 0$ and $c^2 + d^2 = 0$, which implies that S and T take one of the values $\pm i$ at the zero. Since this does not happen, $ac + bd$ and $ad - bc$ are relatively prime.

9.3.9

In particular we deduce

Corollary *Every real polynomial P of degree > 1 can be written in the form*

$$P = ac + bd, \tag{9.61}$$

where b/a and d/c are strongly real rational functions of positive type and where $ad - bc = 1$. In particular P is the sum of two polynomials each having only real zeros.

For polynomials of degree ≤ 1 we can still obtain this form provided that we allow b/a or d/c to be constant (including ∞). We get $P(z)$ from the choices $a = 1, b = P(z), c = 0, d = 1$.

9.4 Critical points of real rational functions

9.4.1

Theorem *Let $Q(z)$ and $T(z)$ be real rational functions with no common poles and set $R(z) = Q(z) + T(z)$. Suppose that*

(i) *Q is a strongly real rational function with a pole at ∞,*
(ii) *T has a zero of order m at ∞.*

Then R has at least $m - 1$ non-real critical points.

Proof We observe firstly that the conditions of the theorem are equivalent to the following representations for Q and T. We have

$$T(z) = \frac{c}{z^m} + \text{higher powers of } \frac{1}{z}, \quad \text{where } c \neq 0 \text{ is real,}$$
$$Q(z) = a + bz + \sum_{k=1}^{N} \frac{c_k}{a_k - z}, \tag{9.62}$$

where a and a_k are real and b and c_k are real and have the same sign, which we may take as positive. We will show that R has at least as many non-real critical points as the function $bz + \frac{c}{z^m}$.

From the representation for R we see that the mapping $w = R(z)$ is 1–1 in the neighbourhood of ∞, mapping this neighbourhood onto a neighbourhood of ∞ in the w-plane. As real values map to real values, the conditions of the theorem imply that the restriction of $R(z)$ to values z in the neighbourhood lying in the upper half-plane H^+ maps to values w in H^+. Thus there is exactly one unbounded component U in H^+ of the set $\{z\colon \Im R(z) > 0\}$. Suppose that $R'(z)$ has exactly χ zeros in U. Then, if ν denotes the number of disjoint curves in H^+ on which R is real and which border U, the Riemann–Hurwitz formula (see chapter 10) shows that $w = R(z)$ is a $(\chi + 1 - \nu)$-fold mapping of U onto H^+; i.e. as z ranges over U, each value $w \in H^+$ is attained exactly $\chi + 1 - \nu$ times, with the usual counting according to multiplicity. This implies that, in addition to the pole at ∞, there are $\chi - \nu$ poles of R on ∂U. Now every pole of T is a finite pole of R and so T has at most $\chi - \nu$ poles on ∂U. However, T has a zero of order m at ∞, and therefore T attains small values of $w \in H^+$ exactly m times near ∞. All unbounded components of the set $\{z \in H^+ \colon \Im T(z) > 0\}$ lie in U, since $\{\Im T(z) > 0 \text{ and } \Im z > 0\} \Rightarrow \{\Im R(z) > 0\}$. As large values of $w \in H^+$ are attained by T the same number of times as small values in each such component, there are at least μ poles of T on ∂U, where (i) if m is even, then $\mu = \frac{1}{2}m$, (ii) if m is odd, then $\mu = \frac{1}{2}(m - 1)$ if $c > 0$ and $\mu = \frac{1}{2}(m + 1)$

if $c < 0$. Thus $\chi - \nu \geq \mu \geq \frac{1}{2}(m-1)$. Finally, as the non-real zeros of R' occur in conjugate pairs, R' has at least 2χ non-real zeros, and therefore at least 2μ non-real zeros.

9.4.2

Corollary *Let $R(z)$ be a real rational function and suppose that the expansion of R at ∞ has the form*

$$R(z) = a + bz + \frac{c}{z^m} + \textit{higher powers of } \frac{1}{z},$$

where $b \neq 0$ and $c \neq 0$. Then R has at least as many non-real critical points as the function $bz + \frac{c}{z^m}$. In particular R has at least $m-1$ non-real critical points.

Since the number of non-real zeros of R' is independent of any Moebius transformation which leaves the real axis fixed (i.e. if ϕ and ψ are two such transformations, then $(\phi \circ R \circ \psi)'$ has the same number of non-real zeros as R'), we can transform our local knowledge at ∞ to local knowledge at a finite point on the real axis, or *vice versa*.

For example, let $S(z)$ be a rational function, whose expansion about the origin takes the form $S(z) = \sigma_0 + \sigma_1 z + \sigma_m z^m + \cdots$, where $m \geq 2$, $\sigma_1 \neq 0$, $\sigma_m \neq 0$, and consider the rational function

$$R(z) = \frac{1}{S\left(\frac{1}{z}\right) - \sigma_0} = \frac{z}{\sigma_1 + \sigma_m / z^{m-1} + \cdots} = \frac{z}{\sigma_1} - \frac{\sigma_m}{\sigma_1} \Big/ z^{m-2} + \cdots$$

expanded about ∞. We obtain the next result.

9.4.3

Corollary *Let $S(z)$ be a real rational function, which has an expansion about the origin in the form*

$$S(z) = A + Bz + Cz^m + \cdots,$$

where $B \neq 0$, $C \neq 0$ and $m \geq 3$. Then the derivative $S'(z)$ has at least as many non-real zeros as the polynomial $B + Cz^{m-1}$ and therefore at least $m-3$ non-real zeros. In particular, we deduce that a real polynomial P of the form

$$P(z) = A + Bz^{m+1} + \cdots,$$

where $A \neq 0$, $B \neq 0$, has at least as many non-real zeros as $A + Bz^{m+1}$ and therefore at least $m-1$ non-real zeros.

9.4.4

Theorem *Let $f(z) = Q(z)e^{P(z)}$, where Q is a real rational function and P is a real polynomial of degree $n \geq 2$. Then f'' has at least $n - 2$ non-real zeros.*

Proof We have $f'/f = Q'/Q + P'$ and note that $Q'/Q \to 0$ as $z \to \infty$. We define $R = z - f/f'$ and note that R has the properties of our theorem with $m = n - 1$. Furthermore $R' = ff''/f'^2$. Since R has at least $n - 2$ non-real critical points, it only remains to show that these are zeros of f''. The alternative is that some of these are fixed points of R. Suppose that there are ρ points in U, which are both fixed points of R and critical points of R. Writing as before $R = z + T$, this would mean that T had ρ additional zeros in U. But then there are corresponding components of $\{\Im T > 0\}$, again lying wholly in U, and hence producing ρ additional poles of T, and therefore of R, on ∂U. We thus find that $\chi \geq \nu + \mu + \rho$, and as ρ of these points are fixed points, we obtain the desired result.

9.4.5

Finally we note the following result by considering $R = z - 1/L$.

Theorem *Let $L = \frac{S}{T}$ be a real rational function, where the polynomials S and T have degrees s and t respectively. Then the rational function*

$$Q = L + \frac{L'}{L}$$

has at least $s - t - 1$ non-real zeros. Furthermore, if $L(x) \to +\infty$ as $x \to +\infty$, then Q has at least $s - t$ non-real zeros.

9.5 Rational functions with real critical points

9.5.1

A real rational function all of whose critical points are real will fail to satisfy the hypotheses of the theorems in the previous section, which give the existence of non-real critical points. We can therefore draw some conclusions about such a function.

Let R be a real rational function with no non-real critical points and consider a component U in H^+ of the level region $\{z : \Im R(z) > 0\}$; applying the Riemann–Hurwitz formula, we deduce that U is simply-connected and the mapping $w = R(z)$ is 1–1 from U onto H^+. Similarly, a component V of $\{z : \Im R(z) < 0\}$ will be simply-connected and R will map V 1–1 onto H^-. Thus the plane will decompose into a finite number of such simply-connected components, each being mapped by R bijectively to one of the two half-planes.

Suppose now that R has a simple pole at ∞. Then by corollary (9.4.2) the expansion of R about ∞ takes the form

$$R(z) = a + bz - \frac{c}{z} + \text{higher powers of } \frac{1}{z}, \tag{9.63}$$

where b and c are $\neq 0$ and have the same sign; thus $bc > 0$. Consider now a point α on the real axis and assume that R has neither a pole nor a critical point at α. Then we can write

$$R(z) = a + b(z-\alpha) + c(z-\alpha)^2 + d(z-\alpha)^3 + \cdots \tag{9.64}$$

near α. The function

$$S(z) = \frac{1}{R(\frac{1}{z} + \alpha) - a} \tag{9.65}$$

has its critical points real with a simple pole at ∞; expanding S at ∞ gives

$$S(z) = \frac{z}{b} - \frac{c}{b^2} + \frac{c^2 - bd}{b^3}\frac{1}{z} + \cdots \tag{9.66}$$

and therefore we obtain $c^2 - bd > 0$. This is equivalent to the inequality

$$3R''(\alpha)^2 - 2R'(\alpha)R'''(\alpha) > 0, \tag{9.67}$$

holding at all real points α, except those at which R has a critical point or a pole. Now consider the function

$$F(x) = \frac{1}{2}x + \frac{R'(x)}{R''(x)}. \tag{9.68}$$

We have

$$F'(x) = \frac{3}{2} - \frac{R'(x)R'''(x)}{R''(x)^2} > 0 \tag{9.69}$$

at all real points other than the poles and critical points of R. It follows that F is a strictly increasing function on the real axis with poles at those zeros of R'' which are not zeros of R'.

9.6 Real entire and meromorphic functions

9.6.1

Some of the methods of this chapter can be extended to certain classes of entire and meromorphic functions. The crucial difference between polynomial/

rational functions and entire/meromorphic functions is the essential singularity at ∞ possessed in general by the latter functions. The extensive literature on these functions may be regarded as a wide and varied study of both the local and global significance of an essential singularity. Usually in order to extend properties of polynomials or rational functions it is necessary to restrict attention to those transcendental functions whose growth is in some way restricted. For entire functions the most important concept is that of a function of finite order.

If f is an entire function, then the **order of** f is given by

$$\alpha = \limsup_{r\to\infty} \frac{\log\log M(r, f)}{\log r}, \tag{9.70}$$

the order being finite when $\alpha \neq \infty$. Here $M(r, f) = \max_{|z|=r} |f(z)|$. For a meromorphic function we have a similar definition with $\log M(r, f)$ replaced by the characteristic function $T(r, f)$.

9.6.2 Strongly real meromorphic functions

Let $F(z)$ be a function meromorphic in the plane; we say that F is **strongly real** if $F(z)$ is real $\Leftrightarrow$ z is real (we exclude $z = \infty$, but include $F = \infty$ as a real value).

9.6.3

Theorem *Let $F(z)$ be a real meromorphic function. The following conditions are equivalent.*

(i) *The zeros and poles of F are real and interspersed on the real axis;*
(ii) *$F(z)$ is strongly real;*
(iii) *$F(z)$ is non-constant and can be written in the form*

$$F(z) = -az + b + \frac{\mu}{z} + \sum_k c_k\left(\frac{1}{z - x_k} + \frac{1}{x_k}\right) \tag{9.71}$$

where b is real, all x_k are real and $\neq 0$, $a \geq 0$, $\mu \geq 0$ and all $c_k > 0$ (or $a \leq 0$, $\mu \leq 0$ and all $c_k < 0$), the series being absolutely convergent for $z \neq x_k$. In this case the series

$$\sum_k \frac{c_k}{x_k^2} \tag{9.72}$$

is convergent.

Proof Since F is real on the real axis, $\overline{F(\bar{z})} = F(z)$ for all z and so in discussing the values of F we may confine ourselves to the real axis and the upper half-plane H^+. It is clear that F is strongly real iff F has no poles in H^+ and $\Im F(z)$ has constant sign in H^+; we will assume that $\Im F(z) < 0$ for $z \in H^+$, i.e. that F is strongly real of negative type. In this case F is locally 1–1 at every point of $\mathbb{R}$ (including the poles), as otherwise the local conformal properties of an analytic function would give the existence of real values of F off the real axis. Thus F has no critical points on $\mathbb{R}$ and in particular every pole of F is real, simple and has a positive residue. At other points $x \in \mathbb{R}$, $F'(x) < 0$, as the condition $\Im F(z) < 0$ for $z \in H^+$ implies that the mapping is orientation reversing near such points $x \in \mathbb{R}$. Thus $F(x)$ is strictly decreasing and in particular between successive poles F decreases from $+\infty$ to $-\infty$, taking every real value, including 0, exactly once. Thus between two successive poles there is exactly one zero. By the same token, between two successive zeros there is exactly one pole. The zeros and poles of F are therefore interspersed. In general there may be infinitely many zeros and poles proceeding in both directions down the axis, giving infinitely many copies of $\mathbb{R}$. Alternatively, there may be infinitely many zeros and poles in one direction, but not in the other. This would give a one-sided stretch of the axis clear of zeros and poles, and F would have a definite limiting value as ∞ was approached from this direction. On the other hand there is clearly no such limit as ∞ is approached from the other direction. The third possibility is that F has only finitely many zeros and poles. In this case, as we shall see, F is rational and we are back in the previously studied case. In particular F has the same limit as ∞ is approached from either direction, and the interspersion is of the circular type.

Conversely, suppose that the zeros and poles of F are interpersed on $\mathbb{R}$. Label the poles x_k with k running from $-\infty$ to $+\infty$, or as appropriate, so that the poles are labelled in the order in which they occur on the axis. Between successive poles x_k and x_{k+1} there is exactly one zero, which we denote by y_k. Assuming for the moment that the number of poles is infinite to the right, consider the product

$$\omega(z) = \prod_k \rho_k \frac{y_k - z}{x_k - z}, \tag{9.73}$$

where $\rho_k = 1$ if $x_k y_k \leq 0$ and $\rho_k = x_k/y_k$ if $x_k y_k > 0$. Clearly $\rho_k = 1$ for at most two terms. Thus for the convergence of the product we require the convergence of

$$\prod_k \frac{1 - z/y_k}{1 - z/x_k} = \prod_k \left(1 + z\frac{\frac{1}{x_k} - \frac{1}{y_k}}{1 - z/x_k}\right) \tag{9.74}$$

and the absolute convergence of the last expression is equivalent to the absolute convergence of

$$\sum_k \frac{\frac{1}{x_k} - \frac{1}{y_k}}{1 - z/x_k}. \tag{9.75}$$

Since $x_k \to \infty$, the denominator $\to 1$ and so we obtain the absolute convergence of the product at all z other than the zeros or poles from the convergence of

$$\sum_k \left(\frac{1}{x_k} - \frac{1}{y_k} \right), \tag{9.76}$$

and this follows immediately from the inequality

$$0 < \frac{1}{x_k} - \frac{1}{y_k} \le \frac{1}{x_k} - \frac{1}{x_{k+1}}. \tag{9.77}$$

Thus $\omega(z)$ is a real meromorphic function with the same poles as F; ω also has the same zeros as F apart possibly from one zero b on the left, if F has only finitely many poles on the left; in this case we add in an extra term $1/(z-b)$ to the product expression for ω. If F has only finitely many poles on the right with one final pole a not followed by a zero, we add into the product expression for ω one extra term $1/(a-z)$; if there is a following zero b, ω is as before, being a finite product in the right terms with a concluding $(z-b)/(z-a)$. With ω defined in this way we see that for $z \in H^+$, $\arg \omega(z)$ is a sum of positive angles with total sum not exceeding π; for $\arg\{(z-y_k)/(z-x_k)\}$ is the angle at z subtended by the segment $[x_k, y_k]$. Thus $0 < \arg \omega(z) < \pi$ for $z \in H^+$, from which it follows that $\Im(\omega(z)) > 0$ for $z \in H^+$.

Furthermore, ω and F have the same zeros and poles, and so F/ω is an entire function with no zeros, real on the real axis and of bounded argument in H^+. F/ω therefore has bounded argument in the plane and so by Liouville's theorem must reduce to a real non-zero constant. F is therefore given, up to a multiplicative real constant, by a product representation, which is clearly strongly real. This proves the equivalence of (i) and (ii). Note also that, if the number of poles of F is finite, then F is rational (since ω is rational).

Consider now a function F of the form (9.71). Assuming the convergence of the series, it is clear that $\Im F(z) < 0$ for $z \in H^+$ and so F is strongly real. Thus to prove that (iii) $\Rightarrow$ (ii) we need to verify that the absolute convergence of the series is equivalent to (9.72). Since

$$c_k \left(\frac{1}{z - x_k} + \frac{1}{x_k} \right) = \frac{c_k z}{x_k(z - x_k)}, \tag{9.78}$$

the equivalence is easily obtained.

It remains to show that a strongly real function F of negative type can be represented in this additive form. To accomplish this we require a lemma, which we give next.

9.6.4

Julia–Carathéodory lemma *Let $\omega(z)$ be analytic in $\mathbb{U}$ and satisfy $|\omega(z)| < 1$ for $z \in \mathbb{U}$. Suppose that ω remains analytic at $z = 1$ and $\omega(1) = 1$. Then $\omega'(1)$ is real and satisfies*

$$\omega'(1) \geq \frac{1 - |z|^2}{|1 - z|^2} \frac{|1 - \omega(z)|^2}{1 - |\omega(z)|^2} \quad (z \in \mathbb{U}). \tag{9.79}$$

Proof As in the proof of the Clunie–Jack lemma 8.1.8 of chapter 8, we easily obtain that $\omega'(1)$ is real and $\omega'(1) \geq 0$. Alternatively, $\omega'(1) = \lim_{z\to 1}((1-\omega(z))/(1-z))$ and $\Re(1 - \omega(z)) > 0$ for $z \in \mathbb{U}$; thus, if $\omega'(1) \neq 0$ and $\arg \omega'(1) = c$, then choosing $\arg(1 - z) = \alpha$, where α is arbitrary in $(-\pi/2, \pi/2)$, it follows that $c + \alpha \in (-\pi/2, \pi/2)$, which is clearly only possible for $c = 0$.

It follows that, if ϕ is analytic in $\mathbb{U}$ with $|\phi| < 1$, $\phi(0) = 0$ and $\phi(1) = 1$, then $\phi'(1) \geq 1$. For writing $\phi(z)/z = \omega$ and applying Schwarz's lemma, we obtain $\omega'(1) \geq 0$, which gives $\phi'(1) \geq 1$. Applying this result to

$$\phi(z) = \frac{\omega(z) - \omega(0)}{1 - \overline{\omega(0)}\omega(z)}, \tag{9.80}$$

we obtain

$$\omega'(1) \geq \frac{|1 - \omega(0)|^2}{1 - |\omega(0)|^2}. \tag{9.81}$$

Finally, choosing $a \in \mathbb{U}$, we apply this result to

$$\Omega(z) = \omega(B(z)), \tag{9.82}$$

where B is Moebius transformation of $\mathbb{U}$ onto $\mathbb{U}$ satisfying $B(0) = a$, $B(1) = 1$. We obtain

$$B'(1)\omega'(1) \geq \frac{|1 - \omega(a)|^2}{1 - |\omega(a)|^2}. \tag{9.83}$$

The desired function B is

$$B(z) = \frac{(1 - a)z + a(1 - \overline{a})}{(1 - \overline{a}) + \overline{a}(1 - a)z}, \tag{9.84}$$

giving

$$B'(1) = \frac{|1-a|^2}{1-|a|^2}. \tag{9.85}$$

9.6.5

To complete the proof of our theorem, assume that $F(z)$ is strongly real with a pole at $z = p$, the residue being $c > 0$. Thus $\Im F(z) < 0$ for $z \in H^+$. Define

$$\phi(w) = p + i\frac{1-w}{1+w}, \qquad \psi(z) = \frac{z+i}{z-i}. \tag{9.86}$$

Then $\phi\colon \mathbb{U} \to H^+$ is bijective with $\phi(1) = p$; $\psi\colon H^- \to \mathbb{U}$ is bijective with $\infty \to 1$. It follows that the mapping $h = \psi \circ F \circ \phi$ maps $\mathbb{U} \to \mathbb{U}$ with $h(1) = 1$. Applying the Julia–Carathéodory lemma to h gives

$$\Re\frac{1+h(w)}{1-h(w)} \geq \frac{1}{h'(1)}\Re\frac{1+w}{1-w} \quad (w \in \mathbb{U}). \tag{9.87}$$

A straightforward calculation gives $h'(1) = 1/c$, and so this inequality is equivalent to

$$-\Im(F(z)) \geq -c\Im\left(\frac{1}{z-p}\right) \quad (z \in H^+). \tag{9.88}$$

In particular, taking $z = i$ gives

$$\frac{c}{1+p^2} \leq -\Im F(i). \tag{9.89}$$

If we now apply this argument repeatedly to the poles x_k with residues c_k we see that the function

$$K(z) = F(z) - \sum_k c_k\left(\frac{x_k}{1+x_k^2} + \frac{1}{z-x_k}\right) \tag{9.90}$$

is meromorphic in the plane and satisfies $\Im K(z) \leq 0$ for $z \in H^+$, the series being convergent, since

$$\sum_k \frac{c_k}{1+x_k^2} \leq -\Im F(i) \tag{9.91}$$

by induction. Moreover K has no poles and so is a real entire function. If K is non-constant, then K is strongly real. However, as we showed earlier, a strongly real function with no poles, and so at most one zero, is rational. It follows that $K(z) = -az + b$, where $a \geq 0$ and b is real. We easily deduce the representation of (iii) for F and the proof of our theorem is complete.

9.6.6

Corollary *If F is strongly real, then*

$$|\Im F(x+iy)| \geq A\frac{y}{(x^2+y^2)} \quad (y>0). \tag{9.92}$$

$$|F(x+iy)| \leq A\frac{x^2+y^2}{|y|} \quad (y \neq 0,\, x^2+y^2 \text{ large}). \tag{9.93}$$

$$\limsup_{r\to\infty} \frac{m(r,F)}{\log r} \leq 1, \tag{9.94}$$

where

$$m(r,F) = \frac{1}{2\pi}\int_0^{2\pi} \log^+ |F(re^{i\theta})|\, d\theta. \tag{9.95}$$

Proof From the representation (9.71) we have, for $y>0$,

$$|\Im F(x+iy)| = y\left(a + \frac{\mu}{x^2+y^2} + \sum_k \frac{c_k}{(x-x_k)^2+y^2}\right), \tag{9.96}$$

and since F is non-constant, at least one of the constants a, μ, c_k is non-zero and therefore the first inequality follows easily. For the second inequality we note that the values $F(i(1+w)/(1-w))$ lie in a half-plane for $|w|<1$, and therefore

$$|F(z)| \leq \frac{A}{1-\dfrac{|z-i|^2}{|z+i|^2}} = A\frac{x^2+(y+1)^2}{4y} \quad (y>0), \tag{9.97}$$

from which the estimate follows. Furthermore the estimate gives

$$\log^+ |F(re^{i\theta})| \leq C + \log r + \log\frac{1}{\sin\theta}, \tag{9.98}$$

and we obtain the third inequality on integration.

9.6.7

Corollary *Let $H(z)$ be a real meromorphic function whose poles are real and simple with the residue at each pole being positive. Then we can write*

$$H(z) = F(z)\phi(z), \tag{9.99}$$

where, if F is non-constant, $F(z)$ is strongly real of negative type and $\phi(z)$ is an entire function.

The decomposition $H = F\phi$ is called the **Levin representation for H** (see Levin [25]). The class of such functions H will be denoted by Ω.

Proof of corollary Let a and b be successive poles of H. Because the residue at each of these poles is positive, $H' < 0$ in (a, b) near the points a and b. Thus in (a, b) H is decreasing from $+\infty$ near a and decreasing to $-\infty$ near b. By continuity H maps (a, b) onto $\mathbb{R}$, and so H has a zero in (a, b). Picking one such zero between every pair of successive poles, we obtain a succession of interspersed zeros and poles. We define F to be the strongly real function of negative type constructed with these zeros and poles, as in the proof of the theorem. H/F is clearly an entire function, since F cancels every pole of H and H cancels every zero of F.

We shall be interested mainly in those elements L of Ω whose Levin representation has the form $L = FP$, where F is strongly real and P is a polynomial.

9.6.8

Lemma *A function $L \in \Omega$ has the form $L = FP$, as above, iff*

$$\alpha = \limsup_{r\to\infty} \frac{m(r, L)}{\log r} < +\infty. \tag{9.100}$$

In this case, if n is the degree of P, we have

$$|n - \alpha| \le 1. \tag{9.101}$$

Proof If $L = FP$, then $\log^+ |L| \le \log^+ |F| + \log^+ |P|$, and so if P has degree n, we obtain with the help of (9.94)

$$\limsup_{r\to\infty} \frac{m(r, L)}{\log r} \le n + 1. \tag{9.102}$$

Conversely, if the above limsup is $= \alpha$ say, then writing $L = F\phi$, where ϕ is entire, we have $\phi = L/F$. Since $1/F$ is strongly real, we obtain

$$\limsup_{r\to\infty} \frac{m(r, \phi)}{\log r} \le \alpha + 1. \tag{9.103}$$

Thus we need to show that this implies that ϕ is a polynomial.

If f is analytic for $|z| \le 1, \ne 0$ on the circle, and B denotes the Blaschke product formed with the zeros of f in $\mathbb{U}$, then $\log|f/B|$ is harmonic in $|z| \le 1$ and so

$$\begin{aligned} \log|f(\rho e^{i\theta})/B(\rho e^{i\theta})| &= \frac{1}{2\pi}\int_0^{2\pi} \Re\frac{1+\rho e^{i(\theta-t)}}{1-\rho e^{i(\theta-t)}} \log|f(e^{it})|\,dt \\ &\le \frac{1+\rho}{1-\rho}\frac{1}{2\pi}\int_0^{2\pi} \log^+|f(e^{it})|\,dt \end{aligned} \tag{9.104}$$

and therefore

$$\log M(\rho, f) \leq \frac{1+\rho}{1-\rho} m(1, f), \tag{9.105}$$

and by continuity this result continues to hold if f has zeros on the circle. Applying this result to $f(z) = \phi(rz)$, we obtain in general

$$\log M(r, \phi) \leq \frac{1+\rho}{1-\rho} m\left(\frac{r}{\rho}, \phi\right), \tag{9.106}$$

and therefore by our assumption, for large r

$$\log M(r, \phi) \leq \frac{1+\rho}{1-\rho}(\alpha + 1 + \epsilon) \log \frac{r}{\rho}. \tag{9.107}$$

It follows that

$$\limsup_{r\to\infty} \frac{\log M(r, \phi)}{\log r} \leq \frac{1+\rho}{1-\rho}(\alpha + 1), \tag{9.108}$$

and letting $\rho \to 0$,

$$\limsup_{r\to\infty} \frac{\log M(r, \phi)}{\log r} \leq (\alpha + 1). \tag{9.109}$$

Thus, if $N > \alpha + 1$, $|\phi(z)/z^N|$ is bounded in the plane. Writing $\phi(z) = P(z) + z^{N+1}\psi(z)$, where P is the Nth partial sum of ϕ, we see that $z\psi(z)$ is bounded in the plane and therefore constant by Liouville's theorem. Thus $\phi = P$ is a polynomial of degree $\leq N$. Since the previous limsup is then equal to the degree of P, we see that ϕ is a polynomial of degree $\leq \alpha + 1$.

9.6.9 Theorem on critical points

Theorem *Let Q be a strongly real rational function of positive type with a pole at ∞, and let $L(z)$ be a function in Ω whose Levin representation has the form $L = FP$, where F is strongly real of negative type and P is a real polynomial of degree n. Assume further that the zeros of L and the poles of Q are distinct. Let A denote the set of poles of L and let*

$$R = Q - \frac{1}{L}. \tag{9.110}$$

Then

(i) *if A is unbounded above and unbounded below, R has at least n non-real critical points,*
(ii) *if A is unbounded above or unbounded below, or empty, then R has at least $n-1$ non-real critical points,*
(iii) *if A is finite, then R has at least $n-2$ non-real critical points.*

9.6.10 We begin with some remarks

(1) *The containment property.* If $z \in H^+$ and $\Im L(z) > 0$, then $\Im R(z) > 0$. It follows that *any component of the set* $\{z \in H^+ : \Im L(z) > 0\}$ *lies in a component of the set* $\{z \in H^+ : \Im R(z) > 0\}$.

(2) Since each pole a of L is real and simple with a positive residue, for $z \in H^+$ near to a, $\Im L(z) < 0$. Therefore there are no poles of L on the boundary of the set $\{z \in H^+ : \Im L(z) > 0\}$. If this set is empty, then L is strongly real. Otherwise, let V be a component of this set. We will show that $L(z) \to \infty$ *as* $z \to \infty$ *in* V.

Let C be a component of ∂V, so L is real on C and C contains no poles of L. C is obtained by analytic continuation of the inverse of L along the real axis, starting from a particular real value corresponding to the image of a point on C; if we meet a branch point of the inverse function, we choose that branch which keeps us on ∂V. C cannot form a loop, as otherwise C would bound a domain in H^+ in which L is analytic with $\Im L = 0$ on the boundary; this clearly contradicts the maximum principle. Thus C is an infinite Jordan curve lying in the closure of H^+. As C divides the plane into two components, one of these domains, D say, lies entirely in H^+. L is analytic in $D \cup C$ and real on C; indeed L maps C injectively onto an interval (a, b) of $\mathbb{R}$, where a and b are asymptotic values of $L(z)$ as $z \to \infty$ along the two arms of C. We shall be showing that $a = -\infty$ and $b = +\infty$. More generally we will show that L *is conformally equivalent in* D *to a real rational function* Φ *in* H^+ *of degree* $\le \frac{1}{2}\rho + 2$, *where* ρ *is the degree of* P, *and satisfying* $\Phi'(x) \neq 0$ *for* x *real.* The proof proceeds in a number of steps:

(3) *The equation* $L(z) = i$ *has at most* $\frac{1}{2}\rho + 1$ *roots in* H^+.

We assume first that L has finitely many poles and so is a rational function. We construct a loop Γ as follows. Γ consists of a large semi-circle Γ_1 in H^+ joining the real points r to $-r$, followed by Γ_2, which is the line segment $(-r, r)$ indented with small semi-circles in H^+ centred at the poles, which are taken to lie in $(-r, r)$. If the radii of these semi-circles are small, then $\Im L < 0$ on them. Hence on Γ_2 we have $\Im(L - i) \le -1$ and so $d(L - i, \Gamma_2) < \frac{1}{2}$. Assuming $\rho \ge 2$, $L(z) = F(z)P(z) \to \infty$ as $z \to \infty$. Hence, if r is large, $d(L - i, \Gamma_1) < d(L, \Gamma_1) + \epsilon$, since $(L - i)/L \to 1$ as $z \to \infty$. But $d(L, \Gamma_1) = d(F, \Gamma_1) + d(P, \Gamma_1) < \frac{1}{2} + \frac{1}{2}\rho + \epsilon$. We obtain, therefore, $d(L - i, \Gamma) \le \frac{1}{2}\rho + 1$ and, since $L - i$ is analytic with all its zeros in H^+ lying inside the region bounded by Γ, we obtain the result from the argument principle.

The case $\rho = 1$ can be dispensed with, because we may always choose the decomposition $L = FP$ in such a way that the degree of P is even. This follows from the fact that the number of zeros of P between two successive poles of L is even, since L decreases from $+\infty$ to $-\infty$ in the interval and one of its zeros is the constructed zero of F. On the other hand a loose zero of P either before the first pole or after the last pole can be absorbed into F by the interspersion theorem.

To prove (3) in the case that L has infinitely many poles, we use an approximation argument. In the representation (9.71) for F we cut the infinite sum off, after N terms say, obtaining a rational function F_N which is strongly real. The corresponding function $L_N = F_N P$ is still in Ω, since each residue at a pole is the same as for L. Thus $L_N - i$ has at most $\frac{1}{2}\rho + 1$ zeros in H^+. Now the zeros of $L - i$ in H^+ lie in the set $I = \{z \in H^+ : \Im L(z) > 0\}$ and L is analytic on the closure of I. Also $L_N \to L$ locally uniformly in I and so on any compact subset of I, $(L_N - i)/(L - i) \to 1$ uniformly as $N \to \infty$. It follows from the argument principle that $L_N - i$ and $L - i$ have equally many zeros in I for large N, and (3) follows.

Returning to (2) we apply the Riemann mapping theorem to obtain a 1–1 mapping $z = \psi(w)$ of $\mathbb{U}$ onto D, which extends continuously to $\mathbb{T}$ with $\psi(1) = \infty$. The function $K(w) = L(\psi(w))$ is analytic in $\mathbb{U}$ and continuous and real on $\mathbb{T} - \{1\}$; furthermore K is injective on $\mathbb{T} - \{1\}$ and has the limiting values a and b as $w \to 1$ from either side on $\mathbb{T}$. We consider the function

$$G(w) = \frac{K(w) - i}{K(w) + i}, \tag{9.111}$$

and observe that $|G(w)| = 1$ on $\mathbb{T} - \{1\}$ and indeed $G(w)$ is a continuous injective mapping of $\mathbb{T} - \{1\}$ onto an arc of $\mathbb{T}$ with a jump discontinuity at $w = 1$. From (2) $G(w)$ has at most $\frac{1}{2}\rho + 1$ zeros in $\mathbb{U}$. If $B(w)$ is the Blaschke product of these zeros, then $V(w) = B(w)/G(w)$ is analytic in $\mathbb{U}$, continuous on $\mathbb{T} - \{1\}$ with $|V(w)| = 1$ and a jump discontinuity at $w = 1$. We will show that

(4) *$V(w)$ is a finite Blaschke product and therefore continuous at $w = 1$.*

Let λ and μ be the one-sided limiting values on $\mathbb{T}$ of $V(w)$ as $w \to 1$ from each side. The limiting values of $V(w)$ as $w \to 1$ from inside $\mathbb{U}$ all lie on the line segment $[\lambda, \mu]$. Indeed from Poisson's formula these limiting values are the limiting values of

$$\int_0^\delta \frac{1 - |w|^2}{|w - e^{it}|^2} V(e^{it}) dt + \int_{-\delta}^0 \frac{1 - |w|^2}{|w - e^{it}|^2} V(e^{it}) dt, \tag{9.112}$$

where $\delta > 0$ is arbitrarily small. In the first term we may write $V = \lambda + \epsilon_1(t)$ where $\epsilon_1 \to 0$ as $t \to 0$; in the second we may write $V = \mu + \epsilon_2(t)$, where $\epsilon_2 \to 0$ as $t \to 0$. Thus we easily see that the limiting values are the limiting values of

$$\lambda \int_0^{\delta} \frac{1 - |w|^2}{|w - e^{it}|^2} dt + \mu \int_{-\delta}^{0} \frac{1 - |w|^2}{|w - e^{it}|^2} dt, \tag{9.113}$$

which are $(1 - t)\lambda + t\mu$ where $0 \leq t \leq 1$. We now choose $\zeta \in \mathbb{U}$ which does not lie on the line segment $[\lambda, \mu]$. Then $V(w) - \zeta$ has finitely many zeros in $\mathbb{U}$, for the function is analytic and does not have 0 as a limiting value on $\mathbb{T}$. Taking the Blaschke product $D(w)$ of these zeros the function

$$M(w) = \frac{V(w) - \zeta}{1 - \overline{\zeta} V(w)} \Big/ D(w) \tag{9.114}$$

is analytic and non-zero in $\mathbb{U}$ and satisfies $|M(w)| = 1$ on $\mathbb{T} - \{1\}$. By the maximum principle $|M(w)| \leq 1$ in $\mathbb{U}$. The same remark applies to $1/M$. Thus $M(w)$ is a constant of absolute value 1. Then (4) follows immediately.

We deduce that $G(w)$ is a ratio of Blaschke products and in particular $G(w) \to 1$ as $w \to 1$ in $\overline{\mathbb{U}}$. Furthermore $G(w)$ is a bijective mapping $\mathbb{T} \to \mathbb{T}$ and therefore $d(G, \mathbb{T}) = \pm 1$. It follows that the number of poles of G in $\mathbb{U}$ does not exceed $\frac{1}{2}\rho + 2$ and the degree of G is at most this quantity. Transforming back to the half-plane, $L(z) \to \infty$ as $z \to \infty$ in $\overline{D}$. Furthermore by mapping $\mathbb{U}$ to H^+ we see that the conformal equivalence of L in D to a rational function in H^+ is as stated.

We may choose C to be a component which minimises $\Im z$ on ∂V, for either $\Im z = 0$ at some points on ∂V or $\partial V \subset H^+$; in the latter case each component of ∂V contains a zero of P and so there are at most finitely many such components. We then see that $V \subset D$ and $L(z) \to \infty$ as $z \to \infty$ in $\overline{V}$.

9.6.11

Lemma *Let K be the set $\{z \in H^+ : \Im R(z) > 0\}$, where R is given by* (9.110) *with $n \geq 2$, and let $\alpha \in H^+$. There is no sequence $\{z_k\} \subset K$ satisfying $z_k \to \infty$ such that $R(z_k) \to \alpha$ as $k \to \infty$.*

Proof We assume the existence of such a sequence. Since $R = Q - 1/L$ and $Q(z) \to \infty$ as $z \to \infty$, we deduce that

$$L(z_k) \to 0 \text{ as } k \to \infty. \tag{9.115}$$

Now $L = FP$, where F is strongly real and P is a polynomial of degree ≥ 2. For $z \in H^+$ we have, by (9.92), $|F(z)| > Ay/r^2$, where A is a positive constant, $y = \Im z$ and $r = |z|$. On the other hand for r large we have $|P(z)| > Br^2$. Thus

$|L(z)| > ABy$ for $z \in H^+$ and r large. The above relation implies therefore that

$$y_k \to 0 \text{ as } z_k = x_k + iy_k \to \infty. \tag{9.116}$$

Now $Q(z)$ is a strongly real rational function of positive type with a pole at ∞, and therefore has the form $Q(z) = az + b + o(1)$ as $z \to \infty$, where $a > 0$ and b is real. Hence $\Im Q(z) = ay + o(1)$ and so

$$\Im Q(z_k) \to 0 \text{ as } k \to \infty. \tag{9.117}$$

Writing $\alpha = \beta + i\epsilon$, we have $\epsilon > 0$; furthermore the hypothesis and the previous relation give

$$\Im(-1/L(z_k)) \to \epsilon \text{ as } k \to \infty. \tag{9.118}$$

It follows that $\Im L(z_k) > 0$ for k large. But then $L(z_k) \to \infty$ as $k \to \infty$ from (2) above, and this contradicts (9.115).

9.6.12

From this result we see that, if $\alpha \in H^+$, the equation $R(z) = \alpha$ has at most finitely many roots in H^+; for, if $\{z_k\}$ is a sequence of such roots, then clearly each $z_k \in K$ and, by the above, no subsequence can tend to ∞. As R is analytic in K, the roots cannot accumulate about a point in K and so $z_k \to \partial K$. But R is real on ∂K and so $R(z_k)$ cannot converge to a non-real value.

9.6.13

More precisely we have

Lemma *Let U be a component of K and suppose that $R'(z)$ has exactly k zeros in U counted according to multiplicity. Then for each $\alpha \in H^+$ the equation $R(z) = \alpha$ has exactly ν solutions in U, where ν is independent of α and satisfies $1 \le \nu \le k+1$.*

Proof Let $a \in U$ be a point such that $R'(a) \ne 0$ and let $b = R(a)$. Then $b \in H^+$ and the inverse function R^{-1} is well defined and 1–1 near b. Let Γ be a curve in H^+ with initial point b and terminal point c. The inverse R^{-1} can be continued along Γ from initial point b giving a curve γ in U. The continuation will reach the point c unless points $z \in \gamma$ approach (i) ∂U, (ii) ∞ or (iii) a critical point of R. Case (i) is ruled out, since R is real on ∂U and so a point of Γ would be real. Case (ii) is ruled out, since by the previous lemma, as $z \to \infty$

in U, R cannot have a limiting value in H^+. Case (iii) can occur only if Γ contains one of the images of a zero of R'; by hypothesis there are at most k such points. Thus R^{-1} can be continued along any curve in H^+ which contains no branch points of R. This immediately yields the result that R is a surjective mapping from U onto H^+. If R were rational, the conclusion of the lemma would be an easy consequence of the Riemann–Hurwitz formula of chapter 10. We will in fact obtain essentially the same formula, but as ∂U is intrinsically more complicated than the corresponding level set for a rational function, some care is needed in giving a modified proof.

For convenience we transform to the unit disc by setting

$$S(z) = \frac{R\left(i\dfrac{1+z}{1-z}\right) - i}{R\left(i\dfrac{1+z}{1-z}\right) + i}. \tag{9.119}$$

S is a meromorphic function, except for a possible essential singularity at $z = 1$, and $|S(z)| = 1$ on $\mathbb{T} - \{1\}$. Let $V = \{z \in \mathbb{U}\colon i(1+z)/(1-z) \in U\}$; then V is a component in $\mathbb{U}$ of the set $\{z\colon |S(z)| < 1\}$. By hypothesis S' has exactly k zeros in V counted according to multiplicity. Furthermore the inverse S^{-1} can be continued along any curve in $\mathbb{U}$, avoiding the at most k branch points, giving a curve in V. We choose ρ in $(0, 1)$ so that ρ is near to 1 and the branch points lie in $|w| < \rho$. Let

$$E = \{z \in V\colon |S(z)| = \rho\}. \tag{9.120}$$

Let $z_0 \in E$ and $w_0 = S(z_0)$. We can continue the inverse S^{-1} from w_0 around the circle $|w| = \rho$ with initial value z_0. Since S attains the value w_0 at most finitely often in $\mathbb{U}$, the continuation will return to the initial branch after a finite number of circuits. This gives a component of E containing z_0, which is a simple closed analytic curve in V. Thus E consists of a union of such curves, which are disjoint, as E contains no critical points of S. Since E is compact, E is a finite union of m such curves $C_1, \ldots, C_m$.

9.6.14

Sub-lemma *Let $\alpha \in C_i$ and $\beta \in C_j$ where $i \neq j$ and let γ be a curve in V joining α to β, which does not meet E, except at α and β. Then*

$$|S(z)| < \rho \quad (z \in \gamma, z \neq \alpha, z \neq \beta). \tag{9.121}$$

Proof If not, then for these z we have $\rho < |S(z)| < 1$ and so the curve $S(\gamma)$ lies in the annulus $\{\rho \leq |w| < 1\}$ with its endpoints on $\{|w| = \rho\}$, say $S(\alpha) = \lambda$

and $S(\beta) = \mu$. We construct a loop Γ in $\{\rho \le |w| < 1\}$ by traversing $S(\gamma)$ from λ to μ followed by an arc of $|w| = \rho$ from μ to λ. We continue S^{-1} along Γ with initial value α; we obtain a terminal value on C_j. From this terminal value we continue the continuation $-n(\Gamma, 0)$ times around $|w| = \rho$. This gives us a new terminal value also on C_j. The total curve, along which we have continued S^{-1}, lies in the annulus and has zero winding number about the origin. It is therefore null homotopic in the annulus. Since S^{-1} can be continued along all curves in the annulus, it follows from the monodromy theorem that S^{-1} has terminal value equal to the initial value. This gives $\alpha \in C_j$, which is a contradiction.

Let D_i denote the Jordan region bounded by C_i $(1 \le i \le m)$. If $z \to 1$ in V, then the limiting values of S lie on $\mathbb{T}$, since the limiting values of R in U at ∞ are real or ∞. It follows that the set $V_\rho = \{z \in V \colon |S(z)| < \rho\}$ contains no boundary points on $\mathbb{T}$. Therefore at least one D_i, say D_1, contains a point of V_ρ : otherwise V_ρ is exterior to all the D_i and so we can approach $\mathbb{T}$ from within V_ρ. We show that

$$V_\rho = D_1 - \bigcup_{i=2}^{m} \overline{D_i}. \tag{9.122}$$

Let X denote the region exterior to C_1 and assume that a point p of V_ρ lies in X. We can join p to 1 in X with a curve and this curve must pass outside V_ρ and so must meet $E - C_1$ at a point $q \in C_2$ say. It then follows that, since C_2 and C_1 are disjoint, $C_2 \subset X$. As $C_2 \subset \mathbb{U}$, the point 1 is exterior to C_2 and so we can join q to 1 in this exterior and in X. Again we must meet a point of E at say $r \in C_3$ and by the sub-lemma the path from q to r lies in V_ρ. As before C_3 is exterior to both C_1 and C_2 and we can repeat the argument. As there are only finitely many C_i, we eventually get a contradiction. It follows that $V_\rho \subset D_1$. We also see that every C_i $(2 \le i \le m)$ lies in D_1 and therefore $\overline{D_i} \subset D_1$ for $2 \le i \le m$. Now by the sub-lemma that part of D_1 exterior to all the C_i $(2 \le i \le m)$ lies in V_ρ; in other words the right-hand side of (9.122) lies in V_ρ. As we cross a C_i out of this region the points z satisfy $|S(z)| > \rho$. Therefore by the sub-lemma, C_i cannot enclose one of the other C_j, for we would obtain a joining curve on which $|S(z)| < \rho$. It follows that $|S(z)| > \rho$ in every D_i $(2 \le i \le m)$ and therefore (9.122) follows.

Thus V_ρ is a domain bounded by the cycle $\Gamma = C_1 - C_2 - \cdots - C_m$, which is homologous to zero in V. The zeros and the critical points of S in V lie in V_ρ. Let ν be the number of zeros of S in V_ρ counted according to multiplicity.

Then by the argument principle

$$k - \nu = d(S'/S, \Gamma) = d(S'/S, C_1) - \sum_{i=2}^{m} d(S'/S, C_i). \qquad (9.123)$$

On the other hand, C_i (assumed positively oriented) has a parametric representation $z_i(t)$ such that $S(z_i(t)) = \rho e^{it}$, where t belongs to an interval whose length is a multiple of 2π. Hence

$$z_i'(t)\frac{S'(z_i(t))}{S(z_i(t))} = i, \qquad (9.124)$$

from which we obtain

$$d(S'/S, C_i) = -1 \quad (1 \le i \le m), \qquad (9.125)$$

since the total change in tangent over a smooth positively oriented loop is 2π. This gives $k - \nu = -1 + (m - 1)$, and so

$$\nu = k + 2 - m \le k + 1. \qquad (9.126)$$

Finally, $\Re((S-c)/S) > 0$ on each C_i for $|c| < \rho$ and therefore, by the argument principle, S attains all values in $\{|w| < \rho\}$ equally often in V_ρ and the lemma follows.

9.6.15

Lemma *Let K be the set $\{z \in H^+ : \Im R(z) > 0\}$, where R is given by (9.110) with $n \ge 2$, and suppose that L has neither zeros nor poles in the interval (a, b), where $L(a) = L(b) = 0$. Then one of the points a and b lies on ∂K. Furthermore, if a is a zero of L of multiplicity r and b is a zero of L of multiplicity s, then the sign of $L^{(r)}(a)L^{(s)}(b)$ is $(-1)^s$.*

Proof If $L'(a) = 0$, then L has a multiple zero at a and so there are points of H^+ near to a at which $\Im L > 0$; by the containment property, at such points we have $\Im R > 0$ and so $a \in \partial K$; similarly, if $L'(b) = 0$. If $L'(a) \ne 0$ and $L'(b) \ne 0$, then $L(x) = (x - a)(x - b)h(x)$, where $h(x) \ne 0$ in $[a, b]$. It follows that $h(a)$ and $h(b)$ have the same sign; moreover, $L'(a)L'(b) = -(a - b)^2 h(a)h(b) < 0$. Thus one of $L'(a)$ and $L'(b)$ is positive. If $L'(a) > 0$, then for $z \in H^+$ near to a we have $\Im L(z) > 0$; as before this gives $a \in \partial K$. To prove the final assertion, we write $L(x) = (x - a)^r (x - b)^s g(x)$, where $g(x)$ has constant sign in $[a, b]$. Then the sign of $L^{(r)}(a)L^{(s)}(b)$ is the sign of $(b - a)^{r+s}(-1)^s g(a)g(b)$. Since $g(a)g(b) > 0$, the result follows.

9.6.16

Corollary *Let c and d be real zeros of L with odd multiplicities r and s respectively, and suppose that $c < d$ and L has no poles in (c, d). Suppose further that every zero of L in (c, d) has even multiplicity. Then $L^{(r)}(c)L^{(s)}(d) < 0$.*

9.6.17 Proof of theorem 9.6.9

Let W denote the union of all components of $K = \{z \in H^+ : \Im R(z) > 0\}$, which have non-empty intersection with $\Lambda = \{z \in H^+ : \Im L(z) > 0\}$. By the containment property, $\Lambda \subset W$. As every component of Λ is unbounded, every component of W is unbounded. Also by 9.6.10 (2), $L(z) \to \infty$ as $z \to \infty$ in Λ and so

$$R(z) \to \infty \text{ as } z \to \infty \text{ in } \Lambda. \tag{9.127}$$

In particular, $R(z)$ attains some large values in H^+ for z near ∞ in W. Also by the rational behaviour of L in Λ, L has at least one zero on the boundary of each component of Λ; therefore R has a pole at each such point, which lies on ∂W.

We assume that R has exactly k critical points in W. By lemma 9.6.13 for $z \in W$, R attains every value in H^+ exactly ν times, where $1 \leq \nu \leq k + 1$. Near any pole R will attain large such values equally often, by the conformality of the mapping. Therefore near ∞ R attains every large value in H^+ at least once. It follows that for $z \in W$, R attains large values in H^+ near finite poles at most k times.

For precision, suppose firstly that L has a zero $\zeta \in H^+$ of order j. Then R has a pole of order j at ζ; hence near ζ the equation $R(z) = w$ has exactly j distinct roots for large $w \in H^+$. In particular, for z near ζ with $\Im Q(z) < 2\Im Q(\zeta)$, we obtain for those large w satisfying $\Im w > 2\Im Q(\zeta)$ that at a root z of $R(z) = w$ we have $\Im R(z) > \Im Q(z)$, and so $\Im L(z) > 0$. Thus each of the j sectors near ζ where $\Im R > 0$ intersects Λ, and therefore each of these regions lies in W. Therefore the equation $R(z) = w$ has for large $w \in H^+$, j roots near ζ in W. It follows that, if α denotes the number of zeros of L in H^+, then the equation $R(z) = w$ has α roots in W for large $w \in H^+$, these roots occurring near the zeros of L.

Secondly, suppose that L has a real zero of order r at a. If r is even, then similar considerations show that the $\frac{1}{2}r$ sectors in H^+ near a in which $\Im R > 0$ are contained in W; therefore the equation $R(z) = w$ has $\frac{1}{2}r$ roots in W near a for large $w \in H^+$. If $r = 2s + 1$, then R takes such values s times if $L^{(r)}(a) < 0$ and $s + 1$ times if $L^{(r)}(a) > 0$.

Now the real zeros of L consist of $n - 2\alpha$ zeros of P and the simple zeros of F (which may coincide with real zeros of P, giving multiple zeros of L). We denote by $|E|$ the number of points in a finite set E. Let I be an open interval containing no poles of L and with endpoints which are poles of L. Let E denote the set of points $a \in I$ at which L has a zero of odd multiplicity, say r, and $L^{(r)}(a) > 0$; let E' denote the set of points $b \in I$ at which L has a zero of odd multiplicity, say s, and $L^{(s)}(b) < 0$. Let t denote the number of zeros of L in I counted according to multiplicity. Then for large $w \in H^+$ the equation $R(z) = w$ has at least $\frac{1}{2}(t + |E| - |E'|)$ roots in W occurring near these zeros. By the preceding corollary 9.6.16, there is at most one point of E' between successive points of E and so $|E'| \leq |E| + 1$. Thus the above equation $R(z) = w$ has at least $\frac{1}{2}(t - 1)$ roots near zeros in I of L. As F has just one zero in I, the quantity $\frac{1}{2}(t - 1)$ is exactly $\frac{1}{2}$(the number of zeros of P in I). Thus in case (i) of the theorem, the equation $R(z) = w$ has at least $\frac{1}{2}n$ roots in W for large w; we deduce that $k \geq \frac{1}{2}n$. Since the number of non-real critical points of R is at least $2k$, we deduce the desired result in this case. In case (ii) we argue similarly except that in the case that I is the infinite interval with one endpoint a pole, there need be no zero of F. Therefore we obtain $k \geq \frac{1}{2}(n - 1)$. Case (iii) gives two such intervals and we obtain $k \geq \frac{1}{2}(n - 2)$. This completes the proof.

9.6.18

Theorem *Let f be a real entire function of finite order all of whose zeros are real. Suppose that f'' has no more than $2p$ non-real zeros for some $p = 0, 1, \ldots$; then f takes the form*

$$f(z) = g(z)\exp(-az^{2p+2}), \tag{9.128}$$

where $a \geq 0$ and where g is an entire function of genus not exceeding $2p + 1$. In particular, if f'' has only real zeros, then f lies in the Laguerre–Pólya class. This implies that f' and all higher derivatives of f have only real zeros.

This result gives a surprising connection between the genus of f and the number of non-real zeros of f''; the result was originally proposed by A. Wiman during lectures in 1911. It appeared firstly in the thesis of M. Ålander of 1914 and was published in Ålander [4]. The work of this section is a modified version of the present author's proof (Sheil-Small [57]) found in 1988. We will prove the result as an application of our theorem 9.6.9 on critical points. We need to establish that under the above hypotheses the function

$$L = \frac{f'}{f} = FP, \tag{9.129}$$

where F is strongly real of negative type and P is a polynomial of degree at most (i) $2p$, (ii) $2p+1$ or (iii) $2p+2$ in the case (i), (ii) or (iii) of theorem 9.6.9.

We shall need Hadamard's canonical representation of an entire function of finite order. A proof can be found in Ahlfors [2]. If $\{a_n\}$ denotes the sequence of zeros of f, then $\exists$ a non-negative integer h such that we can write

$$f(z) = z^{\nu} e^{g(z)} \prod_n \left(1 - \frac{z}{a_n}\right) \exp\left(\sum_{k=1}^{h} \frac{1}{k}\left(\frac{z}{a_n}\right)^k\right), \tag{9.130}$$

where g is a polynomial of degree $\leq h$ and $\nu \geq 0$ is the order of the zero at the origin. The case $h = 0$ in the product corresponds to the sum in the exp term being empty, which can occur only if the product $\prod(1 - z/a_n)$ is convergent. The **genus** of f is the smallest h for which such a representation exists. Logarithmic differentiation gives

$$\begin{aligned} \frac{f'(z)}{f(z)} &= \frac{\nu}{z} + g'(z) - \sum_n \frac{1}{a_n}\left(\frac{1}{1 - \dfrac{z}{a_n}} - \sum_{k=0}^{h-1}\left(\frac{z}{a_n}\right)^k\right) \\ &= \frac{\nu}{z} + g'(z) - \sum_n \frac{1}{a_n} \frac{\left(\dfrac{z}{a_n}\right)^h}{1 - \dfrac{z}{a_n}}. \end{aligned} \tag{9.131}$$

It is easily seen that the necessary and sufficient condition for the absolute and local uniform convergence of this sum at all points $z \neq a_n$ is

$$\sum_n \frac{1}{|a_n|^{h+1}} < +\infty. \tag{9.132}$$

On the other hand with this condition we see that for $|z| = r$ large we have

$$\log^+ \left|\frac{f'(z)}{f(z)}\right| \leq \rho \log r + O(\log r) \text{ as } r \to \infty, \tag{9.133}$$

where $\rho = \max(\deg g - 1, h)$ and h is the genus of the canonical product. We thus obtain

$$\alpha = \limsup_{r\to\infty} \frac{m(r, f'/f)}{\log r} \leq \rho. \tag{9.134}$$

Now, if f has all its zeros real, then $f'/f \in \Omega$, since each pole of f'/f is real and simple with residue the order of the zero of f at that point. Furthermore, by lemma 9.6.8 the Levin representation has the form $L = FP$, where F is strongly real of negative type and P has degree N satisfying $|N - \alpha| \le 1$.

If we now define

$$R(z) = z - \frac{f(z)}{f'(z)} = z - \frac{1}{L(z)}, \tag{9.135}$$

then, since $Q(z) = z$ is trivially strongly real of positive type with a pole at ∞, we can deduce from our theorem 9.6.9 on critical points, that $R(z)$ has at least $N - 2$ non-real critical points. Indeed in cases (i), (ii) and (iii) we obtain a minimum of (i) N, (ii) $N - 1$ and (iii) $N - 2$ non-real critical points. Now

$$R'(z) = \frac{f(z)f''(z)}{f'^2(z)}, \tag{9.136}$$

and therefore the non-real critical points of R are precisely the non-real zeros of f''. It follows that, if f'' has at most $2p$ non-real zeros, then in case (i) $N \le 2p$, in case (ii) $N \le 2p + 1$ and in case (iii) $N \le 2p + 2$. In case (i) we obtain $\alpha \le 2p + 1$. In case (ii) we recall that, if N is odd, we can absorb one zero of P into the strongly real part of the Levin representation; in other words, we can write $FP = F_1 P_1$ where P_1 has degree $N - 1$ and F_1 is strongly real; we then obtain from lemma 9.6.8 that $|N - 1 - \alpha| \le 1$, which gives $\alpha \le 2p + 1$.

Thus we need to prove the next result.

9.6.19

Lemma *Let f be a real entire function of finite order all of whose zeros are real and suppose that $L = f'/f$ is representable in the form*

$$L(z) = F(z)P(z), \tag{9.137}$$

where F is strongly real of negative type and P is a polynomial of even degree $h = 2p$. Assume further that f has infinitely many zeros. Then f has the form (9.127), where $a \ge 0$ and where g is an entire function of genus not exceeding $2p + 1$.

Proof In fact we need to prove that L has the form (9.131), where g' is a polynomial whose degree is either $\le 2p$, or $2p+1$ with the coefficient of z^{2p+1} being negative. Now from theorem 9.6.3 F can be written in the form

$$F(z) = -az + b + \frac{\mu}{z} + \sum_k c_k \frac{z/a_k}{z - a_k} \tag{9.138}$$

where $a \geq 0$, $\mu \geq 0$, $c_k > 0$ and where $\sum_k \frac{c_k}{a_k^2} < +\infty$. Now

$$P(z)c_k \frac{z/a_k}{z - a_k} = \frac{c_k P(a_k) z^{h+1}}{a_k^{h+1}(z - a_k)} + q_k(z), \tag{9.139}$$

where $q_k(z)$ is a polynomial of degree $\leq h$, for each of the two sides has a simple pole at a_k with the same residue there. Since $L = FP \in \Omega$, we see that $c_k P(a_k) > 0$ and so

$$P(a_k) > 0 \text{ for every } k. \tag{9.140}$$

Since P has even degree and $|a_k| \to \infty$ as $k \to \infty$, it follows that $A \geq 0$, where A is the coefficient of z^{2p}. We also see that, if $\mu > 0$, then $P(0) > 0$ and $\mu P(0)$ is the coefficient of $1/z$ for L. The coefficient of z^{2p+1} in L is $-aA \leq 0$. Thus we do indeed obtain the required form provided that we can establish the absolute convergence of

$$\sum_k \frac{c_k P(a_k) z^{h+1}}{a_k^{h+1}(z - a_k)}. \tag{9.141}$$

The convergence of $\sum_k q_k$ to a polynomial of degree $\leq h$ will then follow from the algebraic identities and the given convergence of the sum in the representation of F.

The absolute convergence will be a consequence of the convergence of

$$\sum_k \left| \frac{c_k P(a_k)}{a_k^{h+2}} \right|, \tag{9.142}$$

and since P has degree $\leq h$, $|P(a_k)| \leq M|a_k^h|$ for large k and a suitable constant M, and therefore the required convergence follows from the convergence of $\sum_k \frac{c_k}{a_k^2}$.

9.6.20

This completes the proof of theorem 9.6.18 in the cases that f has infinitely many zeros (apart from a discussion of the Laguerre–Pólya class, which we give below). For case (iii) L is rational and we refer back to theorem 9.4.4. A slight extension of this result is required making use of corollary 9.4.2. If in that theorem we take $n = 2p$ even and consider the case when the leading coefficient of the P of that theorem is < 0, then corollary 9.4.2 gives at least n non-real zeros of f''. This establishes case (iii).

9.6.21 The Laguerre–Pólya class

By definition these are real entire functions f, which have the form

$$f(z) = z^\nu e^{-az^2+bz+c} \prod_n \left(\left(1 - \frac{z}{a_n}\right) \exp\left(\frac{z}{a_n}\right)\right), \qquad (9.143)$$

where $a \geq 0, \nu \geq 0, b$ and c real, a_n real $\neq 0$. For the convergence of the product we require

$$\sum_n \frac{1}{a_n^2} < +\infty. \qquad (9.144)$$

Logarithmic differentiation gives

$$\frac{f'(z)}{f(z)} = \frac{\nu}{z} - 2az + b - \sum_n \frac{1}{a_n} \frac{z/a_n}{1 - z/a_n}. \qquad (9.145)$$

We see immediately that f'/f is either strongly real of negative type or constant, the latter occurring when ν, a, a_n are all zero. Conversely, if f'/f is either strongly real of negative type or constant, then f'/f has the above form and on integration we obtain f in the Laguerre–Pólya class. We have proved the next result.

9.6.22

Theorem *A non-constant real entire function f is in the Laguerre–Pólya class if, and only if,*

$$\Im\left(\frac{f'(z)}{f(z)}\right) \leq 0 \quad (z \in H^+). \qquad (9.146)$$

Note that, if f is in the Laguerre–Pólya class, then both f and f' have all their zeros real. On the other hand, by our theorem, if f is of finite order and both f and f'' have all their zeros real, then f is in the Laguerre–Pólya class. A theorem of Pólya [39] states that the Laguerre–Pólya class is characterised by the property that f and all its derivatives have only real zeros.

9.6.23

Pólya's theorem *Let f be a real transcendental entire function of finite order. Then f is in the Laguerre–Pólya class if, and only if, f and all its derivatives have only real zeros.*

Proof To complete the proof we need to show that, if f is in the Laguerre–Pólya class, then f' is in the Laguerre–Pólya class. The case when $L = f'/f$

is constant is clear, so we assume that L is strongly real of negative type. The result will follow from the next lemma.

9.6.24

Lemma *Let $L(z)$ be a strongly real meromorphic function of negative type and suppose that at each pole of L the residue is ≥ 1. Then the function*

$$L(z) + \frac{L'(z)}{L(z)} \tag{9.147}$$

is strongly real of negative type.

Proof We make use of the product representation from (9.73)

$$L(z) = c \prod_k \rho_k \frac{y_k - z}{x_k - z}, \tag{9.148}$$

where y_k are the zeros of L and x_k are the poles of L, and these are interspersed. If the number of zeros and poles is not infinite in both directions, it may be necessary to add one or two extra terms in the product as explained in the proof of theorem 9.6.3. Differentiating logarithmically gives

$$\frac{L'(z)}{L(z)} = \sum_k \left(\frac{1}{z - y_k} - \frac{1}{z - x_k} \right). \tag{9.149}$$

On the other hand the sum representation of L given in (9.72) is

$$L(z) = -az + b + \frac{\mu}{z} + \sum_k{}' c_k \left(\frac{1}{z - x_k} + \frac{1}{x_k} \right), \tag{9.150}$$

where $\sum'$ denotes the sum taken over all k except the term (if any) where $x_k = 0$, and where c_k is the residue at the pole x_k.

We thus obtain

$$\frac{L'(z)}{L(z)} + L(z) = -az + b + \sum_k{}' \left(\frac{1}{z - y_k} + \frac{c_k}{x_k} + \frac{c_k - 1}{z - x_k} \right) + \frac{\mu - 1}{z} + \frac{1}{z - y_j}, \tag{9.151}$$

the last two terms being necessary only if there is a pole at the origin. Now, since each $c_k \geq 1$ and $\mu \geq 1$, we obtain on taking imaginary parts

$$\Im\left(\frac{L'(z)}{L(z)} + L(z) \right) < 0 \quad (z \in H^+), \tag{9.152}$$

and the lemma follows. If it is necessary to add the extra terms to the product representation for L, this gives extra terms of negative imaginary part.

To complete the proof of Pólya's theorem, we observe that, if $L = f'/f$, where L is strongly real of negative type, then the residue at each pole of L is the order of the zero of f at the same point, and is therefore ≥ 1. Since $f''/f' = L + L'/L$, we obtain $\Im(f''/f') < 0$ in H^+ and so f' is in the Laguerre–Pólya class.

10
Level curves

10.1 Level regions for polynomials

10.1.1

The term **level curve** is used, loosely speaking, to denote a curve on which a function $f(z)$ satisfies a relation

$$F(f(z)) = c \tag{10.1}$$

where $F(w)$ is a real-valued function and c is a constant. In practice it is usually the case that F is one of the three functions $w \mapsto |w|$, $w \mapsto \Re w$ and $w \mapsto \Im w$. We note that in all three cases the equation $F(w) = c$ describes the boundary of a simply-connected domain in the w-plane. In the following discussion we shall adopt this point of view.

Let Ω denote a simply-connected domain and let $P(z)$ be a polynomial of degree $n \geq 1$. Let D be a component of the set $\{z\colon\ P(z) \in \Omega\}$. D is clearly a domain in the z-plane. Choose $a \in D$ such that $P'(a) \neq 0$ and let $b = P(a)$. Let Ω_0 denote the domain Ω with all branch points removed (i.e. points $b = P(a)$ where $P'(a) = 0$). The branch of the inverse function $\phi(w) = P^{-1}(w)$ satisfying $\phi(b) = a$ can be continued along all paths in Ω_0 with initial point b; therefore by the monodromy theorem $\phi(w)$ will return to the initial branch along any null homotopic loop in Ω_0. Note that the values z attained by $\phi(w)$ by continuation along such curves will form a connected set and therefore will lie in D. If $\Omega_0 = \Omega$, so D contains no critical points of P, then ϕ can be continued along all curves so is a single-valued analytic function. It follows that P is a 1–1 mapping of D onto Ω; in particular D is simply-connected. More generally, we have the first result of this chapter.

10.1.2

Theorem *Let $P(z)$ be a polynomial of degree $n \geq 1$ and let $\Omega \neq \mathbb{C}$ be a simply-connected domain. If D is a component of $P^{-1}(\Omega)$, then D is simply-connected and we can write*

$$P(z) = \psi \circ B \circ \tau(z) \quad (z \in D) \tag{10.2}$$

where ψ is a 1–1 conformal map of $\mathbb{U}$ onto Ω, B is a Blaschke product with zeros in $\mathbb{U}$, whose degree is $m+1$ where m is the number of zeros of P' in D counted according to multiplicity, and where τ is a 1–1 conformal map of D onto $\mathbb{U}$. In particular P is a mapping of D onto Ω, which attains each value in Ω exactly $m+1$ times.

Proof We begin by assuming that Ω is a Jordan domain and so is bounded by a Jordan curve Γ. We choose a branch of the inverse function $\phi = P^{-1}$ in Ω which maps into D and then continue this branch around Γ until returning to the initial branch; for simplicity we assume that there are no branch points of ϕ on Γ. We obtain a Jordan curve J in the z-plane with $P(J) = \Gamma$. Let ψ be a 1–1 conformal map of $\mathbb{U}$ onto Ω; then ψ extends continuously to the unit circle $\mathbb{T}$ and gives a homeomorphism of $\overline{\mathbb{U}}$ onto $\overline{\Omega}$. Let $\omega = \psi^{-1}$; the function $\omega(P(z))$ is analytic in D and continuous in $\overline{D}$; for $z \in J$ we have $|\omega(P(z))| = 1$; therefore by the maximum principle $|\omega(P(z))| \leq 1$ for z inside J; it follows that all points z inside J lie in $\overline{D}$, and therefore D is a Jordan domain. Let σ be a 1–1 conformal map of $\mathbb{U}$ onto D; then $\mu = \omega \circ P \circ \sigma$ is a mapping from $\mathbb{U}$ onto $\mathbb{U}$, which extends continuously to a mapping of $\mathbb{T}$ into $\mathbb{T}$; μ has a finite number of zeros in $\mathbb{U}$. Let B denote the Blaschke product with these zeros. Then $\lambda = \mu/B$ is analytic and non-zero in $\mathbb{U}$ and satisfies $|\lambda| = 1$ on $\mathbb{T}$. Since the same is true of $1/\lambda$, it follows from the maximum principle that λ is a constant of absolute value 1. We may assume that $\lambda = 1$ by absorbing a rotation into B. We thus have $B = \omega \circ P \circ \sigma$, which gives the required representation for P in D with $\tau = \sigma^{-1}$. If B has degree ν, then B attains every value in $\mathbb{U}$ exactly ν times, and therefore P attains every value in Ω exactly ν times in D. In addition B' has exactly $\nu - 1$ zeros in $\mathbb{U}$ and so P' has exactly $\nu - 1$ zeros in D.

We consider now the general case that Ω is a simply-connected domain, not the whole plane. As before let ψ denote a 1–1 conformal map of $\mathbb{U}$ onto Ω with $\psi(0) = b \in \Omega$ and write $\omega = \psi^{-1}$. Let $\psi_r(\zeta) = \psi(r\zeta)$, where $0 < r < 1$ and $\zeta \in \mathbb{U}$; ψ_r maps $\mathbb{U}$ onto a Jordan domain $\Omega_r \subset \Omega$ and $P^{-1}(\Omega_r)$ has a component $D_r \subset D$. If $z \in D$, then $P(z) \in \Omega$, so $\omega \circ P(z) \in \mathbb{U}$; therefore, for r close to 1, $P(z) \in \Omega_r$ so $z \in D_r$. Thus $\{D_r\}$ is an expanding sequence of simply-connected domains which exhausts D. It follows that D is simply-connected: to see this, let γ be a loop in D; the domains D_r form an open covering of γ and therefore, by the Heine–Borel theorem, a finite number of D_r cover γ; since $0 < r < s < 1$ implies $D_r \subset D_s$, it follows that, for suitable ρ, $\gamma \subset D_\rho$; now, if $w \notin D$, then $w \notin D_\rho$ and so, because D_ρ is simply-connected, we have $n(\gamma, w) = 0$; thus for all loops γ in D and all points $w \notin D, n(\gamma, w) = 0$; this implies that D is simply-connected (see chapter 2, Problem 2.3.16.3). Since $\Omega \neq \mathbb{C}$, the fundamental theorem of algebra implies that $D \neq \mathbb{C}$; hence we can find a 1–1 conformal mapping $\tau\colon D \to \mathbb{U}$ with $\tau(a) = 0$, where $P(a) = b$. Now for r close to 1, D_r contains all the critical points of P in D; assume there are m such points, counted according to multiplicity. By the first part we have

$$P(z) = \psi_r \circ B_r \circ \tau_r(z) \quad (z \in D_r) \tag{10.3}$$

where B_r is a Blaschke product of degree $m+1$ and where τ_r is a conformal mapping from D_r onto $\mathbb{U}$, which we may assume satisfies $\tau_r(a) = 0$ and $\arg \tau_r'(a) = \arg \tau'(a)$. As $r \to 1$, $\psi_r \to \psi$; also $\tau_r \to \tau$ uniformly on any compact subset $E \subset D$: to see this, observe that, for r near 1, τ_r is defined on E and bounded by 1; therefore $\{\tau_r\}$ converges to χ say for a sequence of $r \to 1$ uniformly on E, and thus locally uniformly on D; χ is analytic on D, and by the well-known application of Hurwitz's theorem is either constant on D or 1–1 on D; now, if $1 > s > r$, then $\tau_s \tau_r^{-1}$ is a 1–1 map from $\mathbb{U}$ into $\mathbb{U}$, which is zero at 0; by Schwarz's lemma the derivative at 0 is at most 1 in absolute value, which gives $|\tau_r'(0)| \le |\tau_s'(0)|$; since $\chi'(0) = \lim_{r\to 1} \tau_r'(0)$, we see that $\chi'(0) \neq 0$, so χ is not constant; finally, to prove $\chi = \tau$, we need to show that χ is a surjective mapping from D onto $\mathbb{U}$; let $\zeta \in \mathbb{U}$ and let $z_r = \tau_r^{-1}(\zeta)$, so $\tau_s(z_r) = \tau_s \tau_r^{-1}(\zeta)$; as above, if $0 < r < s < 1$, then Schwarz's lemma gives $|\tau_s(z_r)| \le |\zeta|$, and so letting $s \to 1$, $|\chi(z_r)| \le |\zeta|$; it follows that the points z_r lie in a compact subset of D and therefore for a suitable subsequence converge to $z \in D$; then $\tau_r(z_r) \to \chi(z)$ as $r \to 1$, so $\zeta = \chi(z)$. To complete the proof of the theorem, it only remains to observe that the Blaschke products B_r of degree $m+1$ converge for a sequence of $r \to 1$ to a Blaschke product B of degree not exceeding $m+1$; we thus obtain the desired representation $P = \psi \circ B \circ \tau$ by letting $r \to 1$; then the number of critical points of B in $\mathbb{U}$ is the same as the number of critical points of P in D, namely m, and so in fact B has degree $m+1$.

10.1.3

Regions D in the z-plane, defined as the components of inverse images of simply-connected domains in the $w(= P(z))$-plane, may reasonably be called **level regions** of the polynomial P. Our theorem states that in such regions P is conformally equivalent, and therefore topologically equivalent, to a finite

Blaschke product in $\mathbb{U}$; the valency of P in D is the degree of the Blaschke product, which in turn exceeds by 1 the number of critical points of P in D.

10.2 Level regions of rational functions

10.2.1

As we have seen, level regions for polynomials are simply-connected. This is not the case in general for the level regions of rational functions. Nevertheless, we can still find relations between the topology of the level region, the valency of the function and the number of critical points within the region. In order to include poles and the point at ∞, a critical point for a rational function is defined as any point at which the function is not locally 1–1. The **order** of the critical point is 1 less than its local valency.

10.2.2

Theorem: Riemann–Hurwitz formula *Let $R(z)$ be a non-constant rational function and let D be a component of the level set $\{z\colon |R(z)| < 1\}$ relative to the extended plane. Let ζ denote the number of zeros of R in D and χ the number of critical points in D, counted according to multiplicity. Finally let $\nu + 1$ denote the number of components relative to the extended plane of the boundary of D. Then R attains in D every value w satisfying $|w| < 1$ exactly ζ times. Furthermore we have*

$$\chi = \zeta + \nu - 1. \tag{10.4}$$

Proof It is clear that D will be bounded by a finite number of level curves on which $|R(z)| = 1$. If D is an unbounded region, then $|R(\infty)| \leq 1$. By considering $R \circ \phi$, where ϕ is a Moebius transformation sending ∞ to a finite point, and sending a finite point c, at which $|R(c)| > 1$, to ∞, we obtain a bounded level region V for $R \circ \phi$ which is homeomorphic to D. Thus it is clear that there is no loss of generality in assuming that D is a bounded region. The boundary of D will then consist of an outer level curve γ_0 together with ν inner curves $\gamma_j\,(1 \leq j \leq \nu)$ on which $|R| = 1$. All the curves are Jordan curves obtained by continuing branches of the inverse function R^{-1} around the unit circle. For simplicity we assume that there are no critical points of R (which, in this case, means zeros of R') on the level curves.

If each γ_j is assumed positively oriented, then the cycle

$$\Gamma = \gamma_0 - \sum_{j=1}^{\nu} \gamma_j \tag{10.5}$$

is homologous to zero relative to D (strictly a domain slightly larger than D). Let ζ denote the number of zeros of R in D and χ the number of critical points, each counted according to multiplicity. Then by the argument principle

$$\zeta = d(R, \Gamma) \quad \text{and} \quad \chi = d(R', \Gamma). \tag{10.6}$$

Now, if $|w| < 1$, then for $z \in \Gamma$, $\Re \frac{R(z)-w}{R(z)} = \Re(1 - w\overline{R(z)}) > 0$, and hence

$$d(R - w, \Gamma) = d(R, \Gamma) = \zeta. \tag{10.7}$$

Therefore in D, $R(z)$ attains every value w in $|w| < 1$ exactly ζ times. This implies that, as z traverses the cycle Γ, each value w on the unit circle is attained exactly ζ times. Indeed we obtain ζ copies of the positively oriented unit circle, for as z traverses γ_0 the circle is traversed positively, say ρ_0 times; and as z traverses γ_j $(1 \le j \le \nu)$ the circle is traversed negatively, say ρ_j times. Thus $\zeta = \sum_{j=0}^{\nu} \rho_j$.

Now we may parametrise γ_0 by $z = z(t)$, where $R(z(t)) = e^{it}$ $(0 \le t \le 2\pi\rho_0)$. We then have

$$z'(t)\frac{R'(z(t))}{R(z(t))} = i, \tag{10.8}$$

and therefore (change in arg dz around γ_0) $+ 2\pi d(R', \gamma_0) - 2\pi d(R, \gamma_0) = 0$. Since the change in tangential angle around a simple smooth curve is 2π, we obtain

$$1 + d(R', \gamma_0) - d(R, \gamma_0) = 0. \tag{10.9}$$

Applying the same reasoning to the remaining γ_j we obtain the relation

$$1 - \nu + d(R', \Gamma) - d(R, \Gamma) = 0. \tag{10.10}$$

This gives

$$1 - \nu + \chi - \zeta = 0. \tag{10.11}$$

Thus

$$\zeta = \sum_{j=0}^{\nu} \rho_j \text{ and } \chi = \zeta + \nu - 1 = \rho_0 - 1 + \sum_{j=1}^{\nu} (\rho_j + 1). \tag{10.12}$$

In the case of a simply-connected region we have $\nu = 0$ and so $\chi = \zeta - 1$, as obtained earlier. Note that in any case

$$\chi \ge \zeta - 1. \tag{10.13}$$

Also note that, as $\rho_j \ge 1$,

$$\zeta \ge \nu + 1 \quad \text{and} \quad \chi \ge 2\nu. \tag{10.14}$$

To deal with the case when the boundary contains critical points we create small indentations in the cycle by the following process. In the continuation process, as we approach a branch point on the unit circle in the w-plane, we indent with a small arc into the unit disc joining two points on the circle on either side of the branch point. In the z-plane this will give us a small crosscut in D joining two portions of the boundary and avoiding the critical point. The amended cycle will still consist of Jordan curves and the above reasoning can be applied with only minor modifications.

10.3 Partial fraction decomposition

10.3.1

Let $R(z)$ be a rational function of degree n with $R = P/Q$, where P and Q are polynomials with no common zeros and where n is the degree of P and Q has degree $< n$. Then for all finite w, with the exception of the values of R at its critical points, the equation $R(z) = w$ has n distinct and simple roots $z_k(w)$. We then have, since $R(z) \to \infty$ as $z \to \infty$,

$$\frac{1}{R(z) - w} = \sum_{k=1}^{n} \frac{1}{R'(z_k)(z - z_k)}. \tag{10.15}$$

It follows that, for at least one k,

$$\Re\left(\frac{R(z) - w}{R'(z_k)(z - z_k)}\right) \geq \frac{1}{n} \tag{10.16}$$

and therefore

$$|z - z_k| \leq n \left|\frac{R(z) - w}{R'(z_k)}\right|, \tag{10.17}$$

which gives an upper bound for the distance from z to a point z_k where $R(z_k) = w$. For example, if $w = 0$, $z = 0$ and $R(0) = 1$, we deduce that $R(z)$ has a zero in the level set $\{|zR'(z)| \leq n\}$; and indeed, more strongly, R has a zero in the set $\{z\colon \Re\frac{-1}{zR'(z)} \geq \frac{1}{n}\}$. The polynomial $R(z) = 1 + z^n$ shows that this is a sharp result.

10.3.2

We also note that, if we take $z = 0$ and $R(0) = 0$, we have

$$\Re\left(\frac{w}{z_k R'(z_k)}\right) \geq \frac{1}{n} \tag{10.18}$$

at a point z_k where $R(z_k) = w$. Thus, if $\phi(w)$ denotes the inverse of R, so ϕ is an algebraic function with branch points at the singularities of R, then the above inequality becomes

$$\Re \frac{w\phi'(w)}{\phi(w)} \geq \frac{1}{n}. \tag{10.19}$$

Our assertion is that this holds at each w, except the branch points, for a suitably chosen branch of the inverse function. It would be interesting to know, for what regions of w we can assert this inequality for the same branch of ϕ. For example, is this the case in a simply-connected region of w which contains no branch points, where we choose that branch of ϕ corresponding to the local inverse of R at $z = 0$? If so, this would be a remarkable result concerning the shape of level curves in regions where R is univalent.

10.3.3 Problem

1. R is a rational function of degree n; α, β and γ are distinct values; a and b distinct points at which $R(a) = \alpha$ and $R(b) = \beta$. Find a region containing a point c at which $R(c) = \gamma$.

We have a solution to this problem when $a = 0$, $b = \infty$; $\alpha = 1$, $\beta = \infty$, $\gamma = 0$; namely the region $\{|zR'(z)| \leq n\}$. We can use this case to solve the general case by finding Moebius transformations A and B such that, if $S = A \circ R \circ B$, then the above solution can be applied to S. B must satisfy

$$B(0) = a, \qquad B(\infty) = b. \tag{10.20}$$

We can take

$$B(z) = \frac{bz + a}{z + 1}. \tag{10.21}$$

A must satisfy

$$A(\alpha) = 1, \quad A(\beta) = \infty, \quad A(\gamma) = 0. \tag{10.22}$$

In this case, assuming α, β and γ are finite, we have

$$A(w) = \frac{\alpha - \beta}{\alpha - \gamma} \frac{w - \gamma}{w - \beta} = \frac{\alpha - \beta}{\alpha - \gamma} \left(1 + \frac{\beta - \gamma}{w - \beta}\right). \tag{10.23}$$

This gives

$$S(z) = \frac{\alpha - \beta}{\alpha - \gamma} \left(1 + \frac{\beta - \gamma}{R\left(\frac{bz + a}{z + 1}\right) - \beta}\right). \tag{10.24}$$

Then $S(0) = 1$, $S(\infty) = \infty$ and $S(z) = 0$ only if $R(\frac{bz+a}{z+1}) = \gamma$. Now $\exists d$ satisfying $S(d) = 0$ and $|dS'(d)| \leq n$. Then

$$zS'(z) = \frac{(\alpha-\beta)(\beta-\gamma)(a-b)}{\alpha-\gamma}\frac{z}{(z+1)^2}\frac{R'\left(\frac{bz+a}{z+1}\right)}{\left(R\left(\frac{bz+a}{z+1}\right)-\beta\right)^2}. \tag{10.25}$$

Writing $c = \frac{bd+a}{d+1}$, we obtain, since $R(c) = \gamma$, and $d = \frac{a-c}{c-b}$,

$$dS'(d) = \frac{\alpha-\beta}{(\alpha-\gamma)(\beta-\gamma)}\frac{(a-c)(c-b)}{a-b}R'(c). \tag{10.26}$$

Thus we can find c satisfying $R(c) = \gamma$ and

$$\left|\frac{(a-c)(c-b)}{a-b}\right||R'(c)| \leq n\left|\frac{(\alpha-\gamma)(\beta-\gamma)}{\alpha-\beta}\right|. \tag{10.27}$$

Note that the level region $\{z\colon |\frac{(a-z)(z-b)}{a-b}||R'(z)| \leq n|\frac{(\alpha-\gamma)(\beta-\gamma)}{\alpha-\beta}|\}$ containing the point c also contains each of the points a and b. Our argument also depends on the fact that the degree of a rational function is invariant under Moebius transformations. Note also that the inequality can be written

$$\left|\left(\frac{R(a)-R(c)}{a-c}\right)\left(\frac{R(b)-R(c)}{b-c}\right)\frac{1}{R'(c)}\right| \geq \frac{1}{n}\left|\frac{R(a)-R(b)}{a-b}\right|. \tag{10.28}$$

The interpretation then is that, if a and b are fixed and c varies over the level set defined by this inequality, then $R(c)$ attains all finite values other than $R(a)$ and $R(b)$. These latter points can also be included, since if $R(c) \to R(a)$, then the inequality can only hold if $c \to a$; therefore the inequality will hold with $c = a$ and the usual limiting interpretation of the left-hand side. Similarly for b.

We can improve on this result by making use of the stronger condition for S

$$\Re\frac{-1}{zS'(z)} \geq \frac{1}{n}. \tag{10.29}$$

We are led to the inequality

$$\Re\left(\frac{a-b}{R(a)-R(b)}\frac{R(a)-R(c)}{a-c}\frac{R(b)-R(c)}{b-c}\frac{1}{R'(c)}\right) \geq \frac{1}{n}. \tag{10.30}$$

10.3.4

Theorem *Let $R(z)$ be a rational function of degree n and let a and b be finite points at which $R(a) \neq R(b)$. Then $R(z)$ attains all values in the plane as z*

ranges over the level set

$$\left\{z : \Re\left(\frac{a-b}{R(a)-R(b)}\frac{R(a)-R(z)}{a-z}\frac{R(b)-R(z)}{b-z}\frac{1}{R'(z)}\right) \geq \frac{1}{n}\right\}. \quad (10.31)$$

This result gives us a level set expressed in terms of two known values of R, namely $R(a)$ and $R(b)$. By letting $b \to a$ we can obtain a level set in terms of $R(a)$ and $R'(a)$, as given next.

10.3.5

Corollary *Let $R(z)$ be a rational function of degree n satisfying $R(0) = 0$ and $R'(0) = 1$. Then $R(z)$ attains all values in the plane as z ranges over the level set*

$$\left\{z : \Re\left[\left(\frac{R(z)}{z}\right)^2 \frac{1}{R'(z)}\right] \geq \frac{1}{n}\right\}. \quad (10.32)$$

10.4 Smale's conjecture

10.4.1

S. Smale [59] (see also M. Shub and S. Smale [58]) has made the following conjecture in connection with the theory of computer complexity. Let $P(z)$ be a polynomial of degree $n \geq 2$. Then for each $z \in \mathbb{C}\ \exists$ a critical point ζ of P such that

$$\left|\frac{P(z)-P(\zeta)}{(z-\zeta)P'(z)}\right| \leq 1. \quad (10.33)$$

The example $P(z) = nz - z^n$ has its critical points at the $n-1$ roots of $z^{n-1} = 1$. At each such point ζ we obtain $P(\zeta) = (n-1)\zeta$ which gives

$$\left|\frac{P(\zeta)}{\zeta P'(0)}\right| = \frac{n-1}{n}. \quad (10.34)$$

This suggests the stronger conjecture

$$\left|\frac{P(z)-P(\zeta)}{(z-\zeta)P'(z)}\right| \leq \frac{n-1}{n} \quad (10.35)$$

for each z and a suitable critical point ζ, which, if true, would be sharp. Smale has shown that there certainly is *some* bound, and he has proved that

$$\left|\frac{P(z)-P(\zeta)}{(z-\zeta)P'(z)}\right| \leq 4 \quad (10.36)$$

for each z and a critical point ζ. The proof is as follows: there is clearly no loss of generality in taking $z = 0$ and assuming that $P(0) = 0$ and $P'(0) = 1$, so $P(z)$ has the form $P(z) = z + a_2 z^2 + \cdots + a_n z^n$. We then need to show that for at least one critical point ζ of P we have

$$|P(\zeta)| \leq 4|\zeta|. \tag{10.37}$$

Let $w = P(z)$ and consider the inverse function $z = P^{-1}(w) = \phi(w)$ say. This is analytic in a neighbourhood of $w = 0$ with the expansion $\phi(w) = w + c_2 w^2 + \cdots$. This function can be analytically continued into the largest disc $|w| < R$ containing no branch points. We then have $R = |P(\zeta)|$, where ζ is a critical point of P. The function $\phi(Rw)/R$ is a normalised univalent function in $|w| < 1$. Therefore by the Koebe $\frac{1}{4}$-theorem we have

$$\left|\frac{\phi(Rw)}{R}\right| \geq \frac{|w|}{4} \quad (|w| < 1) \tag{10.38}$$

from which we easily obtain $|P(\zeta)| \leq 4|\zeta|$, as required.

10.4.2

The conjecture can be expressed in the following alternative form. Let

$$\Phi(z) = \frac{(1 - z)^{n+1} - 1 + (n+1)z}{n(n+1)z^2}, \tag{10.39}$$

which is a polynomial of degree $n - 1$. This generates a symmetric linear form, which we denote by $\Phi(z_1, z_2, \ldots, z_{n-1})$; so $\Phi(z) = \Phi(z, z, \ldots, z)$. For an arbitrary vector $(z_1, z_2, \ldots, z_{n-1})$ with non-zero terms we construct the additional $n - 1$ vectors

$$\frac{1}{z_k}(z_1, \ldots, z_{k-1}, 1, z_{k+1}, \ldots, z_{n-1}) \quad (1 \leq k \leq n-1). \tag{10.40}$$

We consider the n values of the linear form Φ at these points in $\mathbb{C}^{n-1}$. The conjecture is equivalent to the claim that $|\Phi| \leq 1$ for at least one of these values. (The value n here corresponds to the degree of P'.)

Thus for polynomials of degree 2 we have $n = 1$, which gives $\Phi = \frac{1}{2}$. For cubics we have $n = 2$ which gives $\Phi(z) = \frac{1}{2} - \frac{1}{6}z$. The corresponding linear form is identical and we need to show that at least one of the values

$$\frac{1}{2} - \frac{1}{6}z \quad \text{and} \quad \frac{1}{2} - \frac{1}{6z} \tag{10.41}$$

has absolute value ≤ 1. In fact, since either $|z| \leq 1$ or $|1/z| \leq 1$, we see that at least one of the values is $\leq \frac{2}{3}$. Thus the stronger conjecture holds for degrees 2 and 3.

The case $n = 3$. Here we have $\Phi(z) = \frac{1}{2} - \frac{1}{3}z + \frac{1}{12}z^2$, whose corresponding form is

$$\Phi(z, \zeta) = \frac{1}{2} - \frac{1}{6}(z + \zeta) + \frac{1}{12}z\zeta. \tag{10.42}$$

The two associated values are

$$\frac{1}{2} - \frac{1}{6}\left(\frac{1}{z} + \frac{\zeta}{z}\right) + \frac{\zeta}{12z^2} \quad \text{and} \quad \frac{1}{2} - \frac{1}{6}\left(\frac{1}{\zeta} + \frac{z}{\zeta}\right) + \frac{z}{12\zeta^2}. \tag{10.43}$$

Although it is clear that at least one of these three expressions has absolute value $\leq \frac{11}{12}$ (since, for at least one, both variables on which Φ acts lie in the unit disc), the stronger conjecture needs to get the estimate $\frac{3}{4}$, which is not immediately evident. An argument of D. Tischler [64] proves this result as follows: let ζ_1, ζ_2 and ζ_3 be the critical points of P. Then we have

$$\frac{P(\zeta_1)}{\zeta_1} = \frac{1}{2} - \frac{1}{6}\left(\frac{\zeta_1}{\zeta_2} + \frac{\zeta_1}{\zeta_3}\right) + \frac{1}{12}\frac{\zeta_1}{\zeta_2}\frac{\zeta_1}{\zeta_3} \tag{10.44}$$

and similar expressions at ζ_2 and ζ_3. A direct calculation gives

$$\begin{aligned}\sum_{k=1}^{3}\frac{1}{|\zeta_k|^4}\left|\frac{P(\zeta_k)}{\zeta_k} - \frac{1}{2}\right|^2 &= \frac{1}{16}\left(\frac{1}{|\zeta_1|^2|\zeta_2|^2} + \frac{1}{|\zeta_2|^2|\zeta_3|^2} + \frac{1}{|\zeta_3|^2|\zeta_1|^2}\right)\\ &\leq \frac{1}{16}\sum_{k=1}^{3}\frac{1}{|\zeta_k|^4}\end{aligned} \tag{10.45}$$

from which it follows that, for at least one k

$$\left|\frac{P(\zeta_k)}{\zeta_k} - \frac{1}{2}\right| \leq \frac{1}{4}. \tag{10.46}$$

The $\frac{3}{4}$ estimate follows immediately.

10.4.3

A remark on level curves Suppose $P(z)$ has a zero at 0 and consider the smallest value of R such that P is univalent in the level region $D_R = \{|P(z)| < R\}$ containing 0. Then P has a critical point ζ with $|P(\zeta)| = R$. The question arises as to whether P is *convex* in D. For $|w| < R$ we have

$$P(z(w)) = w \tag{10.47}$$

where $z(w)$ denotes the inverse function of P. Then $\exists\rho \leq R$ such that $z(w)$ is convex for $|w| < \rho$, but fails to be convex in larger discs. This means that

$$\Re\left(1 + \frac{wz''(w)}{z'(w)}\right) > 0 \quad (|w| < \rho) \tag{10.48}$$

but not on any larger circle. Now we have

$$z'(w)P'(z(w)) = 1, \qquad z''(w)P'(z(w)) + z'^2(w)P''(z(w)) = 0, \quad (10.49)$$

and therefore

$$1 + \frac{wz''(w)}{z'(w)} = 1 - \frac{P(z(w))P''(z(w))}{P'^2(z(w))}. \quad (10.50)$$

Hence for $z \in D_\rho$ we have

$$\Re\left(1 - \frac{P(z)P''(z)}{P'^2(z)}\right) > 0. \quad (10.51)$$

However,

$$1 - \frac{P(z)P''(z)}{P'^2(z)} = \left(\frac{P(z)}{P'(z)}\right)' = R'(z) \text{ say}. \quad (10.52)$$

Thus the derivative of the rational function $R(z)$ has positive real part in the convex domain D_ρ. It follows from Kaplan's well-known result that $R(z)$ is univalent in D_ρ and maps D_ρ onto a close-to-convex domain. The function $R(z(w))$ is close-to-convex for $|w| < \rho$. In fact $R(z(w)) = wz'(w)$ is *starlike*. Thus P/P' maps D_ρ onto a starlike domain.

10.4.4 The Cordova–Ruscheweyh method [14]

Their general subordination theorem on polynomials implies the weaker version of Smale's conjecture in the case that P has all its critical points on the unit circle. Here is their method:

Suppose that $P(z) = z + a_2 z^2 + \cdots + a_n z^n$ and let $\phi(w) = w + \cdots$ denote the inverse function; this is analytic and univalent for $|w| < R = |P(\zeta)|$, where ζ is a critical point of P. Thus we have

$$P(\phi(w)) = w \quad (|w| < R). \quad (10.53)$$

This gives

$$\phi'(w)P'(\phi(w)) = 1, \qquad \phi''(w)P'(\phi(w)) + \phi'^2(w)P''(\phi(w)) = 0, \quad (10.54)$$

and hence

$$1 + \frac{w\phi''(w)}{\phi'(w)} + \frac{w\phi'(w)}{\phi(w)} \frac{\phi(w)P''(\phi(w))}{P'(\phi(w))} = 1. \quad (10.55)$$

For $0 < \rho < R$ choose a so that $|a| = \rho$ is a point of maximum modulus of ϕ for $|w| \le \rho$. Then by the Clunie–Jack lemma (see chapter 8) we have

$$\Re\left(1 + \frac{a\phi''(a)}{\phi'(a)}\right) \ge \frac{a\phi'(a)}{\phi(a)} = t(a) \ge 1. \tag{10.56}$$

Setting $b = \phi(a)$ the previous two relations give

$$\Re\left(\frac{bP''(b)}{P'(b)}\right) \le \frac{1}{t} - 1 \le 0. \tag{10.57}$$

If ζ_k $(1 \le k \le n-1)$ are the critical points of P, the last relation becomes

$$\sum_{k=1}^{n-1} \Re \frac{1}{1 - \zeta_k/b} \le 0, \tag{10.58}$$

from which it follows that, for at least one k, $\Re(1 - \zeta_k/b) \le 0$ and so $b \le |\zeta_k|$. In fact, more precisely, $|b - \frac{1}{2}\zeta_k| \le \frac{1}{2}|\zeta_k|$. We deduce that $|\phi(w)| \le |\zeta_k|$ for $|w| < R$ and a suitable critical point ζ_k. From Schwarz's lemma we obtain $R \le |\zeta_k|$, i.e.

$$|P(\zeta)| \le |\zeta_k|. \tag{10.59}$$

In particular, if all the critical points lie on the same circle concentric with the origin, then $|P(\zeta)| \le |\zeta|$, as required.

Notice that Smale's conjecture also follows in the case that the function $z \mapsto |P(z)|$ takes on the same value at all the critical points of P.

10.4.5

Let C denote a level curve $\{|P(z)| = |z|\}$ passing through the origin. C is a Jordan curve bounding a domain D, which is a component of $\{|P(z)| < |z|\}$. Smale's conjecture will hold if $D \cup C$ contains a critical point of P. With R defined as in the previous subsection, let Γ_R denote the level curve $\{|P(z)| = R\}$ containing the origin and bounding J_R say. This meets another such level curve at the critical point ζ, and we would like to show that $\zeta \in D \cup C$. Note that Γ_R meets C at points on the circle $\{|z| = R\}$. Now D contains at least one zero a of P; if a is a multiple zero, then a is a critical point of P in D. If a is simple, we choose ρ as large as possible so that the component F of the level set $\{|P(z)| < \rho\}$ which contains a is contained in D. If there is no critical point of P in F, then P is 1–1 in F. Furthermore, there is a point $b \in \partial F \cap C$. We have $|P(b)| = |b| = \rho$. Now if there are no critical points of P in $\overline{F}$, then the boundary of $\overline{F}$ is a simple closed curve lying in $\overline{D}$, and so for $z \in \partial F$ we have $|P(z)| \le |z|$. But also $|P(z)| = \rho$, and therefore $|z| \ge \rho$. Thus ∂F is a simple closed curve lying in $|z| \ge \rho$ and meeting the circle $|z| = \rho$ at b.

Also, for points $z \in C$ near to b, we have $|P(z)| \geq \rho$ and $|P(z)| = |z|$, and so $|z| \geq \rho$. Thus ∂F has a global minimum at b and C has a local minimum at b. In fact in a neighbourhood of b, all points of D in the neighbourhood lie in $|z| > \rho$. It follows that, if Smale's conjecture is false for P, then there is one such minimum on the level set $\{|P(z)| = |z|\}$ corresponding to each zero of P, other than 0.

The question arises as to whether such a minimum point on C corresponds to a critical point of $Q = P(z)/z$ in D, and therefore another zero of Q (and therefore of P) in D, leading to another minimum point and so another critical point of $Q \dots$ etc. In other words, is it true that the number of critical points of Q in D is at least the number of such minimum points on C? If so, Smale's conjecture must hold.

10.4.6

A remark on composition *Let P and Q be non-constant polynomials and D a simply-connected domain with the property that*

$$P(z) \neq Q(\zeta) \quad (z \in D) \tag{10.60}$$

for every critical point ζ of Q. Then $\exists \omega(z)$ analytic in D such that

$$P(z) = Q(\omega(z)) \quad (z \in D). \tag{10.61}$$

Here $\omega(z)$ is locally defined by the equation $\omega(z) = Q^{-1}(P(z))$; the condition implies that ω can be continued along every path in D; from the monodromy theorem ω is a single-valued analytic function in D.

10.4.7 Some tentative conjectures

(1) *Strong conjecture* Let $Q(z)$ be a polynomial of degree m and let E denote the set $\{z\colon |Q(z)| \leq 1\}$. Let $P(z)$ be a polynomial of degree $n > m$ such that Q is a factor of P and suppose that P has all its zeros in E. Then P has at least $n - m$ critical points in E.
(2) *Weak conjecture* Same hypotheses. Then P has at least one critical point in E.

Smale's conjecture is the special case when $P(z)/Q(z) = z$. Note also that the strong conjecture reduces to a weak version of the Gauss–Lucas theorem in the case $m = 1$.

Smale's conjecture can be expressed in the following form: let $Q(z)$ be a polynomial of degree $m \geq 1$ with simple zeros, and for each $r > 0$ let E_r denote the set $\{z: |Q(z)| \le r\}$; then the mapping

$$z \mapsto z + \frac{Q(z)}{Q'(z)} \tag{10.62}$$

maps E_r onto a set $F_r \supset E_r$. This is certainly true for small $r > 0$, when

$$\left| Q\left(z + \frac{Q(z)}{Q'(z)} \right) \right| \approx 2|Q(z)|. \tag{10.63}$$

Thus initially the sets F_r expand faster about the zeros of Q than the sets E_r. On the other hand, for large r

$$\left| Q\left(z + \frac{Q(z)}{Q'(z)} \right) \right| \approx \frac{m+1}{m}|Q(z)| \tag{10.64}$$

and again we easily see that the conjecture is true in this case. Notice that $\exists \rho > 0$ such that $z + Q(z)/Q'(z)$ attains all values in the plane for $|z| \le \rho$: for all large values are attained near the critical points of Q, and only large values are attained near ∞. Thus for large r, F_r covers not only E_r but the whole plane.

Let $R(z) = z + Q(z)/Q'(z)$. The above hypotheses take the form

$$|Q(z)| \le r \Rightarrow \exists \zeta \text{ such that } R(\zeta) = z \text{ and } |Q(\zeta)| \le r. \tag{10.65}$$

Let $Q^{-1} = \phi$ and $R^{-1} = \psi$. Put $w = Q(z)$. Then $|w| \le r$ is to imply that

$$|Q(\psi(\phi(w)))| \le r \tag{10.66}$$

for arbitrary choice of the inverse mapping ϕ and suitable choice of the inverse mapping ψ. Now $w = 0$ gives $\phi(w) = a$, where a is a zero of Q. In this case it is natural to choose the branch of ψ where $\psi(a) = a$. With these choices we have in the neighbourhood of $w = 0$ a well-defined analytic function

$$F(w) = Q \circ \psi \circ \phi(w) \tag{10.67}$$

satisfying $F(0) = 0$. We are required to show that $|F(w)| \le r$ for $|w| \le r$. We have $F'(0) = \frac{1}{2}$ and so, as earlier observed, the conclusion is correct for small w. As long as $F(w)$ remains single-valued, we clearly require $|F(w)| \le |w|$. Now $F(w)$ is single-valued as long as $\psi \circ \phi(w)$ is single-valued. We have

$$\zeta = \psi(\zeta) + \frac{Q(\psi(\zeta))}{Q'(\psi(\zeta))} \tag{10.68}$$

and ζ becomes infinite when $\psi(\zeta)$ is a critical point of Q; $\psi(\zeta)$ branches when $\psi(\zeta)$ is a critical point of R, i.e. $2 - Q''Q/Q'^2$ vanishes at $\psi(\zeta)$. Thus the first branch point of $F(w)$ will occur when $\psi(\phi(w)) = \zeta$ and $2Q'^2(\zeta) = Q''(\zeta)Q(\zeta)$. We have $w = Q(R(\zeta))$ and $F(w) = Q(\zeta)$ and we require that

$$|Q(z)| \le |Q(R(z))| \tag{10.69}$$

for those $z = \psi(\phi(w))$ with $|w| \le |Q(R(\zeta))|$, so $|Q(R(z))| \le |Q(R(\zeta))|$. In other words, we require the rational function $Q \circ R$ to dominate Q in each of the level regions of $Q \circ R$ defined by the zeros of Q, taken up to the critical points of the mapping.

We have the following general formula:

$$Q\left(z + \frac{Q(z)}{Q'(z)}\right) = \sum_{k=0}^{\infty} \frac{1}{k!} Q^{(k)}(z) \left(\frac{Q(z)}{Q'(z)}\right)^k = \sum_{k=0}^{n} \frac{1}{k!} Q^{(k)}(z) \left(\frac{Q(z)}{Q'(z)}\right)^k. \tag{10.70}$$

10.4.8 A Blaschke product approach

Let $P(z) = z + \cdots + a_n z^n$ and let $D = D_c$ denote the component of the level region $\{|P(z)| < c\}$ containing 0. In this region we can write

$$P(z) = c\tau(z) \prod_{k=1}^{m} \frac{\tau(z) - \tau(z_k)}{1 - \overline{\tau(z_k)}\tau(z)} = cB(\tau(z)), \tag{10.71}$$

where $\tau(z)$ is a 1–1 conformal map of D onto $\mathbb{U}$ satisfying $\tau(0) = 0$ and where z_k are those zeros $\ne 0$ of P in D. We have

$$1 = c\tau'(0)B'(0) = c\tau'(0) \prod_{1}^{m} (-\tau(z_k)) \tag{10.72}$$

and so

$$P(z) = \frac{\tau(z)}{\tau'(0)} \prod_{k=1}^{m} \frac{1 - \tau(z)/\tau(z_k)}{1 - \overline{\tau(z_k)}\tau(z)}. \tag{10.73}$$

Now $P'(z)$ vanishes in D when $B'(\tau(z)) = 0$. On the assumption that $m \ge 1$ we wish to show that at some such point we have $|P(z)| \le |z|$. In other words we require that at some ζ satisfying $B'(\zeta) = 0$ we have

$$|B(\zeta)| \le |\tau'(0)||B'(0)||z|, \tag{10.74}$$

where $z = \sigma(\zeta) = \tau^{-1}(\zeta)$. This becomes

$$\left|\frac{B(\zeta)}{B'(0)}\right| \le \left|\frac{\sigma(\zeta)}{\sigma'(0)}\right|. \tag{10.75}$$

Suppose we can in fact show that for at least one critical point ζ of B we have

$$|B(\zeta)| \le |B'(0)||\zeta|. \tag{10.76}$$

Then we may let $c \to \infty$, so D_c approaches a large disc and $|\sigma(\zeta)| \sim |\sigma'(0)\zeta|$. Smale's conjecture will then follow.

This gives us a new conjecture on Blaschke products to consider, which is entirely analogous to the original conjecture, and we may therefore apply similar methods. Smale's inverse function argument will certainly apply to give $|B(\zeta)| \leq 4|B'(0)||\zeta|$ for a suitable critical point ζ. However, his argument can be improved by making use of the fact that the inverse function is bounded.

Let $w = B(z)$ have inverse function at the origin given by $z = \chi(w) = w/B'(0) + \cdots$. Then $\chi(w)$ will remain analytic and univalent in a disc $\{|w| < R = |B(\zeta)|\}$, where ζ is a critical point of B lying in $\mathbb{U}$. Furthermore, $|\chi(w)| < 1$ for $|w| < R$. Therefore, for any x satisfying $|x| \leq 1$, the function

$$\psi(w) = \frac{\chi(w)}{(1 - x\chi(w))^2} \tag{10.77}$$

is univalent for $|w| < R$. It follows that $B'(0)\psi(Rw)/R$ is a normalised univalent function in $|w| < 1$ and we can apply the $\frac{1}{4}$-theorem. We obtain

$$\left|\frac{\zeta}{(1 - x\zeta)^2}\right| \geq \frac{R}{4|B'(0)|} \tag{10.78}$$

and so, since $|x| \leq 1$ is arbitrary, we obtain

$$\left|\frac{B(\zeta)}{\zeta}\right| \leq \frac{4|B'(0)|}{(1 + |\zeta|)^2}. \tag{10.79}$$

10.4.9

These geometrical methods all revolve around the hope that the level curves of a polynomial reflect in their shape the number of critical points inside the curves. For example, the Cordova–Ruscheweyh argument is an attempt to show that the critical point ζ on the curve $|P(z)| = |P(\zeta)|$ bounding a region of univalence containing the origin is a point of maximum modulus on the curve; their argument proves this in the case that all the critical points of P lie on a circle concentric with the origin. The details of their method suggest that we look at the curves where $zP'(z)/P(z)$ is real. If P has degree n, there are at most n such curves, which will be simple and pairwise disjoint provided that the zeros of $(zP'/P)'$ do not lie on these curves. There will be one such curve passing through the origin and also one such curve passing through ∞, except if the mean of the zeros of P is zero. For $zP'/P = 1$ at $z = 0$ and $= n$ at $z = \infty$. These curves contain the zeros and the critical points of P. There is an arc of one such curve joining 0 to a critical point ζ on which zP'/P decreases from 1 to 0. If $|z|$ increases on this curve, then $|P(z)/z|$ decreases, and so $|P(\zeta)/\zeta| \leq |P'(0)| = 1$, and therefore Smale's conjecture holds. To prove this, let $z = z(t)$ denote a parametrisation of the curve and observe that

$$\frac{d}{dt}\log\left|\frac{P(z(t))}{z(t)}\right| = \Re\frac{z'(t)}{z(t)}\left(\frac{z(t)P'(z(t))}{P(z(t))} - 1\right), \tag{10.80}$$

and since $zP'(z)/P(z)$ is real and < 1 on the arc, we see that this expression is negative if, and only if,

$$\Re\frac{z'(t)}{z(t)} > 0, \tag{10.81}$$

i.e. $|z|$ is increasing on the arc. This can be related to the behaviour of P as follows. On the arc we may write

$$\frac{z(t)P'(z(t))}{P(z(t))} = 1 - t, \tag{10.82}$$

from which we obtain

$$\frac{z'(t)}{z(t)}\left(t + \frac{z(t)P''(z(t))}{P'(z(t))}\right) = -\frac{1}{1-t}. \tag{10.83}$$

Thus $|z|$ is increasing on the arc if, and only if,

$$\Re\left(t + \frac{z(t)P''(z(t))}{P'(z(t))}\right) \leq 0. \tag{10.84}$$

10.4.10

Smale's problem is a special case of an interesting class of problems, which we now discuss. Let $Q(z)$ be a polynomial of degree $n \geq 1$ satisfying $Q(0) = 1$ and let $R(z) = \sum_{k=0}^{n} r_k z^k$ be a fixed polynomial of degree $\leq n$. Writing

$$Q(z) = 1 + \sum_{k=1}^{n} q_k z^k = \prod_{k=1}^{n}\left(1 - \frac{z}{z_k}\right), \tag{10.85}$$

we consider the polynomial

$$(R * Q)(z) = r_0 + \sum_{k=1}^{n} r_k q_k z^k \tag{10.86}$$

evaluated at the zeros z_j of Q. Since $q_k = (-1)^k L_k^{(n)}(\frac{1}{z_1}, \ldots, \frac{1}{z_n})$, where $L_k^{(n)}$ is the kth symmetric form in n variables, we obtain

$$\begin{aligned}(R * Q)(z_1) &= r_0 + \sum_{k=1}^{n} r_k(-1)^k L_k^{(n)}\left(\frac{1}{z_1}, \ldots, \frac{1}{z_n}\right) z_1^k \\ &= r_0 - r_1 + \sum_{k=1}^{n-1}(r_k - r_{k+1})(-1)^k L_k^{(n-1)}\left(\frac{1}{z_2}, \ldots, \frac{1}{z_n}\right) z_1^k,\end{aligned} \tag{10.87}$$

and analogous expressions for the other terms $(R * Q)(z_j)$. The generalised Smale problem is to estimate the maximum of the minimum values of the numbers $|(R * Q)(z_j)|$ $(1 \le j \le n)$ over all polynomials Q of degree n, R being kept fixed. Smale's conjecture is concerned with the case when $Q = P'$ and $r_k = 1/(k+1)$. Perhaps the simplest and most natural case of this problem occurs by taking $r_k - r_{k+1} = 1$ $(0 \le k \le n-1)$. Taking $r_0 = 0$, this gives $r_k = -k$, and therefore the problem requires the best estimate for $\min_j |z_j Q'(z_j)|$. Algebraically this amounts to estimating

$$\left| \sum_{k=0}^{n-1} (-1)^k L_k^{(n-1)} \left(\frac{1}{z_2}, \dots, \frac{1}{z_n} \right) z_1^k \right| = \prod_{k=1}^{n-1} \left| 1 - \frac{z_1}{z_{k+1}} \right|, \tag{10.88}$$

etc. In other words we require an estimate for

$$\min_j \prod_{k \ne j} \left| 1 - \frac{z_j}{z_k} \right|. \tag{10.89}$$

The following simple argument solves this problem. Since $Q' = 0$ at a multiple zero of Q, we may assume that the z_j are distinct. The Lagrange interpolation formula then gives

$$\frac{1}{Q(z)} = \sum_{k=1}^{n} \frac{1}{Q'(z_k)} \frac{1}{z - z_k}, \tag{10.90}$$

and therefore, as $Q(0) = 1$,

$$1 = -\sum_{k=1}^{n} \frac{1}{z_k Q'(z_k)}. \tag{10.91}$$

It follows immediately that, for at least one k,

$$|z_k Q'(z_k)| \le n. \tag{10.92}$$

The polynomial $Q(z) = 1 + z^n$ shows that equality can occur.

10.4.11

We conclude the chapter with the following result.

Theorem *Let $P(x) = x + a_2 x^2 + \cdots$ be a polynomial of degree ≥ 2 all of whose critical points are real. Then there is a critical point a of P for which*

$$\left| \frac{P(a)}{a} \right| \le e - 2. \tag{10.93}$$

Proof We may assume that $a_2 \leq 0$ (otherwise consider $-P(-x)$). P has at least one positive critical point: for if all critical points are negative, then P' has positive coefficients, contradicting $a_2 \leq 0$. Let a denote the smallest positive critical point; so $P'(0) = 1$, $P'(a) = 0$ and $P'(x) > 0$ for $0 < x < a$. In particular, $P(x)$ is increasing in $[0, a]$ and so $P(a) > 0$.

Let c_j denote the remaining critical points of P. For each j, either $c_j > a$ or $c_j < 0$. In either case we have

$$\frac{1}{x - c_j} \leq -\frac{1}{c_j} \quad (0 \leq x \leq a). \tag{10.94}$$

From this we obtain

$$\begin{aligned} \frac{P''(x)}{P'(x)} &= \frac{1}{x-a} + \sum_j \frac{1}{x - c_j} \leq \frac{1}{x-a} - \sum_j \frac{1}{c_j} \\ &= \frac{1}{x-a} + \frac{1}{a} + P''(0) \leq \frac{1}{x-a} + \frac{1}{a} \end{aligned} \tag{10.95}$$

for $0 \leq x \leq a$. Integrating this gives

$$P'(x) \leq (1 - x/a)e^{x/a} \quad (0 \leq x \leq a). \tag{10.96}$$

A further integration gives

$$P(a) \leq (e - 2)a, \tag{10.97}$$

as required.

11
Miscellaneous topics

11.1 The *abc* theorem

11.1.1

Mason's theorem *Let a, b and c be polynomials (not all constant) with no common factors and satisfying*

$$a + b = c. \tag{11.1}$$

Then the number of distinct zeros of a, b and c is at least 1 greater than the largest of the degrees of a, b and c.

11.1.2

Mason's theorem is essentially the case $r = 3$ of the following result for rational functions.

Theorem *Let $R(z)$ be a rational function of degree $n \geq 1$ and let w_i $(1 \leq i \leq r)$ be $r \geq 2$ distinct values in the extended plane (i.e. ∞ is a possible value). Then the total number of distinct solutions in the extended plane of the equations*

$$R(z) = w_i \quad (1 \leq i \leq r) \tag{11.2}$$

is at least $(r-2)n+2$.

Proof Assume first that the w_i are finite values $\neq R(\infty)$. Let $Q(w) = \prod_{i=1}^{r}(w - w_i)$; we wish to estimate the number of zeros of the rational function $(Q \circ R)(z)$. Since $Q \circ R$ has degree rn, the total number of zeros counted according to multiplicity is rn. Let z_k $(1 \leq k \leq d)$ be the distinct zeros of $Q \circ R$ and let $\mu(z_k)$ denote the multiplicity of the zero z_k. Then

$$\sum_{k=1}^{d} \mu(z_k) = rn. \tag{11.3}$$

On the other hand, when $\mu(z_k) > 1$, $(Q \circ R)'$ has a zero of multiplicity $\mu(z_k) - 1$ at z_k. Since $Q'(w_i) \neq 0$ for all i, we see that R' has a zero of multiplicity $\mu(z_k) - 1$ at z_k. But R' has at most $2n - 2$ zeros, and therefore

$$\sum_{k=1}^{d} (\mu(z_k) - 1) \leq 2n - 2. \tag{11.4}$$

From these two relations we deduce that $rn - d \leq 2n - 2$, giving $d \geq (r-2)n + 2$, as required.

Next, we consider the case that $w_r = \infty$ and $R(\infty) \neq \infty$. Choose a such that $w_i \neq a$ for $1 \leq i \leq r$ and consider the rational function $S(z) = 1/(R(z) - a)$. Then S has degree n and we require the number of distinct solutions of the equations $S(z) = 1/(w_i - a)\,(1 \leq i \leq r - 1)$ and $S(z) = 0$. By the first part the number is at least $(r - 2)n + 2$, as required.

Finally, we consider the case that $w_r = R(\infty)$. Choose b such that $R(b) \neq w_i$ for $1 \leq i \leq r$ and consider the rational function $T(z) = R(b + \frac{1}{z})$. Then T has degree n and we require the number of distinct solutions of the equations $T(z) = w_i\ (1 \leq i \leq r)$, noting that $T(\infty) \neq w_i\ (1 \leq i \leq r)$ and $w_r = T(0)$. By the previous results we again obtain the lower bound of $(r - 2)n + 2$. This completes the proof.

11.1.3

To prove Mason's theorem we note that there is no loss of generality in assuming that both a and b have degree n, so that c is not identically zero and has degree $\leq n$. Let $R = a/c$. Then writing $w_1 = 0$, $w_2 = \infty$ and $w_3 = 1$, we require the number of distinct solutions of the equations $R(z) = w_i\ (1 \leq i \leq 3)$. Since R has degree n, the above result gives at least $n + 2$ distinct solutions. As ∞ is a solution when c has degree $< n$, we obtain in general at least $n + 1$ distinct finite solutions, which is the required result.

The example $a(z) = (1 + z)^n$, $b(z) = -z^n$ shows that the result is sharp.

11.1.4

As an application one can prove **Fermat's last theorem** for polynomials.

Theorem *There are no polynomials A, B and C with no common factors and not all constant which satisfy identically the equation*

$$A^N + B^N = C^N, \tag{11.5}$$

if $N \geq 3$.

For, if this equation is satisfied, we may assume that A and B have the same degree $n \geq 1$. Then by Mason's theorem the polynomial $(ABC)^N$ has at least $nN + 1$ distinct zeros. On the other hand, it clearly has at most $3n$ distinct zeros. We immediately deduce that $N < 3$. Note that the theorem is true, even if we allow a common factor of A, B and C, provided that when we divide out the highest common factor we are not left with three constants. Obviously $z^N + z^N = 2z^N$. The example $A = z^2 + z$, $B = z + \frac{1}{2}$ and $C = z^2 + z + \frac{1}{2}$ shows that the equation can be satisfied when $N = 2$. Note that rational values for z in this example lead to pythagorean triples; e.g. $z = 1$ gives $3^2 + 4^2 = 5^2$.

11.2 Cohn's reduction method

11.2.1

A. Cohn [13] has given a method for finding the number of zeros of a polynomial P in $\mathbb{U}$. This is based on a step-by-step reduction of the degree making use of the following lemma.

Cohn's lemma *Let $P(z) = \sum_{k=0}^{n} a_k z^k$ be a polynomial of degree n satisfying $|a_0| > |a_n| > 0$ and set*

$$Q(z) = a_n P^*(z) - \overline{a_0} P(z), \tag{11.6}$$

where $P^*(z) = z^n \overline{P(\frac{1}{\bar{z}})}$. Then Q is a polynomial of degree at most $n - 1$, which on the unit circle $\mathbb{T}$ has the same zeros with the same multiplicities as P. Furthermore P and Q have equally many zeros in $\mathbb{U}$.

Proof On $\mathbb{T}$ the rational function $B(z) = P^*(z)/P(z)$ satisfies $|B(z)| = 1$. Also

$$\frac{Q(z)}{P(z)} = a_n(B(z) - \zeta), \tag{11.7}$$

where $\zeta = \overline{a_0}/a_n$ satisfies $|\zeta| > 1$. It follows that $d(B - \zeta, \mathbb{T}) = 0$, and therefore by the argument principle $B - \zeta$ has equally many zeros and poles in $\mathbb{U}$; i.e. P and Q have equally many zeros in $\mathbb{U}$. As $B - \zeta \neq 0$ on $\mathbb{T}$, P and Q have the same zeros with the same multiplicities on $\mathbb{T}$. Finally the coefficient of z^n in Q is $a_n \overline{a_0} - \overline{a_0} a_n = 0$, so Q has degree $< n$. The constant term in Q is $Q(0) = |a_n|^2 - |a_0|^2 < 0$.

11.2.2

Corollary *Let $P(z) = \sum_{k=0}^{n} a_k z^k$ be a polynomial of degree n satisfying $|a_0| < |a_n|$ and define $Q(z)$ by (11.6). Then Q is a polynomial of degree at most $n - 1$,* *which on the unit circle $\mathbb{T}$ has the same zeros with the same multiplicities as P. Furthermore P^* and Q have equally many zeros in $\mathbb{U}$.*

If $a_0 = 0$, the result is trivial. Otherwise apply Cohn's lemma to P^*.

11.2.3

By repeated application of the lemma we have in principle a method for determining the number of zeros of a polynomial in $\mathbb{U}$ by purely algebraic means. We set $P_1 = Q$ and then repeat the method on P_1 obtaining P_2 etc. Of course the method applies provided that at no stage is the absolute value of the product of the zeros $= 1$. If one of the P_j is self-inversive we get $P_{j+1} = 0$. However, by using theorem 7.1.3 of chapter 7 we may then continue the process on P_j'.

It may also be useful to eliminate the zeros on $\mathbb{T}$ by using the division algorithm to find the resultant of P and P^*, and then dividing out this factor.

One may also apply the method to $P(rz)$ with $r > 0$ close to 1, when P has |product of zeros| $= 1$.

11.2.4

Example Find the necessary and sufficient condition on a, b and c so that $az^2 + bz + c \neq 0$ for $z \in \overline{\mathbb{U}}$.

In this case we must have $|c| > |a|$. We have $Q(z) = (a\overline{b} - b\overline{c})z + |a|^2 - |c|^2$, which is $\neq 0$ in $\overline{\mathbb{U}}$ iff

$$|a\overline{b} - b\overline{c}| < |c|^2 - |a|^2. \tag{11.8}$$

11.3 Blaschke products

11.3.1

A **Blaschke product of degree** n is a rational function $B(z)$ of the form

$$B(z) = e^{i\alpha} \prod_{k=1}^{n} \frac{z - z_k}{1 - \overline{z_k} z} \tag{11.9}$$

where α is a real constant and where z_k are complex numbers satisfying $|z_k| < 1$ $(1 \leq k \leq n)$. Note that the zeros and poles of B are conjugate points relative to the unit circle. We have $|B(z)| < 1$ for $z \in \mathbb{U}$, $|B(z)| = 1$ for $z \in \mathbb{T}$ and $|B(z)| > 1$ for $|z| > 1$. On $\mathbb{T}$ we can write

$$B(e^{it}) = e^{i\phi(t)}, \tag{11.10}$$

where $\phi(t)$ is real and $\phi(2\pi) - \phi(0) = 2\pi n$. Furthermore, a differentiation gives

$$\phi'(t) = \frac{e^{it}B'(e^{it})}{B(e^{it})} = \sum_{k=1}^{n} \frac{1 - |z_k|^2}{|1 - \overline{z_k}e^{it}|^2} \geq \sum_{k=1}^{n} \frac{1 - |z_k|}{1 + |z_k|} > 0. \quad (11.11)$$

Thus $\phi(t)$ is a strictly increasing function and

$$\phi'(t) = |B'(e^{it})|. \quad (11.12)$$

In addition we have

$$\phi'(t) \leq \sum_{k=1}^{n} \frac{1 + |z_k|}{1 - |z_k|} \quad (11.13)$$

and therefore by the maximum principle

$$|B'(z)| \leq \sum_{k=1}^{n} \frac{1 + |z_k|}{1 - |z_k|} \quad (z \in \mathbb{U}). \quad (11.14)$$

Note also that, by the argument principle, $B(z)$ and $zB'(z)$ have equally many zeros in $\mathbb{U}$, counted according to multiplicity; therefore *a Blaschke product of degree n has exactly $n - 1$ critical points in $\mathbb{U}$*. We will show that *each critical point of B in $\mathbb{U}$ lies in the convex hull of the set of points* $\{0, z_1, z_2, \ldots, z_n\}$.

Suppose that $z \in \mathbb{U}$, $B'(z) = 0$ and $B(z) \neq 0$. Then

$$0 = \frac{B'(z)}{B(z)} = \sum_{k=1}^{n} \left(\frac{1}{z - z_k} + \frac{\overline{z_k}}{1 - \overline{z_k}z} \right) \quad (11.15)$$

and therefore

$$\sum_{k=1}^{n} \left(\frac{z - z_k}{|z - z_k|^2} + \frac{z_k(1 - \overline{z_k}z)}{|1 - \overline{z_k}z|^2} \right) = 0. \quad (11.16)$$

This gives

$$z \left(\sum_{k=1}^{n} \left(\frac{1}{|z - z_k|^2} - \frac{|z_k|^2}{|1 - \overline{z_k}z|^2} \right) \right) = \sum_{k=1}^{n} \left(\frac{1}{|z - z_k|^2} - \frac{1}{|1 - \overline{z_k}z|^2} \right) z_k \quad (11.17)$$

and solving this we obtain

$$z = \sum_{k=1}^{n} \alpha_k z_k. \quad (11.18)$$

It is immediately verified that $z \in \mathbb{U}$ implies that $\alpha_k > 0\,(1 \leq k \leq n)$ and $\sum_1^n \alpha_k < 1$; thus z is a convex combination of 0 and the elements z_k, as required.

Concerning the critical points of B outside $\mathbb{U}$, we remark that these are precisely the conjugates relative to the unit circle of the critical points of B in $\mathbb{U}$. This follows upon differentiating the identity

$$B(z)\overline{B(1/\overline{z})} = 1. \tag{11.19}$$

Let us take $e^{i\alpha} = 1$, $P(z) = \prod_1^n (z - z_k)$, $P^*(z) = z^n \overline{P(1/\overline{z})} = \prod_1^n (1 - \overline{z_k} z)$. Then

$$B(z) = \frac{P(z)}{P^*(z)} \tag{11.20}$$

and

$$B'(z) = \frac{P'(z)P^*(z) - P(z)P^{*\prime}(z)}{P^{*2}(z)} = \frac{Q(z)Q^*(z)}{P^{*2}(z)} \tag{11.21}$$

where $Q(z) = \prod_1^{n-1} (z - \zeta_k)$, $Q^*(z) = z^{n-1}\overline{Q(1/\overline{z})}$ and ζ_k are the critical points of B in $\mathbb{U}$. If

$$C(z) = \frac{Q(z)}{Q^*(z)} \tag{11.22}$$

is the Blaschke product formed with the points ζ_k, then we have

$$B'(z) = C(z)\left(\frac{Q^*(z)}{P^*(z)}\right)^2 = C(z)R^2(z) \text{ say.} \tag{11.23}$$

On $\{|z| = 1\}$ we obtain

$$|B'(z)| = |R(z)|^2 \tag{11.24}$$

and so writing $R(z) = \sum_0^\infty \rho_k z^k$ we deduce that

$$\sum_{k=0}^{\infty} |\rho_k|^2 = n. \tag{11.25}$$

11.3.2 Non-euclidean lines

A closer analogue of the Gauss–Lucas theorem for Blaschke products has been given by J.L. Walsh [69] in terms of non-euclidean lines in $\mathbb{U}$, which are defined as follows. A **non-euclidean line (*NE* line)** is a circular arc γ in $\mathbb{U}$ with endpoints on $\mathbb{T}$, where the circle Γ, of which γ is an arc, is orthogonal to $\mathbb{T}$; as the radial lines from the origin are orthogonal to $\mathbb{T}$, this definition includes diameters of $\mathbb{U}$. The concept was introduced by Poincaré to model the hyperbolic non-euclidean geometry of Lobachevsky. In general a circle Γ is orthogonal to $\mathbb{T}$ iff there is a Moebius transformation which maps Γ to the real axis and $\mathbb{T}$ onto $\mathbb{T}$. Certainly, if there is such a transformation, then it is conformal and so preserves the orthogonality of the real axis with $\mathbb{T}$; conversely, if Γ is orthogonal to $\mathbb{T}$, then all Moebius transformations which preserve $\mathbb{T}$ map Γ onto orthogonal circles; we need to show that at least one such transformation maps a point of Γ onto ∞; then Γ is mapped onto a line orthogonal to $\mathbb{T}$, which is necessarily a radial line through 0; an additional rotation will then map to the axis. Assuming Γ is not already a line, we perform a rotation first so that the orthogonal circle meets $\mathbb{T}$ at points $e^{i\theta}$ and $e^{-i\theta}$, where $0 < \theta < \pi/2$. The radial lines from 0 through these points will be tangents to the circle, from which we see that the centre of the circle is $\sec\theta$, and the radius is $\tan\theta$. This means that the orthogonal circle is given by the parametric equation

$$z(t) = \frac{1 + e^{it}\sin\theta}{\cos\theta}; \tag{11.26}$$

it is easily verified that with the choice $a = (1 - \sin\theta)/\cos\theta$, the mapping

$$w = i\frac{z - a}{1 - az} \tag{11.27}$$

maps the orthogonal circle to the real axis; since $0 < a < 1$, $\mathbb{T}$ maps to $\mathbb{T}$ and this establishes our assertion.

It follows from the assertion that an *NE* line can be parametrised in the form

$$z(x) = e^{i\alpha}\frac{x + a}{1 + \bar{a}x} \quad (-1 < x < 1), \tag{11.28}$$

where α is real and $|a| < 1$. Writing $b = ae^{i\alpha}$, $w = xe^{i\alpha}$, we see that

$$z = \frac{w + b}{1 + \bar{b}w}, \tag{11.29}$$

where $|b| < 1$ and w traces out a diameter of the unit disc. Thus *NE* lines are the images of diameters of $\mathbb{U}$ under mappings of this form. For a fixed $b \in \mathbb{U}$ we obtain a family of *NE* lines, one corresponding to each diameter, all passing through b but otherwise disjoint. Every point of $\mathbb{U}$ other than b lies on exactly one *NE* line in this b-family. On the other hand every *NE* line passing through b is a member of this family, for it is the image under the above mapping of an *NE* line passing through 0, and it is easily seen that the only *NE* lines passing through 0 are diameters of $\mathbb{U}$ (in our earlier calculations we would obtain $\sec\theta = \tan\theta$ giving $\theta = \pi/2$, $\sec\theta = \infty$). We thus see that each pair of points in $\mathbb{U}$ lies on exactly one *NE* line. Two distinct *NE* lines meet in at most one point in $\mathbb{U}$; indeed, if two orthogonal circles with respect to $\mathbb{T}$ meet at $b \in \mathbb{U}$, then they also meet at $1/\bar{b}$. From this property it follows that an *NE* line l divides $\mathbb{U}$ into two regions each of which is ***NE* convex**: a set E is *NE* convex if each pair of points of E can be joined by an *NE* line lying entirely in E. A region of $\mathbb{U}$ on one side of l has this property, as otherwise there would be a pair of points in the region such that the *NE* segment joining them intersected l in two points, which is impossible. The two regions defined by l will be called ***NE* half-discs**. As an intersection of *NE* convex sets is clearly *NE* convex, it follows that any set $F \subset \mathbb{U}$ has an ***NE* convex hull**, the smallest *NE* convex set containing F. Note that the origin always lies exterior to any circle orthogonal to $\mathbb{T}$. The *NE* convex hull of n points is a 'polygon' whose sides consist of circular arcs. In general the *NE* convex hull K of a set E is the intersection I of the *NE* half-discs containing E, where we include $\mathbb{U}$ as such a half-disc. As I is *NE* convex, $K \subset I$; conversely, suppose that $b \in I - K$; if every *NE* line through b meets K, then after the transformation which maps the *NE* lines through b to diameters through 0, we obtain an *NE* convex set K' such that every diameter meets K'; but then at least one diameter is such that both radii meet K' and therefore $0 \in K'$, which implies that $b \in K$, a contradiction; hence $\exists$ an *NE* line l through b which does not meet K and so K lies on one side of l; this means that E lies on one side of l and so there is an *NE* half-disc containing E which does not contain b; hence $b \notin I$, a contradiction. Thus $I = K$, as asserted.

Consider now a Blaschke product B of degree $n \geq 2$ with zeros $z_1, \ldots, z_n$. Suppose that these zeros lie on one side of an *NE* line. There is an automorphic Moebius transformation $z = \phi(w)$ of $\mathbb{U}$ which maps a diameter of $\mathbb{U}$ onto this *NE* line; then $z_k = \phi(w_k)$, where the w_k lie on one side of this diameter, i.e. in a half disc of $\mathbb{U}$. We have $B(z) = B(\phi(w)) = C(w)$, where $C(w)$ is also a Blaschke product and has its zeros at the points w_k. By our earlier convexity result, $C(w)$ has its critical points in the same half-disc which contains the w_k. It follows that $B(z)$ has its critical points on the same side of the *NE* line as the zeros z_k. As this holds for every *NE* line containing the zeros, it follows that the critical points of B lie in the *NE* convex hull of the set of zeros of B.

11.3.3

Summarising, we have the following result.

Walsh's Blaschke product theorem *Let $B(z)$ be a Blaschke product of degree n with zeros $z_1, \ldots, z_n$ in $\mathbb{U}$. Then $B(z)$ has exactly $n - 1$ critical points in $\mathbb{U}$ and these all lie in the non-euclidean convex hull of the set $\{z_1, \ldots, z_n\}$. In particular the critical points in $\mathbb{U}$ lie in the (euclidean) convex hull of the set $\{0, z_1, \ldots, z_n\}$. The critical points of B outside $\mathbb{U}$ are the conjugates (relative to T) of those in $\mathbb{U}$.*

11.4 Blaschke products and harmonic mappings

In this section we include some results of the author (Sheil-Small [56]) which establish a connection between step functions on $\mathbb{T}$ together with their harmonic extensions into $\mathbb{U}$ and ratios of finite Blaschke products. The general connection is purely algebraic, but it turns out that there are also deeper geometric relations.

11.4.1 The harmonic extension of a step function on $\mathbb{T}$

Let f be a step function on $\mathbb{T}$ defined by

$$f(e^{it}) = c_k \quad (t_{k-1} < t < t_k, 1 \le k \le n), \tag{11.30}$$

where the c_k are complex numbers and where $t_0 < t_1 < \cdots < t_n = t_0 + 2\pi$. The harmonic extension of f to $\mathbb{U}$ is given by

$$f(z) = \frac{1}{2\pi}\int_0^{2\pi}\left(\Re\frac{1+ze^{-it}}{1-ze^{-it}}\right)f(e^{it})dt, \tag{11.31}$$

which can be written in the form $f(z) = \overline{g(z)} + h(z)$, where g and h are analytic in $\mathbb{U}$ and $g(0) = 0$. We then obtain

$$h(z) = \frac{1}{2\pi}\int_0^{2\pi}\frac{f(e^{it})}{1-ze^{-it}}dt = \frac{1}{2\pi}\sum_{k=1}^{n} c_k \int_{t_{k-1}}^{t_k}\frac{dt}{1-ze^{-it}}. \tag{11.32}$$

Differentiating this gives

$$h'(z) = \frac{1}{2\pi}\sum_{k=1}^{n} c_k \int_{t_{k-1}}^{t_k}\frac{e^{-it}}{(1-ze^{-it})^2}dt = \frac{1}{2\pi i}\sum_{k=1}^{n} c_k\left(\frac{1}{z-\zeta_k} - \frac{1}{z-\zeta_{k-1}}\right), \tag{11.33}$$

where $\zeta_k = e^{it_k}$ for $0 \le k \le n$, so that $\zeta_0 = \zeta_n$. Writing $c_{n+1} = c_1$, we obtain

$$h'(z) = \sum_{k=1}^{n}\frac{\alpha_k}{z-\zeta_k} \tag{11.34}$$

where

$$\alpha_k = \frac{1}{2\pi i}(c_k - c_{k+1}) \quad (1 \le k \le n). \tag{11.35}$$

This gives

$$\sum_{k=1}^{n}\alpha_k = 0. \tag{11.36}$$

Similarly, we have

$$g(z) = \frac{1}{2\pi}\int_0^{2\pi}\frac{ze^{-it}}{1-ze^{-it}}\overline{f(e^{it})}dt, \tag{11.37}$$

from which we obtain

$$g'(z) = -\sum_{k=1}^{n} \frac{\overline{\alpha_k}}{z - \zeta_k}. \tag{11.38}$$

Now the relations (11.34/35) imply that

$$h'(z) = \sum_{k=1}^{n-1} \frac{\alpha_k}{z - \zeta_k} - \sum_{k=1}^{n-1} \frac{\alpha_k}{z - \zeta_n} = \sum_{k=1}^{n-1} \alpha_k \frac{\zeta_k - \zeta_n}{(z - \zeta_k)(z - \zeta_n)} \tag{11.39}$$

from which we obtain (using $|\zeta_k| = 1$ for $1 \le k \le n$)

$$\overline{h'\left(\frac{1}{\overline{z}}\right)} = z^2 g'(z). \tag{11.40}$$

Writing $S(z) = \prod_1^n (z - \zeta_k)$, which implies $z^n \overline{S(1/\overline{z})} = S(z) \prod_1^n (-\overline{\zeta_k})$, we have

$$h'(z) = \frac{P(z)}{S(z)}, \qquad g'(z) = \frac{Q(z)}{S(z)}, \tag{11.41}$$

where P and Q are polynomials of degree at most $n - 1$; furthermore

$$\frac{\overline{Q(1/\overline{z})}}{\overline{S(1/\overline{z})}} = \overline{g'(1/\overline{z})} = z^2 h'(z) = z^2 \frac{P(z)}{S(z)} \tag{11.42}$$

from which we obtain

$$z^{n-2} \overline{Q(1/\overline{z})} = P(z) \prod_1^n (-\overline{\zeta_k}). \tag{11.43}$$

It follows that P and Q are polynomials of degree at most $n - 2$; if Q has degree q with q_1 zeros at the origin and q_2 zeros away from the origin, then P has degree p, where $p = n - 2 - q_1$ and P has $n - 2 - q$ zeros at the origin and q_2 zeros away from the origin. The q_2 zeros of P are the conjugates relative to the unit circle of the q_2 zeros of Q.

Thus

$$R(z) = \frac{g'(z)}{h'(z)} = \frac{Q(z)}{P(z)} \tag{11.44}$$

is a rational function of degree at most $n - 2$ satisfying $|R(z)| = 1$ on $\mathbb{T}$; R is therefore a ratio of Blaschke products formed from the q_2 zeros of Q away from the origin together with a zero/pole term z^{q+q_1-n+2}: zero or pole depending on the sign of $q + q_1 - n + 2$.

11.4.2

We now show that the step function $f(e^{it})$ can be reconstructed, up to an arbitrary additive constant, from the following two pieces of information:

(1) an arbitrary polynomial Q of degree at most $n-2$;
(2) a polynomial $S(z)=\prod_1^n(z-\zeta_k)$ with n distinct zeros ζ_k on $\mathbb{T}$.

We define

$$P(z)=z^{n-2}\overline{Q(1/\overline{z})}\prod_1^n(-\zeta_k) \tag{11.45}$$

and set

$$G(z)=\frac{Q(z)}{S(z)},\qquad H(z)=\frac{P(z)}{S(z)}. \tag{11.46}$$

Then we can write

$$G(z)=-\sum_{k=1}^n\frac{\overline{\alpha_k}}{z-\zeta_k}, \tag{11.47}$$

where $-\overline{\alpha_k}=Q(\zeta_k)/S'(\zeta_k)$ for $1\le k\le n$. Since $SG=Q$ has degree at most $n-2$, it follows that

$$\sum_{k=1}^n\alpha_k=0. \tag{11.48}$$

Furthermore, we easily obtain

$$H(z)=\sum_{k=1}^n\frac{\alpha_k}{z-\zeta_k}. \tag{11.49}$$

The step values c_k are defined by $c_1=c_{n+1}=\mu$ and

$$c_k=\mu-2\pi i\sum_{j=1}^{k-1}\alpha_j\quad(2\le k\le n), \tag{11.50}$$

where μ is an arbitrary constant. We immediately deduce the relation (11.35). Assuming that the $\zeta_k=e^{it_k}$ are ordered on $\mathbb{T}$ with $t_0<t_1<\cdots<t_n=t_0+2\pi$, we find by the earlier algebra that the step function

$$f(e^{it})=c_k\quad(t_{k-1}<t<t_k,\ 1\le k\le n) \tag{11.51}$$

extends harmonically into $\mathbb{U}$ as $f=\overline{g}+h$, where $g'=G$ and $h'=H$. Thus up to the arbitrary additive constant μ, f is uniquely determined in $\mathbb{U}$ from (1) and (2).

Observe that the relations $g' = G$ and $h' = H$ imply that the functions G and H defined by (11.46) are primitives in $\mathbb{U}$. This is true in a larger domain than $\mathbb{U}$. Let Γ_k denote the arc (ζ_{k-1}, ζ_k) of the unit circle and let

$$D_k = \mathbb{U} \cup \Gamma_k \cup \{|z| > 1\}, \tag{11.52}$$

so that D_k is the plane cut along the closed complementary circular arc to Γ_k. Let γ be a rectifiable loop in D_k. Then

$$\frac{1}{2\pi i}\int_\gamma G(z)dz = -\sum_{j=1}^{n} \overline{\alpha_j} n(\gamma, \zeta_j) = 0,$$

since $n(\gamma, \zeta_j)$ has the same value for each j (as the ζ_j lie on a connected set in the complement of γ). It follows that G is a primitive in D_k and similarly for H. Thus f has a harmonic extension across Γ_k into the whole of D_k. Up to the additive constant this is unique in $\mathbb{U}$, but may vary in $\{|z| > 1\}$ depending upon across which arc Γ_k we continue f from $\mathbb{U}$. In fact, if we denote by f_k the continuation of f into $\{|z| > 1\}$ across Γ_k, it is easily shown that

$$f_k - f_j = 2(c_k - c_j). \tag{11.53}$$

11.4.3

Summarising, we have

Theorem *Let $f(z) = \overline{g(z)} + h(z)\,(z \in \mathbb{U})$ denote the harmonic extension into $\mathbb{U}$ of a step function on $\mathbb{T}$ given by*

$$f(e^{it}) = c_k \quad (t_{k-1} < t < t_k,\ 1 \le k \le n), \tag{11.54}$$

where $t_0 < t_1 < \cdots < t_n = t_0 + 2\pi$. Then writing

$$S(z) = \prod_{k=1}^{n}(z - e^{it_k}), \tag{11.55}$$

we have for $n \ge 2$ that

$$g'(z) = \frac{Q(z)}{S(z)}, \qquad h'(z) = \frac{P(z)}{S(z)}, \tag{11.56}$$

where P and Q are polynomials of degree at most $n - 2$. Furthermore

$$\frac{g'(z)}{h'(z)} = \frac{Q(z)}{P(z)} = R(z) \tag{11.57}$$

satisfies

$$|R(z)| = 1 \quad (z \in \mathbb{T}) \tag{11.58}$$

and so is a rational function of degree at most $n-2$ *which is the ratio of two Blaschke products. Conversely, if* $Q(z)$ *is a polynomial of degree at most* $n-2$ *and if* $S(z)$ *has the form (11.55), then there exist a polynomial* $P(z)$ *of degree at most* $n-2$ *and a step function of the form (11.54), unique up to an additive constant, such that (11.56/57/58) are satisfied. The harmonic function* f *in* $\mathbb{U}$ *can be continued across any one of the arcs* $(e^{it_{k-1}}, e^{it_k})$ *into* $\{|z|>1\}$ *to give a function harmonic in the domain* D_k *of (11.52).*

11.5 Blaschke products and convex curves

11.5.1 Mappings onto smooth convex curves

Let Γ denote a **smooth convex curve**; by this we mean that Γ can be parametrised in the form $F(e^{it})\,(0 \le t \le 2\pi)$, where $F(z)$ is a conformal mapping of $\mathbb{U}$ onto the convex domain D bounded by Γ, and where F' exists continuously on $\mathbb{T}$ and is $\neq 0$. We consider a mapping of $\mathbb{T}$ onto Γ of the form

$$f(e^{it}) = F\left(e^{i\phi(t)}\right) \quad (0 \le t \le 2\pi), \tag{11.59}$$

where $\phi(t)$ is a real-valued, differentiable function on $\mathbb{R}$ satisfying $\phi'(t) > 0$ and $\phi(t+2\pi) = \phi(t) + 2N\pi$ for every real t, where N is a natural number. Thus f maps $\mathbb{T}$ onto Γ described N times in the positive direction. We denote by $f(z) = \overline{g(z)} + h(z)$ the harmonic extension of f into $\mathbb{U}$.

11.5.2

Theorem *Let* f *as above denote a smooth* N*-fold mapping of* $\mathbb{T}$ *onto the convex curve* Γ*. Assume further that* g' *and* h' *extend continuously to* $\overline{\mathbb{U}}$*. Then* h' *has exactly* $N-1$ *zeros in* $\mathbb{U}$*, say* $z_k\,(1 \le k \le N-1)$*. If* $B(z)$ *is the Blaschke product formed with the* z_k*, then*

$$|g'(z)| < \left|\frac{h'(z)}{B(z)}\right| \quad (z \in \mathbb{U}). \tag{11.60}$$

Furthermore, we can write

$$h'(z) = H(z)Q(z) \quad (z \in \mathbb{U}), \tag{11.61}$$

where H *belongs to the Kaplan class* $K(1, 2N+1)$ *and* $Q(z) = \prod_1^{N-1}((z-z_k) \times (1-\overline{z_k}z))$*.*

Proof Differentiating the relation

$$f(e^{it}) = \overline{g(e^{it})} + h(e^{it}) = F\left(e^{i\phi(t)}\right), \tag{11.62}$$

we obtain

$$e^{it}h'(e^{it}) - \overline{e^{it}g'(e^{it})} = \phi'(t)e^{i\phi(t)}F'\left(e^{i\phi(t)}\right) \neq 0 \tag{11.63}$$

for $0 \leq t \leq 2\pi$. Since Γ is a convex curve bounding the domain D, it is clear from Poisson's formula that $f(\mathbb{U}) \subset D$. The quantity

$$\lambda(t) = \arg\left(e^{it}h'(e^{it}) - \overline{e^{it}g'(e^{it})}\right) = \phi(t) + \arg F'\left(e^{i\phi(t)}\right) \tag{11.64}$$

represents the direction of the outward normal to Γ at the point $f(e^{it})$, and therefore for $z \in \mathbb{U}$

$$\Re\left[(f(z) - f(e^{it}))e^{-i\lambda(t)}\right] < 0. \tag{11.65}$$

Choosing $z = re^{it}$ $(0 < r < 1)$ we can rewrite this inequality as

$$\Re\left\{e^{it}\frac{h(z) - h(e^{it})}{z - e^{it}} \Big/ \left(e^{it}h'(e^{it}) - \overline{e^{it}g'(e^{it})}\right)\right.$$
$$\left. + e^{it}\frac{g(z) - g(e^{it})}{z - e^{it}} \Big/ \left(\overline{e^{it}h'(e^{it})} - e^{it}g'(e^{it})\right)\right\} > 0. \tag{11.66}$$

Letting $r \to 1$ and simplifying we obtain

$$\frac{|h'(e^{it})|^2 - |g'(e^{it})|^2}{\left|e^{it}h'(e^{it}) - \overline{e^{it}g'(e^{it})}\right|^2} \geq 0, \tag{11.67}$$

from which we deduce that

$$|g'(e^{it})| \leq |h'(e^{it})| \quad (0 \leq t \leq 2\pi). \tag{11.68}$$

This inequality together with (11.63) implies that $h'(e^{it}) \neq 0$ for $0 \leq t \leq 2\pi$. Furthermore, (11.63) gives

$$e^{it}h'(e^{it}) = \frac{\phi'(t)e^{i\phi(t)}F'\left(e^{i\phi(t)}\right)}{1 - \overline{e^{it}g'(e^{it})}/e^{it}h'(e^{it})}, \tag{11.69}$$

and, since the denominator of the right-hand side has positive real part and $\phi'(t) > 0$, it follows from the argument principle that the number of zeros of $zh'(z)$ in $\mathbb{U}$ is the degree on $\mathbb{T}$ of the mapping $e^{i\phi(t)}F'(e^{i\phi(t)})$, which is exactly N, as $F'(z) \neq 0$ in $\overline{\mathbb{U}}$. Thus $h'(z)$ has exactly $N - 1$ zeros in $\mathbb{U}$. Forming the Blaschke product B with these zeros, we can prove (11.60) as follows. Let $G(z) = g'(z)B(z)/h'(z)$; then G is analytic in $\mathbb{U}$, extends continuously to $\overline{\mathbb{U}}$ and by (11.68) satisfies $|G(z)| \leq 1$ on $\mathbb{T}$; if (11.60) does not hold, then by the maximum principle $G(z) = c$, where c is constant and $|c| = 1$; this gives $h' = \overline{c}g'B$ and so $\phi'(t)e^{i\phi(t)}F'(e^{i\phi(t)}) = 2ie^{i\psi(t)/2}\Im(e^{it}g'(e^{it})e^{i\psi(t)/2})$, where $\overline{c}B(e^{it}) = e^{i\psi(t)}$; calculating the degree on $\mathbb{T}$ for each side of this equation, we obtain $N = (N-1)/2$, giving the contradictory $N = -1$; thus (11.60) is established.

To complete the proof of the theorem we need to show that $H = h'/Q$ satisfies the definition given in chapter 7, (7.98) of a function in $K(1, 2N+1)$. Clearly H is continuous and $\neq 0$ in $\overline{\mathbb{U}}$ and analytic in $\mathbb{U}$, and therefore it will be sufficient to prove the condition on $\mathbb{T}$. The condition to be proved is

$$d(H, \gamma) > -Nd(z, \gamma) - \frac{1}{2} \tag{11.70}$$

over any positively described arc γ of $\mathbb{T}$. Now from the relation (11.69) we have $d(zh', \gamma) > -\frac{1}{2}$, since the change in the arg of the numerator is positive over γ and the denominator has positive real part. On the other hand, on $\mathbb{T}$ we have

$$e^{it}Q(e^{it}) = e^{iNt} \prod_{k=1}^{N-1} |1 - \overline{z_k}e^{it}|^2, \tag{11.71}$$

and therefore $d(zQ, \gamma) = Nd(z, \gamma)$. The result follows.

11.5.3 Mappings onto general convex curves

In this subsection Γ will denote a general closed convex curve bounding a convex domain D; $F(z)$ is a conformal mapping of $\mathbb{U}$ onto D, and necessarily this extends continuously to $\overline{\mathbb{U}}$, so that $F(e^{it})$ gives a parametric representation of Γ $(0 \leq t \leq 2\pi)$. For a natural number N we write $f \in \Phi\,(\Gamma, N)$ if

$$f(e^{it}) = F\left(e^{i\phi(t)}\right) \quad (0 \leq t \leq 2\pi), \tag{11.72}$$

where $\phi(t)$ is a non-decreasing function on $[0, 2\pi]$ satisfying $\phi(2\pi) - \phi(0) \leq 2\pi N$. The class $\Phi(\Gamma, N)$ is independent of the particular choice of bijective conformal mapping $F\colon \mathbb{U} \to D$; furthermore, by the Helly selection theorem, the class is compact relative to p.p. convergence on the circle. Note that the functions f need not be continuous on $\mathbb{T}$.

As before we continue to denote by $f = \overline{g} + h$ the harmonic extension into $\mathbb{U}$ of f on $\mathbb{T}$. The resulting class of harmonic mappings is compact relative to local uniform convergence in $\mathbb{U}$, there being by the Lebesgue bounded convergence theorem a direct correspondence between local uniform convergence in $\mathbb{U}$ and p.p. convergence on $\mathbb{T}$. Note also that

$$\lim_{r \to 1} f(re^{it}) = f(e^{it}) \text{ p.p.} \quad (0 \leq t \leq 2\pi). \tag{11.73}$$

11.5.4

Making use of a method of smooth approximation we prove the following result.

Theorem *Let $f = \overline{g} + h \in \Phi(\Gamma, N)$. Either f is constant or $\exists N - 1$ points $z_k \in \overline{\mathbb{U}}$ such that*

$$h'(z_k) = 0 \text{ if } z_k \in \mathbb{U} \text{ and } h'(z) \neq 0 \text{ in } \mathbb{U} \text{ if } z \neq z_k \quad (1 \leq k \leq N-1); \tag{11.74}$$

$$|g'(z)| \leq \left|\frac{h'(z)}{B(z)}\right| \quad (z \in \mathbb{U}), \tag{11.75}$$

where $B(z)$ is the Blaschke product formed with the z_k;

$$h'(z) = H(z)Q(z) \quad (z \in \mathbb{U}), \tag{11.76}$$

where H belongs to the Kaplan class $K(1, 2N+1)$ and $Q(z) = \prod_1^{N-1}(z - z_k) \times (1 - \overline{z_k}z)$. If f is non-constant and extends continuously to $\overline{\mathbb{U}}$, then equality cannot occur in (11.75).

Proof For $0 < \rho < 1$ we define

$$f_\rho(e^{it}) = F\left(\rho e^{i\phi_\rho(t)}\right), \tag{11.77}$$

where

$$\phi_\rho(t) = Nt + \frac{1}{2\pi}\int_0^{2\pi}\left(\Re\frac{1+\rho e^{i(t-\theta)}}{1-\rho e^{i(t-\theta)}}\right)(\phi(\theta) - N\theta)d\theta. \tag{11.78}$$

$\phi_\rho(t) - Nt$ is periodic with period 2π and $\phi_\rho(t) \to \phi(t)$ p.p. as $\rho \to 1$. Furthermore, $\phi_\rho(t)$ is infinitely differentiable and

$$\begin{aligned}\phi_\rho'(t) &= \frac{1}{2\pi}\int_0^{2\pi}\Re\frac{1+\rho e^{i(t-\theta)}}{1-\rho e^{i(t-\theta)}}d\phi(\theta) + \frac{2N\pi - \phi(2\pi) + \phi(0)}{2\pi}\Re\frac{1+\rho e^{it}}{1-\rho e^{it}} \\ &\geq N\frac{1-\rho}{1+\rho} > 0.\end{aligned} \tag{11.79}$$

It follows that $f_\rho(e^{it})$ is a smooth N-fold mapping of $\mathbb{T}$ onto the smooth convex curve $\Gamma_\rho = \{F(\rho e^{it})\colon 0 \leq t \leq 2\pi\}$. Since F is continuous in $\overline{\mathbb{U}}$, $f_\rho \to f$ p.p. on $\mathbb{T}$ and locally uniformly on $\mathbb{U}$, where $f_\rho = \overline{g_\rho} + h_\rho$ is the usual harmonic extension in $\mathbb{U}$; thus $g_\rho \to g$ and $h_\rho \to h$ locally uniformly in $\mathbb{U}$. To apply theorem 11.5.2 it only remains to show that the derivatives g_ρ' and h_ρ' extend continuously to $\overline{\mathbb{U}}$. Consider the kth Fourier coefficient of f_ρ given by

$$c_k^\rho = \frac{1}{2\pi}\int_0^{2\pi} e^{-ikt}\sum_{m=0}^{\infty} A_m\rho^m e^{im\phi_\rho(t)}dt, \tag{11.80}$$

where A_m are the coefficients of $F(z)$. Then for $k \neq 0$

$$
\begin{aligned}
c_k^\rho &= \frac{1}{k}\sum_{m=1}^{\infty} m A_m \rho^m \left(\frac{1}{2\pi}\int_0^{2\pi} e^{-ikt} e^{im\phi_\rho(t)}\phi'_\rho(t)dt\right) \\
&= \frac{1}{k^2}\sum_{m=1}^{\infty} m A_m \rho^m \left(\frac{1}{2\pi i}\int_0^{2\pi} e^{-ikt} e^{im\phi_\rho(t)}\big(im\phi'^2_\rho(t) + \phi''_\rho(t)\big)dt\right) \qquad (11.81)
\end{aligned}
$$

etc. Thus we see by integrating by parts q times and making use of the periodicity properties that

$$|k^q c_k^\rho| \le M(\rho, q), \qquad (11.82)$$

where $M(\rho, q)$ does not depend on k. It follows that

$$\sum_{k=1}^{\infty} |k^q||c_k^\rho| \le M(\rho, q+2)\sum_{k=1}^{\infty}\frac{1}{k^2} < \infty, \qquad (11.83)$$

and so all the derivatives of g_ρ and h_ρ have absolutely convergent power series expansions in $\overline{\mathbb{U}}$, which more than proves our assertion.

To complete the proof of the theorem, let $z_k(\rho)$ denote the $N-1$ zeros of h'_ρ in $\mathbb{U}$; for a suitable sequence of $\rho \to 1$, $z_k(\rho) \to z_k$ where $z_k \in \overline{\mathbb{U}}$ for $1 \le k \le N-1$. The corresponding functions B_ρ and Q_ρ converge locally uniformly in $\mathbb{U}$ to B and Q. Clearly h' has zeros at those $z_k \in \mathbb{U}$; also the family $\{h'_\rho\}$ is locally bounded in $\mathbb{U}$, since the family $\{f_\rho\}$ is bounded (f_ρ takes only values in D). Since $K(1, 2N+1) \cap \{k \in \mathcal{A}\colon k(0) = 1\}$ is a compact class, either $H \in K(1, 2N+1)$ or $H \equiv 0$; in the latter case $h' \equiv 0$, which implies $g' \equiv 0$ and so f is constant. If f is non-constant, h' has zeros only at the z_k in $\mathbb{U}$ and the relations (11.75/76) follow.

It remains to consider the case of equality in (11.75) when f is non-constant and continuous in $\overline{\mathbb{U}}$. We then have

$$g'(z) = \frac{h'(z)}{B(z)} \quad (z \in \mathbb{U}), \qquad (11.84)$$

where $B(z)$ is a Blaschke product formed with those $z_k \in \mathbb{U}$. If no $z_k \in \mathbb{U}$, then

$$g(z) = c(h(z) - h(0)), \qquad (11.85)$$

where c is a constant such that $|c| = 1$. This implies that for $z \in \mathbb{U}$, $f(z)$ takes only values on a line segment in D, and therefore certainly cannot map $\mathbb{T}$ onto Γ. It follows that at least one z_k belongs to $\mathbb{U}$, and hence

$$|g'(z)| > |h'(z)| \quad (z \in \mathbb{U}). \qquad (11.86)$$

The Jacobian of f in $\mathbb{U}$ is therefore negative and hence f is locally 1–1 and sense reversing in $\mathbb{U}$. The zeros of $f(z) - f(0)$ in $\mathbb{U}$ are therefore isolated points, which are finite in number, since $f(0) \notin \Gamma$. By the argument principle, the number of such zeros is $-d(f(z) - f(0), \mathbb{T}) \geq 1$ and therefore $d(f(z) - f(0), \mathbb{T}) \leq -1$; but

$$d(f(z) - f(0), \mathbb{T}) = d\big(F\big(e^{i\phi(t)}\big) - f(0), \mathbb{T}\big) \geq 1,$$

since $F(e^{i\phi(t)})$ rotates positively about $f(0)$, which is an interior point of D. Thus equality in (11.75) cannot occur and the proof is complete.

11.5.5 Variations in the argument and tangent of a curve

Let $f(t)\,(0 \leq t \leq 2\pi)$ be the parametric equation of a loop in $\mathbb{C} - \{0\}$ and suppose that $f'(t)$ exists and is continuous and $\neq 0$ for real t, where $f(t)$ is extended by periodicity to $\mathbb{R}$. Our aim is to study the variations of $\arg f(t)$ and $\arg f'(t)$.

We can write $f(t) = R(t)e^{i\phi(t)}$, which gives

$$f'(t) = (R'(t) + iR(t)\phi'(t))e^{i\phi(t)} = \rho(t)e^{i\psi(t)}. \tag{11.87}$$

Thus we wish to compare the variations of ϕ and ψ. Writing

$$\frac{R'(t) + iR(t)\phi'(t)}{|R'(t) + iR(t)\phi'(t)|} = e^{i\omega(t)}, \tag{11.88}$$

we have

$$\psi(t) = \omega(t) + \phi(t). \tag{11.89}$$

Consider first an interval J on which ϕ is increasing; then $\phi' \geq 0$ on J, and so $\Im e^{i\omega(t)} = \sin \omega(t) \geq 0$ on J. If $J = [a, b]$, write $V(\omega; J) = \omega(b) - \omega(a)$. We then have

$$|V(\omega; J)| \leq \pi. \tag{11.90}$$

The same conclusion holds if $\phi' \leq 0$ on J. If $I = [0, 2\pi]$, then $V(\omega; I)$ is a multiple of 2π, and therefore $V(\omega; I) = 0$. It follows that $V(\psi; I) = V(\phi; I)$. Writing $T(\phi; I)$ for the total variation we obtain

$$T(\phi; I) = V(\phi; I) = V(\psi; I) \leq T(\psi; I). \tag{11.91}$$

We will show that the inequality

$$T(\phi; I) \leq T(\psi; I) \tag{11.92}$$

holds in general for the interval $I = [0, 2\pi]$.

Without loss of generality we may assume that $\phi'(0) = 0 = \phi'(2\pi)$ (for if $\phi' \neq 0$, the result is already proved; by a linear shift a zero may be moved to 0).

We further assume, for now, that ϕ' has at most finitely many zeros on I (except possibly for closed intervals of zeros; on any such interval J, $V(\omega; J) = 0$ and the interval can be ignored).We partition $I = [0, 2\pi]$ into sub-intervals $I_1, I_2, \ldots, I_n$ with ϕ alternately increasing and decreasing in successive intervals. Since $\phi' = 0$ at the endpoints of the I_k, $e^{i\omega(t)}$ is real at these points, and so is positive or negative according to the sign of R'. It follows from (11.90) that

$$\left.\begin{aligned}
&V(\omega; I_k) = 0 \text{ if } R' \text{ has the same sign at the endpoints of } I_k;\\
&V(\omega; I_k) = \pi \text{ if } \phi \text{ is increasing and } R' > 0 \text{ at the initial point}\\
&\qquad\qquad \text{and } R' < 0 \text{ at the terminal point of } I_k,\\
&\qquad\quad \text{or if } \phi \text{ is decreasing and } R' < 0 \text{ at the initial point}\\
&\qquad\qquad \text{and } R' > 0 \text{ at the terminal point of } I_k;\\
&V(\omega; I_k) = -\pi \text{ if } \phi \text{ is decreasing and } R' > 0 \text{ at the initial point}\\
&\qquad\qquad \text{and } R' < 0 \text{ at the terminal point of } I_k,\\
&\qquad\quad \text{or if } \phi \text{ is increasing and } R' < 0 \text{ at the initial point}\\
&\qquad\qquad \text{and } R' > 0 \text{ at the terminal point of } I_k.
\end{aligned}\right\} \tag{11.93}$$

Now $\sum_{k=1}^{n} V(\omega; I_k) = V(\omega; I)$ is a multiple of 2π and therefore the number of non-zero terms $V(\omega; I_k)$ is even. We may therefore pair them together taking them in the order in which they partition I. Let I_k and I_l be such a pair. Then R' has the same sign at the terminal point of R_k and the initial point of R_l. We see therefore that

$$\begin{aligned}
V(\omega; I_k \cup I_l) &= 0 \text{ if } \phi' \text{ has the same sign in } I_k \text{ and } I_l,\\
|V(\omega; I_k \cup I_l)| &= 2\pi \text{ if } \phi' \text{ has opposite signs in } I_k \text{ and } I_l.
\end{aligned} \tag{11.94}$$

In the first case, $V(\psi; I_k \cup I_l) = V(\phi; I_k \cup I_l) = V(\phi; I_k) + V(\phi; I_l)$, where the last two quantities have the same sign. We thus obtain

$$T(\psi; I_k \cup I_l) \geq |V(\psi; I_k \cup I_l)| = T(\phi; I_k \cup I_l). \tag{11.95}$$

In the second case

$$\begin{aligned}
T(\psi; I_k \cup I_l) &= T(\psi; I_k) + T(\psi; I_l) \geq |V(\psi; I_k)| + |V(\psi; I_l)|\\
&= |V(\phi; I_k) + V(\omega; I_k)| + |V(\phi; I_l) + V(\omega; I_l)|\\
&\geq |V(\phi; I_k)| + |V(\phi; I_l)|\\
&= T(\phi; I_k \cup I_l),
\end{aligned} \tag{11.96}$$

where we have made use of the fact that $V(\phi; I_k)$ and $V(\phi; I_l)$ have opposite signs and $V(\omega; I_k) = V(\omega; I_l)$. Finally, if $V(\omega; I_j) = 0$, then

$$T(\psi; I_j) \geq |V(\psi; I_j)| = |V(\phi; I_j)| = T(\phi; I_j). \tag{11.97}$$

Adding over these last three inequalities we obtain (11.92).

We now remove the restriction that ϕ' have only finitely many zeros. Since f' is continuous on I, both R' and ϕ' are continuous and in particular ϕ' is bounded on I. Therefore ϕ is a function of bounded variation on I. It follows that, if ψ is a function of bounded variation on I, then so is ω. Therefore the number of intervals I_k on which $|V(\omega; I_k)| = \pi$ is finite (I_k being an interval at whose endpoints $\phi' = 0$, but on whose interior $\phi' \neq 0$). Ordering these I_k on I, if I_k and I_l are successive such intervals, then on the interval J joining I_k to I_l we have $V(\omega; J) = 0$ and therefore R' has the same sign at the endpoints of J. We may therefore argue as above to deduce that $T(\psi; I_k \cup I_l) \geq T(\phi; I_k \cup I_l)$. Thus, as before we obtain (11.92). We have proved the next theorem.

11.5.6

Theorem *If γ is a smooth loop in $\mathbb{C} - \{0\}$, then the total variation of* arg z *on γ does not exceed the total boundary rotation on γ.*

11.5.7

The definition of a function $v(z) = z + a_2z^2 + \cdots$ analytic and of bounded boundary rotation at most $k\pi$ (where $k \geq 2$) is given in chapter 8, subsection 8.8.3. A corollary of our theorem is that *such a function is $\frac{1}{2}k$-valent in* $\mathbb{U}$; for writing $f(t) = v(re^{it}) - w$, where r is close to 1 and $v(z) \neq w$ on $|z| = r$, we have $f'(t) = ire^{it}v'(re^{it})$, and by definition arg $f'(t)$ has total variation $< k\pi$; the same is therefore true of arg $f(t)$ and so the conclusion follows from the argument principle. Notice that this implies that a function of bounded boundary rotation not exceeding 4π is univalent in $\mathbb{U}$.

A second conclusion is that *the function $\int_0^z \frac{v(\zeta)}{\zeta} d\zeta = z + \frac{1}{2}a_2z^2 + \cdots$ is also a function of bounded boundary rotation at most $k\pi$.*

In other words *convolution with the function* $\log \frac{1}{1-z}$ *preserves for each $k \geq 2$ the class of functions of bounded boundary rotation at most $k\pi$.*

11.5.8

We now prove a second inequality. Suppose that ψ satisfies the inequality

$$V(\psi; J) \geq -\pi \tag{11.98}$$

over any interval J. Then

$$V(\phi; I) \leq V(\psi; I). \tag{11.99}$$

To prove this, we need to show that

$$V(\omega; I) \geq 0. \tag{11.100}$$

From the above analysis it is sufficient to consider the case where $V(\omega; I_k) = V(\omega; I_l) = -\pi$. For one of the two intervals I_k or I_l (say I_k) the function ϕ is decreasing, but we then obtain $V(\psi; I_k) < -\pi$, contradicting the hypothesis. Thus (11.99) follows.

11.5.9

From the above we deduce the following.

Theorem *Let $h(z)$ be analytic in $\mathbb{U}$ and suppose that $h'(z)$ has exactly $N-1$ zeros in $\mathbb{U}$, where $N \geq 1$. Suppose further that for r sufficiently close to 1 and for every positively described arc $\gamma \subset \{|z| = r\}$ we have*

$$d(zh', \gamma) > -\frac{1}{2}. \tag{11.101}$$

Then h is N-valent in $\mathbb{U}$.

Proof Let $f(t) = h(re^{it}) - w$, so $f'(t) = ire^{it}h'(re^{it})$; by hypothesis, for r close to 1 $V(\psi; J) > -\pi$ for every interval J. Because h is analytic, ϕ' has only finitely many zeros on I. Thus $V(\phi; I) \leq V(\psi; I)$ provided that $h(z) \neq w$ on $|z| = r$, which will certainly be true for some r close to 1. From the argument principle, the number of zeros of $h(z) - w$ in $\{|z| < r\}$ does not exceed the number of zeros of $zh'(z)$ in the same disc, and this $= N$ if r is close to 1. Thus h is N-valent in $\mathbb{U}$.

11.5.10

Corollary *Let Γ be a closed convex curve. If $f = \overline{g} + h$ is a non-constant function $\in \Phi(\Gamma, N)$, then h is N-valent in $\mathbb{U}$.*

Proof By theorem 11.5.4 the derivative of h has the form $h' = HQ$, where $H \in K(1, 2N+1)$ and $Q(z) = \prod_1^{N-1}(z - z_k)(1 - \overline{z_k}z)$, where $|z_k| \leq 1$. We will show that any function of this form is N-valent in $\mathbb{U}$. For $0 < r < 1$ we consider $h_r(z)$ satisfying $h_r(0) = h(0)$ and

$$h'_r(z) = H(rz)Q_r(z), \tag{11.102}$$

where $Q_r(z) = \prod_1^{N-1}(z - rz_k)(1 - r\overline{z_k}z)$. Let γ be a positively described arc of the unit circle $\mathbb{T}$. Since $H \in K(1, 2N+1)$, we have $d(H(rz), \gamma) > -Nd(z, \gamma) - \frac{1}{2}$; on the other hand, from (11.71), $d(zQ_r, \gamma) = Nd(z, \gamma)$. It follows that

$$d\big(zh'_r(z), \gamma\big) > -\frac{1}{2}, \tag{11.103}$$

from which it follows that $h_r(z)$ is N-valent in $\mathbb{U}$. Now $h_r(z) \to h(z)$ locally uniformly in $\mathbb{U}$ as $r \to 1$. Choose $w \in \mathbb{C}$. For a sequence of $\rho \to 1$, $h(z) \neq w$ for $|z| = \rho$ (because h is non-constant). It follows that on each such circle $|z| = \rho$,

$$\frac{zh'_r(z)}{h_r(z) - w} \to \frac{zh'(z)}{h(z) - w} \text{ uniformly as } r \to 1. \tag{11.104}$$

Integrating this over the circle and applying the argument principle, we deduce that h is N-valent in $\mathbb{U}$.

11.5.11 Problems

1. Let Γ be a closed convex curve and let $f = \overline{g} + h$ be a non-constant function $\in \Phi(\Gamma, N)$. Show that, if $|\zeta| < 1$, the function $\zeta g(z) + h(z)$ is N-valent in $\mathbb{U}$. This result continues to hold if $|\zeta| = 1$, unless f takes only values on a line segment $L \subset \Gamma$. [Hint: if $|\zeta| < 1$, the affine transformation $\zeta\overline{w} + w$ maps Γ onto a closed convex curve.]
2. Let $f \in \Phi(\Gamma, N)$, where Γ is a closed convex curve, and suppose that the Fourier series of f is $\sum_{-\infty}^{\infty} c_k e^{ikt}$. Prove that $\exists\delta(N) > 0$, depending only on N, such that

   $$\sum_{k=0}^{N} |c_k|^2 > d^2(\Gamma)\delta(N),$$

   where $d(\Gamma)$ is the distance from 0 to Γ.

   It can be shown that $\delta(N) > A^{-N}$, where A is an absolute constant (e.g. $A = 2^{11}$). On the other hand, Hall [19] has shown that $\delta(N) \to 0$ as $N \to \infty$. For further details see Sheil-Small [55].
3. Let Γ be a closed convex curve bounding a domain D and let $f \in \Phi(\Gamma, 1)$. Show that either $f(\mathbb{U}) = L$, where L is a line segment contained in Γ, or f is a 1–1 mapping of $\mathbb{U}$ onto a convex domain contained in D; if f is non-constant and continuous on $\mathbb{T}$, then f is a 1–1 mapping of $\mathbb{U}$ onto D.

This is the *Rado–Kneser–Choquet* theorem; for a *hint*: if $f(\mathbb{U}) \neq L$, f is locally 1–1 in $\mathbb{U}$. See Sheil–Small [55] for two separate proofs.

4. Let Γ be a closed convex curve bounding a domain D and let $f \in \Phi(\Gamma, N)$. If f is a continuous mapping of $\mathbb{T}$ onto Γ, show that in $\mathbb{U}$ f attains every value in D at least N times (counting with |multiplicity|). Show further that f attains each value in D at most finitely often.

   The rotation conjecture. We conjecture that f attains no value in D more than N^2 times. Lyzzaik pointed out that this problem is connected with the problem, solved by Wilmshurst, of the number of zeros of a harmonic polynomial; see chapter 2, sections 2.6.10–13. Wilmshurst's example shows that the conjecture would not be true for any smaller upper bound than N^2.

11.6 Blaschke products and convex polygons

11.6.1

Let $c_1, c_2, \ldots, c_n$ be $n \geq 2$ points lying on a closed convex curve Γ and occurring in order as Γ is positively traversed. Equivalently the polygon $\Pi = [c_1, c_2, \ldots, c_n, c_1]$ is convex. We assume that $c_1 \neq c_n$, but otherwise the points need not be distinct. Let f be a step function on $\mathbb{T}$ of the form

$$f(e^{it}) = c_{r(k)} \quad (t_{k-1} \leq t < t_k, 1 \leq k \leq n), \tag{11.105}$$

where $t_0 < t_1 < \cdots < t_n = t_0 + 2\pi$ and where $k \mapsto r(k)$ is a permutation of the numbers $1, 2, \ldots, n$. Thus f takes all the values c_k, though not necessarily in their order on Γ. We begin by proving a lemma.

11.6.2

Lemma *Let F: $\mathbb{T} \to \Gamma$ be a positively oriented homeomorphism. There exist a non-decreasing step function ϕ: $\mathbb{R} \to \mathbb{R}$ and a non-increasing step function ψ: $\mathbb{R} \to \mathbb{R}$ such that*

$$f(e^{it}) = F\left(e^{i\phi(t)}\right) = F\left(e^{i\psi(t)}\right). \tag{11.106}$$

Proof We extend $r(k)$ to the set of integers $\mathbb{Z}$ by the rule

$$r(k) = r(j) \text{ if } j \equiv k \,(\text{mod } n). \tag{11.107}$$

Then $r(k)$ is periodic with period n and r: $\mathbb{Z} \to \{1, 2, \ldots, n\}$ is surjective. Note that $r(k+1) \neq r(k)$ for any $k \in \mathbb{Z}$. We define a function $\rho(k)$ inductively by the following recursion relationship. With $\rho(1) = r(1)$ set

$$\rho(k+1) = \begin{cases} \rho(k) + r(k+1) - r(k) \text{ if } r(k+1) > r(k), \\ \rho(k) + r(k+1) - r(k) + n \text{ if } r(k+1) < r(k). \end{cases} \tag{11.108}$$

This clearly defines $\rho(k)$ for $k \geq 1$; on the other hand, given $\rho(k+1)$ we can recover $\rho(k)$ from the relationship, so $\rho(k)$ is also determined for $k \leq 0$; $\rho(k)$ is a strictly increasing function on $\mathbb{Z}$, which satisfies $\rho(k) \equiv r(k) \pmod{n}$. Also note that

$$\rho(n+1) = \rho(1) + \mu n, \tag{11.109}$$

where $1 \leq \mu \leq n-1$; μ may be regarded as a measure of how much the permutation $r(k)$ differs from the identity permutation: for the identity permutation $\mu = 1$; on the other hand if r reverses the numbers $1, \ldots, n$, then $\mu = n-1$.

We can write $c_k = F(e^{is_k})\,(1 \leq k \leq n+1)$, where $s_1 \leq s_2 \leq \cdots \leq s_{n+1} = s_1 + 2\pi$. We then define

$$\phi(t) = s_{\rho(k)} \quad (t_{k-1} \leq t < t_k) \tag{11.110}$$

for $1 \leq k \leq n$, and $\phi(t_n) = s_{\rho(n+1)}$, where in general, if $j \equiv k \pmod{n}$ and $1 \leq k \leq n$, we have set

$$s_j = s_k + 2\pi \frac{j-k}{n}. \tag{11.111}$$

Then $\phi(t)$ is increasing on $\mathbb{R}$ and $f(e^{it}) = F(e^{i\phi(t)})$. Furthermore

$$\phi(t_0 + 2\pi) - \phi(t_0) = s_{\rho(n+1)} - s_{\rho(1)} = 2\pi\mu. \tag{11.112}$$

Next, we define

$$\sigma(k) = \rho(k) - (k-1)n. \tag{11.113}$$

Then $\sigma(k)$ is strictly decreasing and $\sigma(k) \equiv r(k) \pmod{n}$. Also $\sigma(n+1) = \sigma(1) + (\mu - n)n = \sigma(1) - \nu n$, where $1 \leq \nu \leq n-1$. We set

$$\psi(t) = s_{\sigma(k)} \quad (t_{k-1} \leq t < t_k) \tag{11.114}$$

for $1 \leq k \leq n$ and $\psi(t_{n+1}) = s_{\sigma(n+1)}$. Then $\psi(t)$ is decreasing and $f(e^{it}) = F(e^{i\psi(t)})$. Furthermore

$$\psi(t_0 + 2\pi) - \psi(t_0) = -2\pi\nu. \tag{11.115}$$

Geometrically the number μ measures the number of times we need to traverse Γ in the positive direction in order to pick up the step values in the order of the permutation jumping from each value to the next in the positive direction.

Similarly the number ν makes the corresponding measurement when we interpret the jumps in the negative direction. We shall see that these two numbers, which satisfy $\mu + \nu = n$ and depend only on the permutation $r(k)$, control the zero–pole distribution of the Blaschke products of g'/h'.

11.6.3

Theorem *Let Γ be a closed convex curve and let $c_1, c_2, \ldots, c_n$ be $n \geq 2$ points lying on Γ and occurring in order as Γ is positively traversed, where we assume that $c_1 \neq c_n$. Let f be a step function on $\mathbb{T}$ of the form*

$$f(e^{it}) = c_{r(k)} \quad (t_{k-1} \leq t < t_k, 1 \leq k \leq n), \tag{11.116}$$

where $t_0 < t_1 < \cdots < t_n = t_0 + 2\pi$ and where $k \mapsto r(k)$ is a permutation of the numbers 1, 2, ..., n. Then the harmonic extension to $\mathbb{U}$ of $f = \overline{g} + h$ satisfies

$$\frac{g'}{h'} = \frac{B_2}{B_1}, \tag{11.117}$$

where B_1 and B_2 are Blaschke products. B_1 has degree at most $\mu - 1$ and B_2 has degree at most $\nu - 1$, where μ and ν are the periods arising from the permutation $r(k)$. In particular, g' has at most $\nu - 1$ zeros in $\mathbb{U}$ and h' has at most $\mu - 1$ zeros in $\mathbb{U}$. Furthermore, if the step values c_k are not collinear, g is ν-valent and h is μ-valent in $\mathbb{U}$.

Proof By the lemma, $f \in \Phi(\Gamma, \mu)$ and therefore by theorem 11.5.4

$$|g'| \leq \left|\frac{h'}{B_1}\right|, \tag{11.118}$$

where B_1 is a Blaschke product of degree at most $\mu - 1$. Similarly, $f(\overline{z}) \in \Phi(\Gamma, \nu)$ and therefore

$$|h'| \leq \left|\frac{g'}{B_2}\right|, \tag{11.119}$$

where B_2 is a Blaschke product of degree at most $\nu - 1$. Now B_2 carries the zeros of g' and therefore by the maximum principle the first inequality can be improved to

$$\left|\frac{g'}{B_2}\right| \leq \left|\frac{h'}{B_1}\right|. \tag{11.120}$$

Similarly, B_1 carries the zeros of h' and so the second inequality can be improved to

$$\left|\frac{h'}{B_1}\right| \leq \left|\frac{g'}{B_2}\right|. \tag{11.121}$$

These two inequalities imply (11.117), absorbing a constant of modulus 1 into the Blaschke products. The remaining conclusions also follow from our earlier results.

11.6.4 Topological properties of the harmonic extension

In this subsection we assume that the points c_k $(1 \leq k \leq n, n \geq 2)$ are distinct but otherwise arbitrary points of $\mathbb{C}$, and denote by Π the polygon $[c_1, c_2, \ldots, c_n, c_1]$.

Theorem *Let $f(z) = \overline{g(z)} + h(z)\,(z \in \mathbb{U})$ denote the harmonic extension into $\mathbb{U}$ of a step function on $\mathbb{T}$ given by*

$$f(e^{it}) = c_k \quad (t_{k-1} < t < t_k,\ 1 \leq k \leq n), \tag{11.122}$$

where $t_0 < t_1 < \cdots < t_n = t_0 + 2\pi$. The polygon Π is the set of limit points of $f(z)$ as $z \to \mathbb{T}$ from inside $\mathbb{U}$.

Proof By Schwarz's theorem, $f(z) \to f(e^{it})$ as $z \to e^{it}$ at any point e^{it} of continuity of the boundary function. Thus we need only consider the limiting behaviour of $f(z)$ as z approaches the points e^{it_k}. We will show that as $z \to e^{it_k}$ from inside $\mathbb{U}$, (i) $f(z)$ approaches only points on the line segment $[c_k, c_{k+1}]$, (ii) every such point is a limiting value of f. From Poisson's formula we have

$$f(z) = \sum_{j=1}^{n} c_j \frac{1}{2\pi} \int_{t_{j-1}}^{t_j} \frac{1 - |z|^2}{|1 - ze^{-it}|^2} dt \tag{11.123}$$

and hence, as $z \to e^{it_k}$, all terms in the sum $\to 0$ except the terms $j = k$ and $j = k + 1$. Thus

$$f(z) \to c_k \alpha_k + c_{k+1} \alpha_{k+1} \text{ as } z \to e^{it_k}, \tag{11.124}$$

where α_k and α_{k+1} are respectively the limiting values of

$$\frac{1}{2\pi} \int_{t_{k-1}}^{t_k} \frac{1 - |z|^2}{|1 - ze^{-it}|^2} dt, \qquad \frac{1}{2\pi} \int_{t_k}^{t_{k+1}} \frac{1 - |z|^2}{|1 - ze^{-it}|^2} dt. \tag{11.125}$$

Clearly $0 \le \alpha_k \le 1$ and $0 \le \alpha_{k+1} \le 1$. Furthermore

$$\begin{aligned} \alpha_k + \alpha_{k+1} &= \lim_{z \to e^{it_k}} \frac{1}{2\pi} \int_{t_{k-1}}^{t_{k+1}} \frac{1-|z|^2}{|1-ze^{-it}|^2} dt \\ &= 1 - \lim_{z \to e^{it_k}} \frac{1}{2\pi} \int_{t_{k+1}}^{t_{k-1}+2\pi} \frac{1-|z|^2}{|1-ze^{-it}|^2} dt = 1. \quad (11.126) \end{aligned}$$

Thus the limiting values of $f(z)$ as $z \to e^{it_k}$ are $c_k\alpha_k + (1-\alpha_k)c_{k+1} \in [c_k, c_{k+1}]$. This proves (i). To prove (ii) we need to show that every value on $(0, 1)$ is a limiting value of

$$F(z) = \frac{1}{2\pi} \int_{t_{k-1}}^{t_k} \frac{1-|z|^2}{|1-ze^{-it}|^2} dt \qquad (11.127)$$

as $z \to e^{it_k}$. Now $F(z)$ is the harmonic function in $\mathbb{U}$ arising from the step function defined by $F(e^{it}) = 1\,(t_{k-1} < t < t_k)$ and $F(e^{it}) = 0\,(t_k < t < t_{k-1}+2\pi)$. $F(z)$ remains continuous at all points of $\overline{\mathbb{U}}$, except for the two points e^{it_k} and $e^{it_{k-1}}$, and satisfies $0 \le F(z) \le 1$. Choose α satisfying $0 < \alpha < 1$ and ϵ satisfying $0 < \epsilon < \min(\alpha, 1-\alpha)$. Let $s \in (t_{k-1}, t_k)$ and $t \in (t_k, t_{k-1} + 2\pi)$. Since $F(re^{is}) \to 1$ as $r \to 1$ and $F(re^{it}) \to 0$ as $r \to 1$, for r sufficiently close to 1 we have $F(re^{it}) < \epsilon < \alpha < 1 - \epsilon < F(re^{is})$. From the continuity of F it follows that $\exists \theta = \theta(r) \in (s, t)$ such that $F(re^{i\theta}) = \alpha$. Letting $r \to 1, \theta(r) \to t_k$ and (ii) follows.

This theorem may be interpreted as stating that in a limiting sense f is a mapping of $\mathbb{T}$ onto Π, with the sides of Π attained instantaneously at the jump points. Therefore we might expect that the degree principle would continue to hold for f despite the boundary discontinuities.

11.6.5

Theorem *Let $f(z)$ $(z \in \mathbb{U})$ denote the harmonic extension into $\mathbb{U}$ of a step function on $\mathbb{T}$ given by*

$$f(e^{it}) = c_k \quad (t_{k-1} < t < t_k, 1 \le k \le n), \qquad (11.128)$$

where $t_0 < t_1 < \cdots < t_n = t_0 + 2\pi$. Then $f(\mathbb{U})$ contains all points w about which Π has non-zero winding number.

Proof Consider the parametric representation of Π given by

$$F(e^{it}) = \frac{(t_k - t)c_k + (t - t_{k-1})c_{k+1}}{t_k - t_{k-1}} \quad (t_{k-1} \le t \le t_k) \qquad (11.129)$$

for $1 \le k \le n$. Writing

$$\phi(t) = t_{k-1} \quad (t_{k-1} \le t < t_k, 1 \le k \le n), \tag{11.130}$$

we see that $\phi(t)$ is non-decreasing on $[t_0, t_0 + 2\pi]$ and

$$f(e^{it}) = F\left(e^{i\phi(t)}\right) \tag{11.131}$$

except at the t_k. We now choose a sequence of continuous approximations to ϕ by

$$\phi_j(t) = \begin{cases} t_{k-1} & (t_{k-1} \le t \le t_k - \delta_j), \\ (1/\delta_j)(t_k - t_{k-1})(t - t_k) + t_k & (t_k - \delta_j < t < t_k), \end{cases} \tag{11.132}$$

for $1 \le k \le n$, where $\{\delta_j\}$ is a sequence such that

$$\delta_j \to 0 \text{ as } j \to \infty \text{ and } 0 < \delta_j < \max\{t_k - t_{k-1} \colon 1 \le k \le n\}. \tag{11.133}$$

We define

$$f_j(e^{it}) = F\left(e^{i\phi_j(t)}\right) = \begin{cases} c_k & (t_{k-1} \le t \le t_k - \delta_j), \\ [(t - t_k)(c_{k+1} - c_k)]/\delta_j + c_{k+1} & (t_k - \delta_j < t \le t_k) \end{cases} \tag{11.134}$$

whose harmonic extension into $\mathbb{U}$ is given

$$\begin{aligned} f_j(z) = \sum_{k=1}^{n} \Bigg\{ & c_k \frac{1}{2\pi} \int_{t_{k-1}}^{t_k - \delta_j} \frac{1 - |z|^2}{|1 - ze^{-it}|^2} dt \\ & + \frac{1}{2\pi} \int_{t_k - \delta_j}^{t_k} \frac{1 - |z|^2}{|1 - ze^{-it}|^2} \left(\frac{(t - t_k)(c_{k+1} - c_k)}{\delta_j} + c_{k+1} \right) dt \Bigg\}, \end{aligned} \tag{11.135}$$

from which it is easily seen that

$$f_j(z) \to f(z) \text{ locally uniformly in } \mathbb{U} \text{ as } j \to \infty. \tag{11.136}$$

Furthermore, the functions $f_j(z)$ are continuous in $\overline{\mathbb{U}}$.

Suppose now that $w \notin \Pi$. We show that $\exists R \; (0 < R < 1)$ and j_0 such that

$$f_j(z) \ne w \quad (R \le |z| < 1, j \ge j_0). \tag{11.137}$$

For otherwise, for a suitable sequence of $j \to \infty$ we will have

$$f_j(z_j) = w, \quad z_j \to e^{is}. \tag{11.138}$$

Then for one of the k $(1 \le k \le n)$ we have $t_{k-1} < s \le t_k$ and so from (11.134)

$$w = \lim_{j\to\infty} \left\{ c_k \frac{1}{2\pi} \int_{t_{k-1}}^{t_k-\delta_j} \frac{1-|z_j|^2}{|1-z_je^{-it}|^2} dt \right.$$
$$+ \frac{1}{2\pi} \int_{t_k-\delta_j}^{t_k} \frac{1-|z_j|^2}{|1-z_je^{-it}|^2} \left(\frac{(t-t_k)(c_{k+1}-c_k)}{\delta_j} + c_{k+1} \right) dt$$
$$\left. + c_{k+1} \frac{1}{2\pi} \int_{t_k}^{t_{k+1}-\delta_j} \frac{1-|z_j|^2}{|1-z_je^{-it}|^2} dt \right\}. \tag{11.139}$$

If $s < t_k$, then the second and third terms $\to 0$, which gives

$$w = \lim_{j\to\infty} c_k \frac{1}{2\pi} \int_{t_{k-1}}^{t_k-\delta_j} \frac{1-|z_j|^2}{|1-z_je^{-it}|^2} dt = c_k \in \Pi.$$

This contradicts the hypothesis, so we may assume that $z_j \to e^{it_k}$ as $j \to \infty$. We then see that

$$w = \lim_{j\to\infty} (A_j c_k + B_j c_{k+1}), \tag{11.140}$$

where

$$A_j = \frac{1}{2\pi} \int_{t_{k-1}}^{t_k-\delta_j} \frac{1-|z_j|^2}{|1-z_je^{-it}|^2} dt + \frac{1}{2\pi} \int_{t_k-\delta_j}^{t_k} \frac{t_k-t}{\delta_j} \frac{1-|z_j|^2}{|1-z_je^{-it}|^2} dt,$$

$$B_j = \frac{1}{2\pi} \int_{t_k}^{t_{k+1}-\delta_j} \frac{1-|z_j|^2}{|1-z_je^{-it}|^2} dt + \frac{1}{2\pi} \int_{t_k-\delta_j}^{t_k} \left(1 + \frac{t-t_k}{\delta_j}\right) \frac{1-|z_j|^2}{|1-z_je^{-it}|^2} dt.$$

Then $A_j \ge 0$, $B_j \ge 0$ and

$$A_j + B_j = \frac{1}{2\pi} \int_{t_{k-1}}^{t_{k+1}-\delta_j} \frac{1-|z_j|^2}{|1-z_je^{-it}|^2} dt \to 1 \text{ as } j \to \infty.$$

Hence choosing a suitable sequence of $j \to \infty$ such that $A_j \to A$, we have

$$w = Ac_k + (1-A)c_{k+1},$$

where $0 \le A \le 1$. This gives $w \in \Pi$, contradicting the hypothesis.

For the same $w \notin \Pi$ we now suppose that $f(z) \ne w$ for $z \in \mathbb{U}$. With R and j_0 as in (11.137) we have $f_j(z) \to f(z)$ uniformly on $T_R = \{|z| = R\}$ as $j \to \infty$, and therefore

$$\frac{f_j(z) - w}{f(z) - w} \to 1 \text{ uniformly on } T_R \text{ as } j \to \infty. \tag{11.141}$$

Hence $\exists j_1 \ge j_0$ such that for $j \ge j_1$

$$d(f_j - w, T_R) = d(f - w, T_R) = 0, \tag{11.142}$$

the last equality coming from the degree principle, since $f - w \neq 0$ inside and on T_R. On the other hand, for any such $j \geq j_1$, $f_j(z) - w$ is continuous and $\neq 0$ for $R \leq |z| \leq 1$, and therefore

$$d(f_j - w, \mathbb{T}) = d(f_j - w, T_R) = 0. \tag{11.143}$$

However from (11.134) $f_j(e^{it})$ is a parametric representation of Π, and so this last relation gives $n(\Pi, w) = 0$. Thus $f(\mathbb{U})$ contains all points w about which Π has non-zero winding number. This completes the proof of the theorem.

11.6.6

Theorem *Suppose that the polygon $\Pi = [c_1, \ldots, c_n, c_1]$ is a positively oriented Jordan curve bounding a domain D and let $f(z) = \overline{g(z)} + h(z) (z \in \mathbb{U})$ denote the harmonic extension into $\mathbb{U}$ of a step function on $\mathbb{T}$ given by*

$$f(e^{it}) = c_k \quad (t_{k-1} < t < t_k, 1 \leq k \leq n), \tag{11.144}$$

where $t_0 < t_1 < \cdots < t_n = t_0 + 2\pi$. Then $D \subset f(\mathbb{U})$. Furthermore the following conditions are equivalent (using the notations of 11.4.1):

(1) *f is a homeomorphism of $\mathbb{U}$ onto D;*
(2) *f is univalent in $\mathbb{U}$;*
(3) *$h'(z) \neq 0$ for $z \in \mathbb{U}$;*
(4) *$Q(z)$ has degree $n - 2$ and all its zeros in $\overline{\mathbb{U}}$;*
(5) *$P(z) \neq 0$ for $z \in \mathbb{U}$;*
(6) *$|R(z)| \leq 1$ for $z \in \mathbb{U}$;*
(7) *$|R(z)| < 1$ for $z \in \mathbb{U}$.*

Proof $D \subset f(\mathbb{U})$ follows immediately from the previous theorem. To prove the equivalences: clearly (1)⇒(2); (2)⇒(3) follows from theorem 2.7.14 of chapter 2; (3)⇒(5) follows from (11.41) and (4)⇔(5) follows from (11.43); (5)⇒ that $R(z)$ has no poles in $\mathbb{U}$ from (11.44), and since $|R(z)| = 1$ on $\mathbb{T}$, (6) follows from the maximum principle; (6)⇒(7) from the maximum principle, unless $R(z)$ is a constant c, where $|c| = 1$; in this case $f(\mathbb{U})$ lies on a line segment and so Π is not a Jordan polygon. It remains to prove that (7)⇒(1). Assume (7) and let $a \in \mathbb{U}$; either $|g'(a)| < |h'(a)|$ or g' and h' have a common zero at a of multiplicity m; thus we can write

$$g'(z) = (z - a)^m G(z), \qquad h'(z) = (z - a)^m H(z), \tag{11.145}$$

where G and H are analytic in $\mathbb{U}$, $m \geq 0$ and $|G(a)| < |H(a)|$. Then, expanding f in a harmonic series about the point a, we obtain

$$\frac{(m+1)(f(z)-f(a))}{\rho^{m+1}H(a)e^{i(m+1)\theta}} = 1 + \frac{\overline{G(a)}}{H(a)}e^{-2i(m+1)\theta} + o(1) \qquad (11.146)$$

for $z = a + \rho e^{i\theta}$ and $\rho > 0$ small. Since the right-hand side has positive real part for ρ small, it has zero degree round the circle of centre a and radius ρ. Therefore $f(z) - f(a)$ has an isolated zero at a and

$$m(f(z) - f(a), a) = m + 1 \qquad (11.147)$$

(the multiplicity of the zero of $f(z) - f(a)$ at a is $m + 1$, see chapter 2, section 2.5.1). This result shows that the mapping f is an open mapping on $\mathbb{U}$: for the multiplicity denotes the winding number of the loop $f(C_\rho)$ about $f(a)$, where C_ρ is the circle with centre a, radius ρ (ρ small); for a value w close to $f(a)$ the winding number of $f(C_\rho)$ about w has the same value $m+1 \geq 1$, and therefore by the degree principle f attains the value w. Thus $f(\mathbb{U})$ is an open set.

Now choose $w \notin \Pi$. As we showed in the proof of the previous theorem,

$$n(\Pi, w) = d(f(z) - w, T_R), \qquad (11.148)$$

if R is sufficiently close to 1. On the other hand, since the zeros of $f(z) - w$ are isolated, the argument principle gives, for R close to 1,

$$d(f(z) - w, T_R) = \sum (m(a) + 1), \qquad (11.149)$$

the sum being taken over all zeros $a \in \mathbb{U}$ of $f(z) - w$, where $m(a)$ denotes the multiplicity at a of the zero of h'. Thus

$$\sum (m(a) + 1) = n(\Pi, w) = \begin{cases} 0 & \text{if } w \text{ is exterior to } \Pi, \\ 1 & \text{if } w \in D. \end{cases} \qquad (11.150)$$

Thus the left sum is empty if w is exterior to Π, i.e. $f(z) \neq w$ for $z \in \mathbb{U}$; on the other hand, for $w \in D$ the left sum contains one value a with $m(a) = 0$, i.e. f attains every value in D exactly once. As the mapping is open, $f(z) \neq w$ for $z \in \mathbb{U}$ if $w \in \Pi$. Thus (1) follows.

11.6.7

Putting our results together, we obtain the following result for convex polygons.

Theorem *Suppose that the polygon* $\Pi = [c_1, \ldots, c_n, c_1]$ *is a positively oriented convex curve bounding a domain* D *and let* $f(z) = \overline{g(z)} + h(z)$ $(z \in \mathbb{U})$

denote the harmonic extension into $\mathbb{U}$ of a step function on $\mathbb{T}$ given by

$$f(e^{it}) = c_k \quad (t_{k-1} < t < t_k,\ 1 \le k \le n), \tag{11.151}$$

where $t_0 < t_1 < \cdots < t_n = t_0 + 2\pi$. Then f is a homeomorphism of $\mathbb{U}$ onto D. Furthermore, $g'(z)/h'(z)$ is a Blaschke product of exact degree $n-2$.

Proof By lemma 11.6.2 we have $\rho(k) = r(k) = k$ and so $\mu = 1$. Therefore, by theorem 11.6.3, $h'(z) \neq 0$ for $z \in \mathbb{U}$. Thus f is a homeomorphism of $\mathbb{U}$ onto D by the previous theorem. To prove the exactness of the degree of g'/h' we recall the notations of 11.4.1. We have

$$g' = \frac{Q}{S}, h' = \frac{P}{S}, \frac{g'}{h'} = \frac{Q}{P} = R, z^{n-2}\overline{Q(1/\overline{z})} = cP(z), \tag{11.152}$$

where $|c| = 1$. P and Q have no zeros in common with S, which has its zeros on $\mathbb{T}$. Q has its zeros in $\overline{\mathbb{U}}$ and has degree $n-2$ from the previous theorem. Now, if h' has no zeros on $\mathbb{T}$, then P has no zeros on $\mathbb{T}$ and so Q has all its zeros in $\mathbb{U}$; therefore g' has all its zeros in $\mathbb{U}$ and therefore no zeros in common with h'. Thus R has exact degree $n-2$.

So we see that it is sufficient to show that h' has no zeros on $\mathbb{T}$; to prove this it is enough to show that $|h'(z)| \ge m > 0$ as $|z| \to 1$ from inside $\mathbb{U}$. We make use of Problem 11.5.11.1 in the case $N = 1$: this shows that for each ζ ($|\zeta| = 1$), the function $h + \zeta g$ is univalent in $\mathbb{U}$ and therefore by a classical distortion theorem

$$\left| \frac{h'(z) + \zeta g'(z)}{h'(0) + \zeta g'(0)} \right| \ge \frac{1 - |z|}{(1 + |z|)^3} \tag{11.153}$$

for $z \in \mathbb{U}$ and $|\zeta| = 1$. (We could alternatively use theorem 11.5.4 to deduce that $h' + \zeta g' \in K(1, 3)$ to obtain the same inequality.) Since $|h'(0)| > |g'(0)|$, we obtain, minimising over $\zeta \in \mathbb{T}$,

$$|h'(z)| - |g'(z)| \ge \rho(1 - |z|), \tag{11.154}$$

for $z \in \mathbb{U}$, where $\rho > 0$. This gives

$$|h'(z)| \ge \rho \frac{1 - |z|}{1 - |R(z)|}. \tag{11.155}$$

Now for a Blaschke product R of degree m, we can write

$$\frac{1 + R(z)}{1 - R(z)} = \sum_{k=1}^{m} t_k \frac{1 + \tau_k z}{1 - \tau_k z} + w, \tag{11.156}$$

where $t_k > 0$, $|\tau_k| = 1$ $(1 \le k \le m)$ and $\Re w = 0$. The points $\overline{\tau_k}$ are the m roots on $\mathbb{T}$ of the equation $R(z) = 1$ (similar relations are proved in chapter 7, subsection 7.1.4). Taking real parts we obtain

$$\frac{1-|R(z)|^2}{1-|z|^2} = \sum_{k=1}^{m} t_k \frac{|1-R(z)|^2}{|1-\tau_k z|^2}. \tag{11.157}$$

Now the right-hand expression is continuous in $\overline{\mathbb{U}}$, since $R(z) = 1$ at $\overline{\tau_k}$; it is therefore bounded in $\overline{\mathbb{U}}$ and so

$$\frac{1-|z|^2}{1-|R(z)|^2} \geq \mu > 0. \tag{11.158}$$

Finally from this and (11.155)

$$|h'(z)| \geq \rho\mu \frac{1+|R(z)|}{1+|z|} > \frac{\rho\mu}{2}, \tag{11.159}$$

and the conclusion follows.

11.6.8

Example The following example shows that the mapping f may not be a homeomorphism, if Π is not convex.

Let $\Pi = [1, i, -1, \frac{1}{2}i, 1]$ and take $\zeta_1 = 1, \zeta_2 = i, \zeta_3 = -1, \zeta_4 = -i$. Then

$$f(0) = \frac{1}{2\pi}\sum_{k=1}^{4} c_k(t_k - t_{k-1}) = \frac{1}{4}\sum_{k=1}^{4} c_k = \frac{3}{8}i, \tag{11.160}$$

and this is a point outside the domain bounded by Π. Thus f cannot be 1–1.

11.7 The mapping problem for Jordan polygons

11.7.1

Let $\Pi = [c_1, c_2, \ldots, c_n, c_1]$ denote a positively oriented Jordan polygon with distinct vertices c_k and let D be the domain bounded by Π. The **mapping problem** is to find distinct points $\zeta_k = e^{it_k} \in \mathbb{T}$, labelled in positive cyclic order ($0 \leq k \leq n$, $\zeta_0 = \zeta_n$) such that the harmonic extension to $\mathbb{U}$ of the step function $f(e^{it}) = c_k$ ($t_{k-1} < t < t_k$, $1 \leq k \leq n$) is a homeomorphism of $\mathbb{U}$ onto D. By theorem 11.6.6 the necessary and sufficient condition for this is

$$h'(z) = \sum_{k=1}^{n} \frac{\alpha_k}{z-\zeta_k} \neq 0 \quad (z \in \mathbb{U}), \tag{11.161}$$

where

$$\alpha_k = \frac{1}{2\pi i}(c_k - c_{k+1}) \quad (1 \leq k \leq n), \tag{11.162}$$

taking $c_{n+1} = c_1$. When Π is convex, this condition is satisfied for any ordered set of $\zeta_k \in \mathbb{T}$. Our example shows that for a non-convex Π, the condition may not be satisfied for a given choice of ζ_k. On the other hand, we will see shortly that for some non-convex Π, the mapping problem is soluble. We conjecture, however, that there are Jordan polygons for which the mapping problem is insoluble.

To sharpen the problem, we will consider whether for a given Jordan polygon Π it is possible to solve the mapping problem with a specified dilatation ratio $R = g'/h'$. This is a Blaschke product of degree $\leq n-2$. For convex Π, theorem 11.6.7 shows that R must have exact degree $n-2$, and therefore the mapping problem cannot be solved with a specified dilatation ratio of smaller degree. For example, if we specify that $R(z) = cz$, where $|c| = 1$, then the only convex polygons for which this is possible are triangles, i.e. the case $n = 3$. We will shortly give a complete solution to the mapping problem with this choice of R.

11.7.2

Before doing this, let us consider the problem in general terms. Suppose firstly that we have found an ordered set of ζ_k to solve the mapping problem for a specified R. Then we obtain from (11.38) and (11.162) the following algebraic relationship.

$$\sum_{k=1}^{n} \alpha_k \frac{R(z) - \omega_k}{z - \zeta_k} = 0, \tag{11.163}$$

where $\omega_k = -\frac{\overline{\alpha_k}}{\alpha_k}$. On the other hand, if for the given c_k and R we can find ζ_k ordered on $\mathbb{T}$ with this relation satisfied, then defining h' and g' by (11.34) and (11.38) we obtain $Rh' = g'$, giving $|g'/h'| \leq 1$ in $\mathbb{U}$; by theorem 11.6.6 f is a homeomorphism of $\mathbb{U}$ onto D.

Thus *satisfying identically the algebraic relation* (11.163) *is both necessary and sufficient for solving the mapping problem.*

Note further that for this identity to be satisfied it is necessary that

$$R(\zeta_k) = \omega_k \quad (1 \leq k \leq n). \tag{11.164}$$

Hence, for each k, ζ_k is one of the inverse image values $R^{-1}(\omega_k)$; this gives at most $\deg R$ possible choices, and so at most $n-2$ possible values of ζ_k; in addition, the values ζ_k must be labelled by k in positive cyclic order on $\mathbb{T}$. Thus given Π and the dilatation R, there are at most finitely many solutions to the mapping problem.

11.7.3

Theorem *Let $\Pi = [c_1, c_2, \dots, c_n c_1]$ be a Jordan polygon with distinct vertices c_k bounding a domain D. We can find points $\zeta_k \in \mathbb{T}$, labelled in positive cyclic order, such that the harmonic extension $f = \overline{g} + h$ of the step function on $\mathbb{T}$ is a homeomorphism of $\mathbb{U}$ onto D with $g'(z)/h'(z) = cz$, where c is a constant with $|c| = 1$, if, and only if, the points $\omega_k = -\overline{\alpha_k}/\alpha_k$ occur in positive cyclic order on $\mathbb{T}$. We then have*

$$\zeta_k = \overline{c}\omega_k \quad (1 \le k \le n). \tag{11.165}$$

For the only possible solution of (11.164) is given by $\zeta_k = \overline{c}\omega_k (1 \le k \le n)$. Furthermore, the relation (11.163) is satisfied with this choice of ζ_k. Therefore we have a solution to the mapping problem if, and only if, the points ω_k occur in positive cyclic order on $\mathbb{T}$.

We have already observed that the only possible convex solutions in this case occur when Π is a triangle. On the other hand, if Π is a positively oriented triangle, then it is easily verified that the ω_k are labelled in positive order on $\mathbb{T}$, and therefore the mapping problem has a solution. On the other hand, some non-convex polygons may also have a solution as shown by the example we give next.

11.7.4

Example Let $\Pi = [1, i, -1, \frac{1}{2}i, 1]$ be the quadrilateral polygon of example 11.6.8. Then calculations give $\omega_1 = -i$, $\omega_2 = i$, $\omega_3 = (-3+4i)/5$, $\omega_4 = -(3+4i)/5$ and these are in positive cyclic order on $\mathbb{T}$. Thus the mapping problem can be solved with $R(z) = cz\ (|c| = 1)$.

11.7.5

Nevertheless, the condition that the ω_k occur in order on $\mathbb{T}$ imposes severe geometric restrictions on the possible polygons Π, as the following considerations show.

Writing $\alpha_k = |\alpha_k|e^{ia_k} = (c_k - c_{k+1})/2\pi i$, the condition that Π is a Jordan polygon implies that

$$|a_{k+1} - a_k| < \pi, \qquad a_{n+1} = a_1 + 2\pi. \tag{11.166}$$

Since $\omega_k = -e^{-2ia_k}$, the condition that these occur in order on $\mathbb{T}$ can be written

$$2m_1\pi - 2a_1 < 2m_2\pi - 2a_2 < \cdots < 2m_{n+1}\pi - 2a_{n+1} \tag{11.167}$$

for suitable integers m_k. We deduce that

$$m_{k+1} \geq m_k, \qquad m_{n+1} = m_1 + 3, \tag{11.168}$$

the last relation following from $2m_{n+1}\pi - 2a_{n+1} = 2m_1\pi - 2a_1 + 2\pi$. Therefore, if $k > j$, we obtain

$$2m_k\pi - 2a_k < 2m_{n+1}\pi - 2a_{n+1} = 2m_1\pi - 2a_1 + 2\pi < 2m_j\pi - 2a_j + 2\pi, \tag{11.169}$$

and so

$$a_k - a_j > \pi(m_k - m_j) - \pi \geq -\pi. \tag{11.170}$$

11.7.6

We deduce

Theorem *If* $g'/h' = cz$ $(|c| = 1)$ *for a Jordan polygon* Π, *then* D *is close-to-convex.*

11.7.7

We conclude by solving the mapping problem for a quadrilateral Π with $R(z) = z^2$.

We are required to find $\zeta_k \in \mathbb{T}$, labelled in positive order, such that

$$\sum_{k=1}^{4} \alpha_k \frac{z^2 - \omega_k}{z - \zeta_k} = 0. \tag{11.171}$$

For this to hold, each ζ_k is one of the two square roots of ω_k, and we have

$$\zeta_k = \pm i|\alpha_k|/\alpha_k \quad (1 \leq k \leq 4), \tag{11.172}$$

$$\sum_{k=1}^{4} \alpha_k \zeta_k = 0. \tag{11.173}$$

Thus the problem can be solved if, and only if, both these conditions hold for ζ_k labelled in positive order on $\mathbb{T}$. The two conditions reduce to

$$\zeta_k = \pm\epsilon_k \frac{|c_{k+1} - c_k|}{c_{k+1} - c_k}, \qquad \sum_{k=1}^{4} \epsilon_k |c_{k+1} - c_k| = 0, \tag{11.174}$$

where $\epsilon_k = \pm 1$ for $1 \leq k \leq 4$. Observe that for the second of the two conditions to hold, we must have $\epsilon_k = 1$ for exactly two values of k. Thus the sum of the lengths of two sides of the quadrilateral is equal to the sum of the lengths of the remaining two sides.

To explore the first condition we again write $\alpha_k = |\alpha_k|e^{ia_k}$, so that $|a_{k+1} - a_k| < \pi$ for $1 \le k \le 4$ and $a_5 = a_1 + 2\pi$. We then require

$$(2m_1 + \gamma_1)\pi - a_1 < (2m_2 + \gamma_2)\pi - a_2 < (2m_3 + \gamma_3)\pi - a_3 < (2m_4 + \gamma_4)\pi - a_4 < (2m_5 + \gamma_5)\pi - a_5, \quad (11.175)$$

where the m_k are integers and $\gamma_k = 0$ or 1, depending upon whether $\epsilon_k = 1$ or -1. We deduce that

$$m_{k+1} \ge m_k, \qquad \gamma_5 = \gamma_1, \qquad m_5 = m_1 + 2. \qquad (11.176)$$

We obtain a solution with

$$m_1 = m_2 = 0, \qquad m_3 = m_4 = 1, \qquad m_5 = 2, \qquad \gamma_1 = \gamma_3 = \gamma_5 = 0, \qquad \gamma_2 = \gamma_4 = 1. \qquad (11.177)$$

In other words the mapping problem can be solved for the quadrilateral Π with $R(z) = z^2$ provided that *the two sums of the lengths of opposite sides of* Π *are equal* ($l_1 + l_3 = l_2 + l_4$).

We will show that this condition is also *necessary*, if Π is a convex quadrilateral. The convexity condition implies that $0 < a_{k+1} - a_k < \pi$ for $1 \le k \le 4$. Suppose that we have a quadrilateral where $\gamma_1 = \gamma_2 = \gamma_5 = 0$ and $\gamma_3 = \gamma_4 = 1$. We may take $m_1 = 0$ and $m_5 = 2$. We then have

$$-a_1 < 2m_2\pi - a_2 < (2m_3 + 1)\pi - a_3 < (2m_4 + 1)\pi - a_4 < 2\pi - a_1 \quad (11.178)$$

and writing $r_k = (a_{k+1} - a_k)/\pi$, we have $0 < r_k < 1$ for $1 \le k \le 4$, which implies that $1 < r_1 + r_2 + r_3 < 3$. The above inequality becomes

$$0 < 2m_2 - r_1 < 2m_3 + 1 - r_1 - r_2 < 2m_4 + 1 - r_1 - r_2 - r_3 < 2. \quad (11.179)$$

We have $2m_4 < 1 + r_1 + r_2 + r_3 < 4$, and so $m_4 \le 1$; on the other hand, $2m_2 > r_1 > 0$, so $m_2 \ge 1$. It follows that $m_2 = m_3 = m_4 = 1$, which implies that $r_3 < 0$, contradicting the convexity.

In particular, *there is no solution of the mapping problem for a rectangle* Π *with* $R(z) = z^2$*, except in the case that* Π *is a square.*

11.8 Sudbery's theorem on zeros of successive derivatives

11.8.1

Theorem *Let $P(z)$ be a polynomial of degree $n \geq 2$ and let*

$$\Pi(z) = \prod_{r=0}^{n-1} P^{(r)}(z), \tag{11.180}$$

where $P^{(0)} = P$ and otherwise $P^{(r)}$ denotes the rth derivative of P. Then either Π has exactly one distinct zero or Π has at least $n+1$ distinct zeros.

This result was originally an unpublished conjecture of T. Popoviciu, which was circulated by P. Erdös in his lectures. Sudbery [61] successfully resolved the problem in 1972.

11.8.2 Some algebra of the logarithmic derivative

Let $Q(z)$ be a polynomial whose distinct zeros lie among the d distinct numbers $z_1, \ldots, z_d$. Then we can write

$$\frac{zQ'(z)}{Q(z)} = \sum_{i=1}^{d} \frac{m_i}{1 - z_i/z} = \sum_{j=0}^{\infty} \left(\sum_{i=1}^{d} m_i z_i^j \right) z^{-j}, \tag{11.181}$$

where m_i denotes the multiplicity of the zero z_i of Q, or $m_i = 0$ if z_i is not a zero of Q, the expansion being convergent near ∞. Now for the purposes of the algebra we write $Q(z) = \sum_{-\infty}^{\infty} q_k z^k$ with the understanding that $q_k = 0$ if either $k < 0$ or $k > N$, where N is the degree of Q. We obtain

$$\begin{aligned} \sum_k k q_k z^k &= \sum_k q_k z^k \sum_{j=0}^{\infty} \left(\sum_{i=1}^{d} m_i z_i^j \right) z^{-j} = \sum_{i=1}^{d} m_i \sum_t \left(\sum_{k-j=t} q_k z_i^j \right) z^t \\ &= \sum_{i=1}^{d} m_i \sum_t \left(\sum_{j \geq 0} q_{j+t} z_i^j \right) z^t. \end{aligned} \tag{11.182}$$

Writing $\sigma_j = \sum_{i=1}^{d} m_i z_i^j$ and equating coefficients gives for every integer k

$$k q_k = \sum_{j \geq 0} \sigma_j q_{j+k}. \tag{11.183}$$

In particular we obtain

$$k q_k = \sum_{i=1}^{d} Q_k(z_i) m_i \quad (0 \leq k \leq N), \tag{11.184}$$

where

$$Q_k(z) = q_k + q_{k+1}z + \cdots + q_N z^{N-k} \quad (0 \le k \le N). \tag{11.185}$$

Since $q_N \neq 0$, the case $k = N$ leads to the expected result

$$\sigma_0 = \sum_{i=1}^{d} m_i = N. \tag{11.186}$$

From this we obtain

$$-(N-k)q_k = \sum_{i=1}^{d} m_i S_k(z_i) = \sum_{j=1}^{N-k} \sigma_j q_{j+k} \quad (0 \le k \le N-1), \tag{11.187}$$

where $S_k(z) = q_{k+1}z + \cdots + q_N z^{N-k} = zQ_{k+1}(z)$. Note that this expresses the coefficient q_k linearly in terms of the higher order coefficients $q_{k+1}, \ldots, q_N$ and the powers $z_i, \ldots, z_i^{N-k}$ $(1 \le i \le d)$.

11.8.3

For Sudbery's theorem we consider the corresponding identities for each of the polynomials $P(z), P'(z), \ldots, P^{(n-1)}(z)$, whose degrees are $n, n-1, \ldots, 1$ respectively. We take the z_i as the distinct zeros of $\Pi(z)$ and denote by $m_i(r)$ the multiplicity of z_i as a zero of $P^{(r)}$ ($m_i(r) = 0$ if z_i is not a zero of $P^{(r)}$). Writing $P(z) = \sum_0^n a_k z^k$ we have

$$P^{(r)}(z) = \sum_{k=0}^{n} k(k-1)\ldots(k-r+1)a_k z^{k-r} = \sum_{k=-r}^{n-r} \frac{k(k+1)\ldots(k+r)}{k} a_{k+r} z^k \tag{11.188}$$

for $0 \le r \le n-1$. Writing $\sigma_j(r) = \sum_{i=1}^{d} m_i(r) z_i^j$ we apply the identity (11.183) in the case $N = n - r$. This gives

$$k(k+1)\ldots(k+r)a_{k+r} = \sum_{j\ge 0} \sigma_j(r) \frac{(j+k)(j+k+1)\ldots(j+k+r)}{j+k} a_{j+k+r} \tag{11.189}$$

for $0 \le r \le n-1$. We will assume for now that $d = n$. Thus

$$\sum_{i=1}^{n} m_i(r) = n - r \quad (0 \le r \le n-1). \tag{11.190}$$

For $k = n-1$ this gives

$$-(n-r)a_{n-1} = n \sum_{i=1}^{n} m_i(r) z_i \quad (0 \le r \le n-1), \tag{11.191}$$

taking $a_n = 1$ without loss of generality. This gives us n linear equations satisfied by the n quantities $z_1, \dots, z_n$. These equations imply either that the $n \times n$ matrix $M = (m_{i,j})$ of multiplicities is singular and so has vanishing determinant, or that the values z_i are collinear. For, if M is non-singular, then by solving the equations we find that each z_i satisfies $z_i = t_i a_{n-1}$, where t_i is real and rational. We have

$$m_{i,j} = m_j(i-1) \quad (1 \le i \le n, 1 \le j \le n). \tag{11.192}$$

11.8.4

Sudbery's argument proceeds as follows. Choose m satisfying $1 \le m \le n-1$ and consider the relation (11.189) in the case $k = n-m-r$, where $0 \le r \le n-1$. We obtain

$$\begin{aligned}(n-m-r)(n-m-r-1)\dots(n-m)a_{n-m} \\ = \sum_{j=0}^{m} \sigma_j(r) \frac{(j+n-m-r)\dots(j+n-m)}{j+n-m-r} a_{j+n-m}.\end{aligned} \tag{11.193}$$

Our aim is to show that, for each m $(1 \le m \le n-1)$, we can find a polynomial $\tau_m(x)$ of degree not exceeding m such that $\sigma_m(r) = \tau_m(r)$ for $0 \le r \le n-1$. As we have seen, $\sigma_0(r) = n - r$, which represents σ_0 as a polynomial of degree 1. We proceed by induction on m. In the case $m = 1$ the above expression becomes

$$\begin{aligned}&(n-1-r)\dots(n-1)a_{n-1} \\ &\quad = (n-r)\frac{(n-1-r)\dots(n-1)}{n-1-r} a_{n-1} + \sigma_1(r)\frac{(n-r)\dots(n)}{n-r} a_n.\end{aligned} \tag{11.194}$$

For $0 \le r < n-1$ we obtain

$$a_{n-1} = \frac{n-r}{n-1-r} a_{n-1} + \sigma_1(r) \frac{n}{(n-r)(n-1-r)}, \tag{11.195}$$

taking $a_n = 1$. This gives

$$\sigma_1(r) = -\frac{n-r}{n} a_{n-1}, \tag{11.196}$$

which defines σ_1 as either a polynomial of degree 1 or identically zero. In the case $r = n-1$ we obtain

$$0 = (n-1)! a_{n-1} + \sigma_1 \, n! a_n \tag{11.197}$$

giving $\sigma_1 = -a_{n-1}/n$, which agrees with the above formula for $\sigma_1(r)$ when $r = n-1$. Thus the result is true for $m = 1$. Now assume it is true up to $m-1$, where $m > 1$. The formula (11.193) gives

$$\sigma_m(r)\frac{(n-r)\dots(n)}{n-r}$$

$$= \left\{(n-m-r)(n-m-r-1)\dots(n-m)-(n-r)\right.$$

$$\left.\times\frac{(n-m-r)\dots(n-m)}{n-m-r}\right\}a_{n-m}$$

$$-\sum_{j=1}^{m-1}\sigma_j(r)\frac{(j+n-m-r)\dots(j+n-m)}{j+n-m-r}a_{j+n-m}$$

$$= -\frac{m(n-m-r)\dots(n-m)}{n-m-r}a_{n-m}$$

$$-\sum_{j=1}^{m-1}\sigma_j(r)\frac{(j+n-m-r)\dots(j+n-m)}{j+n-m-r}a_{j+n-m}. \quad (11.198)$$

We thus obtain

$$\sigma_m(r) = -\frac{n-r}{(n-r)\dots n}\left[\frac{m(n-m-r)\dots(n-m)}{n-m-r}a_{n-m}\right.$$

$$\left.+\sum_{j=1}^{m-1}\sigma_j(r)\frac{(j+n-m-r)\dots(j+n-m)}{j+n-m-r}a_{j+n-m}\right]. \quad (11.199)$$

Consider first the range $0 \le r \le n-m$; then this formula can be written as

$$\sigma_m(r) = -m\frac{(n-m)!}{n!}(n-r-m+1)\dots(n-r)a_{n-m}$$

$$-\sum_{j=1}^{m-1}\sigma_j(r)\frac{(j+n-m)!}{n!}(n-r-(m-j)+1)\dots(n-r)a_{j+n-m}. \quad (11.200)$$

Making use of the induction hypothesis, each term is a polynomial in r of degree not exceeding m. Therefore to prove our result by induction, we need to show that this formula for $\sigma_m(r)$ remains valid for $n-m+1 \le r \le n-1$. Suppose that $r = n-m+t$, where $1 \le t \le m-1$. Then in each of the two previous expressions for $\sigma_m(r)$ the term in a_{n-m} is zero, as are those terms in the sum with j in the range $1 \le j \le t-1$. For those terms with $j \ge t$ the two expressions are identical. This establishes the assertion.

Note that the previous formula gives

$$\tau_m(n) = 0 \quad (0 \le m \le n-1). \quad (11.201)$$

In other words, the polynomials to which σ_m are equal at the points $0, 1, \ldots, n-1$ all have a zero at the point n. On the other hand $\sigma_m(n) = 0$ follows from the fact that $P^{(n)}$ is a non-zero constant.

11.8.5

Since the points z_i are distinct ($1 \leq i \leq d$), the $d \times d$ matrix $\{z_i^j\}$ ($0 \leq j \leq d-1, 1 \leq i \leq d$) is non-singular: for otherwise we can find numbers c_j, not all zero, such that

$$\sum_{j=0}^{d-1} c_j z_i^j = 0 \quad (1 \leq i \leq d), \tag{11.202}$$

which gives d distinct zeros of a polynomial of degree $\leq d-1$. It follows that the system of linear equations

$$\sigma_j(r) = \sum_{i=1}^{d} m_i(r) z_i^j \quad (0 \leq j \leq d-1) \tag{11.203}$$

has a unique solution of the form

$$m_i(r) = \sum_{j=0}^{d-1} \sigma_j(r)\zeta_{i,j} \quad (1 \leq i \leq d). \tag{11.204}$$

For $0 \leq r \leq n-1$, $\sigma_j(r)$ is represented by a polynomial of degree at most j and therefore $m_i(r)$ is represented by a polynomial of degree at most $d-1$ for $0 \leq r \leq n-1$. Furthermore, this polynomial has a zero at the point n.

11.8.6

The final part of Sudbery's proof is to show that, for at least one i, the polynomial $r \mapsto m_i(r)$ has degree n. This gives $n \leq d-1$ and so $d \geq n+1$, which is the required result. We can study this problem with the help of the Lagrange interpolation formula. Let us consider a polynomial $u(z)$ of degree not exceeding n, whose values are given at the points $z = 0, 1, \ldots, n$: $u(r) = m_i(n-r)$ for $0 \leq r \leq n$, where m_i is one of our multiplicity functions.

Thus we are given $u(0) = 0$. Let

$$Q(z) = \prod_{k=0}^{n} (z-k). \tag{11.205}$$

By the Lagrange formula $u(z)$ is uniquely determined by the formula

$$u(z) = \sum_{r=1}^{n} \frac{u(r)}{Q'(r)} \prod_{k \neq r} (z - k). \tag{11.206}$$

It follows that, if $u(z)$ does not have degree n, then

$$\sum_{r=1}^{n} \frac{u(r)}{Q'(r)} = 0. \tag{11.207}$$

Now

$$Q'(r) = \prod_{k \neq r} (r - k) = r!(n - r)!(-1)^{n-r}. \tag{11.208}$$

Thus we obtain the relationship

$$\sum_{r=0}^{n-1} (-1)^r \binom{n}{r} m_i(r) = 0 \tag{11.209}$$

for any $m_i(r)$ which does not have full degree n. That this relation can occur non-trivially can be seen from the following example: we consider the case $n = 7$ with $m(0) = 0$, $m(1) = 4$, $m(2) = 3$, $m(3) = 2$, $m(4) = 1$, $m(5) = m(6) = m(7) = 0$; the polynomial $P(z) = 1 + z^5 + z^6 + z^7$ produces this multiplicity structure with the zero at the origin. The above equality does hold in this case, and we find that $m(r)$ has degree 6.

Note further that we can write

$$u(z) = \sum_{r=1}^{n} u(r) \prod_{k \neq r} \frac{z - k}{r - k} \tag{11.210}$$

from which we obtain

$$\begin{aligned} u(N) &= \sum_{r=1}^{n} u(r)(-1)^{n-r} \binom{N}{r} \binom{N - r - 1}{n - r} \quad (N > n), \\ u(-N) &= \sum_{r=1}^{n} u(r)(-1)^r \binom{N + r - 1}{N - 1} \binom{N + n}{N + r} \quad (N > 0). \end{aligned} \tag{11.211}$$

Thus $u(z)$ is integer-valued at the integers, and so has integer coefficients.

11.8.7 Completion of Sudbery's proof

For a complex polynomial P of degree n with more than one zero we can find a zero z_1 of P, whose multiplicity function $m_1(r)$ has the form

$$m_1(r) = \begin{cases} \mu - r & (0 \le r \le \mu), \\ 0 & (\mu + 1 \le r \le n), \end{cases} \tag{11.212}$$

where $1 \le \mu \le n-1$. This follows from the Gauss–Lucas theorem, by choosing z_1 as a vertex of the convex polygon defined as bounding the convex hull of the zeros of P; e.g. we could take z_1 as a zero of maximum modulus. All the zeros of Π lie inside this convex hull; on the other hand, if z_1 has multiplicity μ, then z_1 is not a zero of $P^{(\mu)}$ and therefore lies outside the convex hull of the zeros of $P^{(\mu)}$, which nevertheless contains the zeros of $P^{(r)}$ for $r \ge \mu$. Thus $m_1(r)$ has the form specified.

It only remains to show that the representing polynomial for $m_1(r)$ has exact degree n. This can easily be seen by considering the second difference

$$\Delta_2(m_1(r)) = m_1(r) - 2m_1(r+1) + m_1(r+2) = \begin{cases} 1 & (r = \mu - 1), \\ 0 & \text{otherwise}, \end{cases} \tag{11.213}$$

defined for $0 \le r \le n-2$. It follows that

$$\sum_{r=0}^{n-2} (-1)^r \binom{n-2}{r} \Delta_2(m_1(r)) = (-1)^{\mu-1} \binom{n-2}{\mu-1} \ne 0, \tag{11.214}$$

and therefore $\Delta_2(m_1(r))$ has degree $n-2$, which implies that $m_1(r)$ has degree n.

11.9 Extensions of Sudbery's theorem

11.9.1

Theorem *Let $P(z)$ be a polynomial of degree $n \ge 2$ possessing at least two distinct zeros and let m be a natural number satisfying $1 \le m \le n-1$. Then the number of distinct zeros of P and any m of its derivatives, among $P', \ldots, P^{(n-1)}$, is at least $m+1$. Furthermore, if the m derivatives are successive derivatives $P^{(r)}, \ldots, P^{(r+m)}$, where $r + m \le n-1$, then P and these derivatives have at least $m+2$ distinct zeros.*

Proof Most of the algebra of Sudbery's proof easily extends to this case. Let $r_1, \ldots, r_m$ correspond to the derivatives in question and let z_i $(1 \le i \le d)$ denote the distinct zeros of P and $P^{(r_k)}$ for $1 \le k \le m$; $m_i(r)$ again denotes the multiplicity of the zero z_i as a zero of $P^{(r)}$, where r runs through the values 0 and r_k $(1 \le k \le m)$; $\sigma_j(r)$ is defined as before for these z_i and these values of r.

Simply restricting ourselves to these values of r and using the same algebra, we find that, for $j \geq 1$, there is a polynomial $\tau_j(x)$ of degree at most j, whose values are $\sigma_j(r)$ at $r = 0$ and r_k $(1 \leq k \leq m)$, and furthermore each τ_j has a zero at $x = n$. As before, the $d \times d$ matrix $\{z_i^j\}$ $(0 \leq j \leq d-1, 1 \leq i \leq d)$ is non-singular and we can solve the equations for $m_i(r)$, deducing that, for each i $(1 \leq i \leq d)$ $m_i(r)$ is represented for $r = 0, r_k$ $(1 \leq k \leq m)$ and n by a polynomial of degree at most $d - 1$. It only remains to show that, for at least one i, the representing polynomial for $m_i(r)$ has degree $\geq m$.

We may choose the zero z_1 to have multiplicity function

$$m_1(r) = \max(\mu - r, 0) \quad (r = 0, r_k, n \text{ for } 1 \leq k \leq m), \tag{11.215}$$

where μ satisfies $1 \leq \mu \leq n-1$. We need to show that a polynomial $p(x)$ taking on these values at the specified points necessarily has degree $\geq m$. Writing

$$R(x) = \prod_{k=0}^{m+1} (x - r_k), \tag{11.216}$$

where we have taken $0 = r_0 < r_1 < \cdots < r_m < r_{m+1} = n$, the polynomial of smallest degree taking on these values is given by

$$p(x) = \sum_{k=0}^{m} \frac{m_1(r_k)}{R'(r_k)} \frac{R(x)}{x - r_k}, \tag{11.217}$$

applying the Lagrange interpolation formula. This has degree smaller than $m+1$ if, and only if,

$$\sum_{k=0}^{m} \frac{m_1(r_k)}{R'(r_k)} = 0. \tag{11.218}$$

Now $R'(r_k) = \prod_{j \neq k}(r_k - r_j)$ and so this relation becomes

$$\frac{\sum_{k=0}^{\nu} (\mu - r_k)}{\prod_{j \neq k} (r_k - r_j)} = 0, \tag{11.219}$$

where $r_\nu < \mu \leq r_{\nu+1}$, the values j running from 0 to $m + 1$.

A direct proof that this quantity is non-vanishing would provide a proof that the number of distinct zeros of P and these m derivatives is at least $m + 2$. To prove the theorem we argue as follows. The polynomial $p(x)$ satisfies

$$p(x) = \begin{cases} \mu - x & \text{for } x = r_0, r_1, \ldots, r_\nu, \\ 0 & \text{for } x = r_{\nu+1}, \ldots, r_{m+1} \end{cases} \tag{11.220}$$

and therefore by Rolle's theorem $p'(x) = -1$ has at least one solution in each of the intervals (r_{k-1}, r_k) for $1 \leq k \leq \nu$. It follows that $p''(x) = 0$ has at least $\nu - 1$ solutions in (r_0, r_ν). Similarly $p'(x) = 0$ has at least one solution in each of the intervals $(r_{\nu+k}, r_{\nu+k+1})$ for $1 \leq k \leq m - \nu$; therefore $p''(x) = 0$ has at least $m - \nu - 1$ solutions in $(r_{\nu+1}, r_{m+1})$. It follows that p'' has at least $m - 2$ zeros and therefore p has at least degree m, as required.

Consider next the case when the m derivatives are successive derivatives. There are three cases: (i) $\mu \leq r_1$; (ii) $\mu > r_m$; (iii) $r_1 < \mu \leq r_m$. In case (i) p has $m + 1$ zeros, but is $\neq 0$ at $x = 0$; so p has degree $m + 1$. In case (ii) $p(x) + x = \mu$ has $m + 1$ solutions, but is $\neq \mu$ at $x = n$; therefore p has degree $m + 1$. In case (iii), since the derivatives are successive, $\mu = r_{\nu+1}$ in the earlier reasoning and so $p(x) = \mu - x$ at $x = r_{\nu+1}$. This gives a solution of $p'(x) = -1$ in $(r_\nu, r_{\nu+1})$ and this leads to one extra zero of p''. Thus we obtain $m - 1$ zeros of p'' and so p has degree $m + 1$.

11.9.2

We return now to the case when μ lies strictly between r_ν and $r_{\nu+1}$, so we can write $\mu = (1 - t)r_\nu + tr_{\nu+1}$, where $0 < t < 1$. We then see that

$$p(x) = (1 - t)q(x) + tr(x), \tag{11.221}$$

where $q(x)$ corresponds to the case when $\mu = r_\nu$ and $r(x)$ corresponds to the case when $\mu = r_{\nu+1}$. Thus both q'' and r'' have degree $m - 1$ with all their zeros lying in $(0, n)$. Furthermore, our previous reasoning shows that, for every real t, the polynomial $p = p_t$ is such that p'' has degree at least $m - 2$ with $m - 2$ real zeros. It follows that for every real t, p'' is a polynomial with all its zeros real. This is equivalent to the statement that the rational function

$$R(x) = \frac{r''(x)}{q''(x)} \tag{11.222}$$

takes real values only on the real axis, and therefore is strongly real. Therefore by theorem 9.6.3 of chapter 9, R has interspersed zeros and poles; in other words the $m - 1$ zeros of r'' are interspersed with the $m - 1$ zeros of q'', and all these zeros lie in $(0, n)$. Now for $0 < t < 1$, $p'' = 0$ occurs only when $R < 0$. Thus we wish to show that R takes all its negative values in $(0, n)$. This will be the case if $R(x)$ takes only positive values when x is near ∞. However, this is equivalent to the problem in hand, since it holds iff the leading coefficients of q and r have the same sign. It appears quite possible that this is not necessarily the case, and therefore it may well happen that p does only have degree m.

References

[1] S.S. Abhyankar and T.T. Moh Embeddings of the line in the plane, *J. Reine Angew. Math.* **276** (1975), 149–66.

[2] Lars V. Ahlfors *Complex Analysis*, Second edition, McGraw-Hill, New York (1966).

[3] Lars V. Ahlfors and Leo Sario *Riemann Surfaces*, Princeton Mathematical Ser. no. 26, Princeton University Press (1960).

[4] M. Ålander Sur les zéros extraordinaires des dérivées des fonctions entières réelles *Ark. Mat. Astronom. Fys.* **11** (1916), No. 15.

[5] Hyman Bass, Edwin H. Connell and David Wright The Jacobian conjecture: reduction of degree and formal expansion of the inverse, *Bull. Amer. Math. Soc.* **7** (1982), no. 2, 287–330.

[6] A.F. Beardon *Complex Analysis*, Wiley-Interscience, New York (1979).

[7] G. Birkhoff & S. Mac Lane *A Survey of Modern Algebra*, Macmillan, New York (1953).

[8] B.D. Bojanov, Q.I. Rahman and J. Szynal On a conjecture of Sendov about the critical points of a polynomial, *Math. Z.* **190** (1985), no. 2, 281–5.

[9] Peter Borwein & Tamás Erdélyi *Polynomials and Polynomial Inequalities*, Springer-Verlag, New York, 1995.

[10] D.A. Brannan On coefficient problems for certain power series, *Proceedings of the Symposium on Complex Analysis* (Univ. Kent, Canterbury, 1973), pp. 17–27, London Math. Soc. Lecture Notes Ser., no. 12, Cambridge University Press, London, 1974.

[11] D.A. Brannan, J. Clunie & W. Kirwan On the coefficient problem for functions of bounded boundary rotation, *Ann. Acad. Sci. Fenn. Ser. A.* (1973), No. 523.

[12] J.E. Brown and G. Xiang Proof of the Sendov conjecture for polynomials of degree at most eight, *J. Math. Anal. Appl.* **232** (1999), no. 2, 272–92.

[13] A. Cohn Über die Anzahl der Wurzeln einer algebraischen Gleichung in einem Kreise, *Math. Z.* **14** (1922), 110–48.

[14] A. Cordova and St. Ruscheweyh Subordination of polynomials, *Rocky Mountain J. Math.* **21** (1) (1991), 159–70.

[15] T. Craven, G. Csordas and W. Smith The zeros of derivatives of entire functions and the Pólya–Wiman conjecture, *Ann. of Math.* **125** (1987), 405–31.

[16] Peter L. Duren *Theory of H^p Spaces*, Pure and Applied Mathematics, vol. 38, Academic Press, New York and London (1970).

[17] A.W. Goodman, Q.I. Rahman and J.S. Ratti On the zeros of a polynomial and its derivative, *Proc. Amer. Math. Soc.* **21** (1969), 273–4.

[18] J.H. Grace The zeros of a polynomial, *Proc. Cam. Phil. Soc.* **11** (1902), 352–7.

[19] R.R. Hall On some theorems of Hurwitz and Sheil-Small, *Math. Proc. Cam. Phil. Soc.* **100** (1986), no. 2, 365–70.

[20] W.K. Hayman *Research Problems in Function Theory*, Athlone Press, London (1967).

[21] I.S. Jack Functions starlike and convex of order α, *J. London Math. Soc.* **3** (1971), 469–74.

[22] W. Kaplan Close-to-convex Schlicht functions, *Michigan Math. J.* **1** (1952), 169–85.

[23] J.L. Kelley, I. Namioka and co-authors *Linear Topological Spaces*, D. Van Nostrand Company, Inc., Princeton, NJ (1963).

[24] G.K. Kristiansen Some inequalities for algebraic and trigonometric polynomials, *J. London Math. Soc.* **20** (1979), 300–14.

[25] B.Ja. Levin *Distribution of Zeros of Entire Functions*, Translations of Mathematical Monographs vol. 5, Amer. Math. Soc. (1980).

[26] Z. Lewandowski Sur l'identité de certaines classes de fonctions univalentes. I (French), *Ann. Univ. Mariae Curie-Sklodowska Sect. A.* **12** (1958), 131–46.

[27] Z. Lewandowski Sur l'identité de certaines classes de fonctions univalentes. II (French), *Ann. Univ. Mariae Curie-Sklodowska Sect. A.* **14** (1960), 19–46.

[28] A. Lyzzaik Local properties of light harmonic mappings, *Canad. J. Math.* **44** (1992), 135–53.

[29] A. Magnus On polynomial solutions of a differential equation, *Math. Scand.* **3** (1955), 255–60.

[30] Morris Marden *Geometry of Polynomials*, Mathematical Surveys and Monographs no. 3, Amer. Math. Soc. (1966).

[31] M. Marden Conjectures on the critical points of a polynomial, *Amer. Math. Monthly* **90** (4) (1983), 267–76.

[32] M.J. Miller Maximal polynomials and the Ilieff–Sendov conjecture, *Trans. Amer. Math. Soc.* **321** (1) (1990), 285–303.

[33] M.J. Miller Continuous independence and the Ilieff–Sendov conjecture, *Proc. Amer. Math. Soc.* **115** (1992), no. 1, 79–83.

[34] M.J. Miller On Sendov's conjecture for roots near the circle, *J. Math. Anal. Appl.* **175** (1993), no. 2, 632–9.

[35] T.T. Moh On the Jacobian conjecture and the configuration of roots, *J. Reine Angew. Math.* **340** (1983), 140–212.

[36] Y. Nakai and K. Baba A generalization of Magnus' theorem, *Osaka J. Math.* **14** (1977), 403–9.

[37] S.Yu. Orevkov Rudolph diagrams and analytic realization of the Vitushkin covering (Russian), *Mat. Zametki* **60** (1996), no. 2, 206–24, 319; translation in *Math. Notes* **60** (1996), no. 1–2, 153–64 (1997).

[38] S. Pinchuk A counterexample to the strong real Jacobian conjecture, *Math. Z.* **217** (1994), 1–4.

[39] G. Pólya On the zeros of the derivatives of a function and its analytic character, *Bull. Amer. Math. Soc.* **49** (1943), 179–91.

[40] G. Pólya and I.J. Schoenberg Remarks on de la Vallée Poussin means and convex conformal maps of the circle, *Pacific J. Math.* **8** (1958), 295–334.

[41] G. Pólya and G. Szegö *Aufgaben und Lehrsätze aus der Analysis*, vols. 1 and 2, Springer, Berlin, 1925.

[42] S.P. Robinson *Approximate Identities for Certain Dual Classes*, D. Phil. Thesis, University of York U.K. (1996).

[43] Z. Rubinstein On a problem of Ilyeff, *Pacific J. Math.* **26** (1968), 159–61.

[44] St. Ruscheweyh Duality for Hadamard products with applications to extremal problems for functions regular in the unit disc, *Trans. Amer. Math. Soc.* **210** (1975), 63–74.

[45] St. Ruscheweyh Linear operators between classes of prestarlike functions, *Comment. Math. Helv.* **52** (1977), 497–509.

[46] St. Ruscheweyh On the Kakeya–Eneström theorem and Gegenbauer polynomial sums, *SIAM J. Math. Anal.* **9** (1978), 682–6.

[47] St. Ruscheweyh *Convolutions in Geometric Function Theory*, Les Presses de L'Université de Montréal (1982).

[48] St. Ruscheweyh Geometric properties of the Cesàro means, *Results Math.* **22** (1992), No. 3–4, 739–48.

[49] St. Ruscheweyh and T. Sheil-Small Hadamard products of Schlicht functions and the Pólya–Schoenberg conjecture, *Comment. Math. Helv.* **48** (1973), 119–35.

[50] E.B. Saff and T. Sheil-Small Coefficient and integral mean estimates for algebraic and trigonometric polynomials with restricted zeros, *J. London Math. Soc.* **9** (1974/5), 16–22.

[51] Bl. Sendov On the critical points of a polynomial, *East J. Approx.* **1** (1995) no. 2, 255–8.

[52] T. Sheil-Small On linear accessibility and the conformal mapping of convex domains, *J. Analyse Math.* **25** (1972), 259–76.

[53] T. Sheil-Small On the convolution of analytic functions, *J. Reine Angew. Math.* **258** (1973), 137–52.

[54] T. Sheil-Small The Hadamard product and linear transformations of classes of analytic functions, *J. Analyse Math.* **34** (1978), 204–39.

[55] T. Sheil-Small On the Fourier series of a finitely described convex curve and a conjecture of H.S. Shapiro, *Math. Proc. Cam. Phil. Soc.* **98** (1985), no. 3, 513–27.

[56] T. Sheil-Small On the Fourier series of a step function, *Michigan Math. J.* **36** (1989), 459–75.

[57] T. Sheil-Small On the zeros of the derivatives of real entire functions and Wiman's conjecture, *Ann. of Math. (2)* **129** (1989), no. 1, 179–93.

[58] M. Shub and S. Smale Computational complexity: on the geometry of polynomials and a theory of cost. II, *SIAM J. Computing* **15** (1986), 145–61.

[59] S. Smale The fundamental theorem of algebra and complexity theory, *Bull. Amer. Math. Soc.* **4** (1981), no. 1, 1–36.

[60] Ian Stewart *The Problems of Mathematics*, Second edition, Opus books, Oxford University Press, 1992.

[61] A. Sudbery The number of distinct roots of a polynomial and its derivatives, *Bull. London Math. Soc.* **5** (1973), 13–17.

[62] T.J. Suffridge Starlike functions as limits of polynomials, *Advances in Complex Function Theory, Maryland, 1973/74*, pp. 164–203, Lecture Notes in Mathematics, no. **505**, Springer-Verlag (1976).

[63] G. Szegö Bemerkungen zu einen Satz von J. H. Grace über die Wurzeln algebraischer Gleichungen, *Math. Z.* **13** (1922), 28–55.

[64] D. Tischler Critical points and values of complex polynomials, *J. Complexity* **5** (1989), no. 4, 438–56.

[65] V. Vâjâitu and A. Zaharescu Ilyeff's conjecture on a corona, *Bull. London Math. Soc.* **25** (1993), no. 1, 49–54.

[66] Arno van den Essen Polynomial automorphisms and the Jacobian conjecture, *Algèbre non commutative, groupes quantiques et invariants (Reims 1995)*, pp. 55–81, Sémin. Congr., 2, Soc. Math. France, Paris (1997).

[67] A.G. Vitushkin On polynomial transformations of $\mathbb{C}^n$, *Manifolds–Tokyo (1973) (Proc. Internat. Conf., Tokyo, 1973)*, pp. 415–17, University of Tokyo Press, Tokyo (1975).
[68] J.L. Walsh On the location of the roots of the derivative of a rational function, *Ann. of Math.* **22** (1920), 128–44.
[69] J.L. Walsh The location of critical points of analytical and harmonic functions, Amer. Math. Soc. Colloq. Publ. vol. 34, Amer. Math. Soc., Providence, RI (1950).
[70] E.T. Whittaker and G.N. Watson *A Course of Modern Analysis*, Fourth edition, Cambridge University Press (1935).
[71] A. Wilmshurst *Complex Harmonic Mappings and the Valence of Harmonic Polynomials*, D. Phil. thesis, University of York, U.K. (1994).
[72] A. Wilmshurst The valence of harmonic polynomials, *Proc. Amer. Math. Soc.* **126** (7) (1998), 2077–81.
[73] D. Wright The amalgamated free product structure of $GL_2(k[X_1 \dots, X_n])$ and the weak Jacobian conjecture for two variables, *J. Pure and Appl. Algebra* **12** (1978), 235–51.
[74] A. Zygmund *Trigonometric Series*, vol. I, II, Cambridge University Press (1979).

Index

abc theorem 370
Abel 23
affine mappings/transformations 5, 73
Ålander, M. 343
algebraic function 20
algebraic polynomial 1
algebraic tract 103
analytic continuation lemma 204
analytic polynomial 1
angular separation of critical points and univalence 241, 243
angular separation of zeros
  convolution theorem 254
  Kaplan class condition 248
  sufficient condition 257
  Suffridge condition 249
  third condition 251
annihilation property 129
annular region for zeros 202
apolar polynomials 180
apolarity, weak 182
apolarity theorem 181
approximate identity 147
approximation theorem 258
argument 29
  and tangent of curve 387–90
  of polynomial with zeros on circle 243
argument principle 23, 27, 44, 84
  for Jordan domain 45
  for meromorphic functions 45
asymptotic values 15–29
  of finitely valent mapping 90
  of Pinchuk mapping 93–100

Bernstein's inequality 152, 186
  for trigonometric polynomials 155
Bézout's theorem 4, 84
bisector theorem 200, 215, 216
Blaschke product 60, 193, 231, 233, 351, 373
  and convex curve 382–7
  and harmonic mappings 377–82
  and Smale's conjecture 365
  and step functions on convex curves 394
  Walsh theorem 377
blow-up method 13
Bôcher–Walsh theorem 192
Bojanov–Rahman–Szynal method 213, 226
Bojanov–Rahman–Szynal theorem 215
Bolzano–Weierstrass property 128
Borsuk–Ulam theorem 79
Borwein–Erdélyi inequality 193
  extensions 195–6
bound preserving g.c. operator 170
bound preserving operator 140
boundary inequality 224
boundary rotation of curve 389
bounded boundary rotation 298
  and univalence 389
bounded convolution operators 142
  integral representation 168
bounded operator 139
bounded operator condition 140
bounded variation 159, 221
branch
  of algebraic function 21
  of inverse function 57
  of logarithm 30
  of square root 31
branch point 306
Brannan, D.A. 295
Brouwer's fixed point theorem 48–9
Brouwer's theorem on plane domains 40

Carathéodory's lemma 162
Cauchy–Riemann equations 266
Cauchy's integral formula 128
Cauchy's theorem 24, 31, 36

Cesàro mean 156–7
  of Kaplan class 301
Cesàro polynomial 300
chain 36
change in argument 29
Choquet's theorem 287
circle 73
circular region 178
  zeros and critical points 187
closed convex hull 287
closed under convolution 174
close-to-convex 405
close-to-convex domain 242
close-to-convex function 241–2
closing gap problem 112–5
Clunie, J.G. 171, 266
Clunie–Jack lemma 266
cluster set 58
coefficient 129
  and maximum modulus, zeros on unit circle 239
coefficient bound
  Kaplan class 294
  Suffridge class 260
Cohn's reduction lemma 372
coincidence theorem 47
compact open topology 128
comparison principle 19
complex curve 118
complex form of Bézout's theorem 11
complex polynomial 1
composition remark 363
conical surface 117
conjugate pair 1
conjugate polynomial 153
connected set 30
constant Jacobian 87, 105–24
containment condition, Suffridge convolution theorem 257
containment lemma 266
containment property 335
containment theorem
  for $T(m, \beta)$ 286
  for $T_0(\alpha, \beta)$ 292
continuation of inverse function 57
continuity
  at infinity 83
  of zeros 55
continuous change in argument 29
continuous logarithm 28
convex curve 382
  and step function 392
convex domain 78
convex function 242
convex function condition 275
convex harmonic mapping, univalence theorem 391–2
convex hull 25
convex in direction of real axis 69
convex mapping 260
convex polygon: harmonic univalence criterion 400–1
convex sequence 157
convex set 24
convexity, of level region 360
convexity lemma 162
convexity preserving 141, 174, 279
  functions 144
  g.c. operator 170
  operators and positive trigonometric polynomials 148
convexity theory 286–8
convolution 172
  of harmonic functions 126
convolution condition, Suffridge angular separation 253
convolution containment theorem 280
convolution formula 127
convolution inequality 279
  for $T(m, \beta)$ 284
  for $T_0(\alpha, \beta)$ 291
convolution lemma 269
convolution operator 132
  extension property 141–3
  extension theorem 142
convolution theorem 279
  for $S(\alpha, \beta)$ 297
  for $T(m, \beta)$ 285
  for $T_0(\alpha, \beta)$ 292
  zeros on unit circle 254
Cordova–Ruscheweyh method 361
Craven, T. 306
Craven–Csordas–Smith conjecture 306
  proof when all critical points real 311
  proof when exactly 2 non-real zeros 308–10
critical circle 225
critical point 24, 186, 187
  of Blaschke product 374
  of logarithmic derivative 306; existence of non-real 307
  of rational function 190
  of real polynomial 305
  of real rational functions, theorem on number of non-real 323–4
  of strongly real rational function 321
  on unit circle, univalence theorem 243
critical point inequality 359
critical point theorem 334
critical set 66
cross-cap 123
Csordas, G. 306
cycle 36
cylindrical surface 117

de la Vallée Poussin means 160, 301
degree
  of function on curve 29
  of function on cycle 36
  of function on rectilinear polygon 35
  of harmonic polynomial 3
  of Pinchuk mapping 93
  of polynomial 1
  of real analytic polynomial 4, 83
  of trigonometric polynomial 144
degree estimate, Jacobian conjecture 102–4
degree principle 31, 37
  for circle 34
  for homotopic curves 41
  for rational function 37
  for simply-connected domain 39
  for starlike domain 33
  for triangle 32
Descartes' rule of signs 317
  polynomials with all real zeros 318
Dirichlet means 158
domain 30
dual 172
  of $T(1, 1)$ 271
duality 172
  sums and products 23
duality principle 286
duality theorem
  for $T(1, \beta)$ 275
  for $T(m, \beta)$ 282
  for $T_0(\alpha, \beta)$ 289

elementary mappings 109
ellipse 73
Eneström–Kakeya theorem 281
  Ruscheweyh generalisation 281
entire functions of finite order 327
  with real zeros 343
Erdös, P. 407
essential singularity 44
Euler gamma function 239
extending boundary inequalities 224
extension property 141
exterior of curve 34
extremal distance theorem 221
extremal polynomial 223
  Suffridge 251
extreme point 287
  of Kaplan class 293

Fejér means 147
Fejér–Riesz inequality 129
Fermat's last theorem for polynomials 371
finite limiting value and asymptotic value 19
finite product duality lemma 175, 265, 270, 282
finite valence 84
  and N-fold mapping 390
finite valence condition 390
fixed point 48
Fourier coefficient 129
  of N-fold mapping 391
Fourier series 129
function of bounded boundary rotation 298
fundamental group 42
  of punctured plane 43
fundamental theorem of algebra 1, 22, 26,
    49–50

g.c. kernel 169
g.c. operator 168
Galois 23
gamma function 239
Gauss–Lucas theorem 25, 185, 189, 201
generalised convolution operator 168, 279
genus of entire function 344
global homeomorphism, Jordan domain
    condition 59
global homeomorphism condition 58
Goodman–Rahman–Ratti theorem 207
Goursat method 32
Grace class 174, 253
Grace theorem for rational functions 263
Grace–Heawood theorem 198
Grace's apolarity theorem 181
Grace's theorem 216, 263, 282
  additional forms 183–4
  applied to Ilieff–Sendov conjecture 208–11
Grace–Szegö theorem 173
  circular regions 179
  linear functionals 177
  proof 176–7
graph of surface 101
greatest common divisor 106

Hadamard representation of entire function
    344
harmonic extension 129
  of step function 395; containment theorem
    396; univalence criterion 399
harmonic function 126
  at non-isolated zero 53
  convolution 126
  locally 1–1 64
harmonic mapping and Blaschke product
    377–82
harmonic multiplication operator 135
  class $\mathcal{M}$ 137
harmonic multiplication representation 135
harmonic polynomial 3, 50, 82
  degree 2 73–8
  on critical set 66–72
  theorem on valence 72
  valence near critical set 70–1
  zeros 51, 54

harmonic series 126
Hayman, W.K. 206
Heine–Borel theorem 58
Helly selection theorem 146
highest common factor
  2 110–11
  3 115–17
Hilbert's inequality 130
homeomorphism 57, 81
  condition at infinity 57
  of sphere 80
  of unit disc 60
homologous 36
homology group 42
  of punctured plane 43
homotopic 39
homotopic curves 40
homotopy classes 41
homotopy group 42
homotopy invariant 41
Hurwitz's theorem 65
hyperbolic non-euclidean geometry 375

Ilieff–Sendov conjecture 206
  application of Grace's theorem 208–11
  case of real polynomial 216
  critical circle 225
  extremal distance 221
  extremal polynomial 223
  independent of particular zero 225
  nearest second zero 216
  proof for zeros on unit circle 207
  proof up to degree 5 211–12
  proof when zero at origin 216
  remaining zeros on unit circle 220
  upper bound on distance 215
Ilieff–Sendov problem *see* Ilieff–Sendov conjecture
imaginary zeros 311–17
implicit function theorem 63
inequality
  for analytic polynomials 151–2
  for harmonic polynomials 154
  implies surjectivity 101
instantaneous double reversal 100–1
integral mean theorem, zeros on unit circle 239
integral representation
  of bounded convolution operator 168
  of Kaplan classes 293
  of non-vanishing polynomials 175
interpretation of convolution conditions 270
interspersed zeros and poles
  on unit circle 234
  real rational functions 319
interspersion lemma 232
interspersion of zeros and local maxima 238
interspersion theorem
  real meromorphic function 327
  real rational functions 320
  unit circle 235
inverse function 56
  branch 58
  continuation 57–8
inverse function inequality 356
inverse function theorem 61
invertibility of harmonic multiplication operator 137
inverting tract equation 18–19
inverting transformation 18–19
isolated zeros 8
iterate 49

Jack, I.S. 266
Jacobian 61, 124
Jacobian conjecture 81
  algebraic resolution 124
  condition in polar coordinates 88
  degree of counter-examples 102–4
  examples when true 87
  proof for degree 2 85–6
  proof under additional hypothesis 85
  weak form 104–5
Jacobian determinant 61
Jacobian matrix 61
Jacobian operator, algebraic properties 91
Jacobian problem 89
  geometric transformation 105
Jensen circle 304, 305
Jordan curve 39
Jordan curve theorem 39
Jordan domain 39
Jordan polygon
  and dilatation criterion 404
  mapping problem 402
Julia–Carathéodory lemma 330

Kaplan, W. 241
Kaplan class 244, 277, 382, 385
  coefficient bounds 294
  extreme points 293
  factorisation theorem 246
  integral representation 293
  $K(1, 1)$ 271
  linear functionals 293
Keller Jacobian conjecture 87
kernel 132
knot 120
Krein–Milman theorem 287
Kristiansen, G.K. 240

Lagrange's interpolation formula 2
Laguerre–Pólya class 347
Laurent expansion 126
Lax's theorem 153–4, 186

leading term 88
length of curve 85
level curve 66, 306–7, 350
  geometry 310
  of polynomials 351
level region 306, 352
  and Smale's conjecture 364
  convexity 360
  of rational function 353
Levin representation 332
Levin representation lemma 333
Lewandowski, Z. 242
Lewy's theorem 64
limits of Suffridge's extremal polynomials 258
linear form 181
linear functional 131, 177
  on Kaplan class 293
  on rational function 264–5
linear operator 132
  Grace theorem 205
  on polynomials 203
  on rational functions 264
linear operator lemma 272–3
linearly accessible domain 302
Liouville's theorem 26
Littlewood, J.E. 239
Lobachevsky 375
local multiplicity 56
local uniform convergence 128
locally bounded 128
locally 1–1 24, 82
  function 56
  harmonic functions 64
  polynomial, topology 100
locating critical points 186
location of zeros given critical points 201–2
logarithm 28
logarithmic derivative
  algebra 407
  critical points 306
  existence of non-real critical points 307
logarithmic derivative lemma 187
logarithmic differentiation 23
loop 28
loop lemma 121
Lyzzaik, A. 392

Magnus theorem 109
majorisation 171
mapping problem for Jordan polygons 402
Marden, M. 206
Mason's theorem 370
max–min inequalities 200–1
maximum modulus 238
maximum principle 57
mean 145
meromorphic function 27, 45
monodromy theorem 58
multiple zeros 187
multiplicity 22
  analytic expression 24
  of function at point 44
  of harmonic function with polynomial co-analytic part 47

Nakai and Baba theorem 111
NE convex 376
NE line 375
nearest second zero 216
negative type: strongly real rational function 321
N-fold mapping
  and finite valence 390
  Fourier coefficients 391
  of circle 382–7
non-Euclidean line 375
non-isolated zero 53
non-real critical points of real rational function 323–4
non-real zeros theorem 325
non-separating lemma 119
non-vanishing polynomials 173
norm
  of operator 140
  of self-inversive polynomial 153
normal family 128
null homotopic 41
number of isolated zeros of real analytic polynomial 8
number of zeros 1
  of real analytic polynomial 4
n-valent 66

open mapping 57
order
  of critical point 353
  of entire function 327
  of meromorphic function 327
Orevkov, S. Yu 118
orientation 56

parabola 78
parabolic region 78
parametrisation 27
Parseval's formula 131
partial fraction decomposition of rational function 355
periodic 129
Pinchuk, S. 81, 118
Pinchuk surface 100
Pinchuk's example 90–3
plane topology 39
p-mean preserving operator 169
Poincaré, H. 375

Poisson's formula 128
polar coordinate 28
 Jacobian conjecture 88
polar coordinate form 82
polar derivative 185
pole 27, 44
Pólya–Schoenberg conjecture 260, 275, 276
Pólya and Schoenberg 161
Pólya and Schoenberg's proof 161–5
Pólya and Schoenberg's theorem 161
Pólya's theorem 347
polynomial
 constant on curve 84
 non-vanishing 173
 with all real zeros, Descartes' rule 318
 with zeros on unit circle 231
polynomial mean 145
polynomially invertible 87, 105
Popoviciu conjecture 407
positive convolution operator 142
positive harmonic function 142
positive operator 138
positive trigonometric polynomial 144–51
 and convexity preserving operator 148
 representation 149
 representation theorem 150
positive type, strongly real rational function 321
positivity lemma 157
prime degree 109–10
primitive 31
problems 25–6, 29, 31, 40, 43, 46–7, 55–6, 78–9, 80, 117, 124, 138, 144, 151–2, 155–6, 159– 60, 203, 208, 232, 237–8, 240, 391–2
projection onto plane 89
projective plane 123
properties of Sylvester resultant 12
pseudo-surface 121, 123
pythagorean triples 372

quadratic polynomials 373
quadrilateral, harmonic mapping problem 405

radical 23
Rado–Kneser–Choquet theorem 59, 392
ratio of linear functionals 180
rational function
 convolution containment theorem 265
 critical points 190
 distinct solutions 370
 Grace theorem 263
 in unit disc 193
 linear functional theorem 264–5
 linear operator theorem 264
 proof of extended Grace theorem 269
 strongly real/interspersion theorem 320
 with real critical points 325–6
real analytic polynomial 4, 27, 81
 at infinity 13–19
real critical point 325
real polynomial 1, 304
 critical point 305
 Ilieff–Sendov conjecture 216
 representation as sum of polynomials with real zeros 322
 with imaginary zeros 311–17
real rational function
 critical point theorem 323–4
 interspersed zeros and poles on real axis 319
 representation in terms of strongly real rational functions 321
real zeros, Descartes' rule 318
rectilinear polygon 35
relatively prime degrees 109–10
removable singularity 44
repeated asymptotic value curve 18
repetition
 of asymptotic values 16
 of reduction process 15
 third method 16–17
repetition property 102, 117
representation for harmonic polynomial 3
 of complex polynomial 2, 3
 of linear operator 132–3
residue 43
residue theorem 24, 27
resolving the singularity 13
resultant 124
 and Jacobian 124
reverse inequalities 216
reversed asymptotic value curve 18
Riemann mapping theorem 35, 39
Riemann–Hurwitz formula 353
Robinson, S. 300, 303
Robinson's conjectures 300–1
Rogosinski's coefficient theorem 170
Rogosinski's lemma 157
rotation conjecture 392
Rouché's theorem 48
Rubinstein, Z. 207
rule of signs, Descartes' 317
Ruscheweyh, St. 281
Ruscheweyh theorem for $S(\alpha, \beta)$ 297

$S(\alpha, \beta)$ class 296
Schmeisser, G. 216
Schoenflies theorem 59
Schwarz function 168
Schwarz's lemma 60, 234
second derivative, non-real zeros 325
second dual 172
second zero 216

self-intersections 123
self-inversive polynomial 149, 152, 228
  norm theorem 153
  on unit circle 229
  representation lemma 254
  zeros and critical points 230
Sendov, B. 206
sense preserving 56, 61
sense reversing 56, 61
separated sets 30
separation by a line 186
series representations 125
shift lemma 69
shift operator 135
sign variation 161
simply-connected domain 35, 38, 42
simply-connected surface 89, 117
simultaneous algebraic equations 26
simultaneous equations 81
sine polynomial inequality 156
singularity 27, 44
Smale's conjecture 358
  Blaschke product approach 365
  critical points and minimum points 363
  generalisations 367
  proof for critical points on circle 361–2
  proof for degrees 2, 3, 4 359–60
  real critical points 368
Smith, W. 306
solving algebraic equations 20
sphere 80
starlike, of order $1/2$ 275
starlike approximation lemma 276, 283
starlike domain 33
starlike function condition 275
step function 378
  and convex curve 392
stereographic projection 38, 80
strictly positive operator 138
strongly real meromorphic function 327
  inequality 332
  representation theorem 327
strongly real rational function 319
  critical point 321
  example 321
  positive or negative type 321
strongly real/interspersion theorem 320
subordination 168
subordination inequality 239
subordination principle 234
successive derivatives 413
Sudbery's theorem 407
  extension 413
Suffridge, T.J. 241
Suffridge class 253
  coefficient bound 260
  duality condition 253
Suffridge univalence criterion 243
Suffridge's convolution theorem 254, 261
Suffridge's extremal polynomials 251
  limits 258
Suffridge's theorem
  condition for proving containment 257
  proof of part (c) 255
summability theory 145
surface classification theorem 121
surjectivity, failure of Pinchuk mapping 100
Sylvester determinant 5
Sylvester matrix 10
Sylvester resultant
  evaluation 11
  exact degree 11
  of $P - w$ 11
Sylvester's method of elimination 5
Sylvester's theorem 165
symmetric linear form 181
  Smale's conjecture 359
  Walsh's theorem 182
Szegö's inequality 155

$T(1, 1)$ class 271
$T(1, \beta)$ classification 273–5
$T(2, 2)$ class 302
$T(\alpha, \beta)$ class 269, 289
$T(m, \beta)$ class 282
$T_0(\alpha, \beta)$ class 289
tangent and argument of curve 387–90
tangent vector, Jacobian conjecture 89
Tischler, D. 360
Toeplitz theorem 131
topological argument principle 44
topology of mapping, Jacobian problem 89
total variation of argument of curve 389
trigonometric polynomial 3, 50, 144
trigonometric series 125
two circles theorem 188

uniqueness theorem 2
unit disc, zeros 373
unit disc homeomorphisms 60
unit element 127, 128
univalence
  and angular separation of critical points 243
  and bounded boundary rotation 389
univalence region 197
univalence sector 200
univalent polynomials 241

valence 66, 72, 84
variation diminishing 161
variations in argument/tangent of curve 387–90
Vitushkin, A.G. 118

Walsh, J.L. 305
Walsh's Blaschke product theorem 377
Walsh's theorem on symmetric linear forms 182
Walsh's two circles theorem 188
weak apolarity 182
weak Jacobian conjecture 105
Wilmshurst, A. 52, 392
Wilmshurst's conjecture 52
Wilmshurst's example 52
Wilmshurst's method 8
Wilmshurst's theorem 54
Wiman conjecture 343
winding number 29
  of Jordan curve 46
winding number properties 34
Wright's theorem 115

zero(s) 22, 44
  and critical points, self-inversive polynomials 230
  continuity 55
  exact number 56
  existence of zero path 64
  given critical points 201–2
  of derivatives 407
  of second derivative 325
  on unit circle 231; angular separation condition 248; argument of polynomial 243; bounds on coefficients 239; integral mean/maximum modulus theorem 239; second angular separation condition 249; third angular separation condition 251
zero cycle 36, 41